THE ULTIMATE Tesla Coil Design AND CONSTRUCTION GUIDE

THE ULTIMATE Tesla Coil Design AND CONSTRUCTION GUIDE

Mitch Tilbury

New York Chicago San Francisco
Lisbon London Madrid Mexico City
Milan New Delhi San Juan
Seoul Singapore Sydney Toronto

Library of Congress Cataloging-in-Publication Data on file with the Library of Congress.

McGraw-Hill books are available at special quantity discounts to use as premiums and sales promotions, or for use in corporate training programs. For more information, please write to the Director of Special Sales, Professional Publishing, McGraw-Hill, Two Penn Plaza, New York, NY 10121-2298. Or contact your local bookstore.

3 4 5 6 7 8 9 0 DOC/DOC 0 1 2 1 0 9

ISBN 978-0-07-149737-4
MHID 0-07-149737-4

This book was printed on acid-free paper.

Sponsoring Editor
Judy Bass

Editing Supervisor
Maureen B. Walker

Project Manager
Vageesh Sharma

Copy Editor
Lee A. Young

Proofreader
Aptara Inc.

Production Supervisor
Pamela A. Pelton

Composition
Aptara Inc.

Art Director, Cover
Jeff Weeks

These works are dedicated to my brother Romney. His contributions to the human endeavor are best stated:

All that was great in the past was ridiculed, condemned, combated, suppressed—only to emerge all the more powerfully, all the more triumphantly from the struggle. Let the future tell the truth and evaluate each one according to his work and accomplishments. The present is theirs, the future, for which I really worked, is mine.

—Nikola Tesla

Disclaimer

The author has made every effort to warn the reader of any known dangers involved when working with high voltage. The intent of the material in this text is to assist a hobbyist in designing and operating a Tesla coil while developing insight to safe working practices. Some degree of existing experience is inferred. Electrical design work is by nature interpretive, following the procedures in a design guide are not always reproducible or predictable. Chapters have been included in this guide on electrical safety and designing control circuits that should prevent operator injury if followed. Beware when using surplus parts, whose operating history and material condition is always unknown and should always be treated with caution. The author accepts no legal responsibility for any personal injury or equipment damage encountered by users as a result of following the methods and information contained in this text.

If you are careless, senseless, or otherwise irresponsible you should not engage in this work! This may not be readily apparent as a person would tend to avoid such criticism or awareness. Evaluate your current hobbies or what you do as a profession and how well you do it. Does it involve a degree of creativity? Does it include manual skills and hands-on work? Are you familiar with basic electrical theory and have you applied it to any electrical work? Are you experienced in such matters? If you cannot answer yes to these questions you should engage in another line of interest for your own safety.

Contents

List of Illustrations

List of Tables

Foreword

Mitch Tilbury's *The Ultimate Tesla Coil Design and Construction Guide* covers just about every and any facet of electrical engineering data ever put together in one package. If there is anything of importance to this topic that isn't covered, I certainly can't name it.

The guide is written in chapters divided up into various segments pertaining to the design and construction of Tesla coils and resonant high frequency circuits. Each chapter is provided with a list of references where applicable. The chapters include superb drawings, sketches, and graphs to illustrate the points being discussed. Other features offered are a table of contents, list of tables, and a complete index, something which many authors of such publications often fail to provide. Tilbury even includes a brief, but interesting, history of the life and career of Nikola Tesla.

Want to know something about the theory and construction of spark gaps, mutual inductance, coupling, resonant RLC series circuits, effects of terminal capacitances, current limiting and controls, leakage in inductors and transformers, determining parameters of unmarked transformers, damped oscillations, safety, etc., etc.? It's all there, and then some. The book has a companion Web site, which allows the experimenter to follow a course of planning or to resolve difficult equations using the computer files. (You can find this Web site at www.mhprofessional.com/tilbury/.) Although I did not run any of the programs, the author provides instructions for those who wish to take advantage of the information contained at this Web site.

I characterize this publication as an encyclopedia for coilers at all levels of expertise. Incidentally, there is so much data covered in this book that it defies a review in one reading session.

One thing for sure, this publication is not a set of plans for "How to Build a Tesla Coil" with which experimenters are so familiar. It is, however, a valuable source of information to which experimenters can repeatedly refer for information on just about any facet of planning and building high-frequency apparatus at high potential.

Harry Goldman
President, Tesla Coil Builders Association

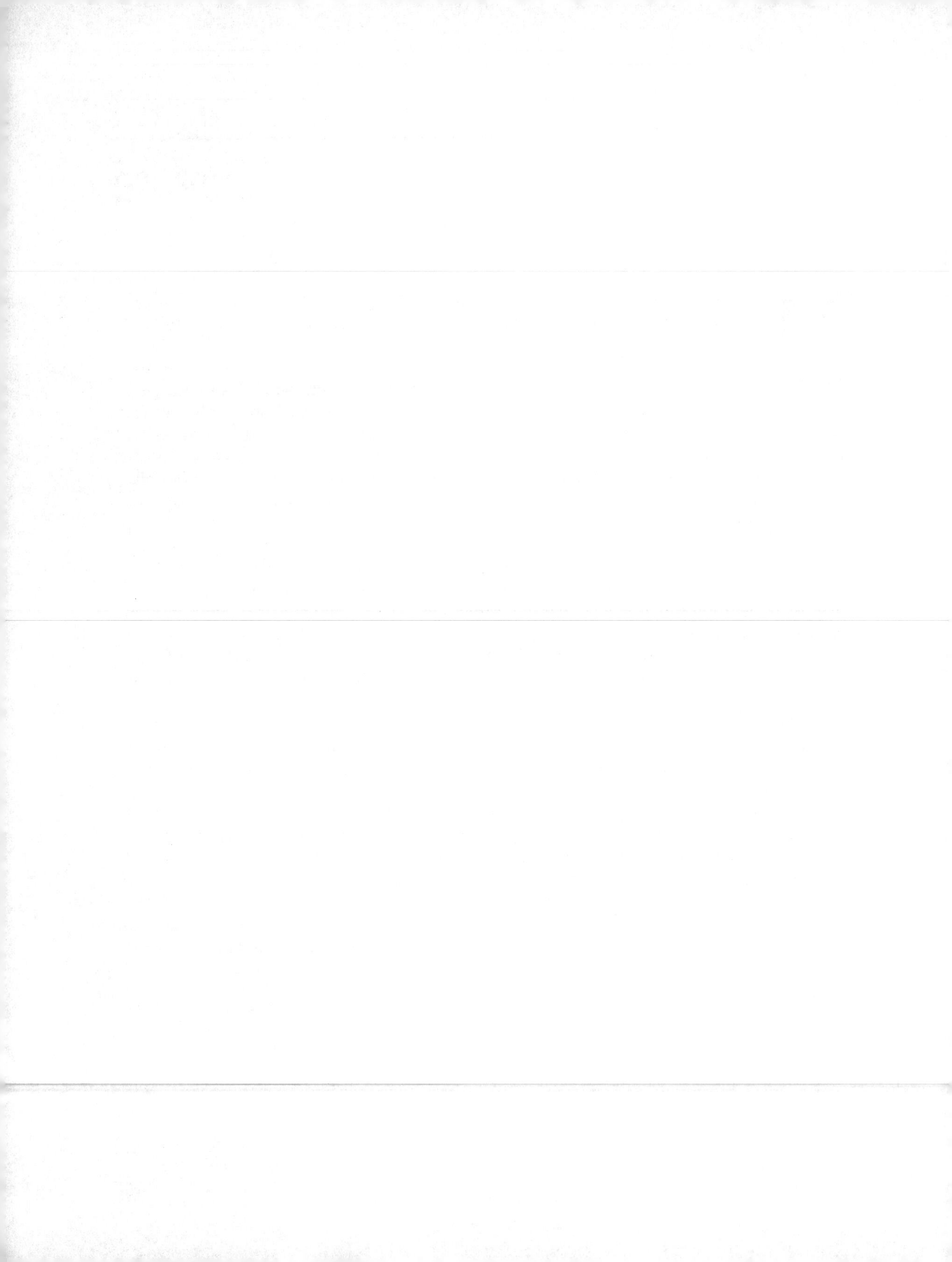

Preface

This effort has not been completely my own. I first became interested in Nikola Tesla in 1987 when I read Margaret Cheney's book *Man Out of Time*. Until then I had no idea who he was or how he impacted the last two centuries. I also had no idea how to generate high voltage. My experience with high voltage was limited to 240 V. It would be another six years before I attempted to build my first Tesla coil. My first attempts were miserable failures. I was having great difficulty obtaining affordable commercial high-voltage parts through mail order sources and area surplus stores. My homemade parts were quickly overcome by the electrical stresses found in an operating coil. I also discovered there was little published information on the subject which was difficult to obtain. Reproducing the coil designs in these published works met with limited success.

It was during these failures that I committed myself to write my own book on Tesla coils someday. I would never have gotten on the right track had I not ran into veteran coiler Ed Wingate of the Rochester Area Tesla Coil Builders (RATCB). I was given Ed's phone number while searching through the surplus stores in Rochester, NY, where he is known for his interest in high–voltage surplus parts. While in town I apprehensively called Ed and he graciously invited me to his laboratory the following day, as I believe he actually perceived my excitement over the phone lines. Ed spent most of the next day showing me how real Tesla coils are built and told me about the ARRL's Hamfests and Harry Goldman's "Tesla Coil Builders Association" (TCBA). I also gratefully received an invitation to Ed's annual Teslathon and have been attending ever since. After Ed's mentoring I was able to obtain quality surplus commercial equipment for a good price. A wealth of coil building information was also becoming available every quarter through the TCBA's newsletter. The coil plans in the newsletter were reproducible and I was soon generating high voltage.

Nikola Tesla designed the first resonant transformer of high frequency and high potential to become known as the Tesla coil. Over the next 100 years this coil found many commercial applications but industry and its engineers never credited Tesla. Although clearly influenced by Tesla these applications were never published, hidden under the guise "proprietary." It fell to the amateur coiler to advance the technique of generating high voltage using a resonant transformer. These techniques have been passed to many generations of coilers and have become accepted practice. Although most of the material in this design guide is my effort to provide conventional formulae and methodology that is referenced to published sources, I would have long ago succumbed to frustration without help from all of the knowledgeable coilers I have met over the years. The ideas, assistance, and parts they have provided are reflected in this guide. I consider myself merely the recorder of these ideas. I would like to express my sincerest gratitude to the following veteran coilers for their invaluable contributions to my high voltage experience: Ed Wingate of the Rochester Area Tesla Coil Builders (RATCB), Tom Vales, Richard Hull of the Tesla Coil Builders of Richmond (TCBOR), Harry Goldman of the Tesla Coil Builders Association (TCBA), John Freau, Tony DeAngelis, Steve Roys, Chris Walton and the many coilers I have had the pleasure of knowing over the years.

Coil building is an amateur endeavor, which places emphasis on the quality and aesthetic appeal of the coil, not just on the spark length being produced. This will always leave challenging opportunities for improvement. I have even seen coilers reproduce antique looking coils of high quality and visual appeal. Without a doubt machinists have built the best coils I have seen. Most of the Tesla coil still requires fabrication of parts and interconnections from stock materials, which makes the machinist a natural coil builder. Coil building has attracted the attention of many tradesman of vocations too numerous to mention.

I am deeply indebted to the late Richard Little Ed.D., being the first to nurture my ability to think on my own.

Last but not least I am indebted to my wife, Jodi, who provides me with as much spare time as I can possibly get away with. This requires an extra effort on her part to maintain our lifestyle. As the high voltage generated increases, so does the noise produced by the spark and she has often bit her tongue while I blasted away in the lab. She is also very generous with my part allowance, as left to my own device I would have long ago propelled us into insolvency. Her efforts are in the spirit of Michael Faraday's wife who contributed portions of her wardrobe so he could insulate the conductors in his solenoids and discover electromagnetism.

Mitch Tilbury

THE ULTIMATE Tesla Coil Design AND CONSTRUCTION GUIDE

CHAPTER 1

Introduction to Coiling

Coiling is the popular term used to describe the building of resonant transformers of high frequency and high potential otherwise known as Tesla coils. Nikola Tesla was the foremost scientist, inventor, and electrical genius of his day and has been unequaled since. If you are unfamiliar with Nikola Tesla see the bibliography in Appendix A for a short narrative and references (5) through (9) for further reading.

Although never publicly credited, Nikola Tesla invented radio and the coil bearing his name, which involves most of the concepts in radio theory. The spark gap transmitters used in the early days of radio development were essentially Tesla coils. The fundamental difference is that the energy is converted to a spark instead of being propagated through a medium (transmitted). The old spark gap transmitters relied on very long antenna segments (approximately 1/4 wavelength) to propagate the energy in a radio wave, the quarter-wave secondary coil is in itself a poor radiator of energy. Tesla coils or resonant transformers of high frequency and high potential have been used in many commercial applications, the only variation being the high voltage is used to produce an effect other than a spark. Although not all commercial applications for Tesla coils are still in use some historical and modern day applications include:

- Spark gap radio transmitters
- Induction and dielectric heating (vacuum tube and spark gap types)
- Induction coils (differ only in the transformer core material being used)
- Medical X-ray devices (typically driven by an induction coil)
- Quack medical devices (violet-ray)
- Ozone generators
- Particle accelerators
- Electrical stage shows and entertainment
- Generation of extremely high voltages with relatively high power levels

Our interest in these coils is of a non-commercial nature or hobby/enthusiast. If you are new to coiling follow along through the chapters to develop an understanding of high voltage and how to generate it utilizing a Tesla coil. Design methodology is explained and taken step by step requiring only an elementary knowledge of electricity. If this is your first coil do not start out drawing 15 kW out of your service box trying to create 15-foot lightning bolts; work with a neon sign transformer and 6-inch sparks. As you progress you will most likely generate lightning bolts and power levels that are limited only by your laboratory space.

The concepts involved in coiling are abstract but not difficult to understand using the illustrations and examples in this design guide. Let's begin.

1.1 Surplus Parts

Throughout this guide, Hamfest is mentioned. If you have never attended one, a Hamfest has a fair-like atmosphere where electronic surplus/salvage vendors and radio enthusiasts come to sell, buy, and trade electrical parts and equipment. The Amateur Radio Relay League (ARRL) sponsors the events. To find the Hamfests nearest you and the dates they occur contact the ARRL headquarters at:

225 Main Street
Newington, CT 06111-1494

Or phone: (860) 594-0200 from 8 AM to 5 PM Eastern time, Monday through Friday, except holidays. They will mail you a Hamfest schedule if requested by phone or mail.

There is also a website address where the Hamfest schedule can be queried for each state and month: http: //www.arrl.com/Hamfests.html.

There are also hundreds of electronic surplus outlets of various sizes located throughout the country. Your local Yellow Pages may list them. Most will attend local Hamfests if not in the phone directory.

1.2 Using Microsoft Excel

Excel spreadsheets were used to automatically perform the calculations used in this guide. You can find these Excel spreadsheets at the companion Web site, which can be found at www.mhprofessional.com/tilbury/. See App. B for a list of Excel files that are used in this book, and the worksheets found in these files. Where the book requests you to open a worksheet, please refer to App. B for the name of the file the worksheet appers in. If you are not familiar with Excel, note that the worksheets are designed to operate with only a few numerical entries into specified cells. All cells requiring an entry in the worksheets are colored blue to separate them from the cells performing calculations. This is as simple as I can make it. Practice with this program will enable any layman to perform complex engineering evaluations. If you do not have Excel try importing or converting the spreadsheet files into a spreadsheet program you do have. If that does not work, all formulae used to derive the design and performance parameters are detailed in the text of each chapter. A spreadsheet can be constructed using a program other than Excel with the formulae and the spreadsheet illustrations contained in the text. A simple notepad and calculator can also be used but will require much more time and patience. Errors are always introduced when performing lengthy hand calculations, which can produce some really poor working coils. Most of the Excel files constructed for this guide contain macros—a series of commands and functions that are stored in a Visual Basic module and are run whenever the task is performed. When you open these files Excel will prompt whether you want to disable or enable the macros. For best results with the worksheet calculations select Enable Macros. When you open these files for the first time a file index will appear. Use this index to help you find the desired worksheet in the file. Some of these worksheets are rather large so use the scroll buttons to navigate around and become familiar with the contents, especially the graphs, which will display the most desired calculation results. The narrative in each section provides detailed instructions on how to perform the calculations in the applicable worksheet. Often the narrative will refer to a colored trace in an Excel graph. As the guide's graphics are monochromatic this is best seen when opening the file and observing it directly in the worksheet graph. If some of the display is lost at the edge of the monitor screen it is because your monitor is smaller than the 17-inch monitor used to construct the files. Simply grab the left-hand edge of the displayed file with the mouse, click the left mouse button, and drag it toward the right edge of the screen. Reposition the displayed file to the left upper corner. You may have to perform this same task on the upper edge of the file. This will expose a tab in the lower right-hand corner used to size the file to the display screen. Reposition as desired and save the changes to customize the file to your monitor.

1.3 Using a Computerized Analog Circuit Simulation Program (Spice)

Spectrum Software's Micro-Cap 8 was used in Chapter 8 to perform Spice simulation of Tesla coil circuits. It is a schematic capture program, meaning that a schematic representation of the circuit is built from which the program generates a spice netlist used to run the simulations. This is an effective tool for designing any electrical circuit including Tesla coils. As this is written, Spectrum Software offers a free evaluation version of Micro-Cap 9 that will run the circuits shown in Chapter 8, which also details how to download and run the program. If you have a Spice circuit simulation program other than Micro-Cap, use the circuit details shown in Chapter 8 to construct a Spice circuit for the program you are using. If you are unfamiliar with these simulators but can download the Micro-Cap 9 program, a few clicks of the mouse will have the circuit up and running, as I have left the settings in the circuit files to operate in that manner. Practice with this program will enable any layman to perform complex engineering evaluations.

1.4 Derivation of Formulae Found in This Guide

General formulae found in this guide were taken directly from the references listed at the end of each chapter. Formulae were modified to obtain expected results if errors were apparent. Math Soft's MathCad 7.0 program was used to derive formula transpositions not found in the references to eliminate any transposition error. Circuit measurements and oscilloscope observations of working circuits were used to verify calculation results. An expected degree of difference exists between calculated values and circuit measurements, ±5% would be within engineering tolerance. Make allowances for these differences. All efforts were made to follow convention. Some of the material, being newly developed, may seem unconventional.

1.5 Electrical Safety, the Human Body Model, and Electrocution

My intention now is to scare you and with this fear develop a healthy respect for the electric currents found in operating Tesla coils. Being over 90% water with electrolytes and salts, your body makes a fair conductor of electricity. The nervous system operates on picoamps of current (1×10^{-12}). The physiological effects on the nervous system with even small currents are listed

Physiological Effect	Current Level at 0 Hz (DC)	Current Level at 60 Hz (AC)
Non-Harmful		
Perception	0 to 4 mA	0 to 1 mA
Surprise	4 to 15 mA	1 to 4 mA
Reflex action	15 to 80 mA	4 to 21 mA
Harmful		
Muscular inhibition	80 to 160 mA	21 to 40 mA
Respiratory block	160 to 300 mA	41 to 100 mA
Death	>300 mA	>100 mA

TABLE 1-1 Electric shock effects.

in Table 1-1 from reference (3). The values in Table 1-1 were used to develop the exponential current vs. frequency characteristics shown in Figure 1-1.

The recipient of a non-harmful shock still controls voluntary movement, which allows them to release their grasp of the source. The recipient of a harmful shock will loose control of their voluntary movement, which keeps them connected to the source (fingers still clutching). The threshold level of these physiological effects increases as the frequency of the current decreases. At first glance it may seem that DC currents are safer than AC currents. However, the DC current has no skin effect and will penetrate to the center of the body where it can do the most harm to the central nervous system. AC currents of 60 Hz can penetrate at least 0.5″below the skin, which is also deep enough to profoundly affect the nervous system. For AC frequencies above 400 Hz the threshold level for dangerous currents decreases to microampere levels; however, the depth of penetration also continues to decrease. At frequencies above the audio range the current remains on the surface of the skin and has difficulty penetrating to the nerves under the skin that control muscular action, respiration and involuntary functions. As there are pain sensors just under the surface of the skin, high frequencies can still cause discomforting shock effects at imperceptible current levels. The reference also notes that most deaths by electrocution resulted from contact with 70 V to 500 V and levels as low as 30 V are still considered potential hazards. Even a small Tesla coil can produce voltages above 100 kV.

Currents in a 60-Hz line frequency are still quite dangerous. The first intentional electrocution in an electric chair was performed using 60-Hz currents at only 2 kV. During the Edison–Westinghouse current wars of the 1890s it was Thomas Edison who promoted the first electric chair and its use in executing William Kemler in New York's Auburn prison to create a fear of AC currents within the public as dramatized in reference (4). It was Nikola Tesla who during this time discovered that frequencies above 2 kHz had a reduced or negligible effect on the nervous system due to skin effect. His public demonstrations during the 1893 Columbia Exposition were quite dramatic. Tesla was reported in the press as being completely aglow in electric fire when he gave the first demonstration of "skin effect"by passing high-frequency,

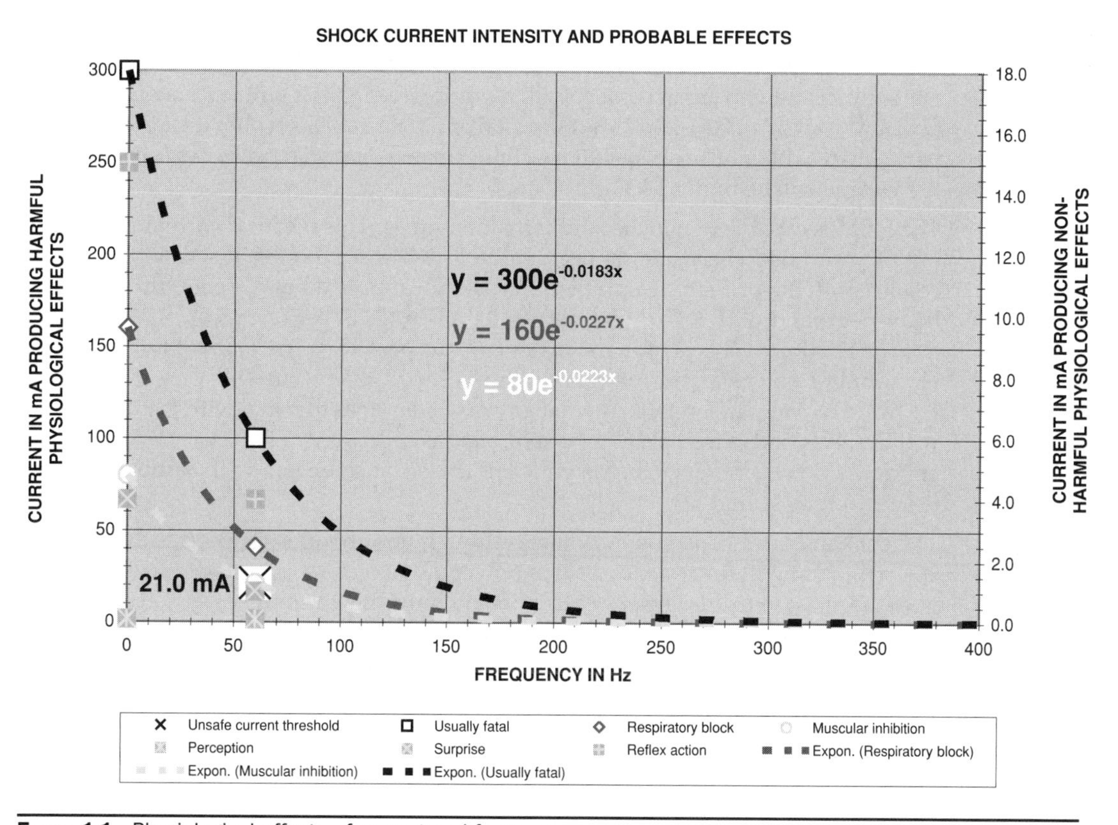

FIGURE 1-1 Physiological effects of current and frequency.

high-voltage currents over his body. These demonstrations were conducted using his "Tesla coil"with an estimated resonant frequency of at least 50 kHz.

With the advent of CMOS devices and their inherent destruction from Electrostatic Discharge (ESD), the electrical characteristics of the human body have been extensively researched. This human body model (HBM) also serves to illustrate the potential hazards found in a Tesla coil and the need for safety. The HBM standard from reference (1) defines the following human body electrical characteristics during a static discharge:

- 100 pF of capacitance.
- 1,500 Ω of resistance.
- 2 to 10 nsec exponential rise time.
- 150 nsec exponential fall time.

The voltage the HBM is charged to ranges from 500 V to 4,000 V. If you accidentally (or intentionally) touch the *primary* circuit in a medium size coil using a 230-V line supplying a 70:1 step-up transformer at full output, the following will theoretically happen:

- You will be unable to physically react within a few nanoseconds (0.000000002 second). Severe electrical shock produces paralysis not to mention the effect of large currents being conducted through a nervous system that operates on picoamps.
- For each second you are in contact with the primary circuit it will reach a peak voltage of (230 V × 1.414) × 70 = 22.77 kV, for a total of 120 times. This is only the instantaneous peak. There will only be about 100 μseconds during these 120 peaks per second where the voltage will be under 1 kV.
- The HBM is capable of nanosecond transition times. This means a current pulse in or out of your body can reach its peak value in 2-10 nsec. For each second you are in contact with the primary circuit the 22.77 kV can supply a peak current through your body-to-ground of: $I = C \times (\Delta v/\Delta t) = 100$ pF × (22.77 kV/2 nsec) = 1,139 A, for a total of 120 times. This is only the instantaneous peak. The current will follow the 60-Hz supply sine wave and the current through you to ground will vary from 0 A to 1,139 A. For comparison look at what an arc welder does to metal with just 100 A. You will draw a steady rms value of current of at least 16.6 kV/1,500 Ω = 11 A. This rms current will probably increase as your HBM resistance lowers with carbonization of tissue. Getting scared yet!
- The peak instantaneous power flowing through your body is (1,139 A × 22.77 kV) = 25.9 MW. You will draw a steady rms power value of 11 A × 16.1 kV = 177 kW. The good news is your line supply will probably limit this to some lower value or some overcurrent protection device (circuit breaker or fuse) in your service box will trip. Pay attention in Chapter 7 to the current limiting and circuit protection sections. Protective devices will generally break a short circuit within one positive or negative alternation of the line, which is less than 8.3 msec.

A good illustration of how dangerous high-voltage 60-Hz currents are can be found in reference (2). First Sergeant Donald N. Hamblin has the distinction of being the only reconnaissance Marine in the Vietnam War with a prosthetic device. Prior to his deploying overseas he was parachute training in Camp Pendleton, California when winds blew his chute toward a high-voltage transmission line. The chute caught on an upper 69-kV line and the First Sergeant swung into a lower 12-kV line, his foot touching the lower line. Observers on the ground described an explosion where the First Sergeant's foot contacted the 12-kV line and his parachute burst into flame, no longer holding him in the lines. The First Sergeant then fell over 50 feet to the ground. All of this happened in a moment. There was not much left of his foot and it was soon amputated (not unusual in high-voltage accidents as the damaged tissue does not heal). The power company was inconsiderate enough to send him a bill for damages. He recovered, learned to function with a prosthetic foot, and served in arduous reconnaissance duty in Vietnam. The point of this illustration is the high-voltage output of the step-up transformer used in a medium or large Tesla coil is operating with the same high-voltage 60-Hz currents. First Sergeant Hamblin was wearing jump boots and the parachute and risers were made of nylon and/or silk. These are all good insulators, which means the First Sergeant did not present a direct short upon contact. If you contact an active primary circuit in an operating Tesla coil you will not be as fortunate! Also keep in mind that Marines are very tough and fit individuals with a lot of self-discipline. The First Sergeant's recovery would not be possible for most people involved in such an accident.

A body 6 feet in height also acts as a grounded 1/4 wavelength antenna resonant (tuned) to: $984 \times 10^6/(6 \times 4) = 41$ MHz. If you're shorter than 6 feet this resonant frequency is higher. Under nominal conditions a person 6 feet in height will absorb the most RF energy when the oscillations are at 41 MHz. Your Q and bandwidth will vary considerably with a variety of factors; therefore the energy absorbed at frequencies above and below this 41 MHz is also dependent upon such factors. Remember that you don't have to touch an RF circuit to get hurt. Your body will receive some amount of any radiated energy, this amount attenuated by the square of the distance from the source. This is typically most harmful around the FM radio broadcast bandwidth. But remember, even some of that 60-Hz transmission line EMF is being received. Tesla coils are usually designed to operate below 500 kHz and a medium-size coil will propagate negligible RF and magnetic fields outside of a 3-meter area so the energy received by an observer is not generally harmful. RF and magnetic field strength are attenuated by the reciprocal of the distance squared (inverse square law) in non-ionized air.

Do not be afraid to continue your electrical investigations, just respect the potential danger involved and use good judgment. Remember to always start out small and work up in small increments. When trying something new, control the current and power levels to the smallest values that will produce the effects you are trying to create, then work up. **NEVER TOUCH THE LINE OR PRIMARY CIRCUIT OF AN OPERATING TESLA COIL!!!** Figure 1-2 illustrates what not to touch.

After Tesla introduced his high-frequency coils and demonstrated their effects at the 1893 Columbian Exposition, electrical shows became quite common. Showman would touch the hundreds of kilovolts being produced by the secondary, or as Tesla first displayed, create a ring of corona around their person. Many science museums still exhibit these effects in public demonstrations. If your Tesla coil is operating above 1 kW it can produce uncomfortable shocks in the secondary. *Even though they are at high frequencies the currents in large coils are dangerous and even deadly*. The highest secondary voltage I have intentionally contacted is about 450 kV at a primary power level of 3 kW and it was quite uncomfortable. If you want to observe the skin effect, try it with a very small coil to start and progress upward with caution. You will quickly find a level that you will not want to visit again.

Other hazards include applying the high-voltage output of a Tesla coil to devices under high vacuum such as X-ray tubes. When the high voltage is applied to an anode in a tube of sufficient vacuum, X-rays are produced and can be lethal. Do not experiment with discharges in high vacuum unless you are absolutely sure of what you are doing. It is not within the scope of this design guide to address discharges in high vacuum. The early X-ray machines (circa 1910s) were essentially Tesla coils driving high-vacuum X-ray tubes. These were even available for home use until regulated by the government.

Do not look at the arc in the spark gap of an operating coil as it produces harmful ultraviolet (UV) rays. The UV is at an intensity comparable to arc welding. Do not stare at the operating spark gap any more than you would stare at an arc welder performing his duties. The spark gap can be safely observed using the same eye protection worn by an arc welder. The pain receptors in the eyes are not very sensitive to UV; however, it is more damaging than infrared and you will not know there is any damage until it is too late. The same damage can be incurred while watching a solar eclipse without proper eye protection (filters). The best course is to not look at it. You can look at the high-voltage spark discharge of the secondary, which is the purpose of building a coil. This spark produces visible light in the violet to white range and low levels of UV, which are not considered harmful.

The minimum safety equipment you should have on hand and safety practices when operating a coil are:

- Fire extinguisher with approved agent for electrical fires (Class C).
- Shoes with thick insulating soles. Tennis shoes have a good inch of non-conducting material between you and ground, which serve well for testing coils. Tesla used special shoes with several inches of cork for the soles, which must have made his already tall appearance seem gigantic.
- Keep one hand in your trouser pocket while circuits are energized, to disable the conductive electrical path from one hand–through the heart–to the other hand. Tesla brought public attention to this safety method.
- Eye protection. There is a potential for anything to come apart when the coil is running. To protect your eyes wear approved industrial eye protection during operation. This is usually required in any industrial or laboratory setting so get into good habits. If the spark gap is to be observed while running the coil use eye protection approved for arc welding.

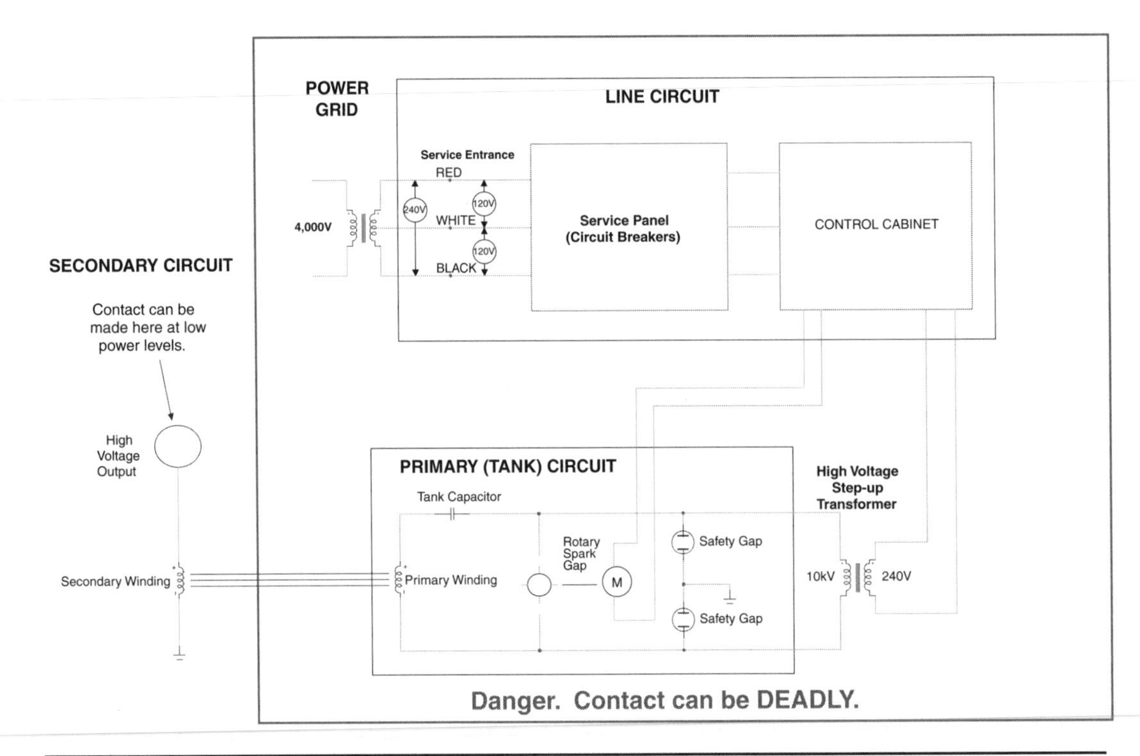

FIGURE 1-2 Dangerous areas in an operating coil.

- Hearing protection. The spark gap will sound like a rapid series of gunshots and as the secondary voltage increases the secondary spark discharge can quickly exceed the sound level produced by the spark gap. Where do you think thunder comes from? While running one of Ed Wingate's large coils the ambient noise level was measured at a safe observation distance by Tom Vales, using a General Radio precision sound level meter. The sound level measured above 124 dB. This is comparable to the loudest jet engines so protect your hearing and wear adequate protection.
- Post a safety observer. If you are just beginning to work in high voltage make sure you have someone around to render assistance if needed. It is good practice to have another person check connections and circuit details before a coil is energized.
- Breathing protection. When cutting, drilling or sanding epoxy, phenolic, plastic, resin laminates and similar materials use a mask to prevent breathing in the fumes and particles. There are a wide variety of materials not addressed in this guide, some quite hazardous. Consult the manufacturer's Material Safety Data Sheets (MSDS) for specific material hazards if they are available. If no information is available use common sense and a mask.

1.6 Derating

Throughout this guide the term derating surfaces. What is derating? Generally manufacturers provide performance parameters for their parts at laboratory temperature (25°C or 77°F), standard air pressure (1 atmosphere pressure or sea level), relative humidity of 50%, and no aging (new or beginning of life). When the part is used in a variety of industrial settings the performance will be degraded or enhanced depending on the environmental conditions of its use. To evaluate these effects on performance the manufacturer will provide a derating methodology to ensure the operating conditions do not overstress the part. Properly applying the derating to the performance parameter will ensure the parts we use in our designs will last as long as the manufacturer intended. A performance parameter is an electrical characteristic of the part such as power handling ability of resistors, working voltage capability of capacitors, or trip current threshold of circuit breakers. The Excel worksheets included on the companion Web site, which can be found at www.mhprofessional.com/tilbury/, will automatically calculate part deratings for the operating environment as explained in the applicable sections. Formulae are included in the text to perform manual calculations if Excel is not used. Derating may also be applied by the manufacturer to increase the service life or reliability of the part.

References

1. ESD Association Standard ANSI/ESD S20.20-1999. Electrostatic Discharge Association, Rome, NY: 1999.
2. D.N. Hamblin and B.H. Norton. *One Tough Marine: The Autobiography of First Sergeant Donald N. Hamblin, USMC*. Ballantine Books: 1993 pp. 179–180.
3. Department of Defense Handbook for Human Engineering Design Guidelines, MIL-HDBK-759C, 31 July 1995, p. 292.
4. Richard Moran. *Executioner's Current*. Alfred A. Knopf, NY: 2002.
5. Cheney, Margaret and Uth, Robert. *Tesla Master of Lightning*. Barnes & Noble Books, NY: 1999.

6. Seifer, Marc J. *Wizard: The Life and Times of Nikola Tesla. Biograph of a Genius.* Birch Lane Press: 1996.
7. Hunt, Inez and Draper, Wanetta. *Lightning in His Hand.* Omni Publication, Hawthorne, CA: 1964.
8. O'Neill, John J. *Prodigal Genius: The Life Story of N. Tesla.* Ives Washburn, NY: 1944.
9. Lomas, Robert. *The Man Who Invented the Twentieth Century: Nikola Tesla, Forgotten Genius of Electricity.* Headline Book Publishing, London: 1999.

CHAPTER 2

Designing a Spark Gap Tesla Coil

First assumption on which to base calculations of other elements is made by deciding on the wavelength of the disturbances. This in well designed apparatus determines the λ/4 or length of secondary wound up. The self induction of the wire is also given by deciding on the dimensions and form of coil hence Ls and λ are given.

Nikola Tesla. Colorado Springs Notes: 1899-1900, p. 56.

Tesla determines the resonant frequency of a new secondary winding.

The classic Tesla coil is based on a spark gap (disruptive discharge) design as shown in Figure 2-8. The resonant primary circuit is typically tuned to the resonant frequency of the secondary circuit. In the primary circuit a capacitor, charged to a high voltage, is in series with the primary winding and the deionized spark gap. The spark gap, once ionized by the capacitor's charge, is used as a switch to produce oscillations in this series resonant LCR circuit. The primary oscillations produced during the discharge of the capacitor are damped as a result of the resistance in the ionized spark gap. The oscillating current in the primary winding is coupled to the secondary winding through the mutual inductance of the air core resonant transformer producing an oscillating current in the secondary winding. This oscillating current produces a high voltage in the secondary winding's resistance and is usually accumulated in a terminal capacitance until the surrounding air is ionized and a spark breaks out.

2.1 Designing the Tesla Coil Using a Spark Gap Topology

To begin a new Tesla coil design open the CH_2.xls file, AWG vs VS worksheet (1). (See App. B.) If this is your first coil project I recommend using a commercial capacitor for the primary tank circuit. Obtaining affordable surplus capacitors sometimes compares to a search for the "holy grail."Building one assumes you can find high-quality materials that will last. A commercial capaci- tor has the advantage of using the best materials, quality engineering and testing in its design. Very thin, hard-to-work with materials are used in the plates and dielectric to yield the most capacitance and dielectric strength per volume. Air and other contaminants are removed from between the plates during construction and some are even filled with insulating oil. You would be hard pressed to build a better capacitor than a team of engineers and technicians familiar with the state-of-the-art that have access to the best materials. Homemade capacitors can be built to perform adequately when carefully constructed.

Referring to Chapter 5 (Capacitors), obtain or build a primary (tank) capacitor suitable for use in the proposed design. When I begin a new design, I start by finding a high-voltage capacitor within the following suggested ranges:

- 0.001 μF to 0.01 μF for a small coil using a current-limited transformer in the 100-W to 1-kW range, e.g., neon sign transformer.
- 0.01 μF to 0.05 μF for a medium coil using a non–current-limited transformer in the 1-kW to 5-kW range, e.g., potential transformer.
- >0.05 μF for a large coil using a non–current-limited transformer in the 5-kW and above range, e.g., distribution transformer.

NOTE: *Veteran coiler Ed Wingate has suggested the often-used term for a distribution transformer—"pole pig" may induce a premature sense of familiarity and lack of caution in new coilers. I concur with his observation therefore the term will not be used again in this guide.*

The design does not have to begin with the capacitor if you have a variety of capacitor values to choose from. It may center on the step-up transformer output voltage and current ratings. Or you may have a desired primary or secondary winding geometry in mind and select the primary capacitance value that produces resonance. However, this will probably not be an option for your first coil as the capacitor and step-up transformer will be the most difficult parts to obtain.

Using the CH_2.xls file, AWG vs VS worksheet (1) and Sections 2.2 through 2.4 complete the design for the selected capacitor. The blue cells (B5 through B46) shown in Figure 2-7 are the required inputs to perform the design calculations. The green cells in column (B) and all cells in columns (C) and (F) are calculations so do not enter any values into these cells. Details for each of these inputs are as follows:

1. *Secondary characteristics*. Enter the desired secondary winding form diameter in inches into cell (B8), magnet wire gauge in cell (B6), number of winding layers in cell (B5), and desired resonant frequency in kHz into cell (B10). The number of turns and total winding height are calculated from these inputs as well as the electrical characteristics of the secondary winding. I recommend beginning with a one-layer winding in the calculations. If an interwinding distance is desired or a wire other than magnet wire is used enter the interwinding separation in cell (B9). Section 2.2 explains these calculations in detail. Enter the value of terminal capacitance in pF into cell (B11). To calculate the terminal capacitance, see Chapter 5. The effects of adding terminal capacitance on the resonant frequency of the secondary are included in the calculations and the calculated resonant frequency with terminal capacitance is shown in cell (F21).
2. *Primary characteristics*. Enter the line frequency in cell (B14). The line frequency is typically 60 Hz (US) and the line voltage either 120 V or 240 V. Enter the step-up transformer's rated output voltage in kV into cell (B15) and rated output current in amps into cell (B16). Enter the estimated *rms* line voltage with the coil running

into cell (B17). The breakdown characteristics of the spark gap will determine this voltage and until the coil is built and tested it can only be estimated. If you have no idea what this value should be start out by entering the maximum line voltage you have available (typically 115–120 V or 230–240 V). After the remaining spark gap and primary characteristics are entered into the worksheet adjust the line voltage in cell (B17) until the calculated step-up transformer peak output voltage in cell (C22) is higher than the calculated spark gap breakdown threshold in cell (B34). This is approximately where the *rms* line voltage will be with the coil running. The power level the coil will operate at is generally dependent on the value of tank capacitance selected and the spark gap break rate (BPS). This is detailed in Section 2.3. Once the coil is running any meter indications can be entered into the worksheet to fine tune and evaluate the performance of the coil in operation.

Enter the step-up transformer's turns ratio in cell (B18). The turns ratio is the ratio of primary turns-to-secondary turns or primary (input) voltage-to-secondary (output) voltage. Transformers will typically have either the turns ratio or input and output voltage labeled somewhere on them. If there is no label see Chapter 4 for details on determining relationships in unmarked transformers. The (1:) is part of the cell formatting so enter only the secondary value of the turns ratio, e.g., a 1:70 ratio is entered as 70. The *rms* output voltage of the step-up transformer is calculated in cell (B22) using the line voltage and turns ratio and the peak output voltage (peak voltage applied to Tesla coil primary tank circuit) in cell (C22). Distribution transformers may pose some confusion. They often operate as step-down transformers to step down the voltage. For use in a Tesla coil they are operated in reverse or as step-up transformers so the turns ratio is reversed from the ratio printed on the case. The printed turns ratio on the case may also refer to the current ratio instead of the voltage ratio as in potential transformers. If no manufacturer's information is available you will have to interpret the ratio. Again refer to Chapter 4 for assistance.

Enter the tank capacitance you have selected in μF into cell (B19). The rated output current and voltage of the step-up transformer and BPS are used to calculate a maximum usable capacitance in cell (F47), which should be greater than or equal to the value of the primary capacitor entered in cell (B19) to optimize the design.

The use of a current-limited transformer makes the design easier but limits optimization once the coil is constructed. Using a primary capacitance larger than the calculated value in cell (F47) will not produce a larger secondary voltage because the primary current is limited. Although current-limited transformers can be connected in parallel to increase the output current, I recommend using a non–current-limited transformer with larger values of primary capacitance. This will enable better optimization of the spark output once the coil is constructed and running as explained in Section 2.6. Enter the DC resistance of the primary winding in cell (B20). This can be calculated for the winding type as detailed in Chapter 4. Enter the separation of the base of the primary winding to the base of the secondary winding in inches into cell (B21). This is explained in detail in Chapter 4 for calculating the mutual inductance of two coaxial coils.

3. *Spark gap characteristics*. Refer to Chapter 6 for details in calculating rotary gap or fixed gap parameters. Enter the distance between the spark gap ends in inches into cell (B25). Enter the applied overvoltage in % into cell (B26). The overvoltage is typically

0%. The calculations can be corrected for the applied temperature–pressure–humidity conditions by entering the correction factor into cell (B27). The correction factor must first be determined using the methodology detailed in Chapter 6. Enter the estimated spark gap ionization time in μs into cell (B28). The number of primary oscillations required to decay to the minimum ionization threshold in seconds entered in cell (B33) is calculated in columns (FV) to (GB) using the methodology in Section 6.11. The calculated number of primary oscillations appears in cell (C29) and the corresponding spark gap operating time period in cell (B29). Unless you are accounting for operating tolerance enter the calculated value in cell (B29) into cell (B28). As the calculations progress this time period can be adjusted. Enter the desired breaks per second (BPS) in cell (B30). If the spark gap is a fixed type the BPS is 120. For a rotary gap the BPS is calculated using the methodology detailed in Chapter 6. The most efficient spark gap that can be built by amateur coilers is the rotary gap. The type of gap selected will depend upon your ability to fabricate parts from stock materials and the availability of these materials.

To characterize a non-synchronous rotary gap performance, enter the phase shift in cell (B31). If you are undecided about the phase shift enter 89.54° (synchronous) to begin the design and we will optimize it later in Section 2.6. For the spark gap characteristics entered the calculated breakdown voltage of the gap during the positive line alternation is shown in cell (B34). The calculated peak output voltage of the step-up transformer in cell (C22) must be larger than the breakdown voltage in cell (B34) or the spark gap will not ionize. If this occurs the gap distance in cell (B25) must be reduced. Enter a 1 into cell (B32) if the electrode material used in the spark gap is copper, brass, aluminum, or silver. Enter a 2 into cell (B32) if magnesium is used. According to reference (5) copper, brass, aluminum, and silver exhibit a linear waveform decrement and magnesium exhibits an exponential waveform decrement. Other materials used could be tungsten or steel. A conditional statement determines whether a linear or exponential gap characteristic is used in the calculations for the selected characteristic entered into cell (B32).

Although magnesium was apparently used in early commercial and experimental spark gaps I would not recommend using it. The plasma in the gap arc can reach several thousand degrees and may ignite the magnesium. If this occurs you cannot put out a magnesium fire, and will have to wait nervously until all of the magnesium material burns away! At this point the surrounding material will certainly be burning. However, it can be extinguished using an agent approved for electrical fires (Class C). Supposedly a fire-extinguishing agent approved for Class D use will extinguish a magnesium fire; however, I doubt it. The U.S. Naval Aviation community has had many fires of this type and develops their own extinguishing agents in a modern laboratory (a division of the Naval Research Laboratory). During fire training it is emphasized that magnesium fires cannot be safely extinguished and to let them burn until out. I will take them on their word as I have not experienced one myself. The danger involved is not worth the risk.

4. *Primary tuning characteristics*. The primary inductance and capacitance determine the frequency of primary oscillations. The primary inductance required to produce oscillations at the resonant frequency of the secondary is shown in cell (B48). Select either an

Archimedes spiral or helically wound primary by entering a 1 or 2 value respectively into cell (B38). If a flat Archimedes spiral (pancake) primary is used enter a value of 1 into cell (B38) and an angle of inclination (θ) of 0° into cell (B43). If the Archimedes spiral primary is not a flat spiral enter a value of 1 into cell (B38) and the desired angle of inclination in degrees into cell (B43). Enter the inside diameter of an Archimedes spiral primary or outside diameter of a helical primary into cell (B40). Enter the total number of turns desired in the primary winding into cell (B42) and the interwinding distance into cell (B41). All of these parameters are explained in detail in Chapter 4.

Vary the number of turns used in the primary in cell (B44) until the calculated primary inductance in cell (B47) is as close to the calculated value in cell (B48) as possible. Expect a typical change of inductance between turns of 5–15 μH. To obtain the maximum theoretical secondary voltage usually requires additional series inductance entered into cell (B46) until the value of calculated inductance in cell (B47) matches the calculated value in cell (B48). An alternate tuning method is to use additional parallel capacitance entered into cell (B45), which is added to the primary capacitance value. The calculated resonant frequency of the primary tank circuit is shown in cell (B49). Remember, the secondary winding with terminal capacitance will determine the resonant frequency. The primary tank circuit is usually tuned to this frequency to couple the most energy to the secondary. Section 2.3 explains these calculations in detail. The primary oscillating current produces a wideband envelope of harmonics. A few turns above or below the selected turn entered into cell (B44) can produce comparable secondary voltages, especially when the primary decrement is low (many primary turns). The lower the primary decrement the less critical the primary tuning becomes.

5. Using the CH_2A.xls file worksheet (see App. B) and Section 2.5, check the performance of the proposed design parameters selected in steps 1 thru 4 before construction begins. Using the Excel files and computer simulations to fine-tune the coil before it is built enables component values outside suggested ranges to be evaluated in the coil's performance, before the time is spent constructing them. In this way, hard-to-find components like high-voltage capacitors can be found first, then the rest of the Tesla coil built around the capacitance value.
6. Using Chapter 7 (Control, Monitoring, and Interconnections), design a control and metering scheme for use with the coil design in the previous steps. The schematic diagram in Figure 2-1 details a medium coil built to produce a 5–6 foot spark. The control, monitoring, and interconnections are shown, which will be similar to any spark gap coil design.
7. Begin construction and have fun.

2.2 Calculating the Secondary Characteristics of a Spark Gap Coil

The secondary winding of many turns produces a high-Q circuit possessing a high self-inductance, a small self-capacitance, and a resistance. It is generally wound as a single layer helix of many turns. The impedance at resonance is equal to the DC resistance of the winding plus the skin and proximity effects of the high-frequency oscillations.

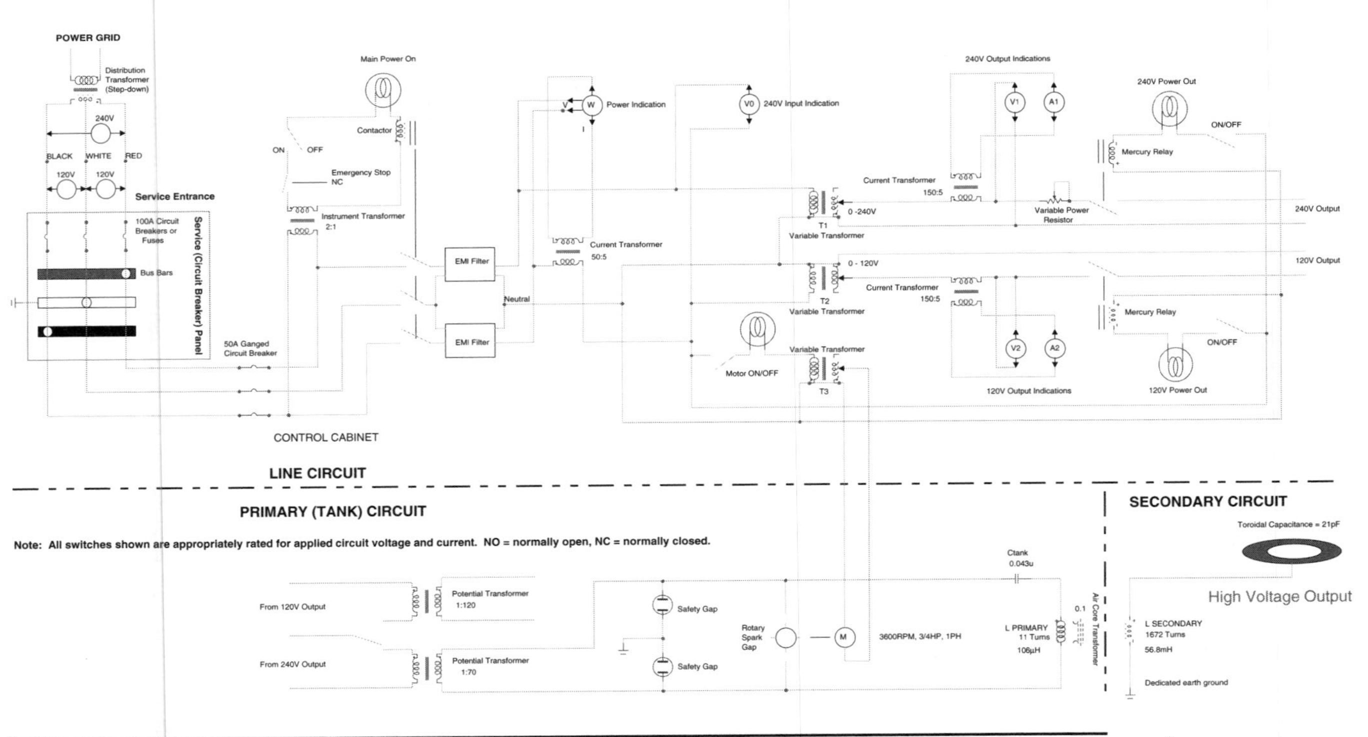

FIGURE 2-1 Details of circuit connections in a spark gap Tesla coil.

Open the CH_2.xls file, AWG vs SECONDARY VOLTAGE worksheet (1). First a quick note about the worksheet calculations. The calculations rely on estimates and approximations of anticipated coil performance. The closer these are to the actual operating coil parameters the more reliable the calculations. Values such as the ionization time entered in cell (B29) require calculation in another worksheet. When directed, perform additional calculations and estimates using the appropriate worksheets for more reliable results. When any value is entered that exceeds the performance limitations of the coil, your only indication will be that some calculations, such as the secondary voltage, will be unrealistically high. Do not be disappointed when your neon sign transformer does not produce a calculated 5.0 MV in the secondary. Back up and review your calculations and find what parameter was not properly defined. The more you work with the calculations the easier they become and the more you will understand about designing Tesla coils. The relationships will soon become intuitive.

The secondary characteristics determine the resonant frequency of the Tesla coil and the frequency the primary oscillations are tuned to; therefore they will be evaluated first. Using a fixed resonant frequency and coil form diameter the remaining secondary parameters can be calculated. The secondary at resonance will act as a quarter wavelength ($\lambda/4$) resonator or antenna. The secondary wire length is therefore proportional to this quarter wavelength:

$$\frac{\lambda}{4} = \frac{c}{fo} \tag{2.1}$$

Where: $\lambda/4$ = One quarter wavelength of resonant frequency in feet = secondary wire length = cell (F14), calculated value.
c = Propagation speed of wavefront in free space (vacuum) = 9.84×10^8 feet/sec, or 2.998×10^8 meters/sec.
fo = Resonant frequency of coil in Hz = cell (B10); enter value in kHz. Converted to Hz in cell (C10) using a 1e3 multiplier.

The wire length per turn in the secondary is calculated:

$$L/T = D\pi + d \tag{2.2}$$

Where: L/T = Length per turn of wire in inches = cell (F13), calculated value.
D = Diameter of secondary coil form = cell (B8), enter value.
d = Diameter of wire with insulation and interwinding distance (if used) = cell (F8), calculated value from NEMA Wire Standard:

$$\text{Wire diameter in inches} = 0.0050 \bullet 1.1229322^{(36-AWG\#)}$$

interwinding distance for close wound magnet wire = insulation thickness. See Figure 2-2. To form a helically wound coil each turn of wire must have a slight angle (pitch) to place it on top of the preceding turn as it is wound up the form. To account for this additional length per turn the diameter of wire with interwinding distance is added to the circumference of the form ($D \times \pi$).

NOTE: *If an interwinding space is desired by entering a value into cell (B9) the interwinding distance will default to the value specified in cell (B9).*

FIGURE 2-2 Interwinding distance in secondary coil.

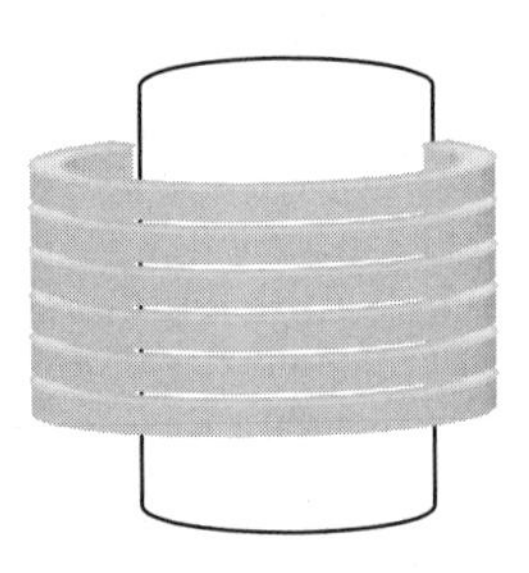

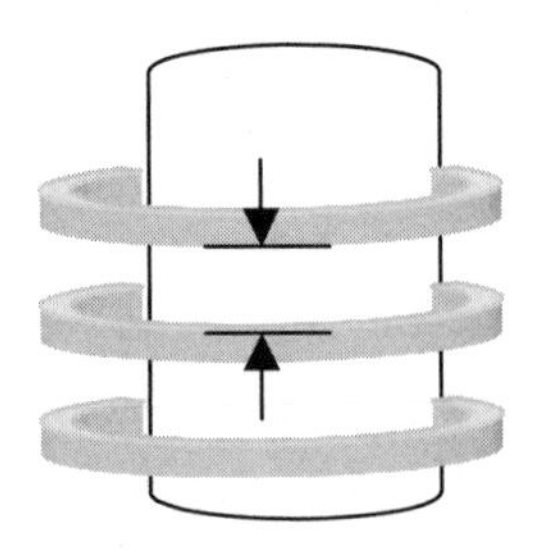

The required number of turns in the secondary can now be determined:

$$N = \frac{\left(\frac{\lambda}{4}\right)}{L/T} \tag{2.3}$$

Where: N = Required number of turns of wire = cell (F11), calculated value.
$\lambda/4$ = One quarter wavelength of resonant frequency in feet = secondary wire length = cell (F14), calculated value from equation (2.1).
L/T = Length per turn of wire in inches = cell (F13), calculated value from equation (2.2).

And the winding height:

$$H = N \bullet L/T \tag{2.4}$$

Where: H = Required height of winding in inches = cell (F15), calculated value.
N = Required number of turns of wire = cell (F11), calculated value from equation (2.3).
L/T = Length per turn of wire in inches = cell (F13), calculated value from equation (2.2).

Now that the physical dimensions of the secondary winding are known the electrical characteristics and can be determined. First calculate the inductance of the coil using the Wheeler formula from reference (8):

$$Ls(\mu\text{h}) = \frac{A^2N^2}{9A + 10H} \tag{2.5}$$

Where: Ls = Inductance of secondary coil in μhenries ± 1.0% = cell (F17), calculated value, converted to mH using a 1e-3 multiplier.
A = Radius of coil form in inches = 1/2 outside diameter entered in cell (B8).
N = Required number of turns of wire = cell (F11), calculated value from equation (2.3).
H = Required height of winding in inches = cell (F15), calculated value from equation (2.4).

Next calculate the self-capacitance of the winding using the selected resonant frequency:

$$Cs = \frac{1}{4\pi^2(fo^2 Ls)} \tag{2.6}$$

Where: Cs = Self-capacitance of secondary coil in farads = cell (F18), calculated value, converted to pF using a 1e12 multiplier.
fo = Resonant frequency of coil in Hz = cell (B10), enter value in kHz.
Ls = Inductance of secondary coil in henries = cell (F17), calculated value from equation (2.5).

The DC resistance of the winding can also be calculated:

$$\mathrm{DC}\Omega = \frac{\frac{\lambda}{4} \bullet \Omega/\mathrm{ft}}{\mathrm{NS}} \bullet (1 + [T_A - 20^\circ\mathrm{C}] \bullet 0.00393) \tag{2.7}$$

Where: $\mathrm{DC}\Omega$ = Total DC resistance of the winding in ohms = cell (F16), calculated value.
NS = Number of layers (strands) of wire used = cell (B5), enter value.
$\lambda/4$ = Total length of wire in winding (one quarter wavelength of resonant frequency) in feet = cell (F14), calculated value from equation (2.1).
T_A = Ambient temperature in °C = cell (B7), enter value. Converted to °F in cell (C7) using the conversion: $^\circ\mathrm{C} \times (9/5) + 32$.
Ω/ft = DC resistance for one foot of selected wire gauge = cell (F7), calculated value = 10.3 Ω/cirmil ft. The wire diameter (d) = $0.0050 \bullet 1.1229322^{(36-AWG\#)}$. The diameter in cirmils = $(d \bullet 1000)^2$.
The 10.3 Ω/cirmil ft DC resistance is for an ambient temperature of 20°C with a temperature coefficient of 0.393%/°C. The DC resistance is adjusted for the ambient temperature entered in cell (B7).

Now that the secondary dimensions and electrical characteristics are defined the skin and proximity effects can be determined. These effects (AC resistance) increase the total resistance beyond the DC resistance calculated in equation (2.7). When only the skin effect is considered a typical formula from reference (1) for calculating the AC resistance is:

$$\mathrm{AC}\Omega = \frac{9.96 \times 10^{-7} \bullet \sqrt{fo}}{d} \tag{2.8}$$

Where: $\mathrm{AC}\Omega$ = Total AC resistance (skin effect only) of the winding in ohms.
fo = Resonant frequency of coil in Hz = cell (B10), enter value in kHz.
d = Diameter of wire = cell (F5), calculated value from NEMA Wire Standard.

However, to properly evaluate the coil performance the proximity effect cannot be ignored. Reference (2) provides a methodology for calculating the total AC and DC resistance in Switch Mode Power Supplies (SMPS), which typically operate in the same frequency range as a Tesla coil. The methodology includes the skin and proximity effects. It uses the Dowell method detailed in reference (3) developed for close wound helical magnet wire with applied sinusoidal AC waveforms, perfect for evaluating the AC effects in our secondary winding. To evaluate these effects the depth of current penetration (skin depth) at the resonant frequency must first be calculated:

$$D\delta = 7.5 fo^{-\left(\frac{1}{2}\right)} \tag{2.9}$$

Where: $D\delta$ = Depth of current penetration in centimeters = cell (F26), calculated value. Converted to inches using a 2.54 divisor.
fo = Resonant frequency of coil in Hz = cell (B10), enter value in kHz.

Next calculate the copper layer factor:

$$F_L = 0.866d \frac{\left(\frac{N}{\text{NS}}\right)}{le} \tag{2.10}$$

Where: F_L = Multiplier used to calculate Q' = cell (F27), calculated value.
d = Diameter of wire in inches = cell (F5), calculated value from NEMA Wire Standard.
N = Required number of turns of wire = cell (F11), calculated value from equation (2.3).
NS = Number of layers (strands) of wire used = cell (B5), enter value.
le = Magnetic path length of winding = height of secondary winding = cell (F15), calculated value from equation (2.4). The magnetic path length in this application is equivalent to the height of the coil.

Now calculate Q', a figure of merit denoting the ratio of the wire diameter (d) and its associated layer factor (F_L) to the depth of penetration ($D\delta$):

$$Q' = \frac{0.866d\,F_L^{\frac{1}{2}}}{D\delta} \tag{2.11}$$

Where: Q' = Ratio of d to $D\delta$ = cell (F28), calculated value.
d = Diameter of wire in inches = cell (F5), calculated value from NEMA Wire Standard.
F_L = Multiplier used to calculate Q' = cell (F27) calculated value from equation (2.10).
$D\delta$ = Depth of current penetration = cell (F26), calculated value from equation (2.9).

The graph in Figure 2-3 is used to determine F_r, a figure of merit denoting the ratio of AC resistance (ACΩ or R_{AC}) to DC resistance (DCΩ or R_{DC}). Find the point on the X-axis (abscissa) that corresponds to the calculated Q' (from equation (2.11)). Move vertically from this point to intersect the line corresponding to the number of layers (NS) used in the winding (black trace for single layer). Note the graph is in a log-log scale and must be read as such. The $R_{\text{AC}}/R_{\text{DC}}$ ratio (F_r) is found directly across this intersection on the Y-axis (ordinate). The F_r is automatically interpolated in the worksheet using the calculations in columns (AS) through (AY), appearing in cell (F29) and used to calculate the total resistance using the formula:

$$Rt = F_r \bullet R_{\text{DC}} \tag{2.12}$$

Where: R_t = Total resistance (ACΩ + DCΩ) due to DC resistance, eddy current (skin), and proximity effects = cell (F30), calculated value. This resistance is combined with the reactance for the total impedance in equation (2.16).

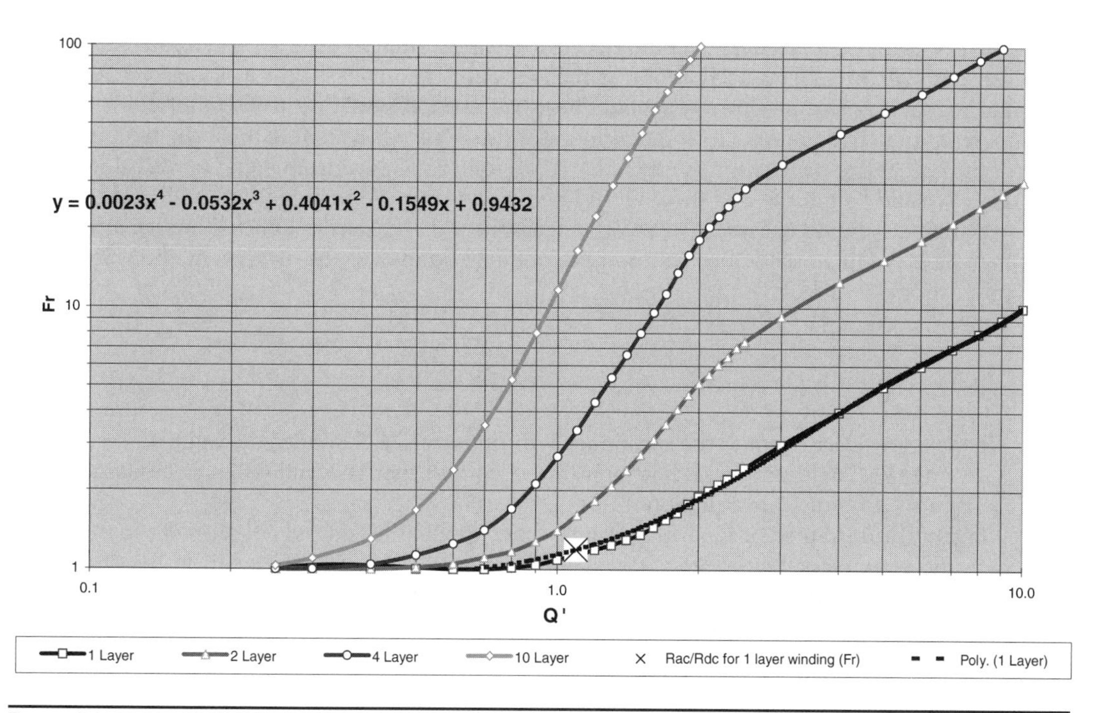

FIGURE 2-3 F_r vs. Q' for sinusoidal waveforms.

F_r = Value of R_{AC}/R_{DC} ratio found using intersection of Q' and NS in Figure 2-3. The fourth-order polynomial curve fit equation shown in the figure is used to calculate F_r in a single layer winding in column (AD).

R_{DC} = Total DC resistance (DCΩ) of the winding in ohms = cell (F16), calculated value from equation (2.7).

Once the total winding resistance at the resonant frequency is determined the quality (Q) of the secondary winding can be calculated:

$$Qs = \frac{\omega o\, Ls}{Rt} \tag{2.13}$$

Where: Qs = Quality of secondary winding (figure of merit) = cell (F32), calculated value.

ωo = Resonant frequency in radians per second = cell (F22), calculated value. $\omega = 2\pi f_o$ where: f_o is the resonant frequency of the secondary in Hz with terminal capacitance = cell (F21).

Ls = Inductance of secondary coil = cell (F17), calculated value from equation (2.5).

Rt = Total resistance of secondary winding in ohms = cell (F30), calculated value from equation (2.12).

The calculated Q considering skin effect only using equation (2.8) is shown in cell (F33) for comparison. Sinusoidal oscillations at the resonant frequency begin in the secondary once the primary oscillations are quenched. These oscillations would continue indefinitely if the Q were infinite. In equation (2.13) the numerator ($\omega o Ls$) is equivalent to the inductive reactance of the coil at resonance ($\omega o Ls = 2\pi f o L$). If the total resistance from equation (2.12) was zero the Q would be infinite and there would be no damping of the secondary oscillations. However, there is always resistance in a resonant circuit and each oscillation is of smaller amplitude than its preceding oscillation. This damping effect is known as the decrement and is inversely proportional to the Q:

$$\delta S = \frac{\pi}{Qs} \tag{2.14}$$

Where: δS = Decrement of the secondary winding = cell (F31), calculated value.
Qs = Quality of secondary winding (figure of merit) = cell (F32), calculated value from equation (2.13).

Whether the decrement is logarithmic, linear, or exponential has less effect on the spark length than the peak secondary voltage developed. The peak voltage developed in the secondary winding will determine the length of the spark and the decrement affects how bright and thick (intensity) the spark is. There are published references to this decrement characteristic being logarithmic, linear, or exponential; the correct decrement being of academic interest only. Reference (4) cites the decrement as logarithmic and reference (5) as either linear or logarithmic, dependent on the type of material used in the spark gap electrodes. These sources infer a linear or logarithmic decrement in the primary circuit, which is independent of the secondary decrement. A comparison between logarithmic and exponential decrement characteristics and the oscillating secondary circuit is performed in Chapter 6. The entire secondary waveform is calculated using the accepted definition of a damped waveform: "each succeeding oscillation will be reduced by the decrement." Both calculations and oscilloscope observations of the secondary waveform indicate the decrement is virtually exponential. An exponential decrement characteristic in the secondary will be inferred throughout this guide.

Observation of operating spark gap coils indicate that adding a terminal capacitance to the top of the coil effectively combines in parallel with the coil's self-capacitance. This changes the resonant frequency of the secondary winding. As more capacitance is added, the lower the resonant frequency becomes. See Section 5.10 for additional details. The resonant frequency of the secondary winding with the terminal capacitance is:

$$fso = \frac{1}{2\pi\sqrt{Ls\,(Ct + Cs)}} \tag{2.15}$$

Where: fso = Resonant frequency of secondary with terminal capacitance in Hz = cell (F21), calculated value.
Ls = Inductance of the secondary winding in henries = cell (F17), calculated value from equation (2.5).
Cs = Self-capacitance of the secondary winding in farads = cell (F18), calculated value from equation (2.6). Converted to farads using a 1e-12 multiplier.

Ct = Terminal capacitance in farads = cell (B11), enter value in pF. Converted to farads in cell (C11) using a 1e-12 multiplier.

The total impedance of the secondary at the resonant frequency is:

$$Zs = \sqrt{Rt^2 + \left(\omega so Ls - \frac{1}{\omega so\,(Cs + Ct)}\right)^2} \tag{2.16}$$

Where: Z_S = AC impedance of secondary circuit at the resonant frequency in ohms = cell (F30), calculated value.

Rt = Total resistance of secondary winding in ohms = calculated value from equation (2.12).

ωso = Resonant frequency in radians per second = cell (F22), calculated value. $\omega so = 2\pi f so$ where: $f so$ is the resonant frequency of the secondary in Hz with terminal capacitance = cell (F21).

Ls = Inductance of the secondary winding = cell (F17), calculated value from equation (2.5).

Cs = Self-capacitance of the secondary winding = cell (F18), calculated value from equation (2.6).

Ct = Terminal capacitance in farads = cell (B11), enter value in pF. Converted to farads in cell (C11) using a 1e-12 multiplier.

2.3 Calculating the Primary Characteristics of a Spark Gap Coil

In contrast to the secondary winding the primary winding of several turns and series tank capacitor produce a low Q resonant circuit. The values of primary inductance and capacitance determine the frequency of oscillations, which typically are close to the secondary resonant frequency. A spark gap is used as a switch to control the repetition period of the primary oscillations. When the spark gap ionizes, a burst of damped oscillations is produced in the primary winding and tank capacitance. The peak primary currents are typically very high and determine how much voltage will be produced in the secondary winding. As the peak primary current increases so does the secondary voltage. The sooner the primary oscillations are damped to the spark gap's deionization threshold, the less power the coil will use in operation.

Throughout the guide $\Delta v/\Delta t$ is used as an expression for the slope of the voltage waveform, which is the change in voltage over a period of time. It may also be expressed as dv/dt. When the current is changing over time the slope is expressed as $\Delta i/\Delta t$. The faster the current or voltage is changing the higher the peak values in the circuit. These relationships can be observed when you pull the plug on an inductive load such as a vacuum cleaner while it is still running. A good sized spark can be seen at the plug end and the outlet as the load and current change abruptly with time resulting in a high voltage being generated (revealed in the spark).

During the design phase of coil building the spark gap characteristics can be estimated as well as certain aspects of coil performance. As these are estimates the actual coil performance may require the step-up transformer to deliver the maximum rated voltage and current. This is a concern only with current-limited types such as neon sign (NST) or plate transformers. To ensure that the primary capacitance value selected is not too large for the step-up transformer's capability the maximum usable primary capacitance is calculated:

$$Cp = \left(\frac{\text{IR}}{\text{VR}} \bullet \frac{1}{\text{BPS}}\right) - Cpt \tag{2.17}$$

Where: Cp = Maximum usable primary capacitance for selected step-up transformer characteristics in farads = cell (F47), calculated value. Converted to μF using a 1e6 multiplier.

IR = Rated output current of step-up transformer in amps *rms* = cell (B16), enter value.

VR = Rated output voltage of step-up transformer in volts *rms* = cell (B15), enter value in kV.

BPS = Breaks Per Second produced by spark gap = cell (B30), enter value. See Section 6.8 for calculating the BPS value.

Cpt = Additional primary tuning capacitance in farads (if used) = cell (B45), enter value in μF. Converted to farads in cell (C45) using a 1e-6 multiplier.

Note the line current drawn by the operating coil can be anticipated using equation (2.31). This *rms* value of primary current is the same as the output current of the step-up transformer. The line current (IL) is the step-up transformer output current (Ip) multiplied by the transformer turns ratio (NP) or Ip $\times$ NP = IL. The calculated *rms* line current is shown in cell (F40). When non–current-limited transformers are used they will deliver whatever current the load demands. For these transformers the maximum usable capacitance calculation is of little utility.

Unless you are including a variable capacitance in the primary circuit for fine tuning adjustment the parameter Cpt is ignored (enter 0 in the worksheet). If a variable capacitance is used in parallel with the primary capacitance its effects are included in all primary calculations.

The step-up transformer has an iron core with very small losses ($\approx$0.1 to 0.5%). The line power input to the step-up transformer is therefore equivalent to its power output to the primary circuit. Presuming the estimated line voltage in cell (B15) is close to the value measured with the coil running the primary circuit power is:

$$Pp = Vp \bullet Ip \tag{2.18}$$

Where: Pp = Primary circuit power drawn from line in watts = cell (F39), calculated value. Converted to kW using a 1e-3 multiplier.

Vp = Applied output voltage of step-up transformer in volts *rms* = cell (B22), calculated value. Converted to kV using a 1e-3 multiplier. The estimated or measured line voltage with the coil running entered in cell (B15) is multiplied by the step-up transformer turns ratio entered in cell (B18) to calculate the applied transformer output voltage (Vp).

Ip = Applied output current of step-up transformer in amps *rms* = cell (F50), calculated value from equation (2.31).

Power transformers are rated using *rms* values. A voltmeter, ammeter, and optional wattmeter are used on the input (line) side of the step-up transformer to monitor the coil's performance.

A wattmeter will measure the same power as calculated in equation (2.18). The volt and ammeter will indicate the *rms* equivalents drawn from the line by the load. The estimated or measured *rms* line voltage with the coil running entered in cell (B15) is multiplied by the step-up transformer turns ratio entered in cell (B18) to calculate the applied transformer output voltage (Vp). This applied *rms* voltage will produce a sinusoidal peak voltage of:

$$Vpk = Vp \bullet 1.414 \tag{2.19}$$

Where: Vpk = Peak voltage output of transformer = cell (C22), calculated value.
Vp = Applied output voltage of step-up transformer in volts *rms* = cell (B22), calculated value. Converted to kV using a 1e-3 multiplier.

A sinusoidal waveform has a positive and negative alternation, each reaching this peak value. The peak-to-peak value is twice the peak value and used as the criterion for the safe working voltage of the capacitor. See Chapter 5 to calculate the applied electrical stresses on the capacitor and determine lifetime and safe operating characteristics. Now that the primary capacitance and step-up transformer are optimized a primary inductance must be selected. To calculate the primary inductance needed to produce oscillations at the resonant frequency of the secondary, the winding topology must first be determined.

If a helically wound primary is used enter the value 2 into cell (B38). If a flat Archimedes spiral primary is used enter a value of 1 into cell (B38) and an angle of inclination (θ) of 0° into cell (B43). If the Archimedes spiral primary is not a flat (pancake) spiral enter a value of 1 into cell (B38) and the desired angle of inclination in degrees into cell (B43).

Because of high-frequency losses and the large current levels required the primary winding is optimized when wound with a conductor comparable to bare copper tubing. An interwinding distance is typically used in the primary to prevent voltage breakdown of the air between the bare windings and allow connection of the series primary circuit to any of the windings (tapped). This distance is measured from the center of the conductor in one winding to the center of the conductor in an adjacent winding as shown in Figure 2-4 and is different from the interwinding distance in the close wound secondary. Enter the value of this interwinding distance in inches into cell (B41).

FIGURE 2-4 Interwinding distance in primary coil.

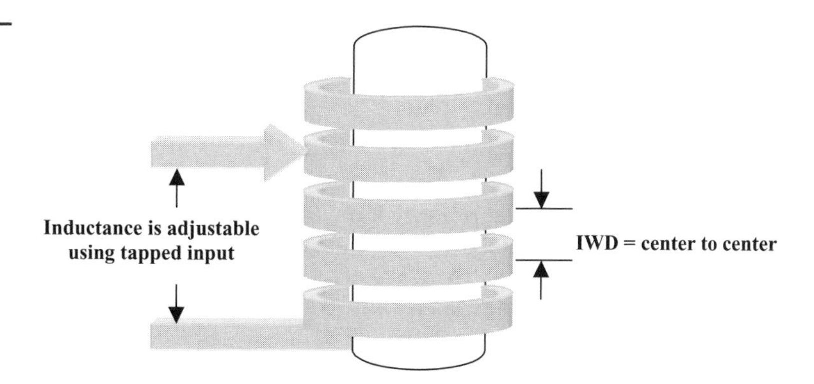

The primary inductance can now be calculated for a helically wound primary:

$$L(\mu h) = \frac{A^2 N^2}{9A + 10H} \tag{2.20}$$

Where: L = Inductance of coil in μhenries = cell (B47), calculated value.
A = Radius of coil form in inches = 1/2 outside diameter entered in cell (B40).
N = Number of turns used to tune coil = cell (B44), enter value.
H = Height (length) of winding in inches = $N \times$ interwinding distance entered into cell (B41).

If a flat or inverse conically wound (Archimedes spiral) primary is used the inductance is:

$$L(\mu h) = \frac{A^2 N^2}{8A + 11W} \tag{2.21}$$

Where: L = Inductance of coil in μhenries = cell (B47), calculated value.
A = Average radius of coil form in inches = {([Outside Diameter (OD) – Inside Diameter (ID)]/2) + ID}/2. Where: ID = cell (B40), enter value. OD = ID + [($N \times$ IWD)/cos θ]. IWD = interwinding distance entered into cell (B41).
N = Number of turns used to tune coil = cell (B44), enter value.
W = Height (width) of winding in inches = calculated value. Where: $W = A/\cos\theta$. θ = angle of incline from horizontal (0°). *Note:* Excel will not calculate the sine, cosine or tangent of angles without first converting to radians. This is done in the worksheet and the cosθ is shown in cell (BC11), sinθ is shown in cell (BC12).

NOTE: *This method agrees with measurements of constructed coils. Do not use for angles of inclination greater than 75°, use the helical formula instead.*

The calculated inductance that appears in cell (B47) is actually performed in columns (BC) through (BJ), rows 3 through 28 for the Archimedes spiral and rows 31 through 56 for the helical winding. These calculations are limited to 25 turns on the primary. To fine tune the primary oscillations to the resonant frequency of the secondary a certain value of inductance will be required:

$$Lp = \frac{1}{4\pi^2 fso^2 (Cp + Cpt)} \tag{2.22}$$

Where: Lp = Required primary inductance for resonance with secondary coil in henries = cell (B48), calculated value, converted to μH using a 1e6 multiplier.
fso = Resonant frequency of secondary with terminal capacitance in Hz = cell (F21), calculated value from equation (2.15).
Cp = Primary capacitance in farads = cell (B19), enter value in μF. Converted to farads in cell (C19) using a 1e-6 multiplier.

Cpt = Additional primary tuning capacitance in farads (if used) = cell (B45), enter value in μF. Converted to farads in cell (C45) using a 1e-6 multiplier.

The primary may not include a fine-tuning provision and the oscillating frequency may be different than the resonant frequency of the secondary. The actual frequency of primary oscillations is:

$$fP = \frac{1}{2\pi\sqrt{(Lp + Lpt)(Cp + Cpt)}} \tag{2.23}$$

Where: fP = Resonant frequency of primary oscillations in Hz = cell (B49), calculated value.

Lp = Calculated primary inductance for turns used in henries = cell (C47), calculated value from equation (2.20) or (2.21). Converted to μH in cell (B47) using a 1e6 multiplier.

Lpt = Additional primary tuning inductance in henries (if used) = cell (B46), enter value in μH. Converted to henries in cell (C46) using a 1e-6 multiplier.

Cp = Primary capacitance in farads = cell (B19), enter value in μF. Converted to farads in cell (C19) using a 1e-6 multiplier.

Cpt = Additional primary tuning capacitance in farads (if used) = cell (B45), enter value in μF. Converted to farads in cell (C45) using a 1e-6 multiplier.

NOTE: *The additional tuning capacitance entered in cell (B45) is added to the primary capacitance in cell (C19). The calculations account for the additional tuning inductance entered in cell (B46) and additional tuning capacitance entered in cell (B45) and are referred throughout the remaining text as Cp and Lp.*

The calculated primary inductance shown in cell (B47) from equations (2.20) and (2.21) is dependent on two additional inputs to calculate the primary characteristics with fine tuning provisions in the design. As already stated our primary design should allow for any winding to be tapped, changing the primary inductance. By entering the desired turn number into cell (B44) the number of turns is changed in equations (2.20) and (2.21) and the calculated inductance in cell (B47). Notice that this will typically produce inductance steps of 5 μH to 15 μH per turn which may not fine tune the primary enough for optimum power transfer to the secondary unless additional tuning capacitance (Cpt) is used or some additional small inductance in series with the primary inductance as shown in Figure 2-5. The calculations are limited to 25 primary turns. If additional series inductance is desired in the design, enter this value into cell (B46). Note that the maximum theoretical secondary voltage cannot be reached without the value in cell (B47) matching the value in cell (B48). This usually necessitates including a variable capacitance or inductance in the primary for fine-tuning.

The calculations will show that a relatively large value of variable capacitance is usually required to accomplish this. About the only way to achieve these values and provide sufficient dielectric strength is with a large commercial RF variable air vane capacitor using many plates (over 20) and its air dielectric. When the capacitor is immersed in a stronger insulator material

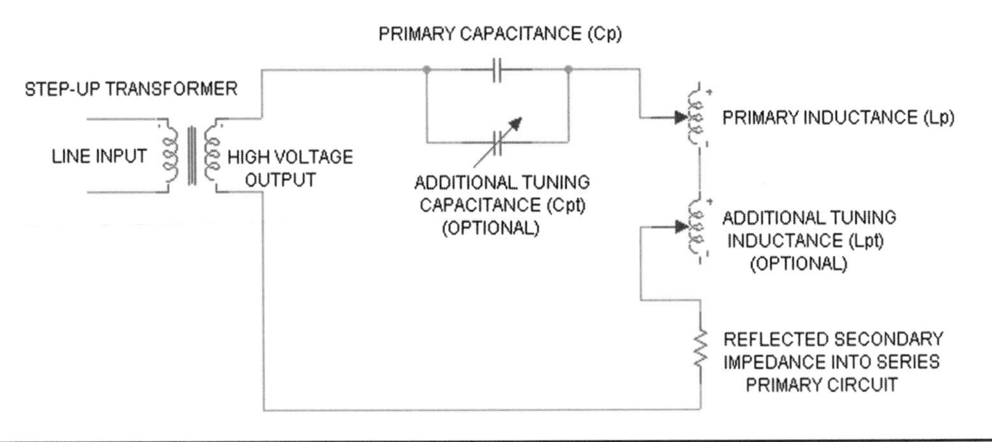

FIGURE 2-5 Additional capacitance or inductance required to fine tune the primary circuit.

such as mineral oil the capacitance is multiplied by the dielectric constant of the insulator (≈2.0–2.6). You may find that this is still not enough capacitance to affect a good tuning range. An easier solution is to use the series inductance. It can be a simple smaller coil that produces 15 μH of total inductance with taps in 0.5-μH increments or a motor driven variable inductor such as those used in commercial RF applications. You will loose a portion of the available primary power in this series inductance as there is no mutual inductance (coupling) with the secondary winding. If a tuning capacitance is used the primary power loss is avoided. Section 5.6 will facilitate tuning capacitance design and calculations. Section 4.2 will facilitate tuning inductance design and calculations.

Do not overemphasize the importance of adding a fine-tuning provision, as it usually will not produce a significant increase in secondary voltage. As long as the primary turn selected produces a primary resonance that is closest to the secondary resonance the course tuning will usually suffice. Build the coil first to see how it runs without fine-tuning and add a fine-tuning provision during the optimization phase discussed in Section 2.6. The tuning calculations are better implemented once the coil is running and the operating parameters are more accurately defined.

A final consideration in the primary winding dimensions is the degree of coupling between the primary and secondary. Pages 184 and 185 of reference (5) indicate this coupling must be less than 20% (coefficient of coupling or $k = 0.20$) for a spark gap primary. Although the secondary in this application was being utilized as a radio transmitter and not generating a spark it is still a good threshold of critical coupling, if not a little high. I have intentionally used 24% (0.24) coupling with no spark output produced in the secondary other than corona losses, interwinding breakdown, and flashover from the secondary-to-primary. When the coupling was decreased to 15% (0.15) the spark output returned to the estimated length.

Section 4.9 details the calculations used to determine the coefficient of coupling and mutual inductance of two coaxial coils. For the primary winding parameters entered the calculated primary radius appears in cell (F58) and primary height in cell (F59). The secondary height and diameter determined in Section 2.2 appear again in cells (F60) and (F61) for comparison. The various form factors are calculated in cells (F62) through (F72) and the mutual inductance in cell (F73) and coefficient of coupling in cell (F74). The mutual inductance and coefficient of coupling between two coaxial coils with an air core is dependent on the geometric relationships of the primary-to-secondary height and primary-to-secondary diameter. Note how the primary height, mutual inductance, and coefficient of coupling changes as the selected turn in cell (B44) is changed. No adjustment to the calculations for the slightly different geometry of the Archimedes spiral was considered necessary as the calculations agreed with measured values of working coils. The calculated mutual inductance and coefficient of coupling can be either positive or negative in value.

These calculations are also used in the worksheet in columns (BK) through (BX) to calculate the mutual inductance (M) and coefficient of coupling (k) for the selected primary winding characteristics and secondary winding characteristics using wire gauges 0000 through 60. Methodology to calculate M also exists in pages 278 and 279 of Reference (9). It is the same as that shown in equation (4.33) except the constant of 0.02505 is replaced by the constant 0.00987. As the Terman methodology used in the worksheet produces reliable results an alternate method is only of comparative interest.

The primary impedance is determined by calculating the resistance of the spark gap, DC resistance of the primary winding, and the reactance of the primary inductance and capacitance. Pages 1–23 of reference (5) detail the methodology for estimating the spark gap resistance. Whether the primary waveform exhibits a linear or exponential decrement characteristic depends on the electrode material (e.g. copper, brass, aluminum or silver is linear, magnesium is exponential). As the data in the reference were based on observation of operating spark gaps it is inferred the primary decrement can be either linear or exponential depending on the type of material used. This is not to be confused with the secondary decrement, which is independent from the primary and examined in detail in Chapter 6. The spark gap characteristics used in the spreadsheet calculations depend on whether a 1 is entered into cell (B32) for a linear characteristic or a 2 is entered for an exponential characteristic. The gap resistance for an exponential characteristic is:

$$Rge = \frac{8Vfo}{\pi Ip} = \frac{8\,(193.04S + 34)}{\pi Ip} \tag{2.24}$$

Where: Rge = Rotary spark gap resistance of exponential electrode material = cell (F52), calculated value.
Vfo = Initial voltage across the ionized gap = linear curve fit formula from Figure 6-35.
S = Spark gap spacing in inches = cell (B25), enter value.
Ip = Peak oscillating current in the primary circuit in amps = cell (F49), calculated value from equation (2.30).

And for a linear characteristic:

$$Rgl = \frac{6Vfo}{\pi Ip} = \frac{6\,(264.16S + 42)}{\pi Ip} \tag{2.25}$$

Where: Rgl = Rotary spark gap resistance of linear electrode material = cell (F51), calculated value.
Vfo = Initial voltage across the ionized gap = linear curve fit formula from Figure 6-35.

S = Spark gap spacing in inches = cell (B25), enter value.
Ip = Peak oscillating current in the primary circuit in amps = cell (F49), calculated value from equation (2.30).

The reactance of the primary capacitance and inductance at the primary oscillating frequency are added to the spark gap resistance and the DC resistance of the primary winding:

$$Zpss = \sqrt{(Rp + Rg)^2 + \left(\omega pLp - \frac{1}{\omega pCp}\right)^2} \tag{2.26}$$

Where: $Zpss$ = AC impedance of primary circuit without reflected secondary at the resonant frequency of primary oscillations in ohms = cell (F44), calculated value.
Rp = DC Resistance of primary winding in ohms = cell (B20), enter value.
Rg = Rotary spark gap resistance for selected material characteristic entered into cell (B33) = calculated value from equation (2.24) or (2.25).
ωp = Resonant frequency of primary in radians per second = cell (F36), calculated value. $\omega p = 2\pi fP$ where: fP is the resonant frequency of the primary oscillations calculated in cell (B49) from equation (2.23).
Lp = Calculated primary inductance for turns used in henries = cell (C47), calculated value from equation (2.20) or (2.21). Converted to μH in cell (B47) using a 1e6 multiplier.
Cp = Primary capacitance in farads = cell (B19), enter value in μF. Converted to farads in cell (C19) using a 1e-6 multiplier.

When the primary is coupled to the secondary an impedance is reflected back into the primary from the secondary. The mutual inductance or coefficient of coupling will determine the value of reflected secondary impedance into the primary circuit. This reflected impedance appears as an additional series resistance in the primary circuit as shown in Figure 2-5. The primary impedance with this reflection is:

$$Zps = \frac{(\omega pM)^2}{Zs} + Zpss \tag{2.27}$$

Where: Zps = AC impedance of primary circuit with reflected secondary at the resonant frequency of primary oscillations in ohms = cell (F42), calculated value.
ωp = Resonant frequency of primary in radians per second = cell (F36), calculated value. $\omega p = 2\pi fP$ where: fP is the resonant frequency of the primary oscillations calculated in cell (B49) from equation (2.23).
M = Mutual inductance of primary and secondary winding in henries = cell (F73), calculated value. Converted to μH using a 1e6 multiplier.
Zs = AC impedance of secondary circuit at the resonant frequency in ohms = calculated value in cell (F30) from equation (2.16).
$Zpss$ = AC impedance of primary circuit without reflected secondary at the resonant frequency in ohms = cell (F44), calculated value from equation (2.26).

The primary Quality factor can now be calculated:

$$Qp = \frac{\omega p Lp}{Zpss} \tag{2.28}$$

Where: Qp = Quality of primary circuit (figure of merit) = cell (F45), calculated value.
ωp = Resonant frequency of primary in radians per second = cell (F36), calculated value. $\omega p = 2\pi f P$ where: $f P$ is the resonant frequency of the primary oscillations calculated in cell (B49) from equation (2.23).
Lp = Calculated primary inductance for turns used in henries = cell (C47), calculated value from equation (2.20) or (2.21). Converted to μH in cell (B47) using a 1e6 multiplier.
$Zpss$ = AC impedance of primary circuit without reflected secondary at the resonant frequency in ohms = cell (F44), calculated value from equation (2.26).

Because of the primary impedance the tank circuit oscillations are damped. The primary tank will produce ringing like oscillations (see Figure 2-9), each successive oscillation will decrease from its preceding oscillation by an amount equivalent to the decrement:

$$\delta P = \frac{\pi}{Qp} \tag{2.29}$$

Where: δP = Decrement of the primary circuit = cell (F46), calculated value
Qp = Quality of primary circuit (figure of merit) = cell (F45), calculated value from equation (2.28).

When the spark gap is deionized the tank capacitor charge will follow the supply voltage changes through the high-impedance output of the step-up transformer. As the spark gap ionizes it forms a low-impedance path for the tank capacitor to discharge this stored energy through the primary winding. The discharge current appears as a damped waveform affected by the primary decrement and electrode material characteristics. The peak value of this oscillating current is:

$$Ip = C\frac{dv}{dt} = Cp\frac{Vp}{\left(\frac{1}{fP}\right)} \tag{2.30}$$

Where: Ip = Peak oscillating current in the primary circuit in amps = cell (F49), calculated value.
Cp = Primary capacitance in farads = cell (B19), enter value in μF. Converted to farads in cell (C19) using a 1e-6 multiplier.
dv = Vp = Peak output voltage of step-up transformer in volts = cell (C22), calculated value from *rms* value in kV in cell (B22) × 1.414. Converted to volts using a 1e3 multiplier.
dt = Time period of primary oscillations = $1/f P$, where $f P$ = Resonant frequency of primary oscillations in Hz = cell (B49), calculated value from equation (2.23).

The *rms* equivalent of this repetitive primary oscillating current waveform is:

$$Ip(rms) = IpEXP - (\delta PfPtD) \bullet \frac{tD}{\left(\frac{1}{\mathrm{BPS}}\right)} \tag{2.31}$$

Where: Ip = Peak oscillating current in the primary circuit in amps = cell (F49), calculated value from equation (2.30).

δP = Decrement of the primary circuit = cell (F46), calculated value from equation (2.29).

fP = Resonant frequency of primary oscillations in Hz = cell (B49), calculated value from equation (2.23).

BPS = Breaks Per Second produced by spark gap = cell (B30), enter value. Refer to Section 6.8. 1/BPS = repetition time period of spark gap ionization in seconds = calculated value in cell (C30).

tD = Time period required for spark gap ionization and primary oscillations to decay to minimum ionization threshold in seconds = cell (B28), enter value in μsec. Refer to Section 6.11. The CH_6A.xls file (see App. B), PRIMARY OSCILLATIONS worksheet (2) calculates *t*D for the applied primary and spark gap characteristics. This methodology is used in columns (FV) to (GB) to calculate the required ionization time period in cell (B29). The required number of primary oscillations is calculated in cell (C29). Unless you are accounting for operating tolerance enter the cell (B29) value into cell (B28).

The remaining calculations will solve for the peak secondary voltage using two methods. The first method uses a transient solution for voltage and current in a series RLC circuit with an applied sinusoidal waveform. The second method calculates the primary-to-secondary impedance ratios (VSWR) to determine the peak secondary voltage. The two methods will produce different solutions to the peak secondary voltage, the actual voltage being close to either solution when the primary oscillating frequency is close to the resonant frequency of the secondary. Actual performance may vary as the calculations assume a pure sinusoidal waveform. The actual waveform may vary, which affects the calculated results. The calculations are intended to present theoretical maximums and relationships for use in optimizing the design. When the number of primary turns used to tune (cell B44) is decreased, the primary oscillating frequency increases above the secondary resonant frequency, and the transient solution becomes more reliable in estimating the secondary voltage. When the number of primary turns used to tune is increased, the primary oscillating frequency decreases below the secondary resonant frequency, and the VSWR solution becomes more reliable in estimating the secondary voltage. For comparison the more traditional primary-to-secondary inductance and capacitance ratio methods of determining the secondary voltage are included. They produce less reliable results than the transient or VSWR solutions.

From pages 157–158 of reference (6) was found formulae for solving the transient (instantaneous) current and voltage in a series RLC circuit with an applied sinusoidal waveform. The peak primary current is calculated in equation (2.30). The peak voltage in the primary winding is:

$$Vpp = Vp \sin(\omega ptp + \delta P) \tag{2.32}$$

Where: Vpp = Peak oscillating voltage in primary winding = cell (F48), calculated value.

Vp = Peak output voltage of step-up transformer in volts = cell (C22), calculated value from *rms* value in kV in cell (B22) × 1.414. Converted to volts using a 1e3 multiplier.

ωp = Resonant frequency of primary in radians per second = cell (F36), calculated value. $\omega p = 2\pi f P$ where: fP is the resonant frequency of the primary oscillations calculated in cell (B49) from equation (2.23).

tp = time period of primary oscillations in seconds = cell (F37), calculated value. $tp = 1/fP$ where: fP is the resonant frequency of the primary oscillations calculated in cell (B49) from equation (2.23).

δP = Decrement of the primary circuit = cell (F46), calculated value from equation (2.29).

NOTE: *α (attenuation factor) which originally appeared in equation (2.32) from reference (5) was substituted with the equivalent primary decrement (δP).*

All necessary parameters in the primary circuit are now known. Next we will calculate the effects the oscillating primary current generates in the secondary.

2.4 Calculating the Resonant Characteristics of a Spark Gap Coil

The primary and secondary circuits are actually two independent oscillating circuits. The low Q primary produces sub-harmonic and harmonic current oscillations over a wide frequency range. This is simulated in Section 8.2. The high-Q secondary is selective to only a narrow bandwidth of the wideband primary oscillations. Only the primary current at this narrow bandwidth is coupled from the primary-through the mutual inductance-to the secondary:

$$Is = \frac{\omega so MIp}{Zs} \tag{2.33}$$

Where: Is = Peak oscillating current in the secondary winding coupled from the primary in amps = cell (B84), calculated value.

Ip = Peak oscillating current in primary circuit in amps = cell (F49), calculated value from equation (2.30).

ωso = Resonant frequency of secondary in radians per second = cell (F22), calculated value. $\omega so = 2\pi fso$ where: fso is the resonant frequency of the secondary in Hz with terminal capacitance = cell (F21).

M = Mutual inductance of primary and secondary winding in henries = cell (F73), calculated value. Converted to μH using a 1e6 multiplier.

Zs = AC impedance of secondary circuit at the resonant frequency in ohms = calculated value in cell (F30) from equation (2.16).

The oscillating secondary current will produce an oscillating voltage in the secondary winding:

$$Vs = -Is\left(\frac{Zs}{\sin\left(\omega sots + \delta S - \tan\phi S\right)}\right) \tag{2.34}$$

Where: Vs = Peak oscillating voltage in the secondary winding = cell (B85), calculated value. Converted to kV using a 1e-3 multiplier.

Is = Peak oscillating current in the secondary winding coupled from the primary in amps = cell (B84), calculated value from equation (2.33).

Zs = AC impedance of secondary circuit at the resonant frequency in ohms = calculated value in cell (F30) from equation (2.16).

ωso = Resonant frequency of secondary in radians per second = cell (F22), calculated value. $\omega so = 2\pi fso$ where: fso is the resonant frequency of the secondary in Hz with terminal capacitance = cell (F21).

ts = Time period of oscillations at the resonant frequency in seconds. $ts = 1/fso$ where: fso is the resonant frequency of the secondary with terminal capacitance calculated in cell (F21) from equation (2.15).

δS = Decrement of the secondary winding = cell (F31), calculated value from equation (2.14).

$\tan\phi S = \frac{\omega so^2 Ls(Cs+Ct)-1}{\omega so(Cs+Ct)Zs}$ = cell (F23), calculated value.

Where: Ls = Inductance of the secondary winding = cell (F17), calculated value from equation (2.5).

Cs = Self-Capacitance of the secondary winding = cell (F18), calculated value from equation (2.6).

Ct = Terminal capacitance in farads = cell (B11), enter value in pF. Converted to farads in cell (C11) using a 1e-12 multiplier.

Note: *From pages 157–158 of reference (6) was found formulae for solving the transient (instantaneous) current and voltage in a series RLC circuit with an applied sinusoidal waveform. The factor α (attenuation factor), which originally appeared in the equation from reference (5), was substituted with the equivalent decrement (δ). Equation (2.34) was transposed from the equation below in Mathcad to solve for the secondary voltage:*

$$I = \frac{V}{Z}\sin\left(\omega t + \delta - \tan\phi\right)$$

Where: I = Current in amps.

V = Applied Voltage.

Z = Impedance of circuit in ohms.

ω = Frequency in radians per second.

t = Time period of oscillations in seconds.

δ = Decrement of circuit.

$$\tan\phi = \frac{\omega^2 LC - 1}{\omega CZ}$$

Where: L = Inductance of series RLC circuit in henries.

C = Capacitance of series RLC circuit in farads.

Equations (2.33) and (2.34) allow for solution of instantaneous voltage and current in the primary and secondary, meaning a waveform can be produced from the calculations. This will be done in Section 2.5.

An alternative method for calculating the secondary voltage is to use the ratio of primary impedance-to-secondary impedance known as the Voltage Standing Wave Ratio (VSWR). First the maximum primary impedance is calculated. At the moment the spark gap is ionized the primary impedance with the reflected secondary impedance was calculated in equation (2.27). The peak primary current at the moment of ionization will decrease by the primary decrement with each succeeding oscillation time period. The primary current oscillations decrease with the primary decrement; therefore the primary impedance must be increasing in a reciprocal manner.

To calculate this increasing primary impedance requires a calculation for each oscillation time period. The maximum primary impedance at the moment of ionization was calculated in cell (F42) and for each wire gauge from 0000 to 60 in column (CB), rows 1 through 69. The calculated impedance in each succeeding column (CC) through (FN) is increased by the reciprocal of the primary decrement $(1/[1 - \delta P])$. A conditional statement is used to determine the maximum primary impedance at the end of the selected spark gap ionization time entered into cell (B28) and shown in column (FO) for each wire gauge from 0000 to 60. For the selected wire gauge entered into cell (B6) the primary impedance at the end of the selected spark gap ionization time entered into cell (B28) is:

$$
\begin{aligned}
Zps(t0)\bullet\left(\frac{1}{1-\delta P}\right) &= Zps(t1)\\
Zps(t1)\bullet\left(\frac{1}{1-\delta P}\right) &= Zps(tn)\\
Zps(tn)\bullet\left(\frac{1}{1-\delta P}\right) &= Zps(te)
\end{aligned}
\tag{2.35}
$$

Where: $Zps(t0)$ = Primary impedance with reflected secondary at moment of spark gap ionization in ohms = calculated value in cell (F42) from equation (2.27) for selected AWG in cell (B6). Also calculated in column (CC) for each wire gauge from 0000 to 60.

δP = Decrement of the primary circuit = cell (F46), calculated value from equation (2.29).

$Zps(tn)$ = Primary impedance with reflected secondary at each succeeding oscillation time period in ohms = calculated value in columns (CD) through (FN) for each wire gauge from 0000 to 60.

$Zps(te)$ = Maximum primary impedance with reflected secondary at end of spark gap ionization time entered into cell (B28) in ohms = calculated value in cell (F43) for selected AWG in cell (B6) and column (FO) for each wire gauge from 0000 to 60.

The maximum primary impedance is now known and the secondary impedance was calculated in equation (2.16). The Voltage Standing Wave Ratio (VSWR) is:

$$\text{VSWR} = \frac{Zp}{Zs} \tag{2.36}$$

Where: VSWR = Maximum Voltage Standing Wave Ratio of tuned primary and secondary circuit = cell (F79), calculated value for selected AWG in cell (B6) and column (FQ) for each wire gauge from 0000 to 60.

Zp = $Zps(te)$ = Maximum primary impedance with reflected secondary at end of spark gap ionization time entered into cell (B28) in ohms = calculated value in cell (F43) for selected AWG in cell (B6) and column (FO) for each wire gauge from 0000 to 60.

Zs = AC impedance of secondary circuit at the resonant frequency in ohms = calculated value in cell (F30) from equation (2.16).

And the coefficient of reflection is:

$$\Gamma = \frac{Zp - Zs}{Zp + Zs} \tag{2.37}$$

Where: Γ = Coefficient of reflection of primary and secondary circuit = cell (F78), calculated value for selected AWG in cell (B6) and column (FP) for each wire gauge from 0000 to 60.

Zp = Maximum primary impedance with reflected secondary at end of spark gap dwell time entered into cell (B28) in ohms = calculated value in cell (F43) for selected AWG in cell (B6) and column (FO) for each wire gauge from 0000 to 60.

Zs = AC impedance of secondary circuit at the resonant frequency in ohms = calculated value in cell (F30) from equation (2.16).

The maximum secondary voltage is calculated:

$$Vs = Vp \bullet \text{VSWR} \tag{2.38}$$

Where: Vs = Peak secondary voltage = cell (F80), calculated value for selected AWG in cell (B6) and column (FR) for each wire gauge from 0000 to 60. Converted to kV using a 1e-3 multiplier.

Vp = Peak output voltage of step-up transformer in volts = cell (C22), calculated value from *rms* value in kV in cell (B22) × 1.414. Converted to volts using a 1e3 multiplier.

VSWR = Maximum Voltage Standing Wave Ratio of tuned primary and secondary circuit = calculated value in cell (F79) from equation (2.36).

NOTE: *The highest calculated secondary voltage for wire gauges 0000 to 60 appears in cell (F84), which is the optimum wire gauge to use. It can be seen as the peak in Chart 1, rows 89 to 144 of the worksheet and shown in Figure 2-6.*

For Tesla coil operation the terminal capacitance should be designed to overcome the dielectric strength of the surrounding air at a value slightly below the maximum voltage generated by the secondary. The dielectric strength of air (before ionization) is approximately 30 kV per cm, or 76.2 kV per inch. When properly designed, the secondary generates its maximum peak voltage, the dielectric (insulating) strength of the air surrounding the terminal capacitor is overcome at some arbitrary point (usually a surface variation or director causing non uniform charge distribution), and a spark breaks out. An air channel originating at this surface variation

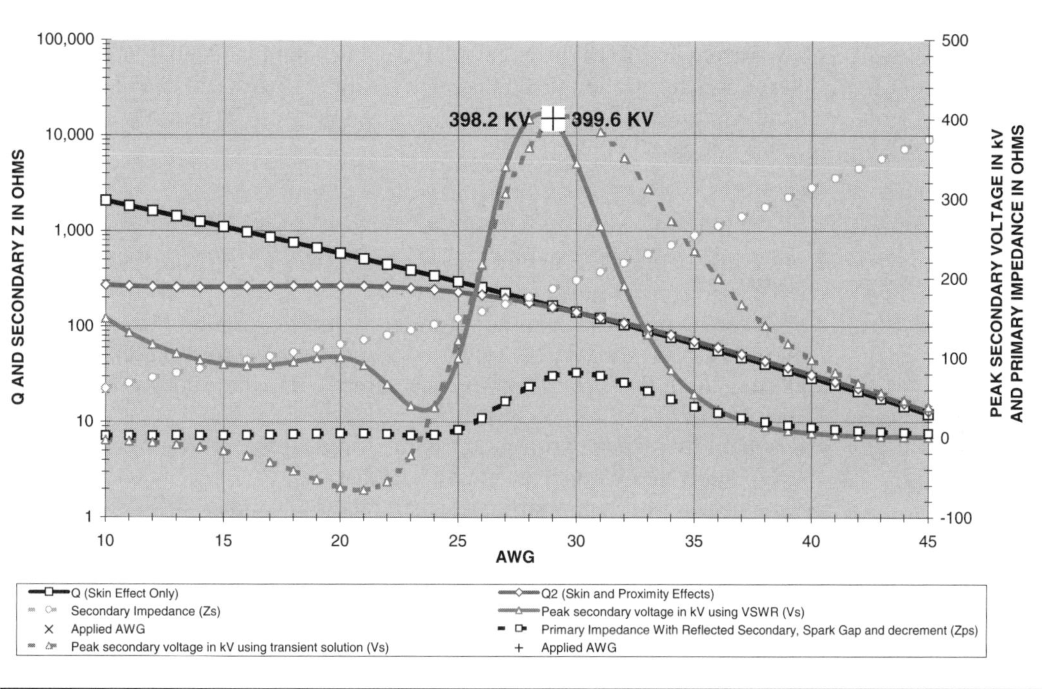

FIGURE 2-6 Optimum wire gauge for selected secondary resonant frequency and coil form diameter.

becomes ionized, creating a low-resistance path. Just as in natural lightning, the higher the voltage the longer the path. It may be more appropriate to call it a "lightning bolt" than the conventional term "spark."

Once the ionized air channel is formed the air is constantly superheated by the discharge. The plasma in the discharge is very hot and lowers the voltage required to maintain the discharge. The spark may appear to grow if a director is used. To characterize the voltage needed to maintain the spark after ionization see Section 6.13. It takes approximately 10 kV to ionize every inch of air in a spark gap coil. The calculated secondary voltage will produce a spark length of:

$$SL = \frac{Vs}{10{,}000} \tag{2.39}$$

Where: SL = Length of high-voltage discharge (spark) in inches = cell (F81), calculated value.

Vs = Peak oscillating voltage in the secondary winding = cell (F80), calculated value from equation (2.38).

The terminal can be made large enough that the high voltage generated in the secondary cannot overcome the insulating threshold of the terminal capacitance and a spark never breaks out. This detunes the coil and standing waves are produced in the secondary. Unless your intention is generating a large electromagnetic pulse (EMP) and electromagnetic interference (EMI), I would not recommend this approach. Sections 5.4 and 5.5 detail terminal capacitance construction and calculate the high-voltage threshold where the spark will break out.

The peak primary current in cell (F49) was calculated using equation (2.30). Using the peak primary current determined in equation (2.33) the peak secondary current is calculated in cell (F82). The *rms* equivalent of the oscillating current in the secondary can be found which would be measurable on an RF ammeter:

$$Is(rms) = Is(pk)EXP - [\delta S fso Tp] \tag{2.40}$$

Where: $Is(rms)$ = *rms* equivalent of oscillating current in the secondary winding = cell (F83), calculated value.

$Is(pk)$ = Peak oscillating current in the secondary winding coupled from the primary in amps = cell (F82), calculated value using equation (2.33).

δS = Decrement of the secondary winding = cell (F31), calculated value from equation (2.14).

fso = Resonant frequency of secondary with terminal capacitance in Hz = cell (F21), calculated value from equation (2.15).

Tp = Time period of rotary spark gap break rate in seconds = 1/BPS. BPS = Breaks Per Second produced by spark gap = cell (B30), enter value. Refer to Section 6.8. Calculated Tp is shown in cell (C30).

The *rms* equivalent of the oscillating voltage in the secondary can also be found:

$$Vs(rms) = Vs(pk)EXP - [\delta S fso Tp] \tag{2.41}$$

Where: $Vs(rms)$ = *rms* equivalent of oscillating voltage in the secondary winding = cell (F85), calculated value. Converted to kV using a 1e-3 multiplier.

$Vs(pk)$ = Peak oscillating voltage in the secondary winding = cell (F80), calculated value using equation (2.38).

δS = Decrement of the secondary winding = cell (F31), calculated value from equation (2.14).

fso = Resonant frequency of secondary with terminal capacitance in Hz = cell (F21), calculated value from equation (2.15).

Tp = Time period of rotary spark gap break rate in seconds = 1/BPS. BPS = Breaks Per Second produced by spark gap = cell (B30), enter value. Refer to Section 6.8. Calculated Tp is shown in cell (C30).

The secondary *rms* voltage can be used to calculate the average power in the secondary as follows:

$$Ps = \frac{Vs(rms)^2}{Zs} \tag{2.42}$$

Where: Ps = Average power in the secondary winding in watts = cell (F86), calculated value. Converted to kW using a 1e-3 multiplier.

$Vs(rms)$ = *rms* equivalent of oscillating voltage in the secondary winding = cell (F85), calculated value from equation (2.41). Converted to kV using a 1e-3 multiplier.

Zs = AC impedance of secondary circuit at the resonant frequency in ohms = calculated value in cell (F30) from equation (2.16).

From page 123 of reference (7) is found an alternate formula for calculating the average power in the secondary:

$$Ps = \frac{Cs \bullet Vs(rms)^2 \omega}{2} \tag{2.42a}$$

Where: Cs = Self-capacitance of secondary coil in farads = cell (F18), calculated value from equation (2.6).
ω appears in the published formula, which is $2\pi f$. This is reduced to simply f, or fso = Resonant frequency of secondary with terminal capacitance in Hz = cell (F21), calculated value from equation (2.15).

However, the peak instantaneous power is much higher:

$$Ps(pk) = Vs(pk) \bullet Is(pk) \tag{2.43}$$

Where: $Ps(pk)$ = Peak instantaneous power in the secondary winding in watts = cell (F87), calculated value. Converted to kW using a 1e-3 multiplier.
$Vs(pk)$ = Peak oscillating voltage in the secondary winding = cell (F80), calculated value using equation (2.38).
$Is(pk)$ = Peak oscillating current in the secondary winding in amps = cell (F82), calculated value using equation (2.33).

The primary-to-secondary inductance ratio has traditionally been used to calculate the peak secondary voltage:

$$Vs = Vp\sqrt{\frac{Ls}{Lp}} \tag{2.44}$$

Where: Vs = Peak oscillating voltage in the secondary winding = cell (B87), calculated value. Converted to kV using a 1e3 multiplier.
Vp = Peak output voltage of step-up transformer in volts = cell (C22), calculated value from *rms* value in kV in cell (B22) × 1.414. Converted to volts using a 1e3 multiplier.
Lp = Calculated primary inductance for turns used in henries = cell (B47), calculated value from equation (2.20) or (2.21). Converted to μH in cell (C47) using a 1e6 multiplier.
Ls = Inductance of secondary coil in henries = cell (F17), calculated value from equation (2.5).

And the calculated secondary voltage using the primary-to-secondary capacitance ratio is:

$$Vs = Vp\sqrt{\frac{Cp}{(Cs + Ct)}} \tag{2.45}$$

Where: Vs = Peak oscillating voltage in the secondary winding = cell (B86), calculated value. Converted to kV using a 1e-3 multiplier.

SECONDARY CHARACTERISTICS

Parameter	Value	Converted
Number of strands **(layers)** of wire **(NS)**	**1**	
Wire gauge **(AWG)**	**29**	
Ambient temperature of wire **(TA)**	25.0°C	77.0 °F
Diameter of coil form **(D)** in inches	**4.50**	11.43cm
Interwinding seperation in inches if close wound magnet wire not used **(S)**	0.0000 in	
Resonant frequency in kHz **(fo)**	**85.50 KHz**	85500Hz
Terminal capacitance in pF **(Ct)**	**21.7 pF**	0.0000000000217 F

PRIMARY CHARACTERISTICS

Parameter	Value	Converted
AC line frequency in Hertz **(Lf)**	60	0.01667
Rated output voltage of step-up transformer in kV **(VR)**	16.8 KV	16800 V
Rated output current of step-up transformer in Amps **(IR)**	0.5000 A	
Measured or estimated (coil running) AC line voltage **(LV)**	200 V	70:1
Turns ratio of step-up transformer **(NT)**	**1:70**	
Tank capacitance in µF **(Cp)**	**0.0430µF**	0.0000000430000 F
Primary winding DC resistance in Ω **(Rp)**	0.0100Ω	
Separation from base of primary to base of secondary in inches **(d1)**	0.0 in	0.00cm
Calculated applied output voltage of step-up transformer in kVrms **(Vp)**	**14.00 KV**	19796 V

SPARK GAP CHARACTERISTICS

Parameter	Value	Converted
Distance between gap ends in inches **(Sg)**	0.160 inches	0.41cm
Applied Overvoltage in percent **(Vo)**	0.0%	
Temperature-Pressure-Humidity Correction Factor **(k)**	1.00	
Spark gap ionization time in µsec **(tD)**	**330.9 µsec**	0.000331 sec
Estimated spark gap ionization time in µsec **(tD)**	**330.9 µsec**	**23**
Spark gap breaks per second **(BPS)**	**460**	0.002174 sec
Phase shift of non-synchronous gap in degrees **(PS)**	89.5 °	472.8143037
Enter: (1 for linear) or (2 for exponential) gap material characteristics	2	
Minimum ionization current **(Imin)**	**1.0 A**	
Calculated breakdown voltage at applied positive alternation in kV **(BVp)**	**13.66 KV**	

Peak applied voltage (Vp) in cell (C22) must be greater than breakdown voltage (BVp) in cell (B34).

PRIMARY TUNING

Parameter	Value	Converted
Enter: (1 for Archemedes spiral) or (2 for helical) wound primary	1	
Enter measured coupling coefficient for Archemedes spiral if known	0.000	
Inside diameter of Archemedes spiral **(ID)** or outside diameter of helical primary in inches **(OD)**	18.0 in	45.72cm
Interwinding distance in inches **(IWD)**	1.000 in	2.54cm
Total number of turns in primary winding **(Ttp)**	13	
Angle of inclination in° if using Archemedes spiral **(θ)**	50.0 °	
Desired primary turn number used to tune **(Tp)**	**11**	
Tuning capacitance in µF if used **(Cpt)**	0.00000 µF	0.0000000000000 F
Tuning inductance in µH if used **(Lpt)**	0.0 uH	0.0000000000000 H
Calculated primary inductance in µH **(Lp)**	**106.48 µH**	0.0001064836971 H
Calculated primary inductance required for resonance in µHenries **(Lp)**	126.46 µH	Lp=1/[4*π^2*(fo^2*Cp+Cpt)]
Primary resonant frequency **(fP)**	**74,378 Hz**	fp=1/[2*π*sqrt([Lp+Lpt]*Cp)]

SECONDARY WINDING CALCULATIONS

Wire diameter in inches **(d)**	0.0113
Wire area in cirmils **(da)**	127
Wire resistance in **Ω / ft**	0.08127
Wire diameter w/insulation **(Dw)**	0.0128
Nominal increase in diameter **(di)**	0.0015
Close wind turns per inch **(T/in)**	78.4
Number of turns of wire **(N)**	2,440
Inter-winding distance **(Di)**	0.0015
Wire length per turn in inches **(LW/T)**	14.1
Wire length total **(Lt)** in feet	2,877.2 ft
Total winding height in inches **(H)**	**31.1**
DC resistance of values entered in Ohms **(Rdc)**	238.4338 Ω
Inductance of values entered in mH **(L)**	**90.91 mH**
Self-Capacitance in pF **(Cs)**	**38.1 pF**
Wavelength in feet **(λ)**	11508.8
Find **ωs**	537212
Resonant frequency of secondary with terminal in Hz **(fso)**	**68,251 Hz**
Find **ωso**	428832
Find **tan ϕ** of secondary oscillations	0.000000

SKIN AND PROXIMITY EFFECT CALCULATIONS

Depth of penetration **(Dδ)**	0.0113
Cu layer factor **(F_L)**	0.76
Q'	0.75
AC resistance factor **(Fr)**	1.0341
Total impedance of values entered in Ohms **(Zs)**	246.5622 **Ω**
Find decrement factor of secondary **(δS)**	0.019869
Quality of resonant circuit (skin and proximity effect) **(Qs)**	**158**
Quality of resonant circuit (skin effect only)	163

PRIMARY CALCULATIONS

Find **ωp**	467331
Find resonant oscillation time period in seconds **(tp)**	0.000013 sec
Find **tan ϕ** of primary oscillations	0.000000
Find Load power in watts **(Pp)**	**4.65KW**
Find Line current in Amps rms **(IL)**	**23.2 A**
Find load resistance in Ohms **(ZP)**	42.2 Ω
Resonant primary impedance with S.G. and reflected sec **(Zps)**	77.8 Ω
Peak primary impedance at end of selected spark gap dwell time **(Zpp)**	4977.3 **Ω**
Resonant primary impedance with S.G., w/o sec **(Zpss)**	2.6198 **Ω**
Find primary Quality factor **(Qps)** with spark gap	18.99
Find decrement factor of primary **(δP)**	**0.165391**
Maximum useable tank capacitance in µF **(Cp)**	**0.0647µF**
Maximum primary winding voltage **(Vpp)**	3259 V
Maximum primary tank current in Amps **(Ip)**	**63.3 A**
Find primary rms current (step-up transformer output) in Amps **(Ip)**	0.332 A
Lineal spark gap resistance in Ohms **(Rgl)**	2.5419 Ω
Exponential spark gap resistance in Ohms **(Rge)**	2.6098 Ω
Primary oscillation frequency-to-resonant frequency multiplier **(γ)**	0.306

FIGURE 2-7 Optimum wire gauge for selected secondary resonant frequency and coil form diameter worksheet calculations.

Conditions:

1. Resonant frequency (fo) is selected and secondary characteristics are calculated from fo, form diameter (D), and AWG.
   Inductance (Ls) and self-capacitance (Cs) assume the wire length is 1/4 λ of fo. Effects of terminal capacitance are included.
   Calculated secondary Q, Rdc and impedance (Zs) includes skin effect and proximity effect using Dowell methodology.
2. Non-synchronous rotary spark gap operation is accounted for using a calculated line supply voltage synchronized to the phase angle entered in (B32). Line supply voltage will approximate synchronous operation and maximum theoretical secondary voltage with 89.54° entered into cell (B32).
   The primary inductance (Lp) is selectable and k is estimated from the calculated primary and secondary dimensions.
   By changing the ID (B40), IWD (B41), Tp (B44) and q (B43) [Archemedes spiral] tune the calculated primary inductance (B47) as close as practical to the required inductance for resonance (B48). If a pancake primary is desired use the Archemedes spiral parameters with an angle of declination (θ) of 0°. If θ exceeds 75° use a helical primary.
   If closer tuning is desired use additional capacitance (B45) in parallel with the tank capacitance or inductance (B46) in series with the primary winding.
3. Primary peak current calculated using I = C(dv/dt) where: C is the tank capacitance (C19), dv the peak applied voltage (C22) and dt the oscillation period (F37). Primary impedance for VSWR solution includes secondary reflected through the mutual inductance. (w * M)^2 / Zs   Primary spark gap calculations use linear(1) or exponential (2) decrement characteristics as selected in (B33). The primary (low Q) and secondary (high Q) decrement is determined by 1/Q where Q = (w*L) / R.
4. Secondary peak current calculated using peak primary current coupled through mutual inductance to secondary.
   Is = [(ω * M) / Zs] * Ip or (ω * M * Ip) / Zs
5. Secondary voltage calculated using transient solution for series R-L-C circuit with applied sinewave.
   E = -I * [R / sin(-ωt - α + ϕ) transposed from solution for transient current: I = E / R * sine(ωt + α - ϕ)
   Primary voltage = E * sin(ωt + α)   Tan ϕ = (ω^2 * LC - 1) / (ωCR)   attenuation factor (α) inferred to = δ
   The transient solution for the secondary voltage is shown as the broken blue trace in chart 1.
6. The maximum primary impedance in cell (F43) at the end of the selected ionization time entered in cell (B29) is calculated using the primary decrement in a reciprocal function. The primary impedance increases by the decrement with each succeeding oscillation.
   The primary and secondary impedance is used to calculate a VSWR. The VSWR times the primary winding voltage is equal to the secondary voltage shown as the solid blue trace in chart 1.
7. **Actual performance may vary as the calculations assume a pure sinusoidal waveform. The actual waveform may vary which varies the calculated results. The calculations are are intended to present theoretical maximums and relationships for use in optimizing the design. It is assumed the applied coupling has not exceeded the critical coupling.**

COUPLING CALCULATIONS

Find primary diameter in inches (**DP**)		35.1 in
Find primary radius in inches (**A**)		17.6 in
Find primary height in inches (**Hp**)		13.3 in
Find secondary radius in inches (**a**)		2.250 in
Find secondary height in inches (**Hs**)		31.1 in
	K1	0.0035
	x1	2.253 in
	x2	15.564 in
	r1	17.700 in
	r2	23.462 in
	D	8.909 in
	k1	31.129 in
	K3	-4.46E-07
	k3	-1.48E+04
	K5	-1.53E-09
	k5	-1.14E+05
Mutual inductance of windings in μH (**M**)		**291.38 μH**
Coefficient of coupling (**k**)		**0.094**

RESONANT CALCULATIONS

Find coefficient of reflection (**Γ**)	0.906
Find Voltage Standing Wave Ratio (**VSWR**)	20.2
Maximum secondary voltage using selected AWG (**Vs**)	**399.6 KV**
Maximum spark length in inches (**SL**)	40.0 in
Find secondary peak current (**Is**)	32.1 A
Find secondary rms current (**Isrms**)	1.7 A
Maximum secondary voltage using optimum AWG (**Vs**)	400.0 KV
Find secondary rms voltage (**Vsrms**)	21.0 KV
Find secondary power (**Ps**)	**0.57KW**
Find voltage-to-current phase shift in radians (**Pm**)	0.006324
Find secondary peak instantaneous power (**Pspk**)	4KW
Find voltage-to-current phase shift in radians (**Pm**)	-1.555

COMPARISON CALCULATIONS

A	B	C / D
Find secondary peak current (**Is**)	32.1 A	Is = [(ωso * M) * Ip] / Zs
Maximum Secondary voltage using transient LCR solution (**Vs**)	**398.2 KV**	Vs=-Is*[Zs/sin(-ωso*t-δS+tan
Maximum secondary voltage using circuit capacitance (**Vs**)	664.9 KV	Vs=Vp*sqrt(Cp/Cs)
Maximum secondary voltage using circuit inductance (**Vs**)	578.4 KV	Vs=Vp*sqrt(Ls/Lp)
Enter theoretical optimum AWG	**27**	Interpret from Chart 1 to find k fo
Coefficient of coupling using optimum AWG (**k**)	0.065	Database lookup from column B>
Maximum secondary voltage using optimum AWG (**Vs**)	306.8 KV	Database lookup from column AR

A B C D E F

FIGURE 2-7 Optimum wire gauge for selected secondary resonant frequency and coil form diameter worksheet calculations. *(continued)*

Vp = Peak output voltage of step-up transformer in volts = cell (C22), calculated value from *rms* value in kV in cell (B22) × 1.414. Converted to volts using a 1e3 multiplier.
Cp = Primary capacitance in farads = cell (B19), enter value in μF. Converted to farads in cell (C19) using a 1e-6 multiplier.
Cs = Self-Capacitance of the secondary winding = cell (F18), calculated value from equation (2.6).
Ct = Terminal capacitance in farads = cell (B11), enter value in pF. Converted to farads in cell (C11) using a 1e-12 multiplier.

Note the primary-to-secondary inductance or capacitance ratio produces voltages that are too high. Another consideration is if the relationship in equation (2.45) is correct, the secondary voltage should increase if the primary capacitance is increased or the secondary self-capacitance or terminal capacitance is decreased. The former would produce an increased output; however, the latter will not. I spent considerable time exploring a way to increase the voltage output of the secondary by decreasing the secondary self-capacitance. The only way to effect this is to change the winding geometry and magnet wire gauge. When a different winding geometry producing less self-capacitance was tested the spark output was severely reduced instead of increased.

Figure 2-6 displays Chart 1 from the worksheet using the input parameters shown in Figure 2-7 and equations (2.1) through (2.38), for magnet wire gauges 10 through 45. Using the graph the optimum secondary winding magnet wire gauge for the selected secondary, primary, and spark gap characteristics is easily found. The calculated secondary voltage corresponding to the optimum wire gauge in the graph is shown numerically in cell (F84). AWG 10 thru 45 were selected for display because anything smaller than 45 gauge is generally too small to wind by hand without breaking it and wire larger than 10 gauge tends to outsize your laboratory space. You can, however, extend the graph to include AWG 0 to 60, which are calculated in columns (J) through (AP) and (FO) through (FR), rows 1 through 69. The traces in the graph were produced using the following methodology:

Column (J) calculates the secondary wire diameter, column (K) the diameter in circular mils, and column (L) the DC resistance per foot for the AWG in column (I).

The secondary winding diameter is fixed to the selected value in cell (B8) but the height will decrease as the wire diameter decreases to produce the selected resonant frequency in cell (B10). The physical length of the wire remains constant for each AWG as defined in equation (2.1). The height is calculated in column (T) using equation (2.4) from the calculated wire length per turn in column (S) and the number of turns in column (P) for each AWG in column (I).

The total DC resistance of the secondary winding for each AWG is calculated in column (U) using equation (2.7).

The inductance of the secondary winding for each AWG is calculated in column (V) using equation (2.5).

The self-capacitance of the secondary winding for each AWG is calculated in column (W) using equation (2.6).

The skin effect in the secondary winding at the selected resonant frequency for each AWG is calculated in columns (X) and (Y) using equation (2.8). The Q for each AWG considering skin effect only is calculated in column (Z) using equation (2.13) and displayed on the graph using the black trace.

The skin and proximity effects at the selected resonant frequency in the secondary winding for each AWG are calculated in columns (AA) through (AE) using equations (2.9) through (2.12). The Q for each AWG considering skin and proximity effects is calculated in column (AF) using equation (2.13) and displayed on the graph using the red trace. Note the difference between the black and red trace. If only the skin effect were considered, the design could never be optimized for different magnet wire gauges. When the proximity effect is ignored the calculations indicate an appreciable voltage increase in the secondary when adding additional winding layers. *This voltage increase does not appear in an operating coil when additional layers are added.* The fact that the voltage does not increase with additional layers illustrates the performance degrading proximity effect, which is taken for granted by most coilers.

The total impedance of the secondary winding with terminal capacitance for each AWG is calculated in column (AI) using equation (2.16).

As the height of the secondary decreases with each smaller magnet wire diameter the coefficient of coupling (k) and the mutual inductance (M) are also changing. Columns (BK) through (BX) calculate the mutual inductance (M) and coefficient of coupling (k) for the selected primary winding and secondary winding characteristics for each AWG using the methodology from Section 4.9. A lookup function displays the calculated coefficient of coupling in column (AJ) and mutual inductance in column (AK).

The total primary impedance with reflected secondary impedance for each AWG is calculated in column (AL) using equation (2.27).

For the selected primary characteristics the peak primary current calculated in cell (F49) using equation (2.30) is applicable to all wire gauges used in the secondary winding. The peak secondary current is calculated for each AWG in column (AN) using equation (2.33). The changing tanø in the primary is calculated in column (AM) and in column (AO) for the secondary.

And finally the peak secondary voltage for each AWG is calculated in column (AP) using equation (2.34).

The broken blue line in the graph displays the calculated peak secondary voltage for selected wire gauges 10 through 45 using the transient solution calculations in equations (2.1) through (2.34). A large [+] coincides with the calculated secondary voltage for the applied AWG entered in cell (B6) and the calculated peak secondary voltage in cell (F84). A solid blue line is also shown displaying the calculated peak secondary voltage for selected wire gauges 10 through 45 using the VSWR calculations in equations (2.35) through (2.38) performed in columns (FO) through (FR). A large [X] coincides with the calculated peak secondary voltage for the applied AWG entered in cell (B6). The calculated secondary impedance is shown in the broken green line. As this is changing for selected wire gauges 10 through 45 the reflected impedance to the primary is also changing. The calculated primary impedance is shown in the broken brown line.

One interesting effect is noticed as the number of layers is changed in cell (B5). The Q will essentially double as a second layer is added (enter 2 in cell B5) by decreasing the DC resistance of the winding to one half the value of a single layer. The secondary voltage depends upon many variables and doubling the Q produces only a small increase in secondary voltage. I confirmed this by adding a second layer onto a coil already constructed and found by the length of the spark the voltage increase was imperceptible.

2.5 Simulating the Waveforms in a Spark Gap Coil Design

A method of simulating the Tesla coil waveforms was needed to evaluate the design before it is built. Using the calculations in Sections 2.2 thru 2.4 an Excel spreadsheet was created to generate the waveforms found in the primary and secondary circuits. They can be used to evaluate the coil's performance for the selected secondary, primary, and spark gap characteristics.

Open the CH_2A.xls file. Two graphs are generated from a table of calculations made from the same values entered in cells (B5) through (B46) as those described in Sections 2.2 through 2.4. The transient solutions for the line supply voltage and primary current calculated in columns (O) and (T) are displayed in the graph in columns (A) through (D), rows (94) to (126). The transient solutions for the secondary voltage calculated in column (V) are displayed as a dark blue trace in the graph in columns (E) through (G), rows (94) to (126). The secondary voltage was also calculated using the primary-to-secondary impedance ratio (VSWR) shown in the graph as a green trace. The calculated waveforms shown in Figure 2-10 are just as they would appear on an oscilloscope with a time period setting of 500 μsec/div. This appeared to be the optimal display setting for most design ranges. Greater detail of the spark gap ionization is shown in the 50 μsec/div graphs in Figure 2-12. The 50 μsec/div graphs appear in the worksheet in columns (A) through (G), rows (128) to (161).

The calculations shown in Table 1, columns (J) through (W) for the transient solution and columns (Y) through (AB) for the VSWR solution, extend to row 1500. To be of any use in design the graphs should display at least 5,000 μs of operation. The file size quickly becomes unmanageable when trying to extend the calculations further and Excel does not like working with it (unless using $\geq$128 MB RAM on the PC). To perform the calculations as described above, a series of nested if-then conditional statements were used. Excel limits nested "if"functions to a maximum of eight, which limits the methodology used to perform the calculations. It is sufficient to perform the calculations and display waveforms for a BPS range of 100 to 1,000, a line supply frequency of 60 Hz, and primary resonant frequencies between 50 kHz and 200 kHz, which is the range of interest. Primary frequencies above 200 kHz reduce the table calculations to less than 5,000 μs of operation. The table calculations can be extended beyond row 1500 to work with frequencies above 200 kHz but you must have the PC capacity to work with the larger file size. The table of calculations was made from the equations shown below.

The time period of the tank oscillations is calculated using the resonant frequency of the selected primary characteristics:

$$tp = \frac{1}{fP} \tag{2.46}$$

Where: tp = Time period of primary oscillations in seconds = cell (K103), calculated value.

fP = Frequency of resonant primary oscillations in Hz = cell (B49), calculated value from equation (2.23).

The damped sinusoidal oscillations found in a Tesla coil have a positive and negative excursion of comparable amplitude for each time period. Cell (K103) is one half the time period, cell (K104) is equal to the time period, cell (K105) is equal to one and one half time periods, with each succeeding cell in column (K), one half time period greater than the preceding cell. In this manner, a positive and a negative peak excursion for each oscillation are calculated. Column (J) rounds off the time period in column (K) to the nearest whole μsec for labeling the horizontal time axis in the graphs.

In addition to the time period of oscillations, the pulse repetition period (PRP) of the spark gap breaks per second (BPS) must also be calculated:

$$\mathrm{PRP} = \frac{1}{\mathrm{BPS}} \tag{2.47}$$

Where: PRP = Pulse Repetition Period of primary oscillations in seconds = cell (C30), calculated value.
BPS = Breaks Per Second produced by spark gap = cell (B30), enter value. See Section 6.8.

An if-then condition in columns (L) and (M) repeats the values in the adjacent column (K) cells until the PRP occurs, then restarts sequential numbering from time zero until the next PRP occurs. In this manner column (K) tracks the cumulative time from zero and column (M) tracks the pulse repetition periods (1/BPS).

In addition to the time period of oscillations and the BPS, the time period of the applied line frequency must also be calculated. To generate a 60-Hz sinusoidal waveform in Excel the instantaneous phase angle of the sinusoidal waveform for each of the time periods in column (K) must be calculated. To calculate the phase angle, the degrees in radians must first be calculated since Excel does not work directly with degrees:

$$\theta = 2\pi Lf\,tp = \omega tp \tag{2.48}$$

Where: θ = Instantaneous phase angle in radians per second = column (N), calculated value.
tp = Cumulative time period of resonant oscillations in seconds = calculated values in column (K) from equation (2.46).
Lf = Supply line frequency in Hz (60 Hz U.S.) = cell (B14), enter value.

The rotary spark gap pulse repetition rate (BPS) may be either synchronous (gap firing is coincident with the line voltage peak) or non-synchronous with the applied line frequency. If there is a synchronous relationship the gap will fire coincident with the line voltage peak occurring at a phase angle of 89.54° (90°). This will repeat during the negative line voltage transient at 270°. In a non-synchronous relationship the gap can fire at any instantaneous point (phase angle) in the sinusoidal line voltage. In this case a phase angle difference exists between the line voltage peak and the firing of the spark gap. To establish the time relationship between the line voltage peak and the firing of the spark gap, enter the line supply phase angle that coincides with time zero (first firing of the spark gap) in cell (B31). For instance, if 89.54° is entered the supply voltage peak coincides with time zero and if 0° is entered the supply voltage is zero at time zero. This angle is added to the phase angle calculated in column (N) to calculate the coincidence of the line voltage and spark gap ionization. This is sufficient to evaluate the full range of non-synchronous operation.

Instantaneous values can now be calculated for an applied 60-Hz sinusoidal line voltage:

$$Vps = Vp \bullet \sin\theta \tag{2.49}$$

Where: Vps = Instantaneous line supply voltage (output of step-up transformer) in volts = column (O), calculated value.
Vp = Peak voltage output of step-up transformer = calculated value in cell (C22) from $1.414 \times rms$ input (in kV) in cell (B22).
θ = Instantaneous phase angle in radians = calculated value in column (N) from equation (2.48) plus the phase angle shift entered in cell (B30).

Because of its low capacitance value (relative to a filter capacitor, e.g., 500 μF) the tank capacitor charge will follow the instantaneous value of the applied primary voltage (line voltage) while the rotary spark gap is deionized. When the rotary spark gap's rotating electrodes come in proximity with their stationary counterparts, the gap ionizes (fires) and the capacitor discharges an oscillating current into the series primary winding. At the moment the gap fires the primary series current is equal to this calculated peak instantaneous current. Each succeeding time period of the oscillation decreases in amplitude (damped) from the preceding oscillation by an amount equivalent to the primary decrement (δP) calculated in cell (F46) from equation (2.29). The instantaneous primary tank current in column (T) uses if-then conditional statements, column (O) and equation (2.30) to calculate the instantaneous current coinciding with the adjacent time period in column (M).

The instantaneous primary winding voltage in column (S) uses if-then conditional statements and equation (2.32) to calculate the instantaneous primary winding voltage that coincides with the adjacent repetitive time period in column (M). When the gap fires the series primary current calculated in column (T) generates a proportional voltage in the primary winding. For each succeeding time period this voltage decreases with the primary current by the primary decrement.

The instantaneous secondary winding current in column (U) uses "if-then" conditional statements and equation (2.33) to calculate the instantaneous secondary winding current that coincides with the adjacent repetitive time period in column (M). As seen in actual transient waveforms in reference (5) and the two generated graphs in the worksheet, the secondary voltage and current are at zero the moment the rotary spark gap ionizes. As the stored energy in the capacitor decreases with the current oscillations and decrement it reaches a point (threshold) where it can no longer sustain the minimum ionization current required by the gap and the gap quenches (deionizes). As the primary oscillations decrease in amplitude (V or $I \times [1 - \delta P]$), the primary impedance and secondary current oscillations increase in a reciprocal manner (V or $I \times 1/[1 - \delta \text{P}]$) and reach their peak coincident with the moment of deionization of the spark gap. For each succeeding time period after deionization the secondary current and voltage decrease by the secondary decrement. In order to make the display match the calculated peak voltage it was necessary to add an offset in cell (B93) to slightly increase the ionization time period by the number of resonant time periods entered in the cell. It is unknown if this time lag actually exists or if there is a small error in the transient methodology.

The instantaneous secondary voltage in column (V) uses if-then conditional statements and equation (2.34) to calculate the instantaneous secondary voltage that coincides with the adjacent repetitive time period in column (M). As the secondary current increases to its peak value, coincident with the moment of deionization of the spark gap, it develops a peak voltage in the winding. The secondary voltage and current are rising while the primary is oscillating because it is being fed energy from the primary through the mutual inductance. Once the primary stops oscillating the secondary has only the stored energy from the primary oscillations and the oscillations begin to decay. For each succeeding time period after deionization, this voltage decreases by the secondary decrement.

The VSWR in column (Z) is calculated using equation (2.36), the secondary impedance calculated in cell (F30) using equation (2.16), and the instantaneous primary impedance in column (Y) using equation (2.35). Conditional statements and equation (2.38) are used to calculate the increasing secondary voltage until the peak is reached. Once the peak is reached the secondary voltage for each succeeding oscillation period decreases by the secondary decrement:

$$V_S = V_P \bullet VSWR \bullet (1 - \delta_S) \tag{2.50}$$

Where: V_S = Instantaneous secondary voltage = column (AA), calculated value.
Vp = Peak voltage output of step-up transformer = calculated value in cell (C22) from 1.414 × *rms* input (in kV) in cell (B22).
$VSWR$ = Instantaneous Voltage Standing Wave Ratio = column (Z), calculated value from equation (2.36).
δS = Secondary decrement = cell (F31), calculated value from equation (2.14).

Figure 2-8 provides a simplified diagram of the operation of the Tesla coil and Figure 2-9 illustrates the associated waveforms and timing relationships. Figure 2-10 shows the primary current, line supply voltage, and secondary voltage waveforms as they would appear on an oscilloscope with a time period setting of 500 μsec/div. The waveforms emulate synchronous performance with a phase shift of 89.54° (90°) entered into cell (B31). Note how the peak primary current at the moment of spark gap ionization coincides with the peak positive alternation of the line voltage at time 0 (0 μs). The waveforms shown in Figure 2-11 emulate nonsynchronous performance with a phase shift of 39° entered into cell (B31). Note how the peak primary current does not coincide with the line voltage peak. Figure 2-12 shows the primary current, line supply voltage, and secondary voltage waveforms as they would appear on an oscilloscope with a time period setting of 50 μsec/div, offering greater detail to the spark gap ionization period. The same primary, secondary, line, and spark gap characteristics from the calculations shown in Figure 2-7 were used to generate the primary and secondary waveforms shown in Figures 2-10 and 2-11.

2.6 Optimizing the Spark Gap Coil Design

To produce the maximum spark length from a spark gap coil, do one or more of the following:

- Increase the primary current by using more tank capacitance. The larger the peak primary current the larger the peak secondary current and resulting secondary peak voltage. If the step-up transformer is current limited the current output must be high enough to supply the increased power demand or the primary current will not

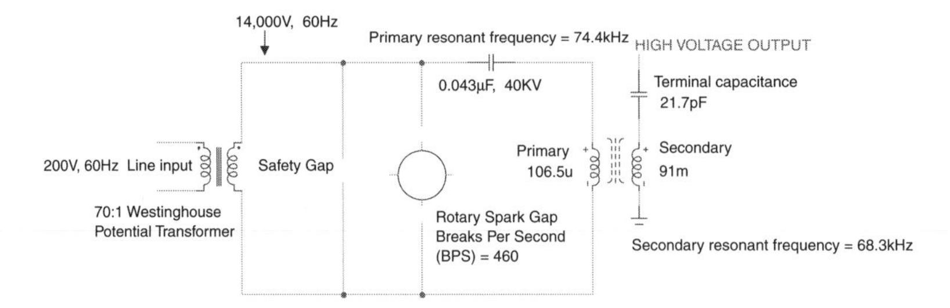

Simplified Rotary Spark Gap Tesla Coil Circuit

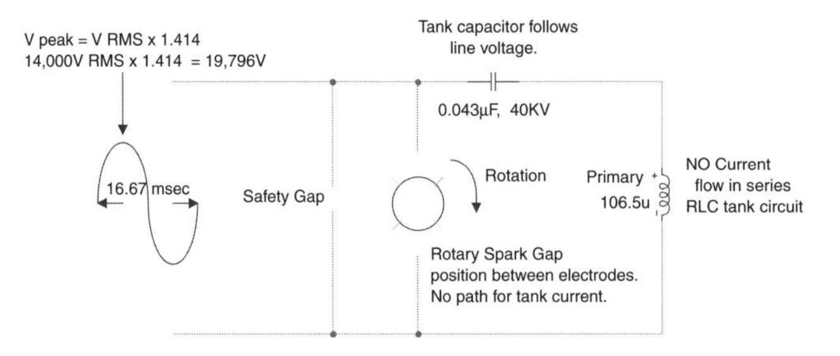

Simplified Primary (Tank) Circuit - Rotary Gap Deionized

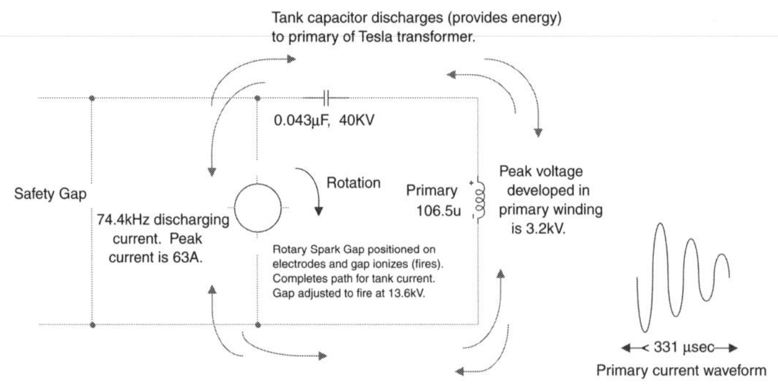

Simplified Primary (Tank) Circuit - Rotary Gap Ionized

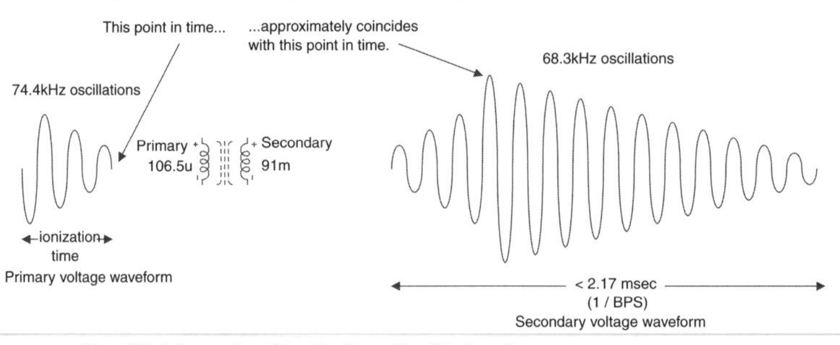

Simplified Secondary Circuit - Capacitor Discharging

FIGURE 2-8 Simplified operation of a spark gap Tesla coil.

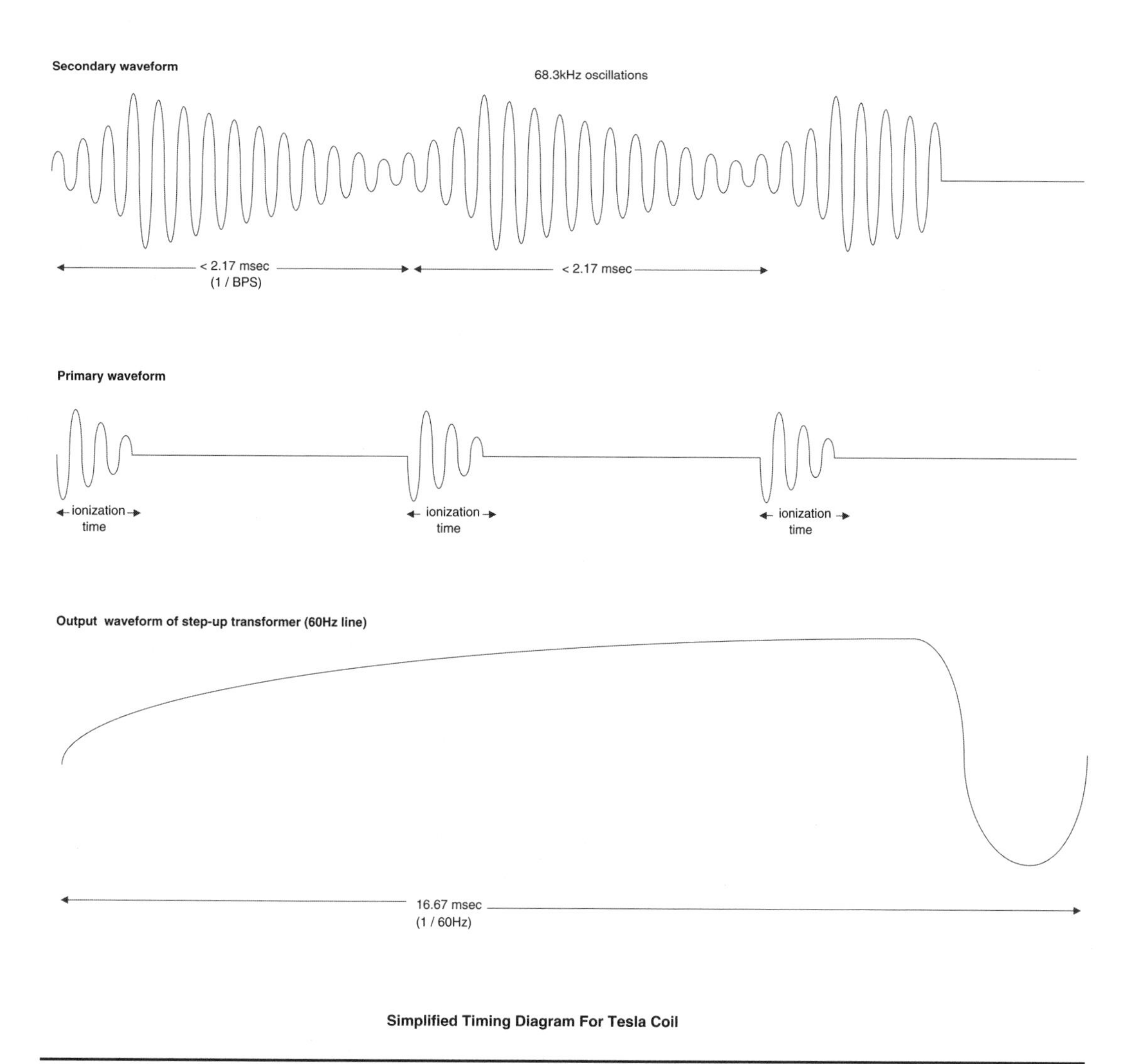

FIGURE 2-9 Waveforms and timing diagram for a spark gap Tesla coil.

increase. To increase the peak primary current for a given step-up transformer output current and voltage rating the capacitance value must be less than or equal to the maximum usable value calculated in equation (2.17). See Section 5.11 for additional details on line voltage, power, and tank capacitance value relationships. If a higher step-up voltage is used the tank capacitor voltage rating must be able to handle it. See Section 5.3 to determine the electrical stresses on the capacitor. The control and monitoring circuits must also be rated for the increased current and power. The primary inductance will have to be retuned for the higher capacitance value by using fewer turns. The decreased primary inductance will lower the primary-to-secondary

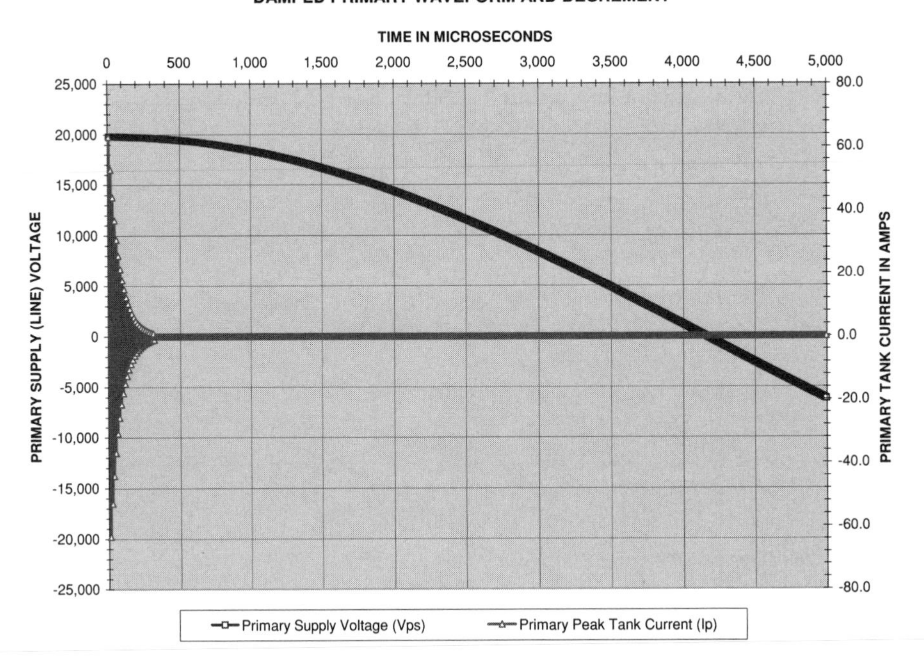

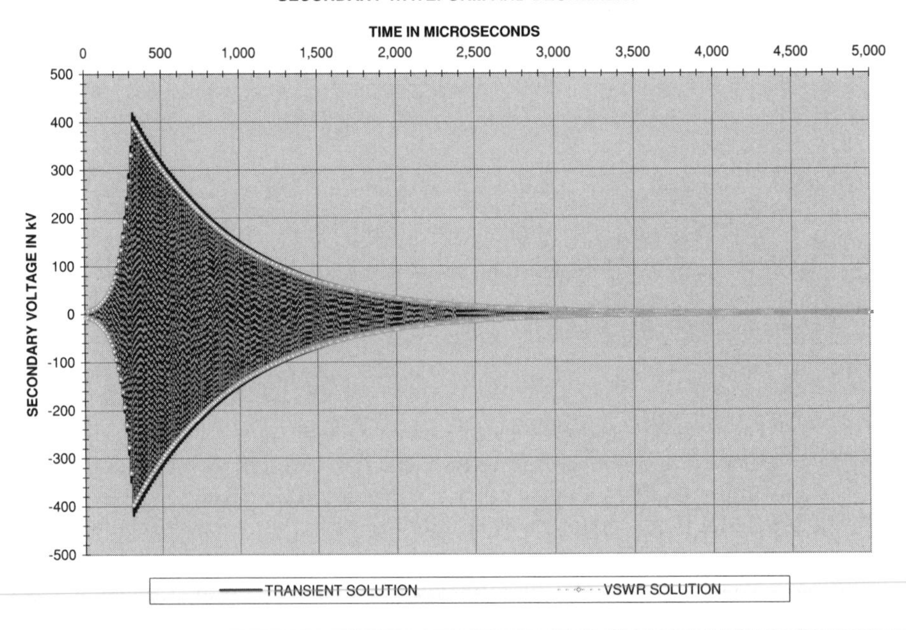

FIGURE 2-10 Simulated primary and secondary waveforms equivalent to 500 μsec/div setting on oscilloscope for a spark gap coil using Excel calculations (phase angle = 90°).

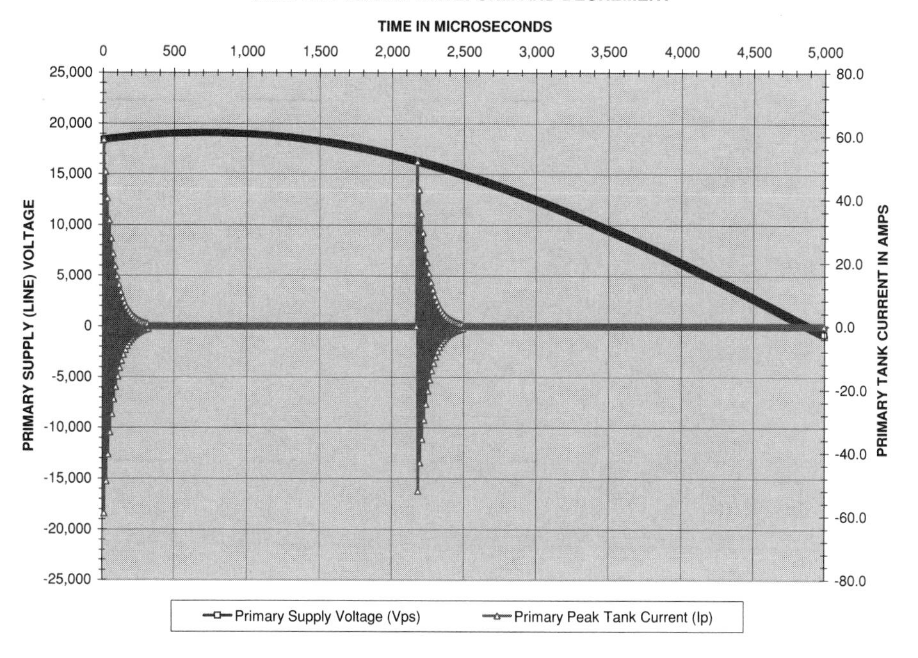

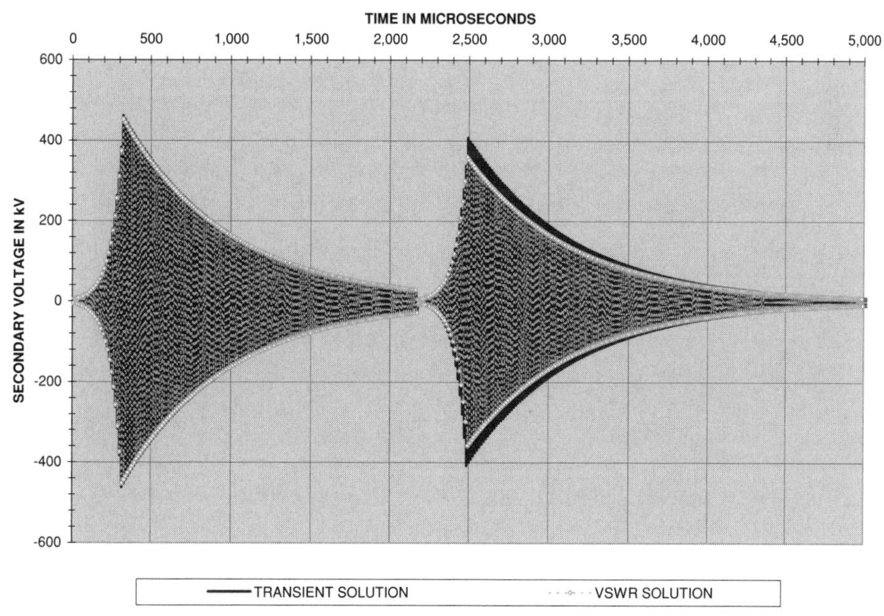

FIGURE 2-11 Simulated primary and secondary waveforms equivalent to 500 μsec/div setting on oscilloscope for a spark gap coil using Excel calculations (phase angle = 39°).

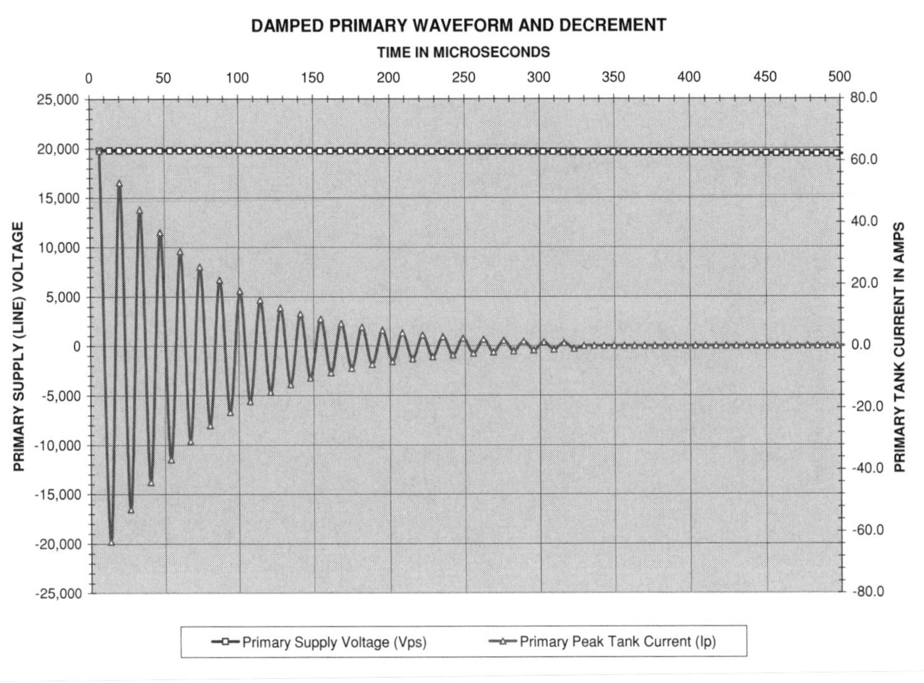

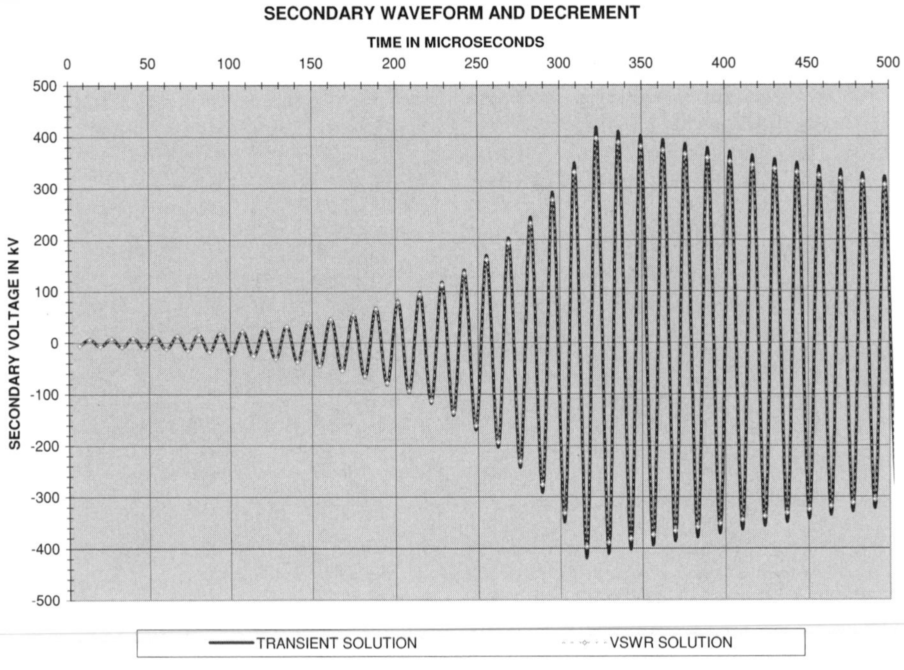

FIGURE 2-12 Primary and secondary waveforms in Figure 2-10 in greater detail.

coupling and mutual inductance but the increased primary current usually results in an overall higher secondary current and voltage. Using fewer primary turns also increases the primary decrement leading to shorter spark gap ionization times. This is generally advantageous and the coil will draw less power from the line than it will when using more primary turns with a lower decrement.

- As illustrated in Figure 2-6 an optimum magnet wire gauge can be selected to produce the highest output voltage in the secondary winding.
- Increasing the secondary coil form diameter will usually produce a higher secondary voltage. The inductance of the winding will also increase. The winding will be physically shorter for a given resonant frequency, which will also increase the coupling and mutual inductance of the primary and secondary. This may result in overcoupling with secondary-to-primary flashover, corona losses, and insulation breakdown. Check the coupling calculations to ensure the coil is not overcoupled (<0.15 to 0.2).
- When considering skin effect only, decreasing the resonant frequency will increase the secondary voltage. However, when considering skin and proximity effects a linear increase in voltage may not result from the decreasing resonant frequency. An optimum resonant frequency may exist for selected coil form dimensions, AWG, and primary characteristics. Another consideration is the higher peak primary current resulting from using higher oscillating frequencies. As the oscillating frequency increases the corresponding time period decreases, allowing higher peak currents. See equation (2.30) for this relationship. The higher peak currents result in higher peak voltages.
- Decrease the secondary winding self-capacitance. Modern coil construction techniques using close-wound magnet wire increases the inductance for the same number of turns, but at the same time increases the interwinding capacitance by minimizing the spacing between turns. In contrast, older methods used a space between turns of double-coated cotton (d.c.c.) wire or a wire with interwinding space and a manually applied insulation (e.g., varnish, beeswax) to achieve high-voltage breakdown strength between turns. A compromise between breakdown strength and interwinding capacitance must be reached in the design of the secondary winding. This is somewhat optimized already by using magnet wire. You will be hard pressed to find wire insulation with higher voltage breakdown strength per unit thickness than that used in modern magnet wire. If an interwinding space is used the self-capacitance is decreased, but so is the inductance. A design compromise may be achieved and the secondary voltage optimized by incorporating an interwinding space in the secondary winding. In Tesla's Colorado Springs notes he emphasized the voltage increase experienced when the self-capacitance of the secondary winding was reduced. The coils used in these experiments had a large spacing between turns. For a given oscillation frequency the winding height will be increased as the interwinding spacing increases. This decreases the primary-to-secondary coupling; however, it can be recovered by redesigning the primary winding. Interwinding spacing in the secondary was used extensively in older coil design but has fallen by the wayside in modern design. Perhaps this technique should be resurrected as it is wide open to experimentation.
- Synchronize the rotary spark gap ionization (firing) with the line voltage peak. This can be done using a synchronous motor to drive the rotary spark gap or by including

a phase shift network in the design to provide a phase lead or lag of the line voltage to the motor. See Section 8.4 on how to produce a phase shift for synchronizing the rotary gap. If a fixed gap is used and the gap spacing adjusted to ionize at the maximum line voltage input it will be automatically synchronous with the line voltage peak.

- Increase or decrease the gap distance between the rotating and stationary electrodes. This changes the spark gap ionization and quenching characteristics. A smaller gap ionizes at a lower voltage. A larger gap ionizes at a higher voltage requiring higher step-up transformer output voltage and capacitor voltage ratings. The primary decrement (δP) is an indication of how many primary oscillations will result from each rotary spark gap firing. As δP increases, fewer primary oscillations result. If the required gap ionization time exceeds the rotary spark gap dwell time the gap will continue to ionize after the rotary electrode passes the stationary electrode, which can be observed as an excessively long arc as the electrodes pass each other. For low decrements the long ionization time may exceed the dwell time by a large factor, which leads to inefficient operation and wasted power. To resolve this problem use fewer primary turns to increase δP and shorten the required ionization time or use larger diameter electrodes. Remember to retune the primary when using a different tank capacitance. Additional capacitors can be connected in parallel with the tank capacitor increasing the primary capacitance and allowing small adjustments. To find the optimum setting may require several test firings and adjustments. Remember to disconnect the line power and discharge the tank capacitance (see Section 5.12) before making adjustments to the spark gap and primary winding. The electrode material and diameter used in the gap can also be experimented with.
- Increase the breaks per second (BPS) in the primary circuit. This will be limited by the tank capacitance value. See Section 5.11 for the BPS and tank capacitance value relationships. If the BPS is greater than about 300 a synchronous motor/gap system will not produce appreciable increases in secondary voltage. At rates higher than this the gap will fire at or very near the peak line input no matter what the phase angle relationship between the peak line voltage and the rotary gap ionization. Although a fixed gap can be easily adjusted to fire at the line peak (120 BPS) the quenching characteristics of the rotary gap and its higher BPS rate are superior and will usually produce a larger spark output and a smoother running coil. Fixed gaps can be run in series with a rotary gap for even better performance and higher power handling capability.

  An unfortunate side effect is the higher the BPS the more power the coil draws from the line. This may not be an issue but should be considered for efficient design. I added twice as many rotating electrodes to a 460 BPS gap for a 920 BPS rate. The coil had a high primary decrement to facilitate a short required ionization time. The only noticeable feature using the 920 BPS rate was a brighter secondary spark; however, the coil drew twice as much power and the spark length was not appreciably greater. Break rates higher than around 300 BPS are generally of no use if your intent is producing the longest spark with the least amount of power. If you use a variable BPS provision, this can be easily experimented with to find the optimum BPS.

- If a selection of primary capacitance values is available reduce the number of primary turns used and retune to resonance by increasing the tank capacitance. An additional capacitance can be attached in parallel to the existing capacitance. This procedure can be evaluated using the worksheets and calculations; however, actual performance limitations can be determined only in an operating coil. Therefore build the coil and experiment with decreasing the number of primary turns used or add tank capacitance until the primary is tuned to the resonant frequency of the secondary. As this is repeated an optimum combination will be discovered for the secondary effect you are trying to create.

  In his Colorado Springs experiments Tesla used a 50-ft diameter primary winding of 1 turn for an inductance of about 56 μH. This produced a very large primary decrement. It also enabled a break rate approaching 4,000 BPS, which allowed him to efficiently pump a lot of power into the primary. Experiment with the primary decrement.

- Increase the coefficient of coupling (k) between the primary and secondary windings. This also increases the mutual inductance (M) between the primary and secondary. A flat Archimedes spiral primary will produce the lowest coupling. A helical coil primary produces the highest coupling. An inverted conical coil (Archimedes spiral with inclination) primary produces a coefficient of coupling somewhere in between for the same outside diameters. Decreasing the primary diameter to place it in closer proximity with the secondary will also increase coupling; but possibly at the expense of increased corona losses, flashover between the secondary and primary, and secondary winding insulation failure. A test run will determine your design's limitations. As your coupling approaches 0.20 it will be too tight to prevent insulation failure and flashover. Experience indicates that 0.10 to 0.15 is optimal for this type of Tesla coil.

  When conservation of energy is considered, the primary power must equal the secondary power plus any power lost in the transformer core. Note in Figure 2-7 that the average secondary power in cell (F86) is 0.79 kW and average primary power in cell (F39) is 4.65 kW. This indicates about 17% of the primary power is coupled to the secondary with the remaining power either lost in the air core or spread out in the frequency spectrum.

  I watched a demonstration by Richard Hull of the Tesla Coil Builders of Richmond (TCBOR), where he top loaded (increased the terminal capacitance) a Tesla coil until the spark could no longer break out. He continued top loading until a bright violet corona could clearly be seen surrounding the entire space between the Archimedes spiral primary and helically wound secondary. Thanks are extended to coiler Steve Roys for allowing Richard to push the operating envelope of his coil, which held up nicely. This demonstrated the energy loss in the air core, even if it cannot be seen. If you can see it, the coupling is too tight and some redesign is necessary, but don't assume you are ever transferring all of the primary power to the secondary.

  A Tesla coil is an air core resonant transformer. Using a core material other than air to increase the permeability between the primary and secondary is not an accepted method of increasing the coefficient of coupling. If another material was used it would have to encompass the entire volume of the primary and secondary to provide mutual coupling, which makes air ideal when you think of it.

2.7 Calculation Accuracy

The accuracy of the calculations performed in this chapter is dependent on the factors listed below:

1. Section 4.2 states that the Wheeler formula for calculating inductance is within ±1%. If a flat Archimedes spiral primary is used the accuracy decreases to within ±5%.
2. Section 4.9 states that the formula for calculating M or k is within ±0.5%.
3. Generally the value marked on the case of mica capacitors is within a ±5% manufactured tolerance. This means the actual value can be up to +5% greater or −5% less than the printed case value. Other high-voltage capacitor types are generally within ±10% to ±20% manufactured tolerance. The capacitance can be measured reducing the calculation error to the measurement error of the device used to measure the capacitance. Many low-cost digital capacitance meters provide a measurement error no greater than 1%.
4. Measuring the load current and voltage requires costly high-voltage measuring apparatus and is generally limited to calculations. The line current and voltage can easily be measured using analog metering. The meter indications are generally within ±5% of the actual value resulting from manufactured tolerance and interpretational error when reading the indications. This could be reduced to ±1% if digital metering is used to measure the line characteristics. Remember that digital devices are less robust than their analog counterparts when subjected to high-voltage transients.
5. The rated output voltage of a high-voltage step-up transformer often disagrees with the actual output. This is a result of manufacturing tolerances. If the exact primary-to-secondary turns ratio is known the output voltage and current error can be eliminated. Current-limited neon sign transformers (NST) have a manufacturing tolerance estimated at ±5%. Non–current-limited potential transformers have a ±0% manufacturing tolerance and are tested after assembly to ensure the turns ratio is exactly as marked on the case. Other transformer types will fall somewhere between this range. The calculations depend on the measured or estimated line input and the turns ratio of the step-up transformer to determine the output voltage, therefore the calculated output voltage accuracy is dependent upon the accuracy of the turns ratio.

Any additional error cannot be characterized from the references providing the formulae used in the calculations. Open the CH_2.xls file, CALCULATION ERROR worksheet (2) to determine the maximum calculation error. (See App. B.) The resulting cumulative error in the calculations will be within:

$$Te = \pm Le \pm Me \pm Ce \pm Lme \pm NTe \tag{2.51}$$

Where: Te = Maximum calculation error in percent = cell (E3), calculated value.
Le = Maximum calculated inductance error using the Wheeler formula (Section 4.3) = ± 1% entered into cell (B3).

Me = Maximum calculated mutual inductance error (Section 4.9) = ± 0.5% entered into cell (B4).
Ce = Maximum calculation error introduced from manufactured tolerance of high-voltage tank capacitor = cell (B5), enter value in percent.
LMe = Maximum calculation error introduced from manufactured tolerance of line voltage and current metering = cell (B6), enter value in percent.
NTe = Maximum calculation error introduced from manufactured tolerance of step-up transformer turns ratio = cell (B7), enter value in percent.

The actual circuit values are within +9.5% and −9.5% of the calculated results when the tank capacitance is measured with a ±3% digital meter, ±5% line voltage and current metering and no error introduced using a potential step-up transformer with a 70:1 turns ratio. As the accuracy of the input parameters improves, so does the accuracy of the calculated results.

One final note on accuracy. In a conversation I had with John Freau, he voiced concern over obtaining any accuracy in calculations. He observed that once the coil generates enough voltage to initiate a spark, the rapid discharge will change the terminal capacitance characteristics, operating frequency, and perhaps other operating parameters. I agree with his observation—however, the objective is only to initiate a spark output. Once the spark is initiated it is only of academic interest what the coil does. The coil eventually recovers from any spark discharge effects and resumes its initial operating parameters to create another spark discharge, and does this at a rate equal to the BPS.

References

1. *Reference Data for Radio Engineers*. H.P. Westman, Editor. Federal Telephone and Radio Corporation (International Telephone and Telegraph Corporation), American Book. Fourth Ed.: 1956, p. 129.
2. Unitrode Power Supply Design Seminar SEM-900, p. M2-4, pp. M8-1 thru M8-10.
3. P.L. Dowell. *Effects of Eddy Currents in Transformer Windings*. Proceedings IEE (UK), Vol. 113, No. 8, August 1966.
4. Terman F.E. *Radio Engineer's Handbook*. McGraw-Hill: 1943. Section 3: Circuit Theory. Pp. 135–172.
5. Dr. J. Zenneck, Translated by A.E. Seelig. *Wireless Telegraphy*. McGraw-Hill: 1915.
6. *Reference Data for Radio Engineers*. H.P. Westman, Editor. Federal Telephone and Radio Corporation (International Telephone and Telegraph Corporation), American Book. Fourth Ed.: 1956, pp. 151–155.
7. Nikola Tesla. *Colorado Springs Notes: 1899–1900*. Nikola Tesla Museum (NOLIT), Beograd, Yugoslavia: 1978.
8. Wheeler formula found in: H. Pender, W.A. Del Mar. *Electrical Engineer's Handbook*. third Ed. John Wiley & Sons, Inc.: 1936. Air Core Inductors pp. 4-17 to 4-18. This reference cites the following source: Proc. I.R.E., Vol. 16, p. 1398. October 1928.
9. U.S. Department of Commerce, National Bureau of Standards, *Radio Instruments and Measurements, Circular 74*. U.S. Government Printing Office. Edition of March 10, 1924, reprinted Jan. 1, 1937, with certain corrections and omissions.

CHAPTER 3

Resonance

A very curious feature was the sharpness of tuning. This seems to be due to the fact that there are two circuits or two separate vibrations which must accord exactly.

Nikola Tesla. Colorado Springs Notes: 1899-1900, p. 43.

Tesla observes resonance of two tuned circuits.

When a changing voltage is applied to a circuit that contains reactive components the circuit will exhibit resonant effects. Since all electrical parts and circuits contain at least some small amount of parasitic capacitance and inductance the circuit will exhibit resonant response to variable-frequency excitation. Understanding resonant effects is the key to designing tuned or oscillating circuits.

3.1 Resonant Effects in Series RLC Circuits

Unless otherwise noted reference (1) is used as the source for resonant formulae in Sections 3.1 and 3.2. The resonant response for the series RLC circuit formed by the secondary winding detailed in Chapter 2 is shown in Figure 3-1. Note this is only for the winding and ignores skin and proximity effects and the terminal capacitance. At the resonant frequency of the circuit marked by the large [X] the impedance is at minimum (black trace) and the line current is at maximum (red trace). At frequencies below resonance the phase angle is leading, indicating a predominately capacitive circuit. At resonance the phase angle is at 0° (blue trace). At frequencies above resonance the phase angle is lagging, indicating a predominately inductive circuit. Similar resonant effects are also found in power distribution systems where power factor correction (capacitance) is used to offset a lagging phase angle.

The secondary winding of a Tesla coil is a good example of a lightly damped (low decrement) series RLC resonant circuit. To evaluate the resonant response of a series RLC circuit open the CH_3.xls file, RLC RESONANCE worksheet (2). (See App. B.) The calculation methodology is described below.

From reference (3) is found Thomson's equation discovered by Sir William Thomson (Lord Kelvin) for calculating the resonant frequency of condenser circuits that are *not* extremely damped (low decrement as in the secondary). The entire formula is:

$$fo = \frac{1}{2\pi\sqrt{LC}} \bullet \frac{1}{\sqrt{1 + \left(\frac{\delta}{2\pi}\right)^2}} \tag{3.1}$$

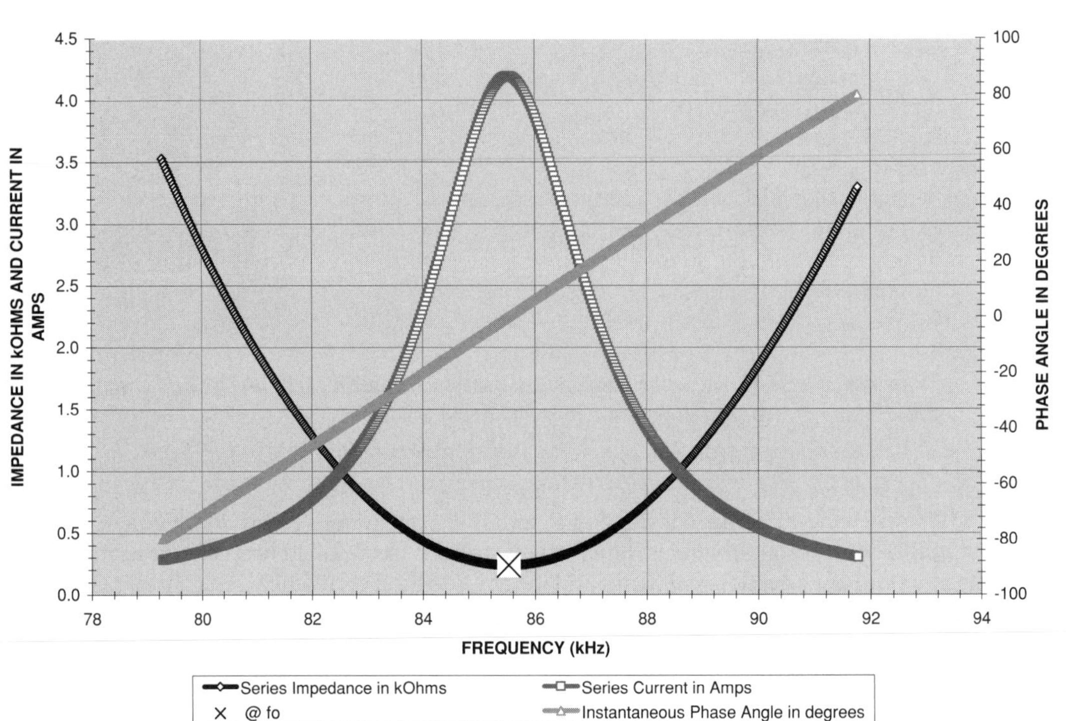

FIGURE 3-1 Series RLC circuit resonant response.

Where: fo = Resonant frequency of series RLC circuit in Hz = cell (H3), calculated value. Converted to kHz in cell (G3) using a 1e-3 multiplier.

C = Capacitance of circuit in farads = cell (B5), enter value in μF. Converted to pF in cell (C5) using a 1e6 multiplier and farads in cell (D5) using a 1e-12 multiplier.

L = Inductance of circuit in henries = cell (B3), enter value in μH. Converted to mH in cell (C3) using a 1e-3 multiplier and henries in cell (D3) using a 1e-6 multiplier.

δ = Decrement of coil = cell (G14), calculated value from $\delta = \pi / Q$ and $Q = (\omega \bullet L)/R$.

This can be shortened to the more convenient and familiar form for high-Q circuits with negligible loss in accuracy:

$$fo = \frac{1}{2\pi\sqrt{LC}}$$

The resonant time period is:

$$t = \frac{1}{fo} \tag{3.2}$$

Where: t = Time period of resonant frequency in seconds = cell (H7), calculated value. Converted to μsec in cell (G7) using a 1e6 multiplier.

fo = Resonant frequency of series RLC circuit in Hz = cell (H3), calculated value from equation (3.1).

Table 1 in columns (K) through (T), rows 21 to 522 generates the calculated values displayed in Chart 1. Figure 3-1 shows Chart 1 as it appears in columns (A) through (D), rows 21 to 55 of the worksheet. The resonant frequency calculated in cell (H3) is used in the center of the table in row 272 to calculate the same range of frequencies above and below resonance in column (K). Each preceding cell from (K271) to (K22) is decreased by the value entered into cell (B7). Each succeeding cell from (K273) to (K522) is increased by the value entered into cell (B7). Column (L) converts column (K) to kHz for use in the graph display.

The frequency in cycles per second (hertz) in column (K) is converted to radians per second in column (M):

$$\omega = 2\pi f \tag{3.3}$$

Where: ω = Frequency of interest in radians per second = calculated value in column (M) for each corresponding value in column (K).

f = Frequency in Hz = calculated value in column (K).

Below the resonant frequency the circuit is more capacitive ($-\theta$ with current leading the voltage) and above resonance it is more inductive ($+\theta$ with current lagging the voltage). These reactance effects are accounted for by including the phase angle ($+\theta$) in the calculations:

$$\theta = \omega \bullet t \tag{3.4}$$

Where: θ = Phase angle in radians = calculated value.

ω = Frequency of interest in radians per second = calculated value in column (M) from equation (3.3).

t = Time period of resonant frequency in seconds = cell (H7), calculated value from equation (3.2).

The magnitude (real) is calculated in column (N) using the cos(θ) and the phase angle in radians (imaginary) is calculated in column (O) using sin(θ). The phase angle in degrees is:

$$\theta(\text{deg}) = \left(\frac{180}{\pi}\right)\theta(\text{rad}) \tag{3.5}$$

Where: θ (deg) = Phase angle in degrees = calculated value in column (P).

θ (rad) = Phase angle in radians = calculated value in column (O) from equation (3.4).

The reactance of the circuit inductance is:

$$X_L = \omega L \tag{3.6}$$

Where: X_L = Reactance (impedance) of inductance in ohms = calculated value in column (Q) at corresponding frequency in column (M).

ω = Frequency of interest in radians per second = calculated value in column (M) from equation (3.3).

L = Inductance of circuit in henries = cell (B3), enter value in μH. Converted to mH in cell (C3) using a 1e-3 multiplier and henries in cell (D3) using a 1e-6 multiplier.

The reactance of the circuit capacitance is:

$$X_C = \frac{1}{\omega_C} \tag{3.7}$$

Where: X_C = Reactance (impedance) of capacitance in ohms = calculated value in column (R) at corresponding frequency in column (M).

ω = Frequency of interest in radians per second = calculated value in column (M) from equation (3.3).

C = Capacitance of circuit in farads = cell (B5), enter value in μF. Converted to pF in cell (C5) using a 1e6 multiplier and farads in cell (D5) using a 1e-12 multiplier.

The total impedance (resistance plus reactance) of the series RLC resonant circuit is:

$$|Zs| = \sqrt{Rs^2 + \left(\omega L - \frac{1}{\omega C}\right)^2} = Rs + \sin\theta\left(\omega L - \frac{1}{\omega C}\right) \tag{3.8}$$

Where: Zs = Impedance of series RLC resonant circuit in ohms = calculated value in column (T) at corresponding frequency in column (M). Converted to kΩ using a 1e-3 multiplier.

Rs = Resistance of circuit in ohms = cell (B4), enter value.

$\sin\theta$ = Resultant phase angle (imaginary) of reactive components in series RLC resonant circuit in radians calculated in column (O).

ω = Frequency of interest in radians per second = calculated value in column (M) from equation (3.3).

L = Inductance of circuit in henries = cell (B3), enter value in μH. Converted to mH in cell (C3) using a 1e-3 multiplier and henries in cell (D3) using a 1e-6 multiplier.

C = Capacitance of circuit in farads = cell (B5), enter value in μF. Converted to pF in cell (C5) using a 1e6 multiplier and farads in cell (D5) using a 1e-12 multiplier.

The line current drawn from the source is the same current in the series RLC circuit:

$$Is = \frac{V}{Zs} \tag{3.9}$$

Where: Is = Line (series) current in amps = calculated value in column (S) at corresponding frequency in column (M).

V = Voltage applied to circuit = cell (B6), enter value.

Zs = Impedance of series RLC resonant circuit in ohms = calculated value in column (T) using equation (3.8) at corresponding frequency in column (M).

Instantaneous values (vectors) of an AC waveform contain two components: the magnitude, which is real (r) and the phase angle, which is imaginary (j) and indicates the magnitude's position in a 360° arc of rotation. Think of the rotor position in a generator. To generate a sinusoidal wave in the stator, the rotor must travel in a 360° arc of rotation, which also corresponds to the value 2π. Any calculation using 2π produces a positional (phase) vector in radians. Figures 3-2 through 3-4 were constructed using the calculations in CH3.xls file, Rectangular worksheet. Note in Figure 3-3 as the sinusoidal wave completes the first positive alternation from time zero it corresponds to 180° or π. After the first positive and negative alternation, the arc of rotation has completed 360° or 2π. As the second positive alternation is completed 3π is reached. A selected point of interest on the rotor actually travels in a circle, best depicted in Figure 3-2 but is often more intuitive as shown on an oscilloscope, illustrated in Figure 3-3 for two complete rotations. The large [X] in both figures correspond to this same point of interest. Figure 3-4 displays the series RLC resonant circuit in its respective rectangular quadrants.

The calculated instantaneous values used to plot Figure 3-2 are also known as rectangular coordinates. The value at [X] is correctly expressed as: 0.62 j0.79 (the r is inferred in the value

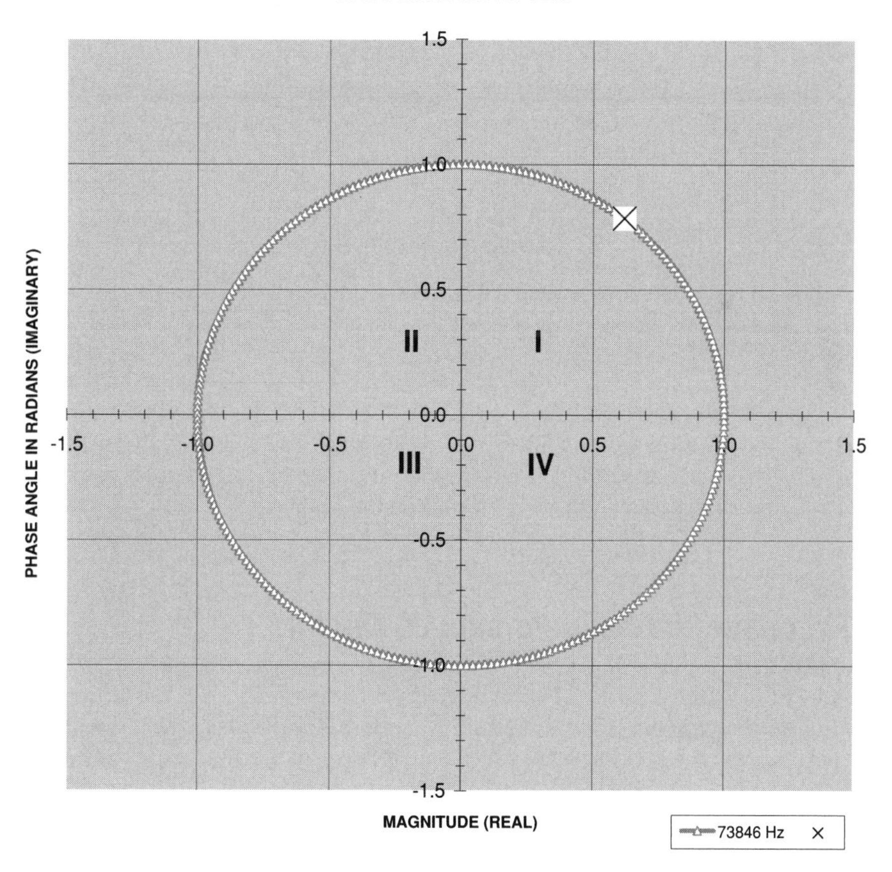

FIGURE 3-2 Periodic sinusoidal waveform in rectangular form showing magnitude (real) and phase angle (imaginary) characteristics.

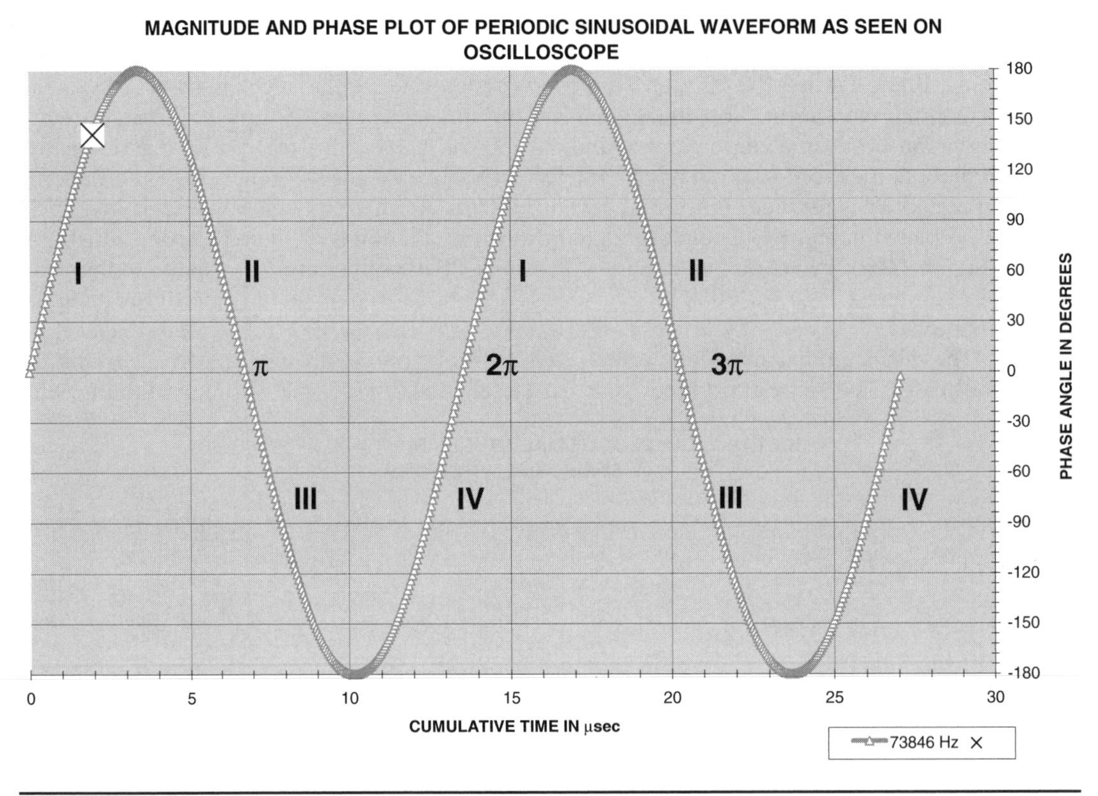

FIGURE 3-3 Magnitude (real) and phase angle (imaginary) of periodic sinusoidal waveform as displayed on oscilloscope (time).

0.62 and the j is expressed in the value 0.79). The CH_10.xls file, POLAR-RECTANGULAR worksheet (6) performs polar-to-rectangular and rectangular-to-polar conversions and notes on each type of expression to assist in conversion. (See App. B.)

3.2 Resonant Effects in Parallel LC Circuits

The resonant response for a parallel LC circuit is shown in Figure 3-5. At the resonant frequency of the circuit marked by the large [X] the impedance is at maximum (black trace) and the line current is at minimum (red trace). The branch currents in the inductive (blue trace) and capacitive (green trace) branches are also displayed in the figure. At frequencies below resonance the phase angle is lagging, indicating a predominately inductive circuit. At resonance the phase angle is at 0°. At frequencies above resonance the phase angle is leading, indicating a predominately capacitive circuit.

To evaluate the resonant response of a parallel *LC* circuit open the CH_3.xls file, RLC RESONANCE worksheet (2). The same worksheet used for the series RLC resonant circuit

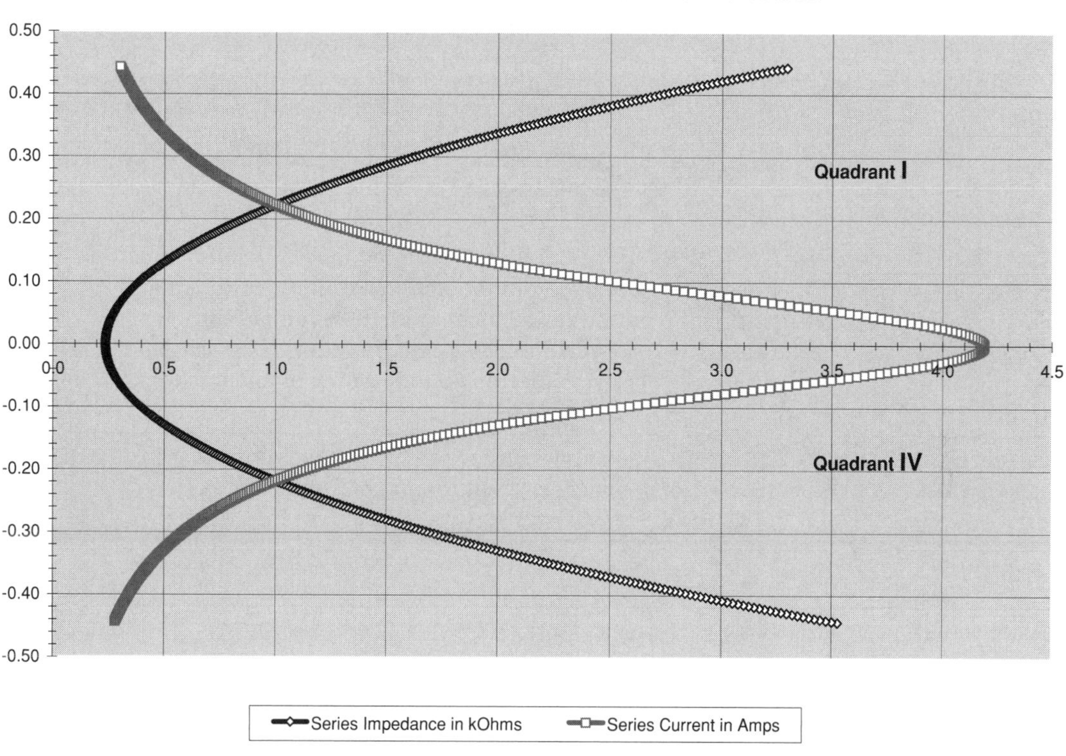

FIGURE 3-4 Series RLC circuit resonant response in rectangular form showing real characteristics.

is also used for the parallel *LC* circuit calculations. The calculation methodology is described below.

The resonant frequency of the parallel LC circuit is calculated in column (W) using equation (3.1) and the same methodology as the series RLC circuit. The time period calculated in cell (H20) is also the same methodology as the series RLC circuit.

Table 2 in columns (W) through (AI), rows 21 to 522 generate the calculated values displayed in Chart 2, Figure 3-5 shows Chart 2 as it appears in columns (F) through (J), rows 21 to 55 of the worksheet. The resonant frequency calculated in cell (H16) is used in the center of the table in row 272 to calculate the same range of frequencies above and below resonance in column (W). Each preceding cell from (W271) to (W22) is decreased by the value entered into cell (B21). Each succeeding cell from (W273) to (W522) is increased by the value entered into cell (B21). Column (X) converts column (W) to kHz for use in the graph display. The frequency in column (W) is converted to radians per second in column (Y) using equation (3.3). The magnitude (real) is calculated in column (Z) using the cos(θ) and the phase angle in radians (imaginary) is calculated in column (AA) using the sin(θ). The phase angle (θ) in degrees is calculated in column (AB) using equation (3.5).

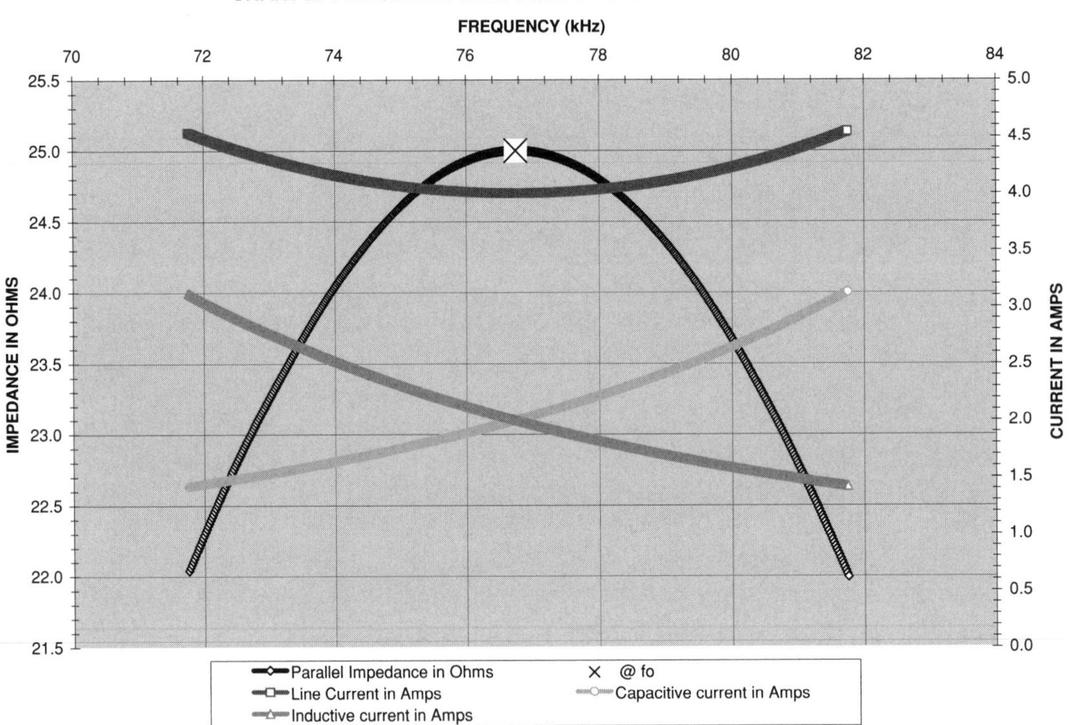

FIGURE 3-5 Parallel LC circuit resonant response.

The parallel *LC* circuit has an inductive and capacitive branch to the line current. The reactance of the inductive branch is:

$$ZL = RL + X_L = RL + \sin\theta(\omega L) \tag{3.10}$$

Where: ZL = Reactance (impedance) of inductive branch in ohms = calculated value in column (AC) at corresponding frequency in column (Y).

X_L = Reactance (impedance) of inductance in ohms = ωL.

RL = Resistance in inductive branch in ohms = cell (B17), enter value.

$\sin(\theta)$ = Resultant phase angle of reactive components in parallel LC resonant circuit in radians calculated in column (AA).

ω = Frequency of interest in radians per second = calculated value in column (Y) from equation (3.3).

L = Inductance of circuit in henries = cell (B16), enter value in μH. Converted to mH in cell (C16) using a 1e-3 multiplier and henries in cell (D16) using a 1e-6 multiplier.

The reactance of the capacitive branch is:

$$ZC = RC - X_C = RC - \frac{\sin\theta}{\omega C} \tag{3.11}$$

Where: ZC = Reactance (impedance) of capacitive branch in ohms = calculated value in column (AE) at corresponding frequency in column (Y).
X_C = Reactance (impedance) of capacitance in ohms = $1/(\omega C)$.
RC = Resistance in capacitive branch in ohms = cell (B19), enter value.
$\sin(\theta)$ = Resultant phase angle of reactive components in parallel LC resonant circuit in radians calculated in column (AA).
ω = Frequency of interest in radians per second = calculated value in column (Y) from equation (3.3).
C = Capacitance of circuit in farads = cell (B18), enter value in μF. Converted to pF in cell (C18) using a 1e6 multiplier and farads in cell (D18) using a 1e-12 multiplier.

The line current splits in the parallel LC circuit. The portion of line current in the inductive branch is:

$$IL = \frac{V}{ZL} \tag{3.12}$$

Where: IL = Line current in the inductive branch in amps = calculated value in column (AD) at corresponding frequency in column (Y).
V = Voltage applied to circuit = cell (B20), enter value.
ZL = Impedance of inductive branch in ohms = calculated value in column (AC) using equation (3.10).

The portion of line current in the capacitive branch is:

$$IC = \frac{V}{ZC} \tag{3.13}$$

Where: IC = Line current in the capacitive branch in amps = calculated value in column (AF) at corresponding frequency in column (Y).
V = Voltage applied to circuit = cell (B20), enter value.
ZC = Impedance of capacitive branch in ohms = calculated value in column (AE) using equation (3.11).

The total impedance (resistance plus reactance) of the inductive and capacitive branches in the parallel LC resonant circuit is found using the product-sum method for parallel resistance:

$$Zp = \frac{ZL \bullet ZC}{ZL + ZC} \tag{3.14}$$

Where: Zp = Impedance of parallel LC resonant circuit in ohms = calculated value in column (AH) at corresponding frequency in column (Y).
ZL = Reactance (impedance) of inductive branch in ohms = calculated value in column (AC) using equation (3.10) at corresponding frequency in column (Y).
ZC = Reactance (impedance) of capacitive branch in ohms = calculated value in column (AE) using equation (3.11) at corresponding frequency in column (Y).

The line current drawn from the source is the sum of the inductive and capacitive branch currents in the parallel LC circuit:

$$Ip = \frac{V}{Zp} \tag{3.15}$$

Where: Ip = Line current in amps = calculated value in column (AG) at corresponding frequency in column (Y).
V = Voltage applied to circuit = cell (B20), enter value.
Zp = Impedance of parallel LC resonant circuit in ohms = calculated value in column (AH) using equation (3.14) at corresponding frequency in column (Y).

3.3 Determining the Resonant Frequency of Two Tuned Circuits in a Spark Gap Coil

A resonant transformer consists of two tuned LCR circuits known as the primary and secondary. The secondary circuit determines the resonant frequency of the Tesla coil. The secondary winding forms a series LCR circuit with a self-inductance, self-capacitance, and resistance. The resistance generally increases as the resonant frequency increases due to skin and proximity effects. If a terminal capacitance is used atop the secondary to increase the voltage reached before the spark breaks out, this capacitance increases the winding's self-capacitance, decreasing the resonant frequency of the secondary (Section 5.10). The series LCR circuit forms the secondary impedance, which varies over a selected frequency range. The secondary has a high Q with a corresponding narrow bandwidth frequency response. The secondary impedance is at a minimum value at the resonant frequency and increases sharply above and below resonance due to the high Q of the circuit as described in Section 3.1. The high Q results in a low decrement and the oscillations produced in the secondary take much longer to dampen than those in the primary.

The primary circuit is the source of energy for the resonant transformer. It is treated as a separate tuned circuit linked to the secondary through the mutual inductance of the primary and secondary windings. The primary winding also has a self-inductance, self-capacitance, and resistance; however, the self-capacitance and resistance of the primary winding are of little interest as the much larger tank capacitance and spark gap resistance are predominant in the circuit. The primary circuit also forms a series LCR circuit using the primary winding inductance, tank capacitance, and spark gap resistance. The impedance of the primary LCR circuit is predominantly the resistance of the spark gap during ionization. As seen in equations (2.24) and (2.25) this resistance is affected by the voltage developed across the gap, electrode material, gap spacing, and peak oscillating current. The primary has a low Q with a wide bandwidth frequency response. The low Q results in a high decrement, which quickly dampens the primary oscillations to the deionization threshold of the spark gap. As the decrement value increases, fewer current oscillations develop in the primary as the capacitor discharges its stored energy. The primary is effectively a wide bandwidth current generator. Sufficient current can be coupled over a wide frequency range to produce a high voltage in the secondary therefore the primary oscillations do not have to be tuned exactly to the resonant frequency of the secondary. However, Tesla discovered that the voltage increase in the secondary is highest when the primary and secondary are tuned to the same frequency.

The relationship between Q and bandwidth is:

$$Q = \frac{1}{BW} \tag{3.16}$$

To evaluate the primary and secondary resonance open the CH_3.xls file, SPARK GAP COIL RESONANCE worksheet (1). (See App. B.) This worksheet gives a graphical display of the frequency response of the primary and secondary reactive components. The worksheet is a modification of the worksheets used in Chapter 2. Table 1 in columns (I) through (R), rows 105 to 929 is used to calculate the values displayed in Chart 1. Figure 3-6 shows Chart 1 as it appears in columns (A) through (F), rows 90 to 123 of the worksheet. The table of calculations is constructed as follows:

As the primary is the source of energy in the coil it is chosen as the reference point for resonance. Cell (I485) in the center of column (I) is equal to the calculated resonant frequency in cell (B49) in Hz. The calculated frequencies in cells (I105) to (I484) decrease from the resonant frequency in cell (I485) by the value entered into cell (I96). The calculated frequencies

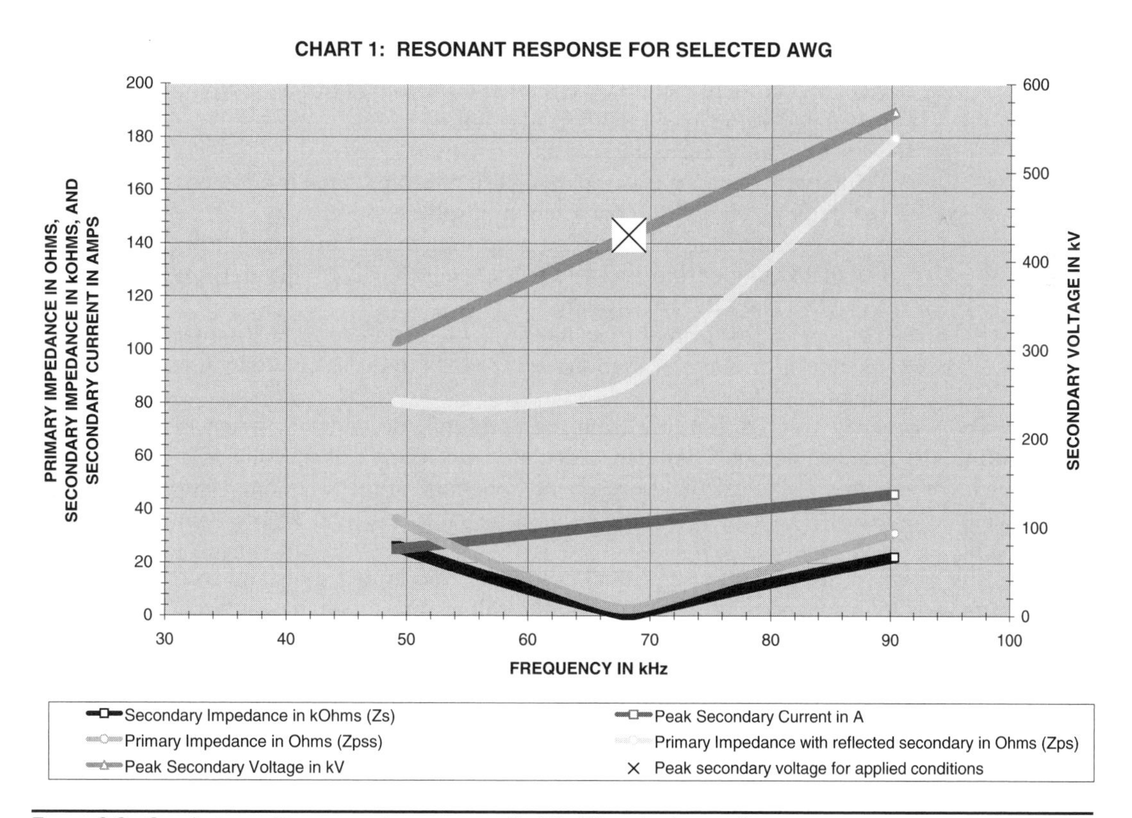

FIGURE 3-6 Spark gap coil's resonant response with 22-pF terminal capacitance.

in cells (I486) to (I929) increase from the resonant frequency in cell (I485) by the value entered into cell (I96). The frequency resolution in Hz in cell (I96) can be changed to increase the frequency range of the calculations in Table 1. This also extends the data displayed in the graph. The frequency in column (I) is converted to radians per second (ω) in column (K). The primary impedance with reflected secondary impedance is calculated in column (L) using equation (2.27) and without reflected secondary impedance in column (M) using equation (2.26).

Columns (N) and (O) are used to calculate Q′ and Fr to determine the skin and proximity effects in the secondary as detailed in equations (2.7) through (2.12) with the secondary impedance at the corresponding frequency in column (I) calculated in column (P) from equation (2.16). The secondary current is calculated in column (Q) from a variation of equation (2.33):

$$Is = \frac{\omega MIp}{Zs} = \frac{\omega M \bullet Cp(\frac{dv}{dt})}{Zs} \tag{3.17}$$

Where: Is = Peak oscillating current in the secondary winding coupled from the primary in amps = column (Q), calculated value.
ω = Frequency in radians per second = column (I), calculated value.
M = Mutual inductance of primary and secondary winding in henries = cell (F73), calculated value. Converted to μH using a 1e6 multiplier.
Z_S = AC impedance of secondary circuit at the resonant frequency in ohms = calculated value in cell (F30) from equation (2.16).
Ip = Peak oscillating current in primary circuit in amps = $Cp \times (dv/dt)$.
Cp = Primary capacitance in farads = cell (B19), enter value in μF. Converted to farads in cell (C19) using a 1e-6 multiplier.
dv = Vp = Peak output voltage of step-up transformer in volts = cell (C22), calculated value from rms value in kV in cell (B22) $\times$ 1.414. Converted to volts using a 1e3 multiplier.
dt = Time period of primary oscillations = 1/fP, where fP = Resonant frequency of primary oscillations in Hz = cell (B49), calculated value from equation (2.23).

The secondary voltage resulting from the oscillating secondary current is calculated in column (R) from equation (2.34). The calculated peak secondary voltage in column (R), which corresponds to the calculated resonant frequency of the secondary with terminal capacitance in cell (F21) using equation (2.15), appears in cell (I99) and as a large [X] in the graph.

The effects of using a 22-pF terminal capacitance on a medium-sized coil are shown in Figure 3-6. At resonance the primary exhibits minimum impedance as seen in the green trace. At resonance the secondary also exhibits minimum impedance as seen in the black trace. The resonant frequencies of both the primary and secondary are very close, with a calculated peak secondary voltage of 429 kV. When the terminal capacitance is increased to 50 pF the secondary resonance decreases well below that of the primary with a corresponding decrease in peak voltage to 291 kV as shown in Figure 3-7. To increase the secondary voltage the primary would have to be retuned using more turns (decreases resonant frequency).

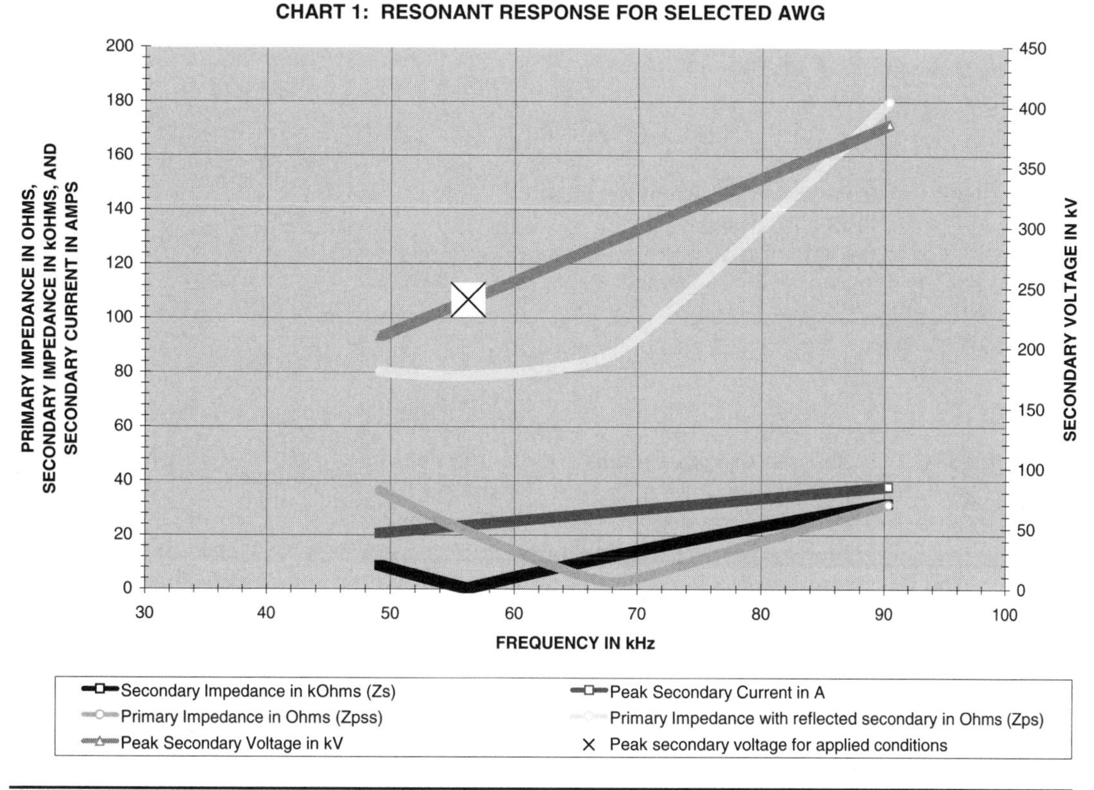

FIGURE 3-7 Spark gap coil's resonant response with 22-pF terminal capacitance increased to 50-pF.

3.4 Other Resonant Calculations

In the periodic sinusoidal oscillations the current is changing over time and expressed as a differential equation ($\Delta i/\Delta t$). In resonant calculations there are two domains of interest: the time and frequency domains. The time-domain is equivalent to running a transient analysis using a Spice simulation as explained in Chapter 8. The frequency-domain is equivalent to running an AC analysis in Spice. The two domains are linked by the Fourier transform.

A Fourier transform resolves the time domain into the frequency domain or a time-domain waveform into a frequency-domain spectrum. The term ω used in the calculations has the following relationships:

$$\omega = \frac{2\pi}{t} \quad f = \frac{1}{t} \quad \therefore \omega = 2\pi f \tag{3.18}$$

Where: ω = Frequency of interest in radians per second.
f = Frequency of interest in hertz (cycles per second).
t = Time period of oscillations in seconds.

A Fourier series is used to evaluate the entire frequency spectrum. Reference (4) states that the current in a periodic symmetrical sinusoidal waveform at the fundamental (resonant) frequency can be found using:

$$I = Ip \bullet \sin(\omega t) \tag{3.19}$$

Where: I = Instantaneous current in amps.
Ip = Peak current in amps.
ω = Resonant frequency in radians per second.
t = Time period of resonant oscillations in seconds.

The phase angle (θ) in degrees can be calculated using the sine function as shown in equations (2.48) and (2.49) which are combined below:

$$\sin\theta = \sin(2\pi ft) = \sin(\omega t) \tag{3.20}$$

Where: θ = Instantaneous phase angle in radians per second.
ω = Resonant frequency in radians per second.
t = Time period of resonant oscillations in seconds.
f = Frequency of resonant oscillations in Hz.

And the magnitude (voltage, current) can be calculated using the cosine function.

$$\cos\theta = \cos(2\pi ft) = \cos(\omega t) \tag{3.21}$$

From reference (2) is found a technique to calculate the impedance at resonance for lightly damped (high Q) parallel LC resonant circuits in the frequency domain using a Laplace transform:

$$\omega o = \sqrt{\frac{1}{LC}} \tag{3.22}$$

Where: ωo = Resonant frequency in radians per second.
C = Capacitance of resonant circuit in farads.
L = Inductance of resonant circuit in henries.

And at resonance the gain is:

$$A = R\sqrt{\frac{C}{L}} \tag{3.23}$$

Where: A = Gain of resonant circuit.
R = Resistance of resonant circuit in ohms.
C = Capacitance of resonant circuit in farads.
L = Inductance of resonant circuit in henries.

The secondary impedance to the source can be found using a variation of equation (3.23):

$$Z = \sqrt{\frac{L}{C}} \tag{3.24}$$

Where: Z = secondary impedance to the source at resonance in ohms.
C = Self-capacitance of secondary and terminal (if used) in farads.
L = Inductance of secondary in henries.

From page 111 and 112 of reference (3) is found the relationship of the primary and secondary circuits at resonance:

$$LpCp = LsCs \quad \text{and} \quad \frac{Lp}{Ls} = \frac{Cs}{Cp} \tag{3.25}$$

Where: Lp = Primary inductance in henries.
Cp = Primary capacitance in farads.
Ls = Secondary inductance in henries.
Cs = Secondary capacitance in farads.

Spark Gap DEIONIZED
primary-to-source impedance @ 60Hz = 61.7 kOhms

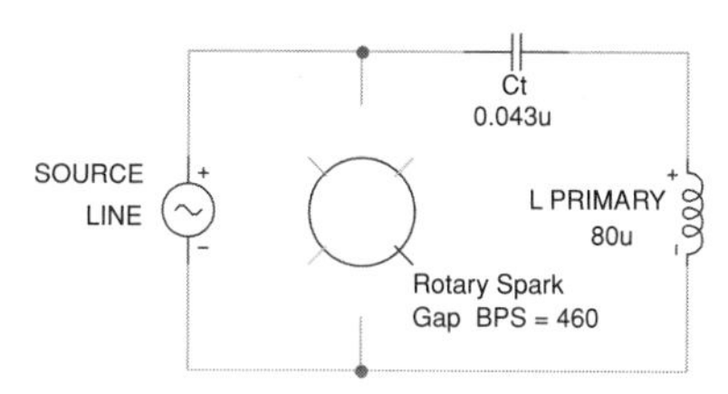

When the spark gap is deionized the 60Hz line frequency encounters a capacitive reactance of:
1 / 2*pi*f*C = 1 / 6.28*60*0.043µ = 61.7 kOhms.
and also encounters a series inductive reactance of:
2*pi*f*L = 6.28*60*80µ = 0.03Ohms.
The total line impedance is 61.7 kOhms. The approximate line current during the 2.15ms the gap is deionized is:
I = E / Z = 16.8kVrms / 61.7kOhms = 272mA.

If the tank capacitor (Ct) and rotary spark gap's position in the circuit are switched, the line impedance during deionization becomes even greater.

Spark Gap IONIZED
primary impedance @ fo (79kHz) = 22 Ohms

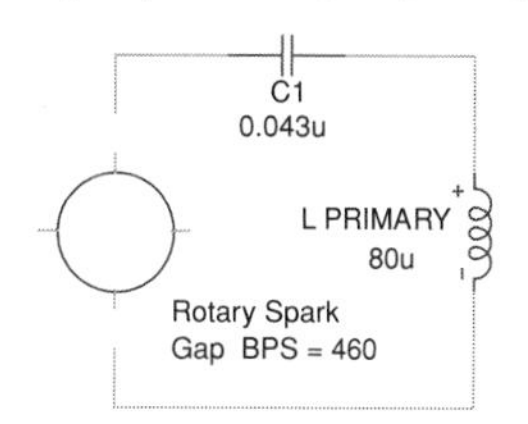

When the spark gap is ionized the 60Hz line frequency still encounters the total line impedance of 61.7 kOhms. However, the impedance the tank circuit encounters at the resonant frequency of 79kHz with the tank capacitor as the source is approximately 2 Ohms. When the secondary impedance is reflected into the primary this impedance becomes approximately 22 Ohms during the 25µs of ionization time.

Spark Gap DEIONIZED
secondary-to-source impedance @ 60Hz = Meg Ohms

PRIMARY Z
61.7 kOhms to 60Hz

SECONDARY Z
Meg Ohms

The secondary is connected to the primary by the mutual inductance (coupling) when current is flowing in the primary. While the spark gap is deionized there is negligible current flow in the primary and several Meg Ohms of secondary impedance to the 60Hz source.

Spark Gap IONIZED
secondary-to-source impedance @ 60Hz = 35.4 kOhms
secondary impedance @ fo (79kHz) = 221 Ohms

PRIMARY Z
22 Ohms @ resonance
61.7 kOhms to 60Hz

SECONDARY Z
221 Ohms @ resonance
35.4 kOhms to 60Hz

The secondary is connected to the primary by the mutual inductance (coupling) when current is flowing in the primary. While the spark gap is ionized there is intense current flow in the primary and 35.4 kOhms of secondary impedance to the 60Hz source.

When the primary current stops after 25µs the secondary is isolated from the primary by the Meg Ohm impedance-to-source. The secondary now becomes it's own source with a 221 Ohm impedance to the resonant frequency of 79kHz.

FIGURE 3-8 Spark gap coil's resonant impedances as seen by primary and by source (line).

Figure 3-8 illustrates the impedance seen by the line (source) in an operating Tesla coil. The values are typical for operating coils.

References

1. Terman F.E. *Radio Engineer's Handbook*. McGraw-Hill: 1943. Section 3: Circuit Theory, pp. 135–172.
2. Dan Mitchell and Bob Mammano. *Designing Stable Control Loops*. Unitrode 2001 Power Supply Design Seminar SEM-1400, p. 5-5.
3. Dr. J. Zenneck, Translated by A.E. Seelig. *Wireless Telegraphy*. McGraw-Hill: 1915. Page 5.
4. H. Pender, W.A. Del Mar. *Electrical Engineer's Handbook*. Third Ed. John Wiley & Sons, Inc.: 1936. Fourier Series, pp. 3-06 to 3-09.

4 CHAPTER

Inductors and Air Core Transformers

...the consideration shows why with a large distributed capacity a very high pressure cannot be obtained on the free terminal. All the electrical movement set up in the coil is taken up to fill the condenser and little appears on the free end. This drawback increases, of course, with frequency and still more with the emf.

Nikola Tesla. Colorado Springs Notes: 1899-1900, p. 74.

Tesla observes the effects of distributed capacitance in the winding using different interwinding distance.

When a 30-foot strand of 24 AWG magnet wire is held straight at 3 feet above ground the inductance is approximately 15.4 μH. If this wire is close wound around a 1-inch diameter form a helical winding of approximately 112 turns will result with an inductance of 107.8 μH. Winding the coil into a helix with no interwinding space increases the inductance of the wire by 7 times ($107.8\,\mu H/15.4\,\mu H = 7$). The winding height of 2.46 inches produces the optimum winding height-to-diameter ratio of 2.46 according to reference (15). If the diameter of the form is changed to 2.46 inches and the 30′ wire is wound 46 turns for a 1-inch winding height, the resulting inductance is 151.9 μH or about 10 times the inductance of the straight wire ($151.9\,\mu H/15.4\,\mu H = 10$). This illustrates how a wire's inductance is increased when wound in a coil.

4.1 Calculating Inductor Parameters

The only resemblance a Tesla coil has to a transformer is the mutual inductance between the primary and secondary coil windings. Therefore when designing the primary and secondary windings, they can be thought of as separate inductors. The worksheets described below can be found in the CH_4.xls file. (See App. B.) Section 4.8 details the calculations used for determining magnet wire parameters.

The applied inductance of a magnetic device is:

$$L = \frac{0.4\pi\mu N^2 A_e}{\ell_e \times 10^8} \tag{4.1}$$

Where: L is the inductance in henries.
μ is the permeability of the core material used.
N is the number of turns of the inductor.
A_e is the cross-sectional area of the core in cm^2.
ℓ_e is the core magnetic path length in cm.

The permeability (μ) of the core is given by:

$$\mu = \frac{B}{H} \tag{4.2}$$

Where: μ is the permeability of the core material used.
B is the applied magnetic flux density in the core from the magnetizing winding (usually the primary).
H is the applied magnetic field strength in the core from the magnetizing current in the primary.

For core materials other than air the manufacturer of the core material will test its properties and provide μ values for the material with the applied AC and DC magnetization, temperature, high frequency, aging, and manufacturing tolerance effects on μ. For Tesla coils the core material is air and its μ will be defined by its applied B density and H field; therefore, equation (4.2) is not yet of any use. An Excel worksheet was prepared for each type of inductor to calculate the inductance and detailed in Sections 4.2 thru 4.5.

4.2 Helical Inductors

When an inductor is wound in a helix on a cylindrical form as shown in the HELIX (magnet wire) worksheet (3) and Figure 4-1 the inductance is found using the Wheeler formula from reference (12):

$$L(\mu\text{h}) = \frac{A^2 N^2}{9A + 10H} \tag{4.3}$$

Where: L = Inductance of coil in μhenries = cell (D34), calculated value. When H is greater than A this formula is within ± 1.0% accuracy.
A = Radius of coil form in inches = 1/2 outside diameter entered in cell (B6).
N = Number of turns used in coil = cell (D7), enter value.
H = Height (length) of winding in inches = cell (B25), enter value.

NOTE: *When determining the height compare it to the dimension in cell (D27). This will indicate the minimum height needed in a closely wound secondary for the selected number of turns entered in cell (D7).*

Using the dynamics of the worksheet the coil's diameter, number of turns, and height can be changed to find the optimum design for a desired inductance value before construction begins.

The DC resistance of the winding is displayed in cell (D25) and adjusted for the temperature entered in cell (D8). The resistance at 20°C is calculated:

$$\text{DCresistance@20°C} = (L_N \times N) \times \Omega/f \tag{4.4}$$

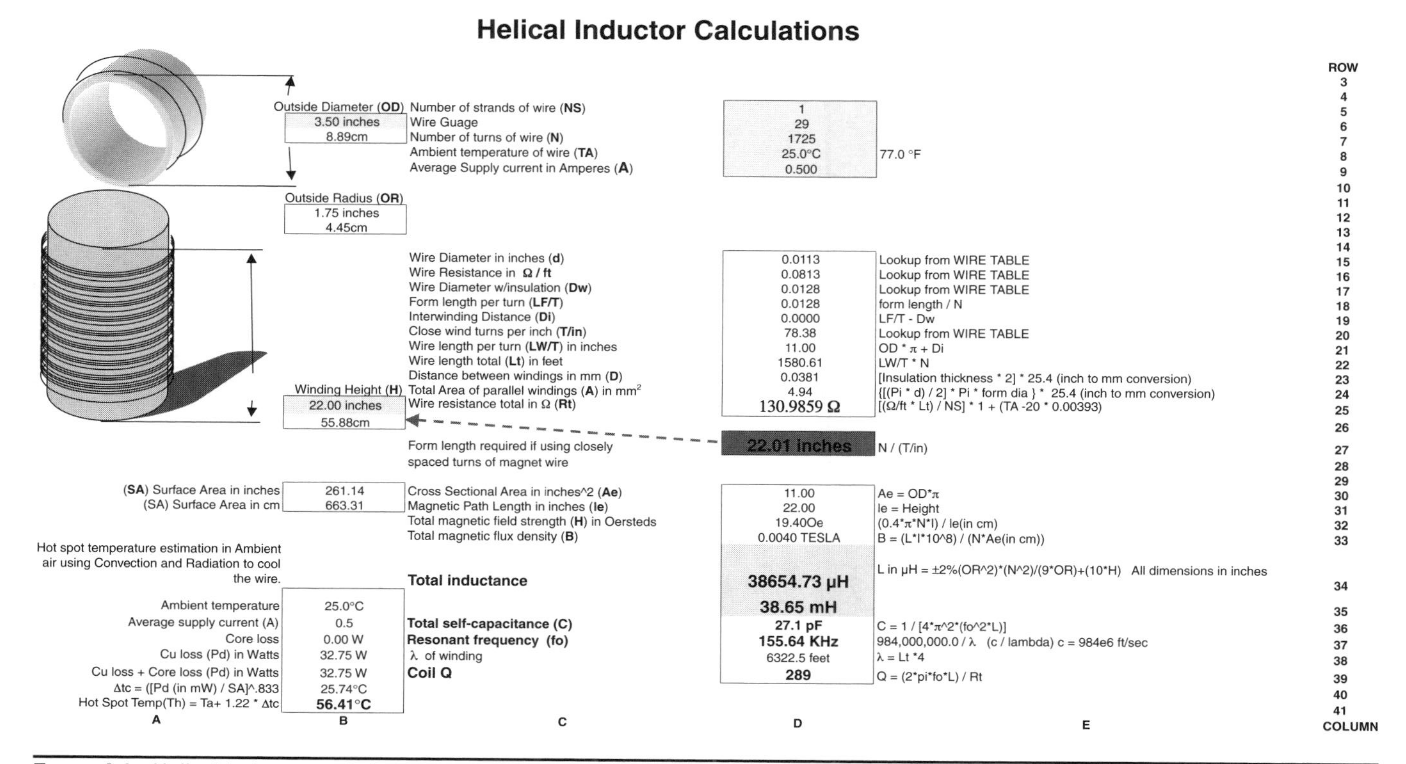

FIGURE 4-1 Helical inductor worksheet calculations.

Where: L_N = Length per turns in feet (LW/T in worksheet) = winding form circumference = $\pi \times$ form diameter entered in cell (B6).
N = Number of turns used in coil = cell (D7), enter value.
Ω/ft = DC resistance for one foot of selected wire gauge from NEMA wire table (see Section 4.7).

NOTE: *$L_N \times N$ = total length of winding in feet.*

The copper used in magnet wire has a temperature coefficient of 0.393%/°C, so the formula for adjusting the winding resistance for temperature is:

$$\text{DCresistance} = \text{DCresistance@20°C} \times 1 + [(T_A - 20°\text{C}) \times 0.00393] \tag{4.5}$$

Where: DCresistance@20°C = calculated DC resistance for application at 20°C from equation (4.4).
T_A = Ambient temperature in °C entered in cell (D8), cell (E8) displays T_A in °F.

If a bifilar (2 strands) or multiple layer application is desired the winding resistance is also adjusted for the number of strands entered in cell (B5) by dividing the resistance by the number of strands used, e.g., if two strands are used the resistance is 1/2, if three strands are used it is 1/3, if four are used it is 1/4.

Once the winding's wire length is known the resonant frequency can be calculated:

$$fo = \frac{c}{\lambda} \tag{4.6}$$

Where: fo = Resonant frequency of coil in Hz = cell (D37), calculated value. Converted to kHz using a 1e-3 multiplier.
c = Speed of light in free space = 984,000,000 feet/sec or 299,800,000 meters/sec.
λ = Electrical wavelength of coil. The secondary of a Tesla coil is a series LCR circuit which is resonant to 1/4 the physical wavelength or length of the wire, therefore 4 × the physical wire length is the electrical wavelength. The physical length of the coil wire = $L_N \times N$.

NOTE: *From reference (2) the velocity of light in free space is 2.998×10^8 meters per second or 186,280 miles per second or 9.84×10^8 feet per second. The permeability (μ) to a magnetic field component is $4\pi \times 10^{-7}$ or 1.257×10^{-6} henry per meter. The permittivity (e) to an electric field component is 8.85×10^{-12} or $(36\pi \times 10^9)^{-1}$ farad per meter. The characteristic impedance $(\mu/e)^{0.5}$ is 376.7 ohms or 120π ohms.*

The interwinding (self) capacitance of the coil is:

$$C = \frac{1}{4\pi^2(fo^2 L)} \tag{4.7}$$

Where: C = Self-capacitance of coil in farads = cell (D36), calculated value. Converted to pF using a 1e12 multiplier.
fo = Resonant frequency of coil in Hz = cell (D37), calculated value from equation (4.6).
L = Inductance of coil in henries = cell (D34), calculated value from equation (4.3).

Using the calculated resonant frequency the Q or quality of the coil can be calculated:

$$Q = \frac{\omega o\, L}{R} \tag{4.8}$$

Where: ωo = Resonant frequency in radians or $2\pi fo$.
R = DC winding resistance = cell (D25), calculated value.
L = Inductance of coil in henries = calculated value in cell (D34) from equation (4.3), converted to henries.

In an inductor the current flow encounters the DC resistance and dissipates power, which produces a temperature rise. From reference (3) the surface area of a cylinder is determined first to calculate this temperature rise:

$$\text{area}_{\text{surface}} = 2\pi A(A + H) \tag{4.9}$$

Where: $\text{area}_{\text{surface}}$ = Inductor surface area in square cm (cm^2) = cell (B31), calculated value. Converted to inches in cell (B30).
A = Radius of coil form in inches = cell (B12), calculated value from 1/2 outside diameter entered in cell (B6).
H = Height (length) of winding in inches = cell (B25), enter value.

The core and winding temperature will increase and a "hot spot"will develop somewhere within the core (inductor). Reference (4) suggests the following methodology for calculating the hot spot temperature:

$$\text{Hotspot} = \left[\frac{\text{copperloss} + \text{coreloss}}{\text{area}_{\text{surface}}}\right]^{0.833} 1.22 + T_A \tag{4.10}$$

Where: Hotspot = The hottest temperature in °C reached in any portion of the inductor, = cell (B41), calculated value.
$\text{area}_{\text{surface}}$ = Inductor surface area in square cm = calculated value in cell (B31) from equation (4.9).
copperloss = Power dissipated in the winding = $I^2 \times R$ = cell (B38), calculated value. I is the DC (average) value of current through the winding and found using the formulas shown in Table 5.1 and value entered in cell (D9). R is the DC winding resistance calculated in cell (D25).
coreloss = Power dissipated in the core. The air core in a Tesla coil will dissipate power but due to the large surface area its contribution to self-heating is essentially zero.
T_A = Ambient temperature in °C entered in cell (D8), cell (E8) displays T_A in °F.

Insulation Type	NEMA Standard	Maximum Rated Temperature (Thermal Class)
Enamel, Polyurethane, Formvar	MW-1, (2), (3), (15), (18), (19), (29)-C	105°C
Polyurethane, Nylon	MW-28, (75)-C	130°C
Polyester, Polyurethane Nylon-155	MW-5, (79), (80)-C	155°C
Polyester imide Nylon	MW-30, (76), (77), (78)-C	180°C
Glass fiber, Polyester amide imide, Polytetrafluoroethylene (Teflon)®	MW-43, (44), (35), (36)-C	200°C
Polyimide	MW-16, (20)-C	220°C
Teflon is a registered trademark of the Dupont Corporation.		

TABLE 4-1 Maximum temperature ratings of magnet wire insulation.

It is important that the core's operating temperature and higher hot spot temperature not exceed the manufacturer's rated Curie temperature for the core. Above this Curie temperature the core will lose its magnetizing ability, which may be permanently lost even after the temperature falls below the Curie point again. In our Tesla coil application the air core does not exceed this limitation but the magnet wire's insulation can. Therefore keep the hot spot temperature below the manufacturer's maximum rated temperature to ensure that the wire insulation is not damaged thermally, subjecting it to corona losses or voltage breakdown between windings. Table 4-1 lists several maximum temperature ratings for typical magnet wire insulation found in reference (5).

Using Ampere's Law the applied magnetic field strength (H) of the coil is calculated:

$$H = \frac{0.4\pi NI}{le} \tag{4.11}$$

Where: H = Magnetic field strength in coil in oersteds = cell (D32), calculated value.
N = Number of turns used in coil = cell (D7), enter value.
I = Average (DC) value of current applied to magnetizing winding (see Table 4-1) = cell (D9), enter value.
le = Magnetic path length of core in inches = height of winding = cell (D31), and cell (B25), enter value.

Using Faraday's Law the applied magnetic flux density (B) of the coil is calculated:

$$B = \frac{LI \times 10^8}{NAe} \tag{4.12}$$

Where: B = Magnetic flux density in coil in teslas = cell (D33), calculated value.
L = Inductance of magnetizing coil (primary) in henries = calculated value in cell (D34) from equation (5.3), converted to henries.
N = Number of turns used in coil = cell (D7), enter value.
I = Average (DC) value of current applied to magnetizing winding (see Table 4-1) = cell (D9), enter value.
Ae = Cross-sectional area of coil form in square inches = $\pi \times$ outside diameter (OD), OD = cell (D30), enter value.

NOTE: *The calculated Ae and le are approximations. The magnetic flux tends to concentrate on the inside corners of the magnetic path (ID), and will not be equally distributed among the form.*

As B and H are now defined the μ of coil can now be calculated using equation (4.2). If the coil is wound with an interwinding space, use the HELIX (bare wire) worksheet (4) to calculate the above parameters. This worksheet modifies the calculations used for close wound magnet wire so a desired number of windings can fill the selected winding length chosen with evenly spaced distance between the windings. This would be most advantageous in a primary winding. This worksheet can also be used for any non–magnet wire type insulated wire. The interwinding distance in this case would be 2 × insulation thickness. Bare wire with a separation between windings can also be applied. Use worksheet (3) HELIX (magnet wire) when constructing a close-wound secondary.

4.3 Helical Solenoids

When an inductor is wound in a helix on a cylindrical form in *more than one layer* as shown in Figure 4-2, the inductance is found using another of Wheeler's formulas from reference (12) and calculated using HELIX (solenoid) worksheet (5):

$$L(\mu h) = \frac{0.8A^2N^2}{6A + 9H + 10C} \tag{4.13}$$

Where: L = Inductance of coil in μhenries = color cell (D32), calculated value. When A, H, and C are of relatively equal values the accuracy is within ± 1.0%.
A = Distance from center of coil form to center of winding (center of winding is midway between inside and outside of layers or average diameter of winding without the coil form) in inches:

$$A = \text{OR} + \left(\frac{\text{NL} \bullet \text{Dw}}{2}\right)$$

Where: NL = Number of winding layers in color cell (D9), enter value.
OR = Radius of outside diameter of coil form in inches = 1/2 OD entered in color cell (B6).

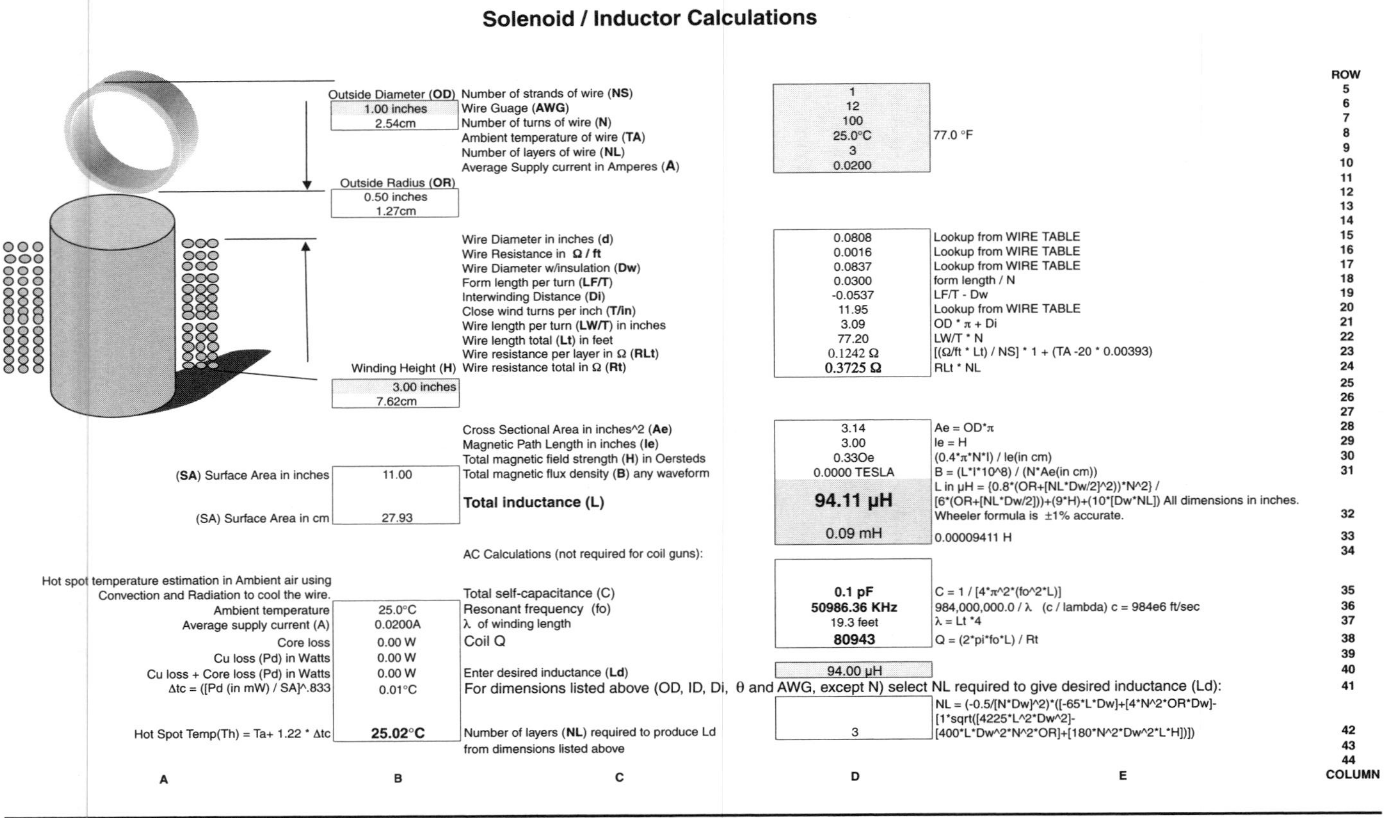

FIGURE 4-2 Solenoid winding worksheet calculations.

Dw = Diameter of magnet wire with insulation = cell (D17) from AWG wire table.
C = Height of layers in inches = NL•Dw.
N = Number of turns used in coil = color cell (D7), enter value.
H = Height (length) of winding in inches = color cell (B25), enter value.

The methodology shown in equations (4.4) thru (4.12) also applies to this worksheet and is not repeated. Becoming familiar with worksheet (3) and the helix methodology in Section (4.2) before beginning another worksheet should provide assistance in performing the calculations.

4.4 Archimedes Spiral Inductors

When an inductor is wound in a flat spiral or an inverse conical (Archimedes spiral) form as shown in Figure 4-3, the inductance is calculated using another of Wheeler's formulas from reference (12) and calculated using Archimedes (mag wire) worksheet (6):

$$L(\mu\text{h}) = \frac{A^2 N^2}{8A + 11W} \tag{4.14}$$

Where: L = Inductance of coil in μhenries = cell (D28), calculated value. The accuracy is within ± 5.0% for a flat spiral when W > (A/5).
A = Average radius of coil form in inches:

$$A = \frac{\left[\frac{\text{OD}-\text{ID}}{2} + \text{ID}\right]}{2}$$

Where: OD = Outside Diameter = cell (B3), enter value.
ID = Inside Diameter = cell (B8), enter value.
N = Number of turns used in coil = cell (D5), enter value.
W = Height (width) of winding in inches = cell (B19), calculated from: $W = A/\cos\theta$.

NOTE: *θ = angle of incline from horizontal (0°). Excel will not calculate the sine, cosine or tangent of angles without first converting to radians. This is done in the worksheet and the $\cos\theta$ is shown in cell (B34), $\sin\theta$ is shown in cell (B35). Do not use for angles greater than 75°, use the Helical formula.*

Using the dynamics of the worksheet the coil's inside and outside diameter and number of turns can be changed to find the optimum design for a desired inductance value before construction begins. The methodology shown in equations (4.4) thru (4.12) applies to this worksheet also and is not repeated. Becoming familiar with worksheet (3) and the helix methodology in Section (4.2) before beginning another worksheet should provide assistance in performing the calculations.

Use ARCHIMEDES (mag wire) worksheet (6) when constructing a close wound secondary as found in an Oudin coil. Use ARCHIMEDES (bare wire) worksheet (7) when constructing

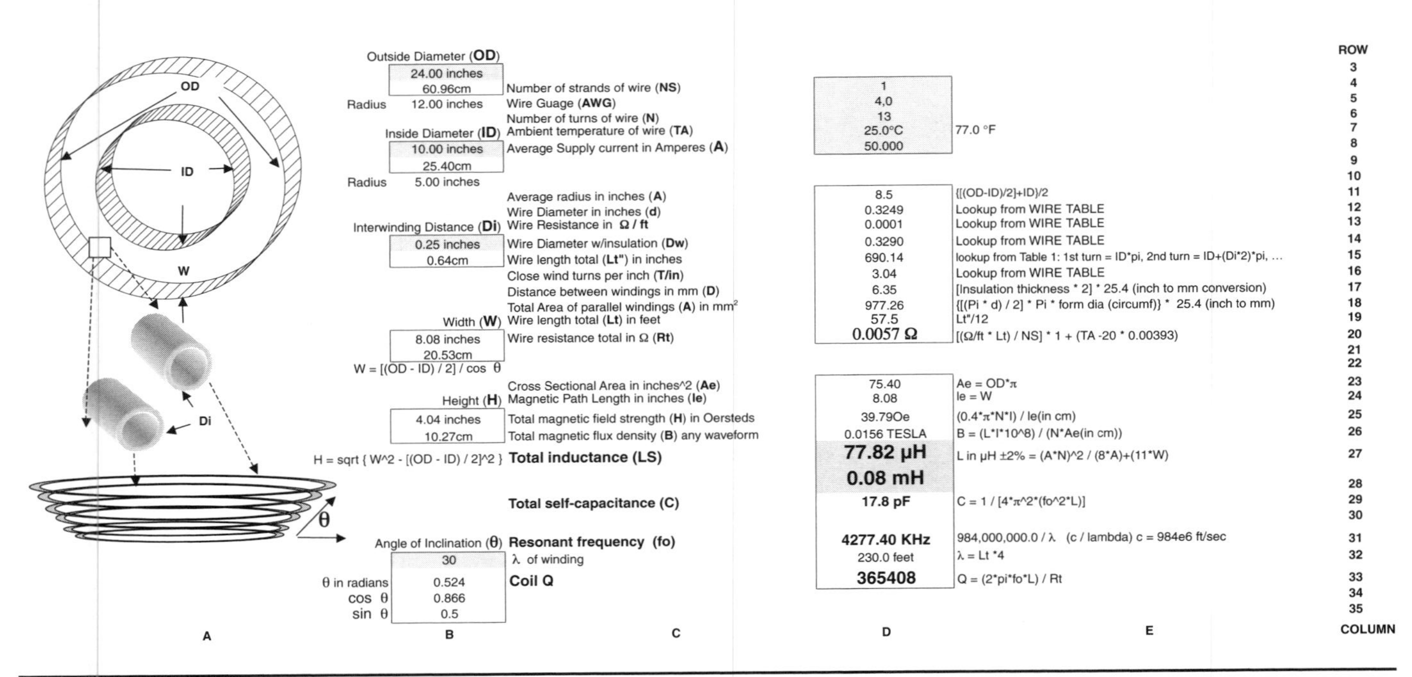

FIGURE 4-3 Inverse conical winding worksheet calculations.

primary windings that use an interwinding space. If insulated wire other than magnet wire is used enter the interwinding distance + (insulation thickness × 2) in cell (B14).

4.5 Toroidal Inductors

When an inductor is wound on a toroidal form (air core) as shown in Figure 4-4, the inductance is found using another of Wheeler's formulae from reference (12) and calculated using TOROID (air core) worksheet (8):

$$L(\mu\text{h}) = 0.01257 \bullet N^2 \bullet \left(\text{AR} - \sqrt{\text{AR}^2 - \text{AT}^2}\right) \tag{4.15}$$

Where: L = Inductance of coil in μhenries = cell (D28), calculated value.
N = Number of turns used in coil = cell (D5), enter value.
AR = The average radius or distance from center of toroid to center of winding cross section in cm calculated in cell (D10):

$$AR = \frac{\left[\frac{\text{OD}-\text{ID}}{2} + \text{ID}\right]}{2}$$

Where: OD = Outside Diameter in cm = cell (B5), enter value in inches in cell (B4). Converted to cm using 2.54 multiplier.
ID = Inside Diameter in cm = cell (B19), enter value in inches in cell (B18). Converted to cm using 2.54 multiplier.

AT = The equivalent radius of winding turn calculated in cell (D17):

$$\text{AT} = \frac{2\left(\frac{\text{OD}-\text{ID}}{2} + H\right)}{2\pi}$$

Where: H = Height of toroidal form in cm = cell (B26), enter value in inches in cell (B25). Converted to cm using 2.54 multiplier.

NOTE: *The toroidal form in the worksheet uses a rectangular cross section, therefore the rectangular circumference is (height × width × 2), the radius of a circle is (circumference/π × 2) and the equivalent radius of a rectangular cross section is (rectangular circumference/π × 2).*

When the core is not air and its permeability is known the inductance is calculated in TOROID (ferrite core) worksheet (9):

$$L(H) = \frac{0.4\pi\mu Ae\, N^2}{le \times 10^8} \tag{4.16}$$

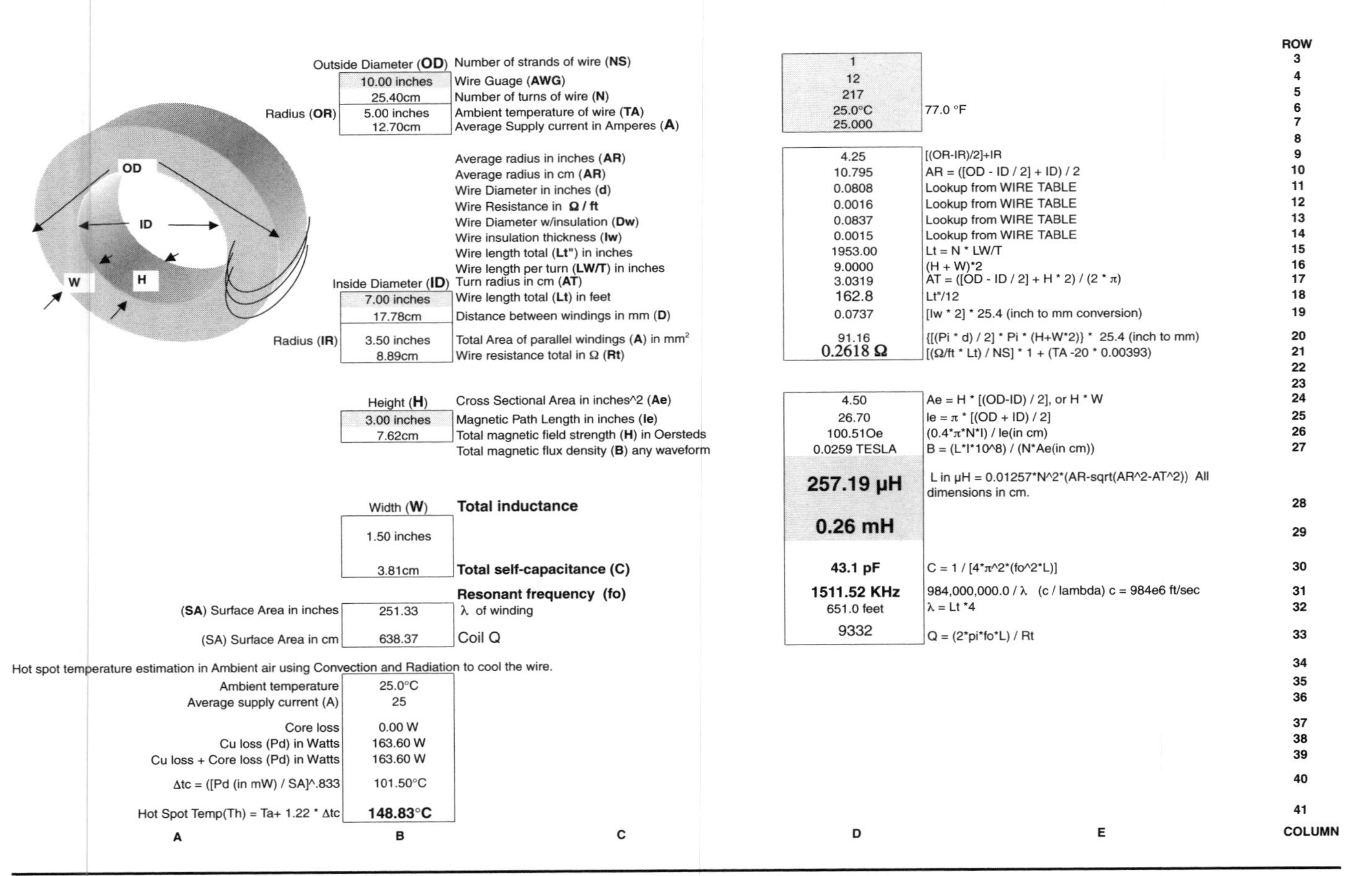

ROW	A	B	C	D	E
3	Outside Diameter (**OD**)		Number of strands of wire (**NS**)	1	
4		10.00 inches	Wire Guage (**AWG**)	12	
5		25.40cm	Number of turns of wire (**N**)	217	
6	Radius (**OR**)	5.00 inches	Ambient temperature of wire (**TA**)	25.0°C	77.0 °F
7		12.70cm	Average Supply current in Amperes (**A**)	25.000	
8					
9			Average radius in inches (**AR**)	4.25	[(OR-IR)/2]+IR
10			Average radius in cm (**AR**)	10.795	AR = ([OD - ID / 2] + ID) / 2
11			Wire Diameter in inches (**d**)	0.0808	Lookup from WIRE TABLE
12			Wire Resistance in **Ω / ft**	0.0016	Lookup from WIRE TABLE
13			Wire Diameter w/insulation (**Dw**)	0.0837	Lookup from WIRE TABLE
14			Wire insulation thickness (**Iw**)	0.0015	Lookup from WIRE TABLE
15			Wire length total (**Lt"**) in inches	1953.00	Lt = N * LW/T
16			Wire length per turn (**LW/T**) in inches	9.0000	(H + W)*2
17	Inside Diameter (**ID**)		Turn radius in cm (**AT**)	3.0319	AT = ([OD - ID / 2] + H * 2) / (2 * π)
18		7.00 inches	Wire length total (**Lt**) in feet	162.8	Lt"/12
19		17.78cm	Distance between windings in mm (**D**)	0.0737	[Iw * 2] * 25.4 (inch to mm conversion)
20	Radius (**IR**)	3.50 inches	Total Area of parallel windings (**A**) in mm²	91.16	{[(Pi * d) / 2] * Pi * (H+W*2)} * 25.4 (inch to mm)
21		8.89cm	Wire resistance total in Ω (**Rt**)	0.2618 Ω	[(Ω/ft * Lt) / NS] * 1 + (TA -20 * 0.00393)
22					
23					
24	Height (**H**)		Cross Sectional Area in inches^2 (**Ae**)	4.50	Ae = H * [(OD-ID) / 2], or H * W
25		3.00 inches	Magnetic Path Length in inches (**Ie**)	26.70	Ie = π * [(OD + ID) / 2]
26		7.62cm	Total magnetic field strength (**H**) in Oersteds	100.51Oe	(0.4*π*N*I) / Ie(in cm)
27			Total magnetic flux density (**B**) any waveform	0.0259 TESLA	B = (L*I*10^8) / (N*Ae(in cm))
28	Width (**W**)		**Total inductance**	**257.19 µH**	L in µH = 0.01257*N^2*(AR-sqrt(AR^2-AT^2)) All dimensions in cm.
29		1.50 inches		**0.26 mH**	
30		3.81cm	**Total self-capacitance (C)**	**43.1 pF**	C = 1 / [4*π^2*(fo^2*L)]
31			**Resonant frequency (fo)**	**1511.52 KHz**	984,000,000.0 / λ (c / lambda) c = 984e6 ft/sec
32	(**SA**) Surface Area in inches	251.33	λ of winding	651.0 feet	λ = Lt *4
33	(SA) Surface Area in cm	638.37	Coil Q	9332	Q = (2*pi*fo*L) / Rt
34	Hot spot temperature estimation in Ambient air using Convection and Radiation to cool the wire.				
35	Ambient temperature	25.0°C			
36	Average supply current (A)	25			
37	Core loss	0.00 W			
38	Cu loss (Pd) in Watts	163.60 W			
39	Cu loss + Core loss (Pd) in Watts	163.60 W			
40	Δtc = ([Pd (in mW) / SA]^.833	101.50°C			
41	Hot Spot Temp(Th) = Ta+ 1.22 * Δtc	**148.83°C**			

FIGURE 4-4 Toroidal winding worksheet calculations (air core).

Where: L = Inductance of coil in henries = cell (D29), calculated value. Converted to μH using a 1e-6 multiplier.
μ = Permeability of core material = cell (D7), enter value.
Ae = Cross-sectional area of coil form in square inches = form Height (H) entered in cell (B25) × ([Outside Diameter (OD) entered in cell (B4) – Inside Diameter (ID) entered in cell (B18)]/2) = Height × Width = cell (D25), calculated value.
N = Number of turns used in coil = cell (D5), enter value.
le = Magnetic path length of core in inches = π × ([Outside Diameter (OD) – Inside Diameter (ID)]/2) = cell (D26), calculated value.

NOTE: *The calculated Ae and le are approximations. The magnetic flux tends to concentrate on the inside corners of the magnetic path (ID) and will not be equally distributed among the toroidal form. In the absence of manufacturer's data this is as close as it gets. If a core is used which has Ae and le measured data, copy the worksheet and replace the formula in cells (D25) and (D26) with the measured values. Likewise, if the μ is unknown, typical values are 10 to 500 for powdered iron (ceramic) cores and 1,000 to 15,000 for ferrite cores. Other materials such as a hollow plastic form would have a low value (<10, 1.0 suggested).*

Using the dynamics of the worksheet the coil inside and outside diameter, height, μ, and number of turns can be changed to find the optimum design for a desired inductance value before construction work is done. The methodology shown in equations (4.4) thru (4.12), apply to this worksheet also and are not repeated. Becoming familiar with worksheet (3) and the helix methodology in Section (4.2) before beginning another worksheet should provide assistance in performing the calculations.

4.6 Effect of Coil Form Geometry

A method to determine the optimum winding geometry was derived as follows: Open the CH_4A.xls file, HvsD worksheet (8). (See App. B.) This worksheet will perform calculations and graph the performance of a helically wound inductor using closely spaced turns of magnet wire. First, enter the known values listed below:

- Known values: wire gauge in cell (B3), number of strands in cell (B2), ambient temperature in cell (B4), coil form diameter in cell (B5), resonant frequency in cell (B6), interwinding space if not using magnet wire in cell (B7).

The worksheet will calculate the following parameters from the known values:

- No frequency effects: number of turns in column (L), winding resistance in column (O), inductance in column (P), self-capacitance in column (Q), winding Q in column (R), winding decrement (δ) in column (T).
- Skin effect only: winding Q in column (AH), winding δ in column (AI).
- Skin and proximity effects: winding Q in column (AN), winding δ in column (AO).

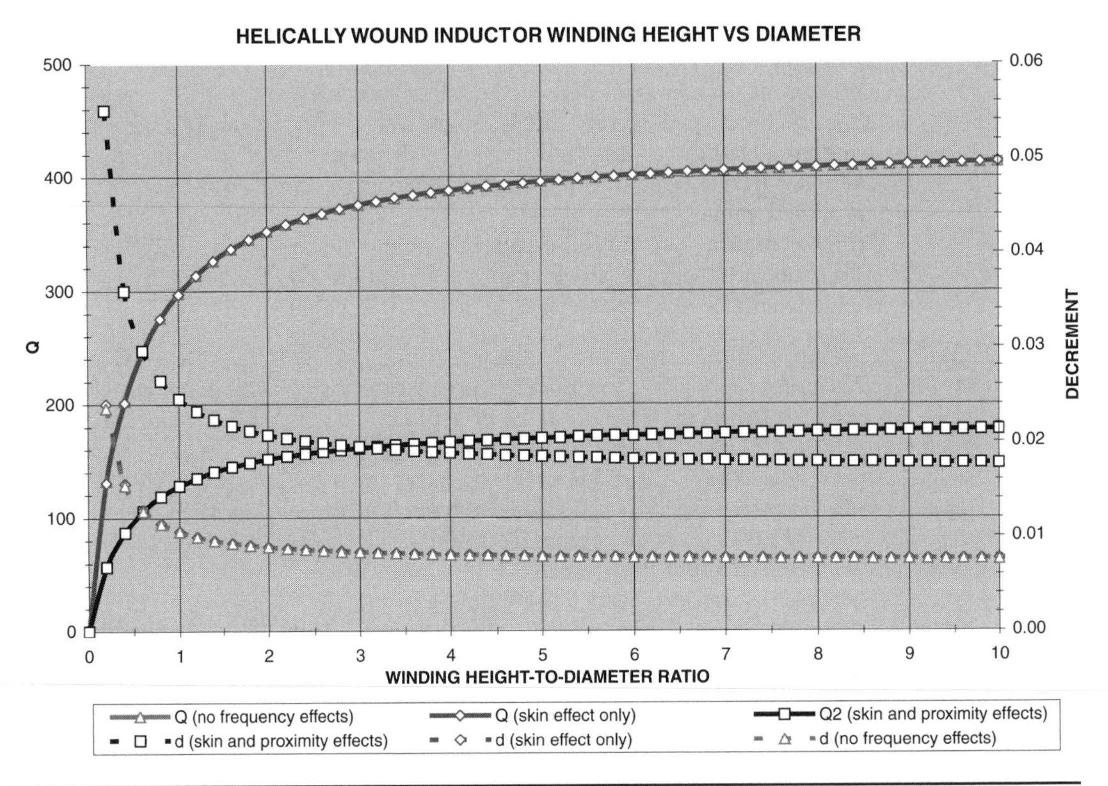

FIGURE 4-5 Optimum H-to-D ratio for close wound magnet wire, case 1.

For the fixed coil diameter entered in cell (B5), the winding height is increased in column (J) in 0.05-foot (0.6″) increments. This produces an increasing coil height (H)-to-coil diameter (D) ratio shown in column (K). Section 4.2 details the equations used for the calculated values. The coil's performance (Q and δ) can now be graphed using the (Y) axis and compared to the H-to-D ratio in the (X) axis. The skin effect and both the skin and proximity effects on the winding are also displayed on the graph.

The results shown in Figures 4-5 thru 4-7 illustrate that the optimum height-to-diameter geometry ratio, with respect to the coil's *Q*, is probably between 3:1 and 7:1. Case 1 uses a coil form diameter of 3.0 inches, resonant frequency of 225 kHz, with 1 strand of 28 AWG magnet wire. Further increases in the *H*-to-*D* ratio produce only marginal increases in *Q* as shown in the slope of the trace in Figure 4-5. In order to display the calculated results in this manner, the resonant frequency had to remain fixed. This is not unusual, as the completed Tesla coil's primary tank is usually tuned to a specific resonant frequency, determined by the tuned secondary circuit. Several different winding geometries can be used in the secondary winding for a selected resonant frequency. The object of the calculations is to illustrate the winding geometry (*H*-to-*D* ratio) that produces the optimum winding *Q* and decrement.

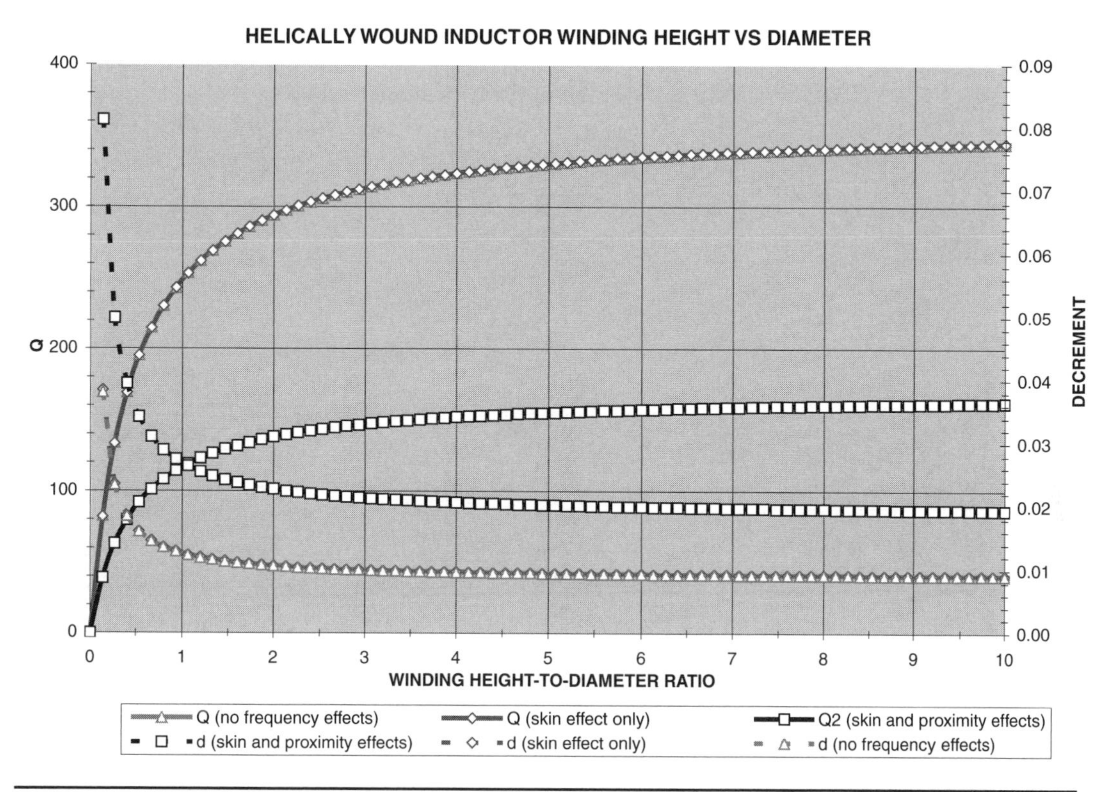

FIGURE 4-6 Optimum *H*-to-*D* ratio for close wound magnet wire, case 2.

Case 2 uses a coil form diameter of 4.5 inches, resonant frequency of 125 kHz, with 1 strand of 28 AWG magnet wire. Again the optimum *H*-to-*D* ratio is somewhere between 3:1 and 7:1, as further increases produce only marginal increases in *Q* as shown in the slope of the trace in Figure 4-6.

Case 3 uses a coil form diameter of 12.0 inches, resonant frequency of 75 kHz, with 1 strand of 24 AWG magnet wire. The optimum *H*-to-*D* ratio is somewhere between 3:1 and 7:1, as further increases produce only marginal increases in *Q* as shown in the slope of the *Q* traced in Figure 4-7.

It can be seen in Figures 4-5 through 4-7 that considering only the skin effect (red trace) in the winding is not much different than considering no frequency effects at all (blue trace). When both the skin and proximity effects are considered (black trace) the reduced *Q* and decrement becomes apparent. Referring to equation (2.14), the higher the secondary *Q*, the lower the decrement of the oscillations produced in the tuned secondary circuit. For a fixed coil form diameter the impedance increases as the height is increased, by increasing the winding resistance. When a second winding layer is added in cell (B2) the *Q* effectively doubles and δ decreases by half. However, a second winding will not double the secondary voltage as the proximity effect degrades performance. Several other coil form diameters and wire gauge combinations were tried, all producing similar results. When an interwinding space is used

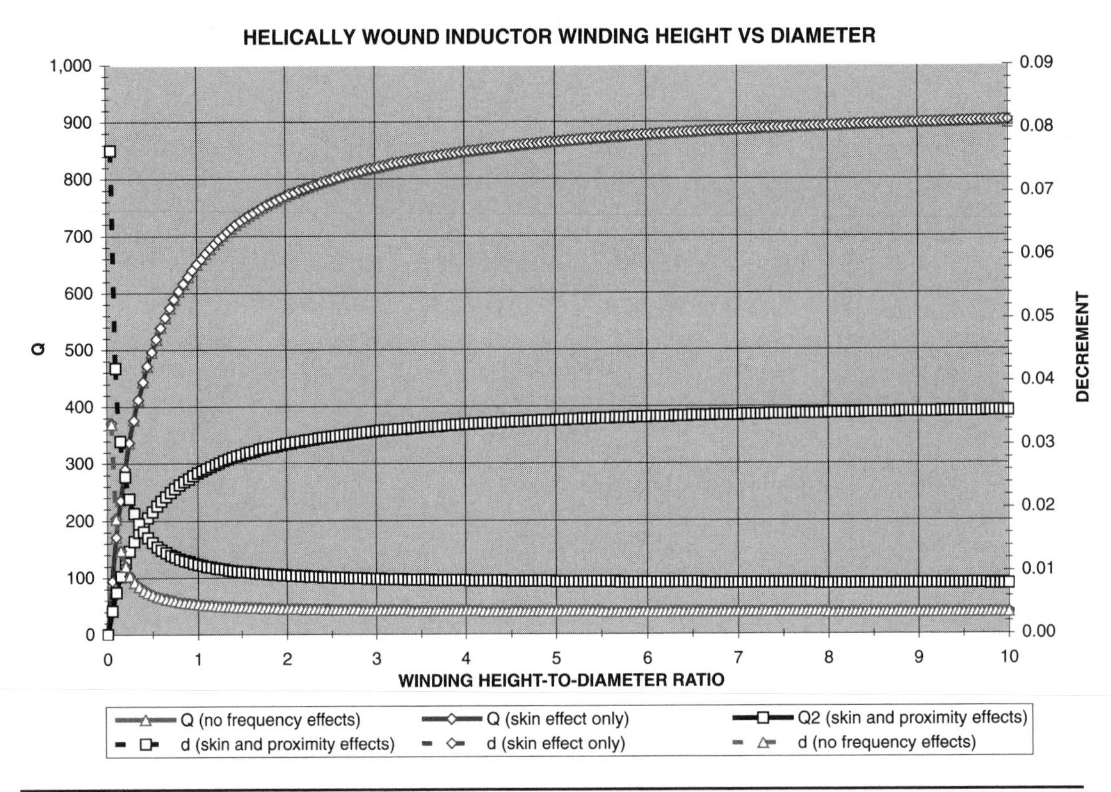

FIGURE 4-7 Optimum *H*-to-*D* ratio for close wound magnet wire, case 3.

by adding a small separation in cell (B7) the curve follows the same characteristics, but with a dramatically lower *Q*.

Winding geometry, resistance, *Q*, inductance, and self-capacitance are all design compromises. All factors affect each other, but an optimum design can be found using careful analysis. The height of the secondary winding affects the primary-to-secondary coupling, which is another design consideration. A winding geometry *H*-to-*D* ratio between 3:1 and 7:1 will produce optimum *Q* and δ in a close-wound secondary. The calculations in this section alone will not optimize the design. The interaction of all the characteristics detailed in Chapter 2 must be considered simultaneously in order to optimize the design. If you already have an operating coil try adding a second winding layer on the secondary. When the spark length does not double it will become apparent that there is much more to consider than merely the skin effect, secondary *Q*, and δ.

4.7 Calculating Magnet Wire Parameters

Magnet wire is used primarily for the secondary coil, but can be used in other applications. The National Electrical Manufacturer's Association (NEMA) Wire Table worksheets contained in several of the Excel files, including the CH_4.xls file, were constructed using reference

(6). Calculations for the maximum bare wire diameter in inches, maximum overall diameter with insulation, the dielectric strength minimum breakdown voltages, resistance per foot, and weight in pounds per 1,000 feet for magnet wire AWG sizes 0000 (4,0) thru 60 are performed as follows:

The bare wire diameter for each American Wire Gauge (AWG) number is calculated using the NEMA conversion in Section 1.6.7.1 of the reference:

$$n_{\text{AWG}} = 0.0050 \bullet 1.1229322^{(36-n)} \tag{4.17}$$

Where: n_{AWG} = Nominal bare wire diameter in inches for each AWG number.
0.0050 = Nominal base diameter in inches for 36AWG.
1.1229322 = The ratio of the diameter of any AWG number to the (smaller) diameter of the next larger AWG number = $^{39}\sqrt{92}$.
36 = The AWG number of the base diameter.
n = The number designation for the desired AWG size. For sizes 4/0 (0000) to 2/0 (00) AWG, n is a negative number from −3 to −1 respectively with 0 AWG = 0.

The wire area in circular mils is calculated using this bare wire diameter (d):

$$\text{circular mils} = d^2 \tag{4.18}$$

The resistance in ohms per foot is calculated using information provided in the MWS industries wire catalog for copper wire (OFHC and ETP copper). This was used since the NEMA manual contains no resistance per foot data in their tables. Stated in reference (7), "The resistance of 1 foot of annealed copper wire @20°C and 1 mil in diameter is 10.3 ohms." This 10.3 Ω per circular mil foot (cirmil ft) produces a resistance of:

$$R = \frac{kl}{A} \tag{4.19}$$

Where: R = Resistance in ohms.
k = 10.3 ohms for 1 foot, 1mil diameter or 10.3 Ω/cirmil ft.
l = Length of wire in feet.
A = Area of wire in cir mils.

The density (weight) was calculated using information provided in reference (7) for copper wire (OFHC and ETP copper). Stated in the MWS catalog data, "The density of OFHC and ETP copper is 0.323 pounds per cubic inch." The area of one foot of magnet wire in cubic inches can be calculated using a formula for determining the volume of a circular cylinder found in reference (8):

$$A^3 = \pi r^2 L \tag{4.20}$$

Where: A^3 = Area of one foot of magnet wire in cubic inches.
r = Radius of wire in inches or the diameter/2.
L = Length of wire = 12 inches.

And finally the weight per foot is calculated:

$$lB/\text{ft} = 0.323 A^3 \tag{4.21}$$

Where: A^3 = Area of one foot of magnet wire in cubic inches.

Because the Excel lookup function will not work with consecutive zero characters, using the lookup function for the larger wire diameters requires entering the wire gauge (AWG) into the Excel worksheets as follows:

AWG	Enter into cell
0000	4,0
000	3,0
00	2,0
0	0

Reference (15) details the origins of the AWG numbering system (Nikola Tesla is cited as one of the authorities consulted although they spell his name Nicola in the credits): "The Washburn & Moen gage was established in 1830 for use in the manufacture of iron and steel wire. The American Steel and Wire Company, upon absorbing the Washburn & Moen Company, adopted its steel wire gage and gave it the name of the new company. In 1912 the U.S. Bureau of Standards made a complete study of wire gages. This report showed that the great majority of steel wire was being made in accordance with the American Steel and Wire gage and that this gage was quite well adapted to the purpose. It recommended this gage therefore as the 'steel wire gage' for the United States and the American Institute of Electrical Engineers adopted this name as standard."

Also from the reference: "The American Wire Gage (A.W.G.) is used exclusively in America for wires intended for use as electrical conductors. It was devised by the Brown & Sharpe Manufacturing Company in 1857 and is often referred to as the B. & S. gage. The American Institute of Electrical Engineers adopted this gage as standard under the name 'American Wire Gage' and this is the term now used by manufacturers of electrical wire in designating their output. Copper is made the basis of comparison, and it is said to have 100 per cent conductivity when its purity and density are such that 1 foot of copper wire having a diameter of 1 mil (0.001 inch) has a resistance of 10.371 Ω at a temperature of 20°C. Ordinary commercial drawn or rolled copper has a conductivity of 96 to 99.5 percent of this value, and electrolytic copper may have a conductivity of over 100 percent. It is usual to specify a conductivity of about 98 percent when selecting cables for heavy currents and important service."

This would indicate that as early as the original copyright of 1905, annealed copper wire had a standard resistance of 10.3 Ω/cirmil ft. at 20°C. For electrical equipment manufactured in the United States it can be assumed the AWG method is used and determines the diameter of the conductor. This becomes significant when evaluating antique and surplus electrical devices.

4.8 Skin Effect in Inductors

When direct current is applied to a straight conductor it distributes itself evenly throughout the wire's cross-sectional area. When an alternating current is applied to a straight conductor, eddy currents develop and the current will tend to flow on the surface. As the AC frequency is increased it becomes increasingly difficult for the current to penetrate into the center of the conductor, which flows along the conductor surface (skin). This increases the effective resistance of the wire and is called skin effect. Chapter 7 details skin-effect calculations in straight conductors. When a straight conductor is wound into a coil, the resulting proximity effects of the adjacent windings will produce even further losses. To calculate these losses references

(9) and (10) were used to prepare the CH_4.xls file, SKIN EFFECT worksheet (2) shown in Figure 4-8 and described below:

Enter the ambient temperature in °C (T_A), wire gauge (AWG), frequency (f), magnetic path length (le), number of turns in winding (N), number of layers (N_L) the winding occupies on the core, number of strands (N_S) in the winding, and the length per turn in feet (ft/T) in cells (B5) thru (B13) respectively. The CH_4.xls inductor worksheets provide a calculated le, explained in Sections 4.1 thru 4.5.

The number of winding layers (N_L) will generally be a single layer (1) in Tesla coil construction. Multiple layers (for solenoids) can be estimated using the calculations provided in rows 29 to 33. Enter the wire gauge in cell (B29) and number of turns into cell (B30). Cell (E29) automatically returns the outside diameter of the wire with its insulation thickness from the NEMA wire table and is used to calculate the number of layers for bobbin (winding form used in E cores) or toroid cores from the following formula:

$$\text{Bobbin } N_L = \frac{NdN_S}{B} \tag{4.22}$$

Where: Bobbin N_L = Number of layers required = cell (E32), calculated value.
d = Diameter of wire with insulation in inches = cell (E29), calculated value from wire gauge entered in cell (B29).
N = Number of turns used in winding = cell (B30), enter value.
N_S = Number of strands of wire used in winding = cell (B31), enter value.
B = Winding breadth of bobbin core in inches = cell (B32), enter value.

$$\text{Toroid } N_L = \frac{NdN_S}{\text{ID}\pi} \tag{4.23}$$

Where: Toroid N_L = Number of layers required = cell (E33), calculated value.
ID = Inside diameter of toroidal core in inches = cell (B33), enter value.

For magnet wire the wire diameter (d) is automatically converted to cm in cell (E6) and inches in cell (F6) from the wire gauge entered in cell (B6) using equation (4.17). This is representative of the single build magnet wire currently being manufactured and does not necessarily match old or outdated wire tables (e.g., double coated cotton or dcc). It will, however, approximate the diameter for any other type of wire for the gauge entered.

Skin effect (D) is essentially the inability of current to penetrate from the periphery toward the center of a conductor as the frequency is increased. This is a direct result of eddy currents established in the conductor from the changing AC flux (see Figure 4-9). The eddy currents reinforce current flow on the conductor's "skin,"decreasing exponentially as they move toward the center. Proximity effects result from high-frequency current-carrying conductors being in proximity to each other. Figure 4-10 illustrates the proximity effect in two conductors with opposing and aiding (same) current directions. While the proximity effect is dependent on the number of winding layers and layer construction, the skin effect is affected by the applied frequency:

$$D = \sqrt{P\pi\mu f} \tag{4.24}$$

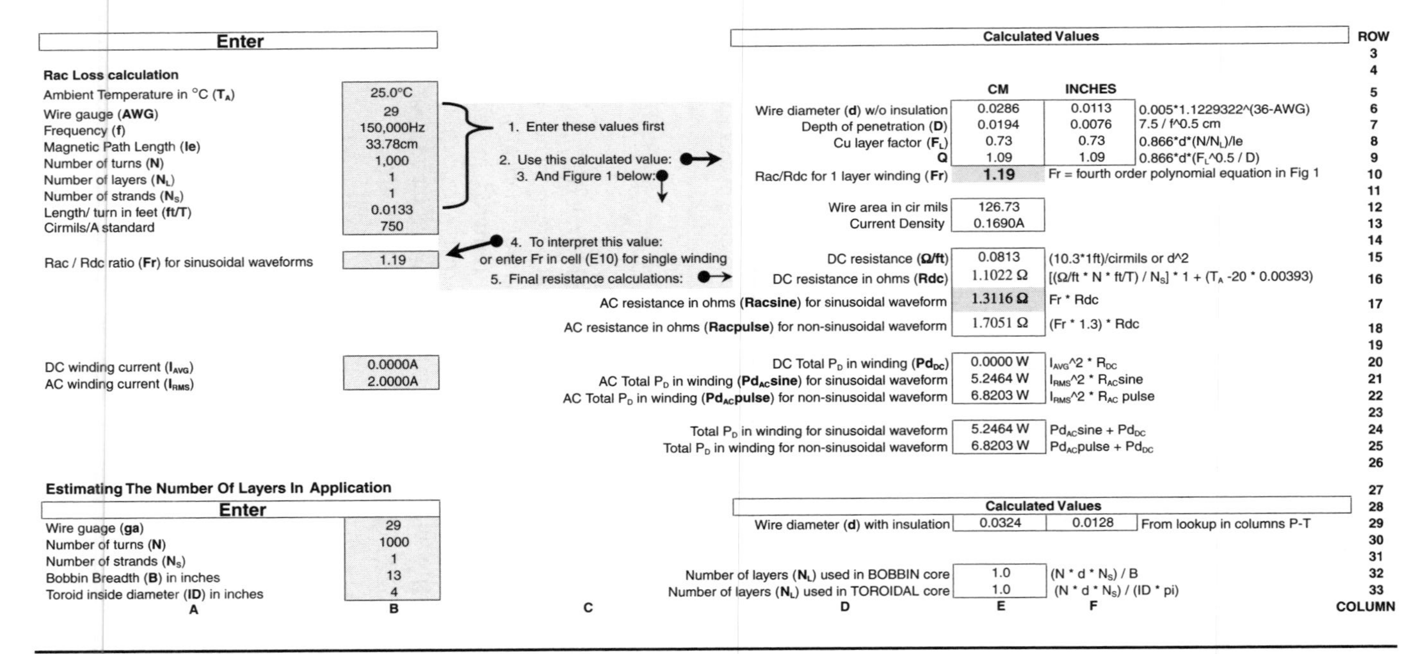

FIGURE 4-8 Skin effect worksheet calculations.

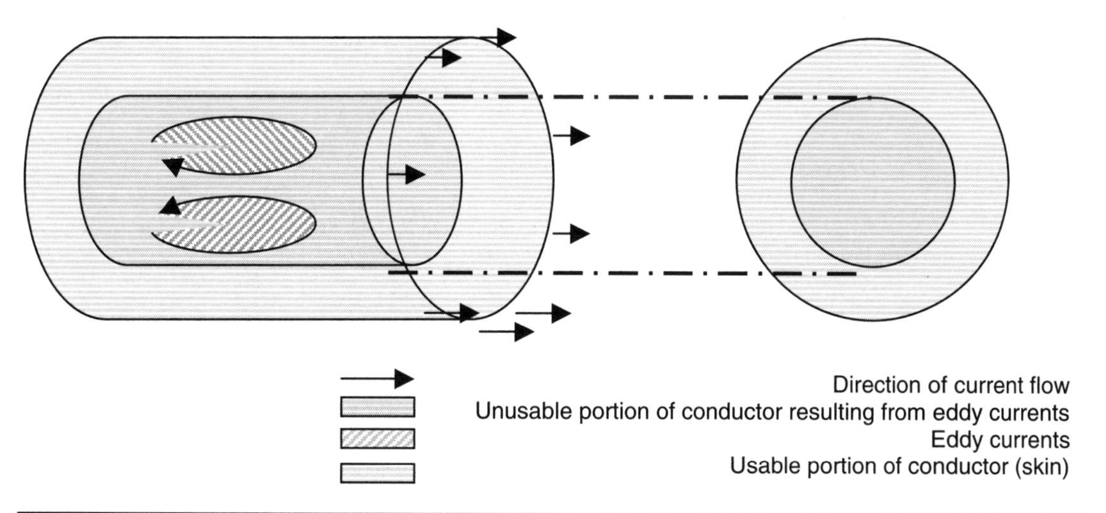

FIGURE 4-9 Illustration of eddy current (skin) effect.

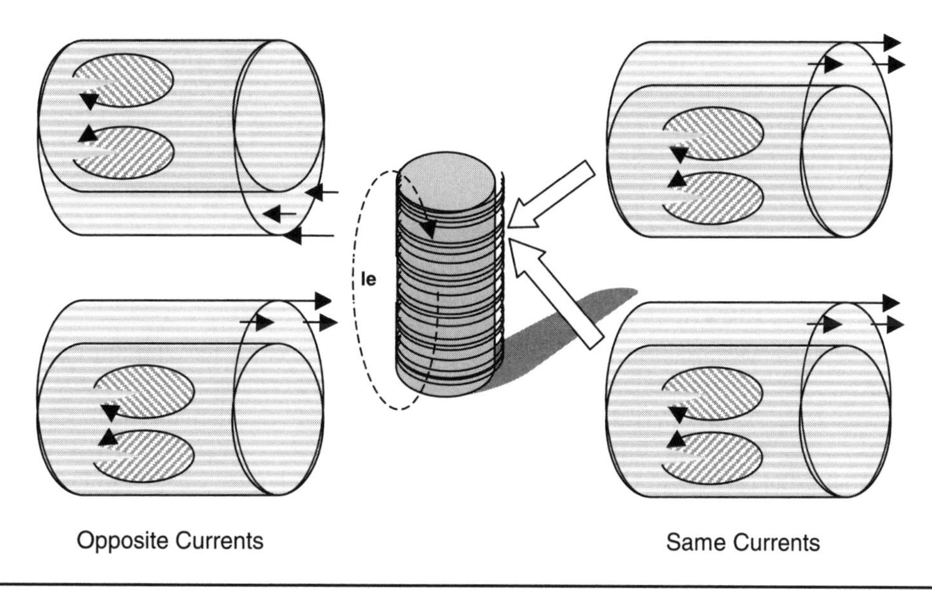

FIGURE 4-10 Illustration of proximity effect.

Where: D = Depth of current penetration in meters.

P = Resistivity of material in μohm/cm. For copper the *Handbook of Chemistry and Physics* lists this value as 1.7241e-6 @20°C. For other temperatures P = P@20°C · 1+0.0042 · T-20°C where T = ambient temperature.

μ = Permeability of conductor in ohm-meters. For copper the *Handbook of Chemistry and Physics* lists this value as 1.72e-8.

f = Applied frequency in hertz.

Reference (9) and the worksheet use a simplified form of this calculation:

$$D = 7.5 f^{-0.5} \tag{4.25}$$

Where: D = Depth of current penetration in centimeters = cell (E7), calculated value. Cell (F7) converted to inches.
f = Applied (resonant) frequency in hertz = cell (B7), enter value.

A copper layer factor (F_L) is determined:

$$F_L = 0.866d \frac{\left(\frac{N}{N_L}\right)}{le} \tag{4.26}$$

Where: F_L = Multiplier used to calculate Q' = cell (E8), calculated value.
d = Diameter of wire in cm = cell (E6), calculated value from wire gauge entered in cell (B6). Column (F) converts these calculations to inches.
N = Number of turns used in winding = cell (B9), enter value.
N_L = Number of layers used in constructing the winding on the core = cell (B10), enter value.
le = Magnetic path length of core in cm (winding breadth of toroid) = cell (B8), enter value. The CH_4.xls inductor worksheets provide a calculated le, explained in Sections 4.1 thru 4.5 and shown in Figure 4-10.

A figure of merit (Q') denoting the ratio of the wire diameter (d) and its associated layer factor (F_L)-to-depth of penetration (D) is then calculated:

$$Q' = \frac{0.866d(F_L)^{0.5}}{D} \tag{4.27}$$

Where: Q' = Ratio of d to D = cell (E9), calculated value.
d = Wire diameter in inches = cell (F6), calculated value.
F_L = Multiplier used to calculate Q' = calculated value in cell (E8) from equation (4.26).
D = Depth of current penetration in centimeters = calculated value in cell (E7) from equation (4.25).

Using the calculated Q' from equation (4.27), refer to Figure 2-3. The data used to generate the graph in the figure are originally found in reference (9) and *was developed for sinusoidal waveforms*, which makes it perfect for evaluating Tesla coils. Locate the calculated Q' on the X-axis. Moving up, intersect the line corresponding to the number of layers used in the winding (black trace for single layer). Note the graph is in a log–log scale and must be read as such. The ratio of AC resistance-to-DC resistance or R_{AC}/R_{DC} ratio (F_r) is found directly across this intersection point on the Y-axis. Enter this value into cell (B15). The fourth-order polynomial curve fit formula shown in Figure 2-3 is used in cell (E10) to calculate F_r for a single-layer winding. This number can be entered in cell (B15) when using a single-layer winding, eliminating the graph interpretation. This ratio is used to calculate the AC resistance using the formula:

$$R_{AC} = F_r \bullet R_{DC} \tag{4.28}$$

Where: R_{AC} = AC resistance (R_{AC}) due to eddy current (skin) effects.
F_r = value of R_{AC}/R_{DC} ratio found using intersection of Q' and N_L in Figure 2-3.
R_{DC} = DC resistance of winding = calculated value in cell (E15) from NEMA wire table. Adjusted for temperature using: $1 + [(T_A - 20°C) \times 0.00393]$, where T_A is the temperature entered in cell (B5).

The total AC resistance for sinusoidal waveforms attributed to eddy current (skin) and proximity effects is displayed in cell (E17). If a pulsed, square wave, or duty cycled waveform (such as a coil gun or solid state coil) is applied to the primary, use the F_r calculation error noted in reference (9): for duty cycles >0.4 a correction factor of 1.3 is applied to the calculated sinewave value of F_r. The correction factor will increase R_{AC} by an additional 30% for switched waveforms. The higher R_{AC} for non-sinusoidal waveforms is displayed in cell (E18).

The cells in row 13 refer to the NEMA wire table in computing the current density of the wire. Enter circular mils per ampere standard 375, 750, 1,000 or any other standard value in cell (B13) and the calculated current density is displayed in cell (E13).

The total power dissipated in the magnetic windings is the sum of the AC and DC power dissipation:

$$P_{\text{Dtotal}} = P_{\text{Dac}} + P_{\text{Ddc}} \tag{4.29}$$

Where: P_{Dtotal} = Total power dissipation from AC and DC resistance = cell (E24), calculated value.
P_{Dac} = Power dissipation attributable to AC resistance = calculated value from equation (4.30).
P_{Ddc} = Power dissipation attributable to DC resistance = calculated value from equation (4.31).

From the AC power dissipation:

$$P_{\text{Dac}} = I_{\text{RMS}}^2 \bullet R_{AC} \tag{4.30}$$

Where: P_{Dac} = Power dissipation attributable to AC resistance = cell (E21), calculated value.
I_{RMS} = rms value of current through windings = cell (B21), enter value. See Section 10.5 for rms equivalents of waveforms.
R_{AC} = AC resistance from equation (4.28).

And the DC power dissipation:

$$P_{\text{Ddc}} = I_{\text{DC}}^2 \bullet R_{DC} \tag{4.31}$$

Where: P_{Ddc} = power dissipation attributable to DC resistance = cell (E20), calculated value.
I_{DC} = DC (average) value of current through windings = cell (B20), enter value. See Section 10.5 for average equivalents of waveforms.
R_{DC} = DC resistance of winding = calculated value in cell (E15) from NEMA wire table. Adjusted for temperature using: $1 + [(T_A - 20°C) \times 0.00393]$, where T_A is temperature entered in cell (B5).

The combined I^2R losses of the R_{AC} and R_{DC} components for a non-sinusoidal waveform are shown in cell (E25). For convenience Figure 2-3 is included on the worksheet in Rows 39 thru 86. The cells shaded in yellow instruct the user through the steps detailed above.

4.9 Mutual Inductance

The mutual inductance (M) is affected by the proximity of the primary winding to the secondary and the geometries of both the primary and secondary. For example, a loosely wound helical primary will have a higher mutual inductance than a loosely wound flat Archimedes spiral (pancake) for the same number of turns and outside diameter. As the magnetic flux (B) from the primary induces (couples) its magnetic field to the secondary, the magnetic field strength (H) in the secondary will have a higher density as it is moved closer to the primary and the more perpendicular the field is to the conductor (winding). If you visualize two helically wound primary coils surrounding a 3-inch diameter secondary as shown in Figure 4-11, one primary OD of 12″and the other with an OD of 6″, the 6″ diameter primary will have a higher M and coefficient of coupling (k) with the secondary. As the height of the primary is increased, M and k increase.

Reference (13) contains methodology to calculate the mutual inductance of two coaxial coils. Both concentric (centers of both coils aligned) and non-concentric types are included as shown in Figure 4-12. Open the CH_4A.xls file, MUTUAL INDUCTANCE worksheet (1) shown in Figure 4-13. (See App. B.) To calculate the mutual inductance of two coaxial concentric coils, the primary and secondary geometries and number of turns must first be known. To calculate the required height and number of turns for a specified resonant frequency enter the following secondary characteristics into the worksheet:

- Desired resonant frequency in kHz into cell (B6). Converted to Hz in cell (C6).
- Desired coil form diameter in inches into cell (B3). Converted to cm in cell (C3).

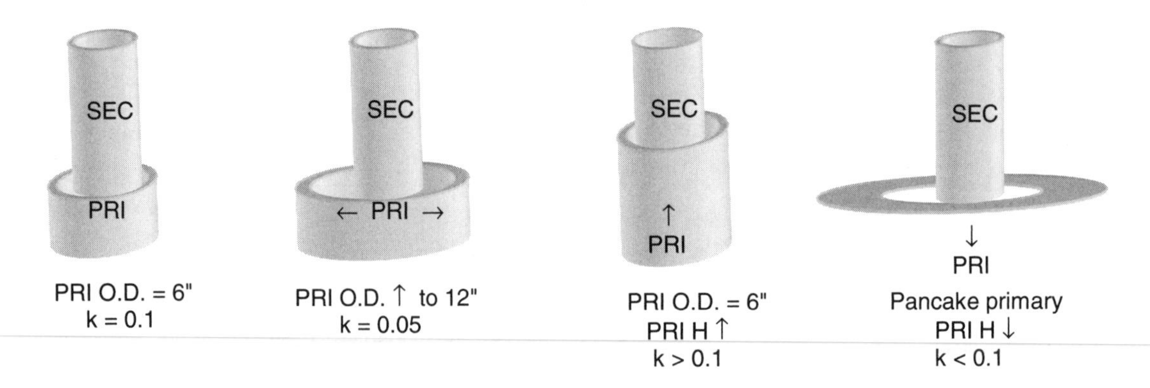

FIGURE 4-11 Mutual inductance of two windings.

FIGURE 4-12 Illustration of parameters required for calculating the mutual inductance for coaxial coils.

Coaxial Concentric Coils

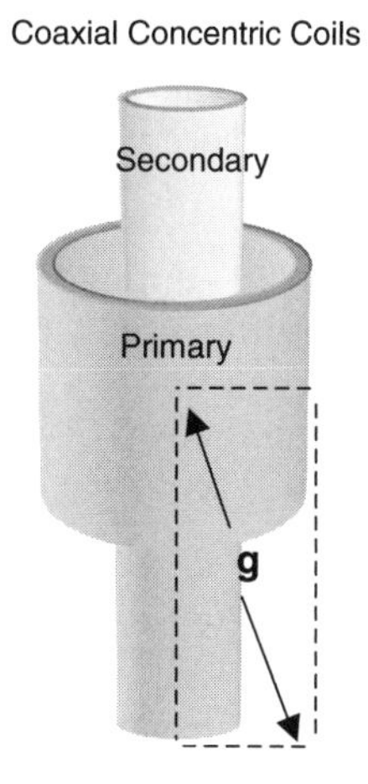

Coaxial Non-Concentric Coils

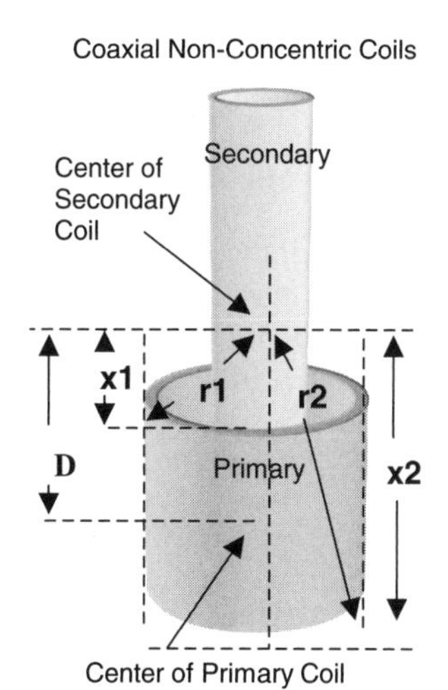

Secondary Winding Characteristics

A	B	C
Outside Diameter of secondary winding in inches (**ODs**)	4.50 inches	11.43cm
Wire Gauge (**AWG**)	29	
Interwinding seperation in inches if close wound magnet wire not used (**S**)	0.0000 in	
Resonant frequency in Hz (**fo**)	85.50 KHz	85500Hz

Primary Winding Characteristics

A	B	C
Enter: (1) for Archemedes spiral or (2) for helical wound primary	1	
Interwinding Distance in inches (**IWD**)	1.000 in	2.54cm
Total number of turns in primary winding (**Ttp**)	13	
Angle of inclination in° if using Archemedes sprial (θ)	50.0 °	
Desired primary turn number used to tune (**Np**)	**11**	
Inside Diameter of Archemedes spiral (**ID**) or Outside Diameter of helical primary in inches (**OD**)	18.0 in	89.19cm
Separation from base of primary to base of secondary in inches (**d1**)	0.0 in	0.00cm

Calculated k and M can be either positive or negative in value.

Coaxial Concentric Coils **Coaxial Non-Concentric Coils**

From: Terman F.E. Radio Engineer's Handbook. McGraw-Hill Book Co., 1943: Section 2, Pp. 71-73.

D	E	ROW
Secondary Winding Calculations		
Radius of secondary winding in inches (**a**)	2.250 in	3
Wire Diameter in inches (**d**)	0.0113	4
Wire Area in cirmils (**da**)	127	5
Wire Diameter w/insulation (**Dw**)	0.0128	6
Nominal increase in diameter (**di**)	0.0015	7
Close wind turns per inch (**T/in**)	78.4	8
Number of secondary turns (**Ns**)	**2440**	9
Inter-winding Distance (**Di**)	0.0015	10
Wire length per turn in inches (**LW/T**)	14.1	11
Wire length total (**Lt**) in feet	2877.19 ft	12
Height of secondary winding in inches (**Hs**)	**31.129 in**	13
Inductance of Values Entered in mH (**Ls**)	**90.91 mH**	14
		15
Primary Winding Calculations		16
Outside Diameter of primary winding in inches (**ODp**)	35.1 in	17
Radius of primary winding in inches (**A**)	17.556 in	18
Height of primary winding in inches (**Hp**)	**13.312 in**	19
Calculated Primary Inductance in µH (**Lp**)	**106.5 uH**	20
		21
		22
Coaxial Concentric Coil Characteristics		23
Center of secondary winding to imaginary base of primary winding in inches (**g**)	23.46 inches	24
Mutual inductance of windings in µH (**M**)	**284.17 µH**	25
Coefficient of coupling (**k**)	0.091	26
Coefficient of coupling for Hp>Hs (**k**)	0.038	27
		28
		29
Coaxial Non-Concentric Coil Characteristics		30
K1	0.0035	31
x1	2.253 in	32
x2	15.564 in	33
r1	17.700 in	34
r2	23.462 in	35
D	8.909 in	36
k1	31.129 in	37
K3	-4.46E-07	38
k3	-1.48E+04	39
K5	-1.53E-09	40
k5	-1.14E+05	41
Mutual inductance of windings in µH (**M**)	**291.38 µH**	42
Coefficient of coupling (**k**)	0.094	43
D	E	COLUMN

FIGURE 4-13 Mutual inductance worksheet for coaxial coils.

- Desired wire gauge of magnet wire (AWG) into cell (B4).
- If an interwinding distance is desired between the windings, enter the distance in inches from the edge of one winding to the adjacent edge of the next winding (as shown in Figure 2-2) into cell (B5).

Using the methodology shown in equations (2.1) through (2.5), the calculated secondary inductance is shown in cell (E14). The required number of turns is shown in cell (E9) and the winding height is shown in cell (E13).

To calculate the primary inductance, required height, and number of turns to tune the primary to the resonant frequency of the secondary enter the following into the worksheet:

- If a helically wound primary is used enter the value 2 into cell (B10). If a flat Archimedes spiral primary is used enter a value of 1 into cell (B10) and an angle of inclination (θ) of 0° into cell (B13). If the Archimedes spiral primary is not a flat (pancake) spiral enter a value of 1 into cell (B10) and the desired angle of inclination in degrees into cell (B13).
- If a helically wound primary is used enter the Outside Diameter (OD) into cell (B15). If an Archimedes spiral primary is used enter the Inside Diameter (ID) into cell (B15).
- Desired total number of primary turns into cell (B12).
- Desired number of primary turns to tune to the resonant secondary frequency into cell (B14).
- Desired interwinding distance in inches from the center of one winding to the center of the next winding (as shown in Figure 2-4) into cell (B11).
- For vertically oriented windings such as shown in Figure (4-12), if the base of the secondary winding is elevated above the base of the primary winding, enter this separation in inches into cell (B16). If horizontally oriented windings are used, orthogonally rotate (turn 90°) the image to a vertical orientation and enter the separation into cell (B16).

Using the methodology shown in equations (2.20) and (2.21) the calculated primary inductance is shown in cell (E20) and the winding height is shown in cell (E19). The calculated inductance is derived from the calculations in columns (K) through (R), rows (14) through (40) for an Archimedes spiral and rows (44) through (70) for a helically wound primary. The calculations are limited to 25 primary turns. If more turns are needed to tune the primary to the resonant secondary frequency a larger tank capacitance is needed instead of more primary turns.

Now that the primary and secondary height and number of turns are known the mutual inductance can be calculated for two coaxial concentric coils:

$$M(\mu\text{H}) = 0.0501\frac{a^2 NpNs}{g}\left[1 + \frac{A^2a^2}{8g^4}\left(3 - 4\frac{(0.5Hp)^2}{a^2}\right)\right] \tag{4.32}$$

Where: M = Mutual inductance of coaxial concentric primary and secondary winding in μH.
a = Radius of secondary coil form in inches = calculated value in cell (E3) from: 0.5 × OD entered in cell (B3).
A = Radius of primary coil form in inches = calculated value in cell (E18) from: 0.5 × calculated OD in cell (E17). A conditional statement is used in cell (E17) to determine the OD for an Archimedes spiral or helically wound primary.
Np = Number of primary turns = cell (B14), enter value.
Ns = Number of secondary turns = calculated value in cell (E9).
Hp = Height of primary in inches = calculated value in cell (E19).
g = Hypotenuse of imaginary right triangle formed from base of secondary coil and outer edge of the primary (see Figure 4-12):

$$g = \sqrt{A^2 + (0.5 \bullet Hs)^2}$$

Where: Hs = Height of secondary in inches = calculated value in cell (E13).

NOTE: *Page 281 of reference (16) lists the formula as shown in equation (4.32) except the constant of 0.0501 is replaced by the constant 0.01974.*

The mutual inductance for two coaxial non-concentric coils can also be calculated:

$$M(\mu\text{H}) = 0.02505 \frac{a^2 A^2 Np Ns}{4(0.5Hp \bullet 0.5Hs)}(K1k1 + K3k3 + K5k5) \tag{4.33}$$

Where: M = Mutual inductance of coaxial concentric primary and secondary winding in μH.
a = Radius of secondary coil form in inches = calculated value in cell (E3) from: 0.5 × OD entered in cell (B3).
A = Radius of primary coil form in inches = calculated value in cell (E18) from: 0.5 × calculated OD in cell (E17). A conditional statement is used in cell (E17) to determine the OD for an Archimedes spiral or helically wound primary.
Np = Number of primary turns = cell (B14), enter value.
Ns = Number of secondary turns = calculated value in cell (E9).
Hp = Height of primary in inches = calculated value in cell (E19).
Hs = Height of secondary in inches = calculated value in cell (E13).
And the following form factors (see Figure 4-12):

$$D = (0.5Hs - 0.5Hp) + d1$$

Where: D = calculated value in cell (E36).
$d1$ = Separation from base of primary to base of secondary in inches = cell (B16), enter value.
$k1$ = Hs = Height of secondary in inches = calculated value in cell (E13).

$$K1 = \frac{2}{A^2}\left(\frac{x2}{r2} - \frac{x1}{r1}\right)$$

Where: $K1$ = calculated value in cell (E31).
$x1$ = Distance from center of secondary coil to the upper edge of primary = calculated value in cell (E32) from: $x1 = D - (0.5Hp)$
$x2$ = Distance from center of secondary coil to the lower edge of primary = calculated value in cell (E33) from: $x2 = D + (0.5Hp)$
$r1$ = Hypotenuse of imaginary right triangle formed from center of secondary coil and upper edge of the primary = calculated value in cell (E34) from: $r1 = \sqrt{x1^2 + A^2}$
$r2$ = Hypotenuse of imaginary right triangle formed from center of secondary coil and lower edge of the primary = calculated value in cell (E35) from: $r2 = \sqrt{x2^2 + A^2}$

$$K3 = 0.5\left(\frac{x1}{r1^5} - \frac{x2}{r2^5}\right)$$

Where: $K3$ = calculated value in cell (E38).

$$k3 = a^2 0.5Hs\left(3 - \frac{4(0.5Hs)^2}{a^2}\right)$$

Where: $k3$ = calculated value in cell (E39).

$$K5 = -\frac{A^2}{8}\left[\frac{x1}{r1^9}\left(3 - \frac{4x1^2}{A^2}\right) - \frac{x2}{r2^9}\left(3 - \frac{4x2^2}{A^2}\right)\right]$$

Where: $K5$ = calculated value in cell (E40).

$$k5 = a^4 0.5Hs\left(\frac{5}{2} - 10\frac{(0.5Hs)^2}{a^2} + 4\frac{(0.5Hs)^2}{a^4}\right)$$

Where: $k5$ = calculated value in cell (E41).

NOTE: *Pages 278 and 279 of reference (16) list the formula as shown in equation (4.33) except the constant of 0.02505 is replaced by the constant 0.00987.*

The coefficient of coupling (k) can also be calculated now that the mutual inductance is known:

$$k = \frac{M}{\sqrt{L_P L_S}} \tag{4.34}$$

Where: k = Coefficient of coupling between primary and secondary windings (less than 1.0) = calculated value in cell (E26) for coaxial concentric coils and cell (E43) for coaxial non-concentric coils.

M = Mutual inductance of primary and secondary winding in μH = calculated value in cell (E25) for coaxial concentric coils from equation (4.32) and cell (E42) for coaxial non-concentric coils from equation (4.33).

L_P = Inductance of primary coil in μhenries = calculated value in cell (E20) from equations (2.20) and (2.21).

L_S = Inductance of secondary coil in μhenries = calculated value in cell (E14) from equations (2.1) through (2.5).

Two graphs are also contained in the worksheet. The mutual inductance and coefficient of coupling are calculated for concentric coaxial coils in columns (S–U), rows (16–40) for an Archimedes spiral primary and rows (46–70) for a helical primary of 1 to 25 turns. The calculations are shown in the graph in columns (A–D), rows (88–128) and in Figure 4-14 for the applied parameters in Figure 4-13. The mutual inductance and coefficient of coupling are also calculated for non-concentric coaxial coils in columns (W–AI), rows (16–40) for an Archimedes spiral primary, and rows (46–70) for a helical primary of 1 to 25 turns. The calculations are shown in the graph in columns (A–D), rows (46–86), and in Figure 4-15 for the applied parameters in Figure 4-13. The graphs aid in determining the coupling for a desired number of turns when tuning the primary.

No adjustment to the calculations for the slightly different geometry of the Archimedes spiral was considered necessary as the calculations agreed with measured values of working coils. The calculated k and M can be either positive or negative in value. Although the calculation accuracy was not specified in the reference for single layer coils, the accuracy for multiple layer coils using the same methodology provides a specified accuracy of better than ±0.5% and it is assumed the same accuracy can be expected in single layer coils.

The mutual inductance can also be measured as shown in Figure 4-16. To perform the calculations open the CH_4A.xls file, AIR CORE RELATIONSHIPS worksheet (12). (See App. B.)

The mutual inductance and coefficient of coupling affect the magnetizing inductance. Therefore the magnetic flux density (B) calculation in equation (4.12) also depends on M and k. The magnetizing inductance is the inductance of the winding the current is being applied to (primary), the other winding in the transformer being the one the current is being transferred to (secondary). The magnetizing inductance is decreased by the coefficient of coupling:

$$L(h) = Lp(1 - k) \tag{4.35}$$

Where: L = Inductance of primary (magnetizing) coil in henries.

Lp = Inductance of primary coil in henries.

k = Coefficient of coupling.

Since k is always less than 1.0, the magnetizing inductance and B will always be less than that calculated without k.

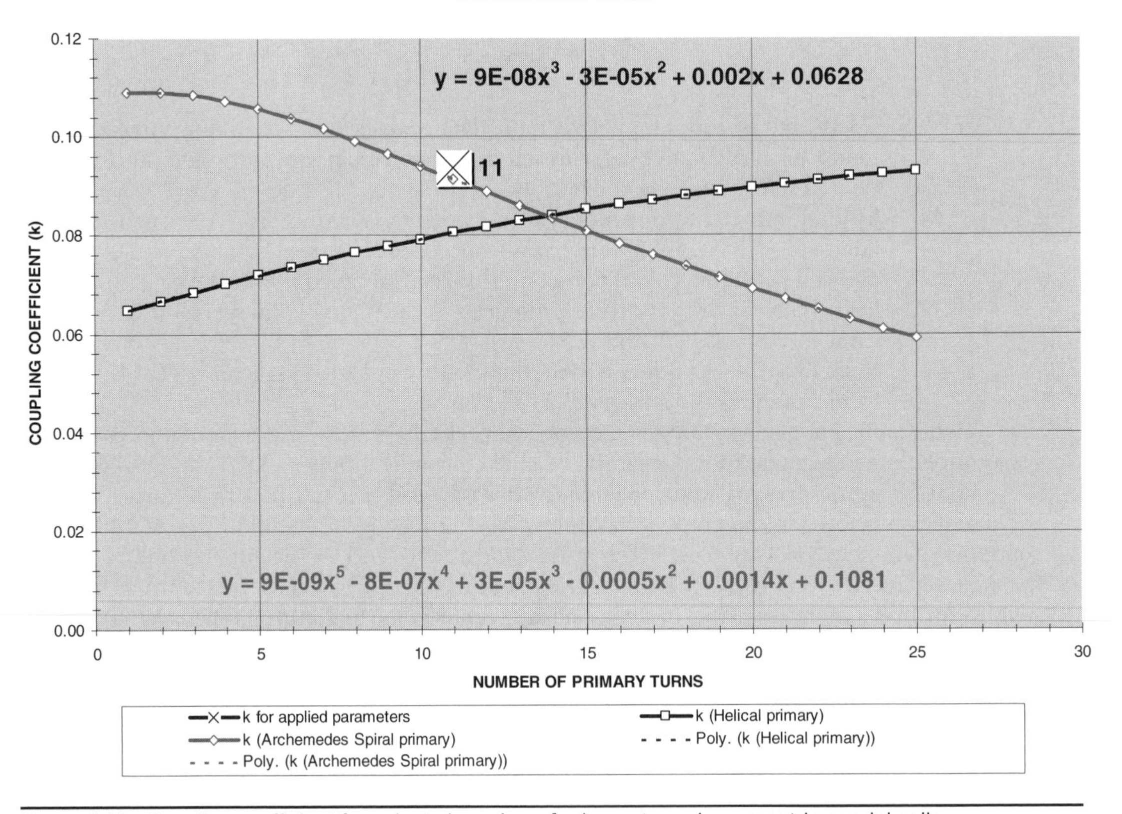

FIGURE 4-14 Coupling coefficient for selected number of primary turns in concentric coaxial coils.

This effect is not very pronounced in spark gap coils with a few primary windings and high decrement. The primary and secondary are coupled only for a very brief period of the total operating time. This produces an almost negligible effect on the primary inductance in most applications. The effect becomes more pronounced in vacuum tube coils, which are typically coupled tighter for nearly 50% of the duty cycle. This effect was pronounced enough for Tesla to observe it in his Colorado Springs experiments even with his primitive measurement systems. This was likely due to the large scale of his coil (50-ft. diameter primary and secondary with an 8-ft. diameter extra coil) and high break rates (BPS up to ≈4,000).

For coaxial coils that are wound on the same coil form such as the primary and grid winding in a tube coil, the mutual inductance can be determined using methodology from reference (14). Figure 4-17 shows two coaxial coils wound on the same form and their associated parameters. A tube coil's primary and grid windings will be used to illustrate the calculation methodology; however, it can be applied to any similar arrangement in another application. In such an application the primary will be the magnetizing winding with the grid being the other winding mutually coupled to the magnetizing winding.

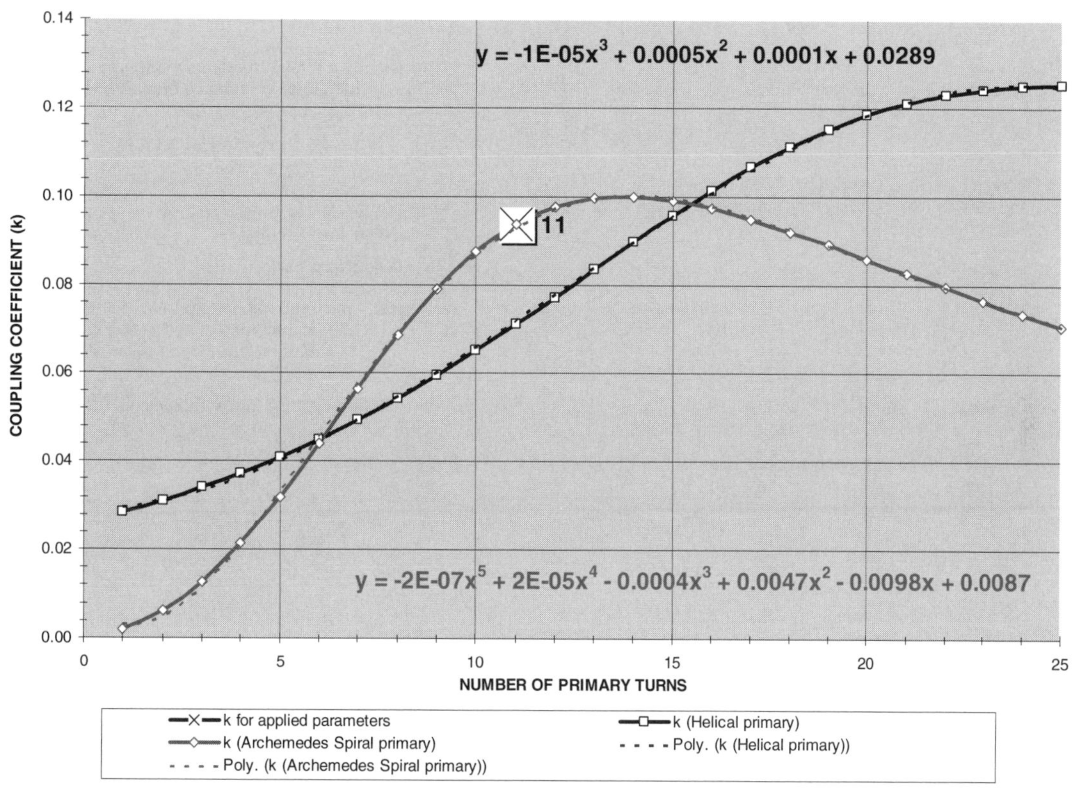

FIGURE 4-15 Coupling coefficient for selected number of primary turns in non-concentric coaxial coils.

Figure 4-18 shows the worksheet construction in the CH_4A.xls file (see App. B), k for $A = 0.2$ worksheet (3) used to calculate the mutual inductance (M) and coefficient of coupling (k) for two coaxial coils wound on the same form with a primary-to-grid height ratio of 0.2. There are several dimensions used in the calculations.

Only six parameters are required to perform the calculations:

- The outside diameter (OD) in inches of the coil form is entered in cell (K4).
- Distance between the primary and grid windings (Hn) in inches entered into cell (K5).
- Number of primary turns entered in cell (K8).
- The primary interwinding distance entered in cell (K9).
- Number of grid winding turns entered in cell (K12).
- The grid interwinding distance entered in cell (K13).

The primary height (Hp) is calculated in cell (K18) and grid height (Hg) is calculated in cell (K19). The primary-to-grid height ratio (A) is shown in cell (K20). There are worksheets for

To measure mutual inductance (M) proceed as follows:

1. Refer to the transformer below with its measured inductances shown:

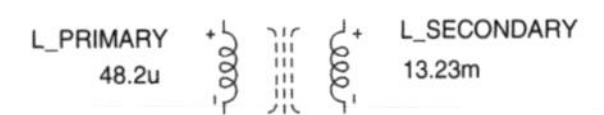

2. Connect windings in series as shown below and measure the resulting total inductance:

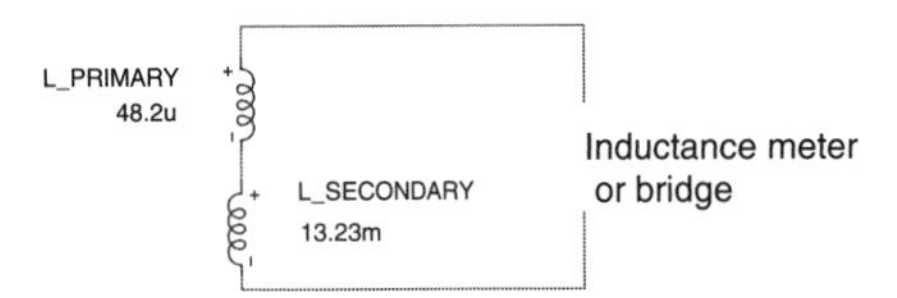

e.g. measures 14.45mH

3. Reverse the winding connections on one of the coils and again measure the resulting total inductance:

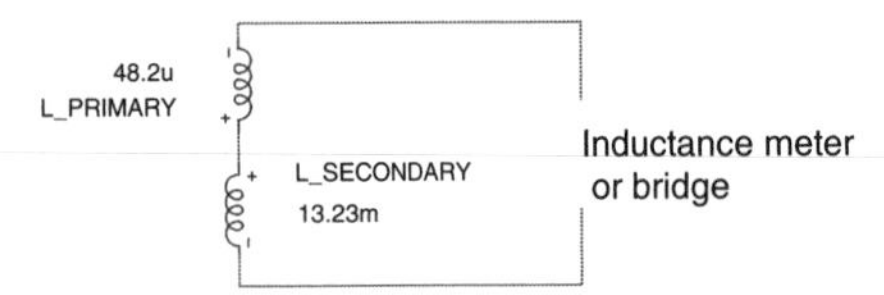

e.g. measures 14.08mH

4. The mutual inductance (M) is one fourth the difference between the two measured values:

M = (L2 - L3) / 4

Where: L2 = total inductance measured in Step 2
L3 = total inductance measured in Step 3

e.g. (14.45mH - 14.08mH) / 4 = 92.5μH

5. The coefficient of coupling (k) can be found from the measured values:

k = M / sqrt(Lp x Ls)

Where: Lp = inductance of primary winding
L3 = inductance of secondary winding
M = Mutual inductance (from step 2 - 4)

e.g. 92.5μH / sqrt(48.2μH x 13.23mH) = 0.116

Terman, F.E. Radio Engineers' Handbook. McGraw-Hill:1943, pp. 906-910.

FIGURE 4-16 Measuring the mutual inductance of two windings in an air core transformer.

FIGURE 4-17 Coaxial coils wound on same coil form.

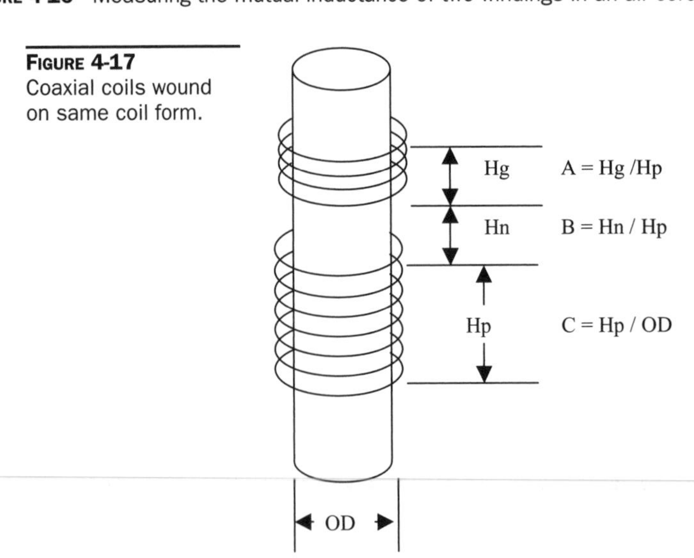

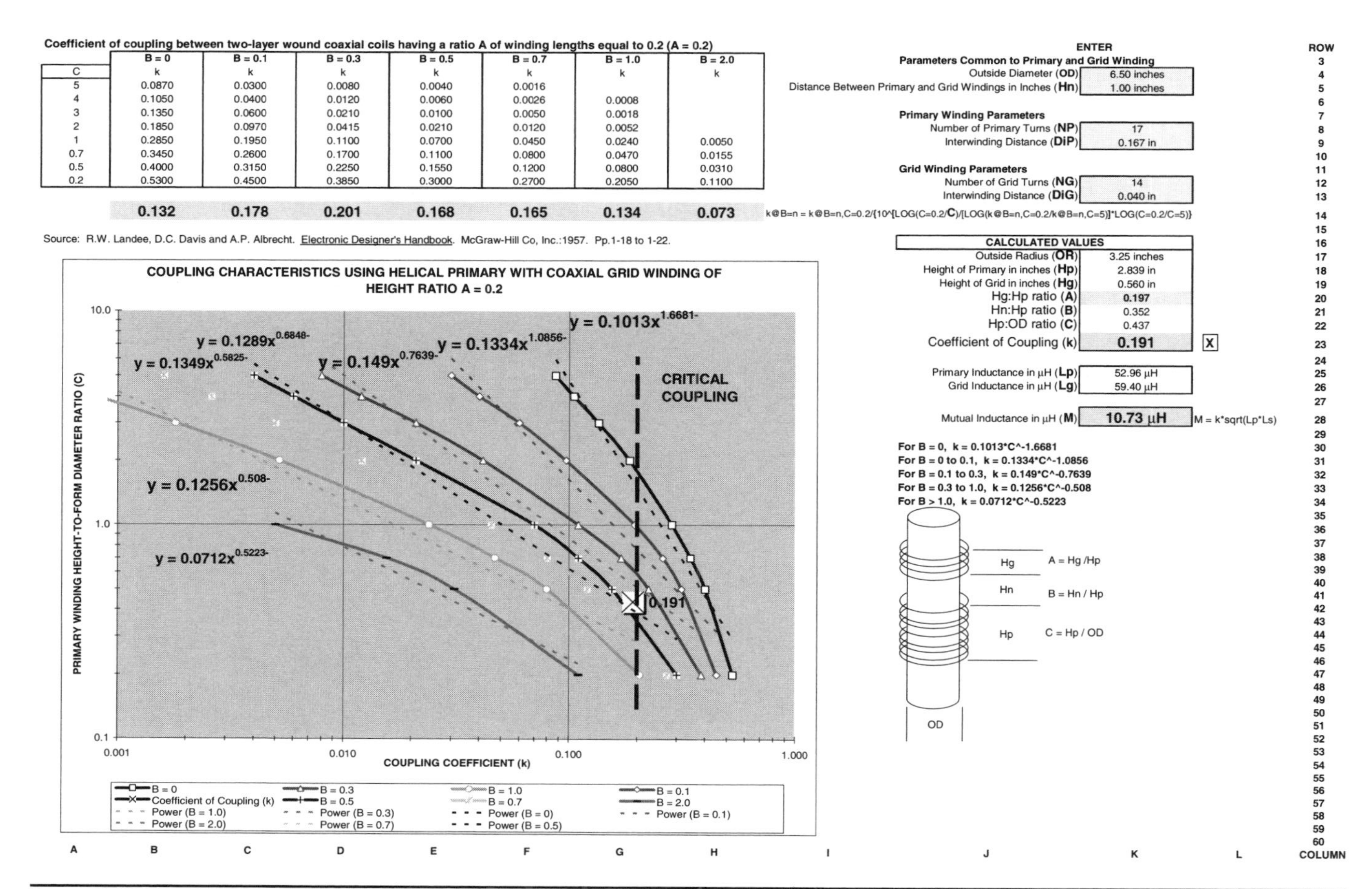

Coefficient of coupling between two-layer wound coaxial coils having a ratio A of winding lengths equal to 0.2 (A = 0.2)

C	B = 0 k	B = 0.1 k	B = 0.3 k	B = 0.5 k	B = 0.7 k	B = 1.0 k	B = 2.0 k
5	0.0870	0.0300	0.0080	0.0040	0.0016		
4	0.1050	0.0400	0.0120	0.0060	0.0026	0.0008	
3	0.1350	0.0600	0.0210	0.0100	0.0050	0.0018	
2	0.1850	0.0970	0.0415	0.0210	0.0120	0.0052	
1	0.2850	0.1950	0.1100	0.0700	0.0450	0.0240	0.0050
0.7	0.3450	0.2600	0.1700	0.1100	0.0800	0.0470	0.0155
0.5	0.4000	0.3150	0.2250	0.1550	0.1200	0.0800	0.0310
0.2	0.5300	0.4500	0.3850	0.3000	0.2700	0.2050	0.1100
	0.132	0.178	0.201	0.168	0.165	0.134	0.073

FIGURE 4-18 Worksheet calculations for coaxial coils wound on same coil form and primary-to-grid winding height ratio of 0.2.

primary-to-grid height ratios of 0.1, 0.2, 0.3, 0.5, 0.7, and 1.0. Selecting the proper worksheet ensures greater accuracy. If the calculated value of A is closer to another worksheet's *A* value, switch to that worksheet and reenter the six parameters. For the values entered in Figure 4-18, $A = 0.197$ so worksheet (3) for $A = 0.2$ is appropriate for the calculations.

The primary-to-grid winding separation (*Hn*)–to–primary winding height (*Hp*) ratio is designated (B) in the methodology and calculated in cell (K21). The primary winding height (*Hp*)–to–winding form diameter (OD) ratio is designated (C) in the methodology and calculated in cell (K22). The reference provides table values of the coupling coefficient for selected values of dimensions B and C and is shown in columns (A) through (H), rows 3 to 12. These table values are used to construct the graph shown below the table. The coefficient of coupling (k) can now be determined from the graph for the calculated *B* and *C* dimensions. The curve fit formulae shown in the graph for selected B dimensions were used with a conditional statement to calculate k for the applied dimensions in cell (K23). The curve fits are crude but eliminate manual graph interpretation. The calculated value of k is plotted on the graph by a large [X]. If the automated calculation is considered inadequate, k can be manually interpreted. Equation (4.3) is used to calculate the inductance of the close wound helical primary and grid windings in cells (K25) and (K26).

The mutual inductance can now be calculated:

$$M = k\sqrt{L_P L_G} \tag{4.36}$$

Where: M = Mutual inductance of primary and grid windings in μH = calculated value in cell (K28).

k = Coefficient of coupling between primary and grid windings (less than 1.0) = calculated value in cell (K23).

L_P = Inductance of primary coil in μhenries = calculated value in cell (K25) from equation (4.3).

L_G = Inductance of grid coil in μhenries = calculated value in cell (K26) from equation (4.3).

For determining *k* without the Excel worksheets calculate the A, B and C dimensions as explained above and interpret *k* for the appropriate value of *A* from Figures 4-19 through 4-24.

Worksheet (1) in the file can also be used to estimate *M* and *k* if the form diameters entered for each winding are the same. There may, however, be some disparity between the two methods. A dashed line appears at $k = 0.2$ in the worksheet graphs and in Figures 4-19 through 4-24 denoting the critical coupling threshold for spark gap coils. The worksheets can be used to design any type of coil; however, the critical coupling threshold of $k = 0.2$ applies only to spark gap coils. Tesla coils using an extra coil (magnifiers), tube coils, and solid-state coils can be coupled tighter than traditional spark gap coils and critical coupling in these applications can exceed the 0.2 marked on the graphs.

4.10 Leakage Inductance

The mutual inductance is the degree of coupling the primary and secondary share. The leakage inductance is the inductance not shared between the primary and secondary. The mutual inductance couples the primary energy to the secondary. Since all the power must be accounted for (law of conservation), the power not transferred to the secondary is dissipated

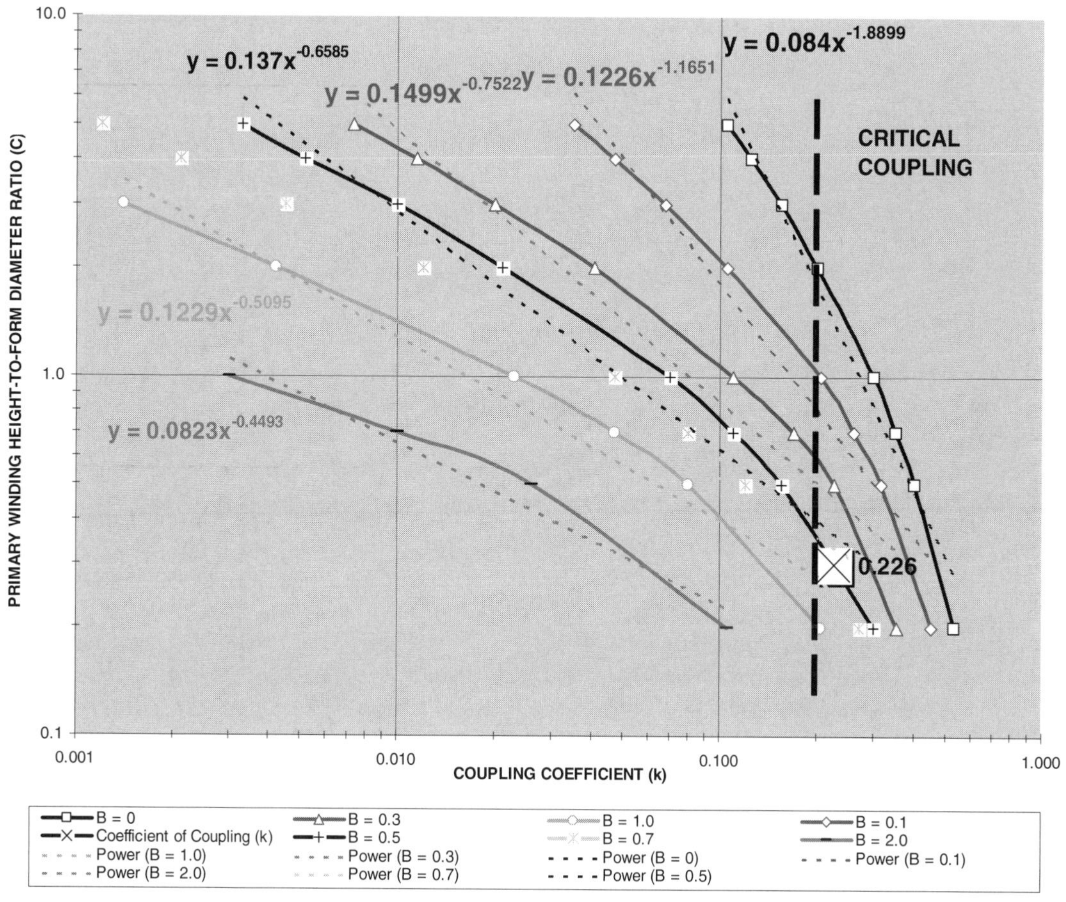

FIGURE 4-19 Coupling coefficients for $A = 0.1$.

in the leakage inductance. To perform these calculations for an iron cored power transformer open the CH_4A.xls file, IRON CORE RELATIONSHIPS worksheet (11). (See App. B.) Looking at the example in Figure 4.25, the primary inductance is 35.2 mH with a maximum leakage inductance of 600 μH. This produces a coupling coefficient of:

$$k = 1 - \left(\frac{\text{Leakage}}{Lp}\right) \tag{4.37}$$

Where: k = Coefficient of coupling = cell (F3), calculated value.

Leakage = Measured leakage inductance of primary (magnetizing) winding in henries with secondary winding shorted together = cell (B4), enter value in μH. Converted to H using a 1e-6 multiplier.

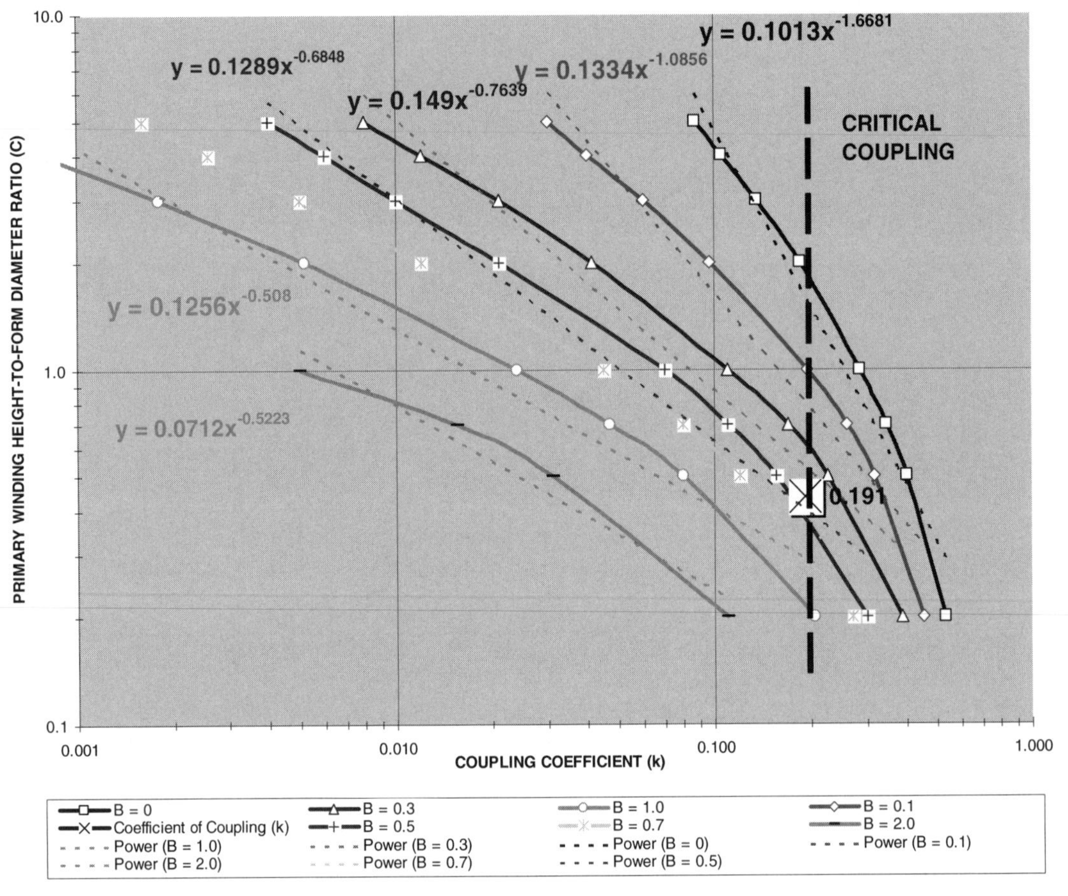

FIGURE 4-20 Coupling coefficients for $A = 0.2$.

$Lp =$ Inductance of primary (magnetizing) winding in henries = cell (B3), enter value in mH. Converted to H using a 1e-3 multiplier.

From the example iron core transformer shown in Figure 4-25 the calculated k of 0.983 from actual measurements ensures that 98.3% of the energy drawn by the primary winding from the source will be delivered to the load in the secondary. The remaining 1.7% will be dissipated in the leakage inductance. In an iron core transformer the coupling is usually very high, on the order of 0.95 to 1.0. Air core transformers have much lower coupling due to the decreased magnetic permeability (μ) of air compared with the higher μ of iron. The coupling is also reduced in the air core's primary and secondary windings as they are not in close proximity of each other as in the iron core power transformer. With an air core it takes considerably more primary power to transfer a desired power to the secondary. Typically more than half the primary power is lost in the leakage inductance of an air core transformer.

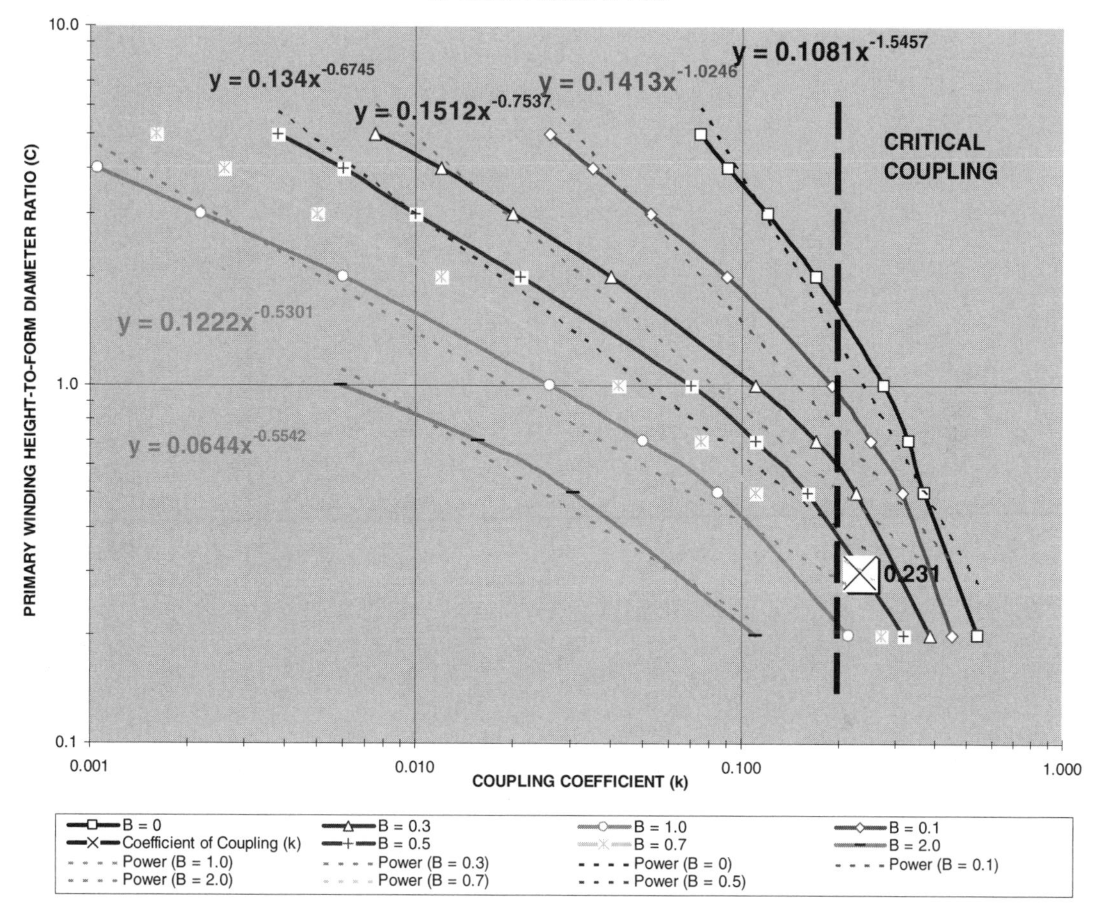

FIGURE 4-21 Coupling coefficients for $A = 0.3$.

The coefficient of coupling and leakage inductance share the following relationships:

$$k = 1 - \frac{M}{Lp} \quad \text{and} \quad M = -(k-1)Lp \tag{4.38}$$

Where: M = Mutual inductance of primary and secondary windings in henries = cell (F4), calculated value (also see Section 4.9). Converted to μH using a 1e6 multiplier.

Lp = Inductance of primary (magnetizing) winding in henries = cell (B3), enter value in mH. Converted to H using a 1e-3 multiplier.

k = Coefficient of coupling = cell (F3), calculated value from equation (4.37).

To measure the leakage inductance of a transformer, refer to Figure 4-25.

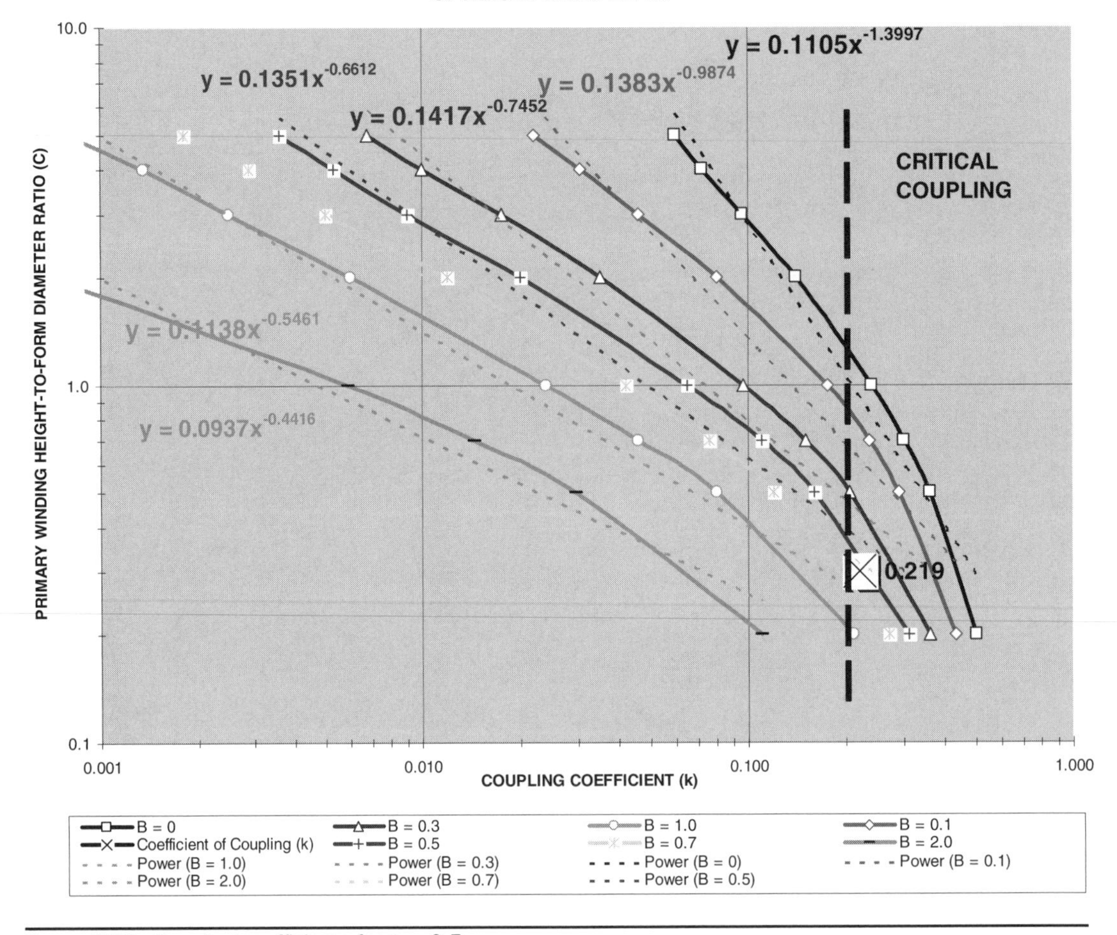

FIGURE 4-22 Coupling coefficients for $A = 0.5$.

4.11 Relationships of Primary and Secondary Windings in Iron Core Transformers

To perform these calculations open the CH_4A.xls file, IRON CORE RELATIONSHIPS worksheet (11). The turns ratio (N) defines the relationships in an iron core transformer and can be calculated using the measured primary and secondary inductance:

$$N = \sqrt{\frac{Ls}{Lp}} \tag{4.39}$$

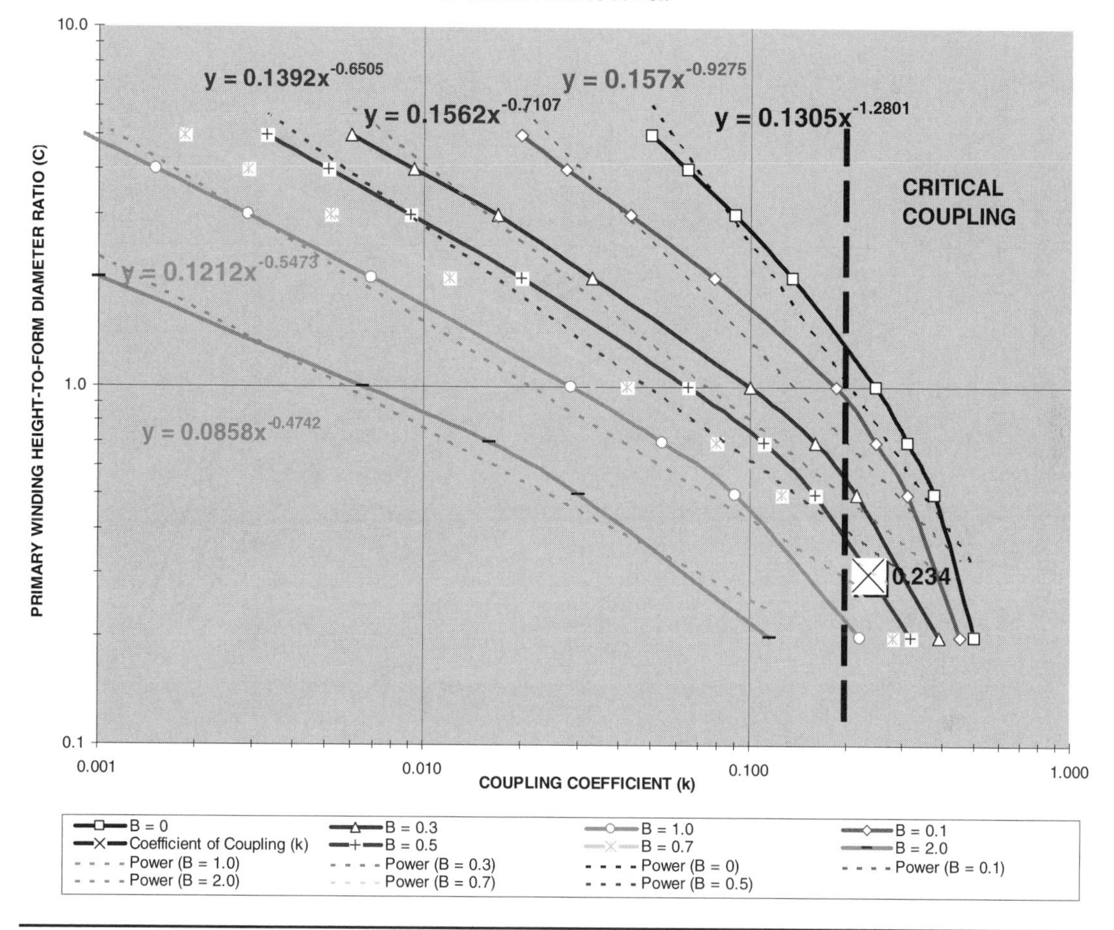

FIGURE 4-23 Coupling coefficients for $A = 0.7$.

Where: N = Primary-to-secondary turns ratio = cell (F7), calculated value.
Lp = Inductance of primary (magnetizing) winding in henries = cell (B3), enter value in mH. Converted to H using a 1e-3 multiplier.
Ls = Inductance of secondary winding in henries = cell (B7), enter value.

Once N is determined the relationship of current, voltage, and impedance in an iron core transformer is:

$$N = \frac{Np}{Ns} = \frac{Vp}{Vs} = \frac{Is}{Ip} = \sqrt{\frac{Zp}{Zs}} \tag{4.40}$$

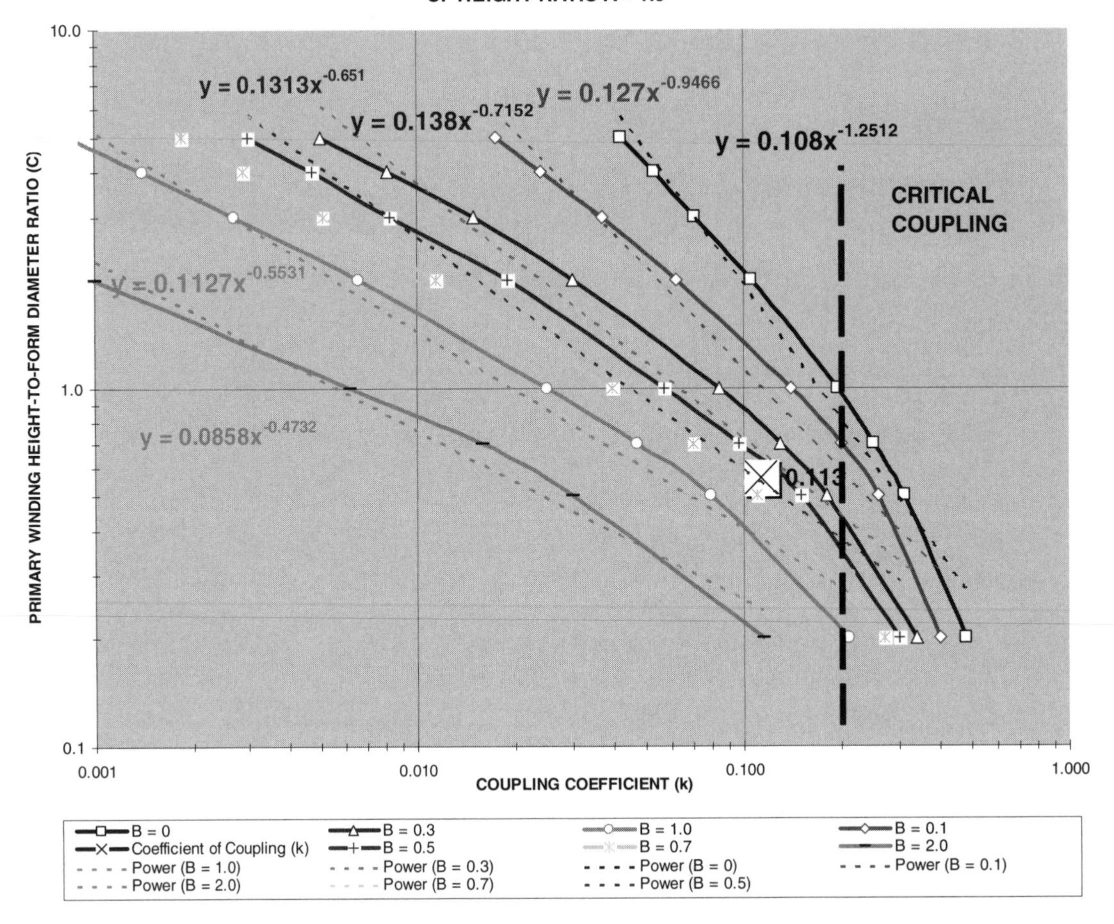

FIGURE 4-24 Coupling coefficients for $A = 1.0$.

Where: N = Turns ratio between primary (magnetizing winding) and secondary winding = cell (F7), calculated value.

Ls = Inductance of secondary winding in henries = cell (B7), enter value.

Lp = Inductance of primary (magnetizing) winding in henries = cell (B3), enter value in mH. Converted to H using a 1e-3 multiplier.

Np = Number of turns used in the primary winding.

Ns = Number of turns used in the secondary winding.

N = Primary-to-secondary turns ratio = cell (F7), calculated value from equation (4.39).

Vp = Voltage applied to the primary winding = cell (B8), enter value.

To measure leakage inductance proceed as follows:

1. Refer to the iron core transformer below with its measured inductances shown (Westinghouse 70:1 potential transformer):

L PRIMARY 35.2m L SECONDARY 172.48

2. To find the maximum leakage inductance, short the secondary winding as shown below and measure the primary inductance:

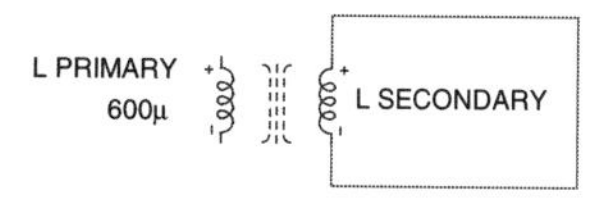

e.g. measures 600µH

3. To find the leakage inductance at a specific load impedance, short the secondary winding as shown below and measure the primary inductance:

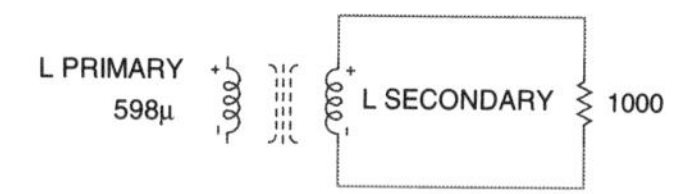

e.g. measures 598µH

4. The leakage inductance is the measured primary inductance. If using an inductance meter, the frequency of the internal oscillator is the frequency of the inductance measurements. If a different frequency is required the technique in steps 5 thru 7 will measure the primary inductance at a different frequency.

To measure inductance (L) proceed as follows:

5. Construct the circuit shown below:

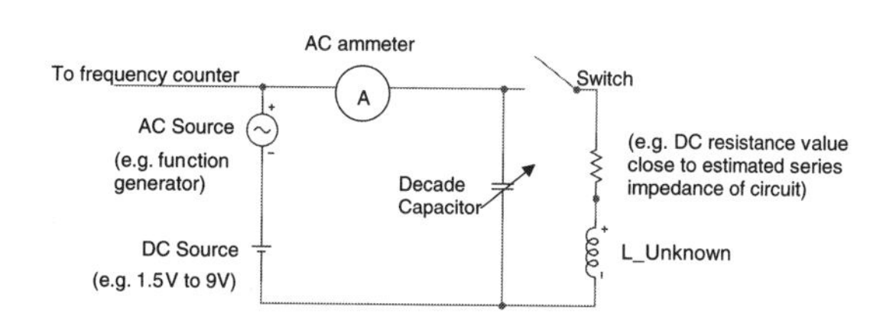

6. Adjust the decade capacitor until the AC current measured by the AC meter is the same with the switch opened and closed. This is the circuit's point of resonance.

7. The inductance is calculated as follows:

wL = (1/2) x [1/(wC)] and, L = 1/[2(w^2C)]

Where: L = inductance value of unknown coil in henries
w / (2*pi) = frequency of AC source (f) as measured with frequency counter. w = 2*pi*f
C = capacitance of decade capacitor at resonance

Terman, F.E. Radio Engineers' Handbook. McGraw-Hill:1943, pp. 906-910.

FIGURE 4-25 Measuring the leakage inductance of two windings in an iron core transformer.

Vs = Voltage developed on the secondary winding = cell (F8), calculated value.

Ip = Magnetizing current applied to the primary winding in amps = cell (B9), enter value.

Is = Current induced in the secondary winding from the primary in amps = cell (F9), calculated value.

Zp = Impedance of the primary circuit into the transformer (line) = cell (F10), calculated value.

Zs = Impedance of the secondary circuit out of the transformer (load) = cell (F11), calculated value.

To reflect a load or other resistance value from the secondary into the primary winding:

$$R_{\text{ref}} = \frac{Rs}{N^2} \tag{4.41}$$

Where: R_{ref} = Secondary resistance reflected to the primary (magnetizing winding) in ohms = cell (F15), calculated value.
N = Primary-to-secondary turns ratio = cell (F7), calculated value from equation (4.39).
Rs = Resistance in secondary circuit (load) in ohms = cell (B15), enter value.

To reflect a capacitance value from the secondary into the primary winding:

$$C_{ref} = \frac{Cs}{N^2} \tag{4.42}$$

Where: C_{ref} = Capacitance reflected from the secondary winding to the primary winding in farads = cell (F14), calculated value. Converted to μF using a 1e6 multiplier.
N = Primary-to-secondary turns ratio = cell (F7), calculated value from equation (4.39).
Cs = Capacitance in secondary circuit = cell (B14), enter value in μF. Converted to farads using a 1e-6 multiplier.

When the turns ratio is known and your inductance meter's range cannot measure the primary inductance value, but can measure the secondary:

$$Lp = \frac{Ls}{N^2} \tag{4.43}$$

Where: Lp = Inductance of primary (magnetizing) winding in henries = cell (F18), calculated value. Converted to mH using a 1e3 multiplier.
N = Primary-to-secondary turns ratio = cell (B20), enter value. Also equal to the step-up ratio of primary voltage-to-secondary voltage. The 1: is part of the cell formatting so enter only the secondary value of the turns ratio.
Ls = Measured inductance of secondary winding in henries = cell (B19), enter value.

And conversely when the turns ratio is known and the inductance meter's range cannot measure the inductance value of the secondary, but can measure the primary:

$$Ls = \frac{Lp}{\left(\frac{1}{N}\right)^2} \tag{4.44}$$

Where: Ls = Inductance of secondary winding in henries = cell (F19), calculated value.
N = Primary-to-secondary turns ratio = cell (B20), enter value. Also equal to the step-up ratio of primary voltage-to-secondary voltage. The 1: is part of the cell formatting so enter only the secondary value of the turns ratio.
Lp = Measured inductance of primary winding in henries = cell (B18), enter value in mH. Converted to H using a 1e-3 multiplier.

To determine the voltage induced in the secondary from the primary winding:

$$Vs = \frac{Vp}{N} \tag{4.45}$$

Where: Vs = Voltage developed in the secondary winding = cell (F8), calculated value.
N = Primary-to-secondary turns ratio = cell (F7), calculated value from equation (4.39).
Vp = Voltage applied to the primary winding = cell (B8), enter value.

To determine the current induced in the secondary from the primary winding:

$$Is = IpN \tag{4.46}$$

Where: Is = Current induced in the secondary winding from the primary = cell (F9), calculated value.
N = Primary-to-secondary turns ratio = cell (F7), calculated value from equation (4.39).
Ip = Magnetizing current in the primary winding = cell (B9), enter value.

4.12 Determining Transformer Relationships in an Unmarked Iron Core Transformer

At the last Hamfest you picked up what looks like a nice vacuum tube plate transformer (has porcelain high-voltage bushings) but there are no markings on it. Now what?

A variable autotransformer also known as a Variac (General Radio) or other trade name will come in handy. Though no longer in business, General Radio invented the variable autotransformer. A very interesting paper detailing the history of the company and its inventions, "A History of the General Radio Company"written by Arthur E. Thiessen in 1965 for the company's 50th anniversary celebration can usually be found on an Internet search. The variable voltage output of the variable autotransformer can be connected to any winding in the unmarked transformer. Reading the voltage on any of the other windings as the autotransformer is slowly increased will enable you to determine the transformer turns ratio. Make sure the autotransformer is increased very slowly to avoid damaging the voltmeter or danger caused by an internal short in the transformer. An even safer method is to use a signal generator. A small 1.0 V sine wave input equal to the line frequency will be increased by the primary-to-secondary turns ratio, the ratio being determined by voltage measurements on each winding. Even if the turns ratio is as high as 1:200, the other windings will be no higher than 200 V. Use equation (4.40) to calculate the unknown turns ratio from the voltage measurements ($N = Vp/Vs$).

4.13 Relationships of Primary and Secondary Windings in Air Core Transformers

To perform these calculations open the CH_4A.xls file, AIR CORE RELATIONSHIPS worksheet (12). In an air core transformer the turns ratio does not determine the reflected impedance from one winding to another. This is dependent on the mutual inductance as previously defined in equation (2.27):

$$R_{\text{ref}} = \frac{(\omega M)^2}{R_S} \tag{4.47}$$

Where: R_{ref} = Resistance of secondary circuit reflected through the mutual inductance, into the primary circuit in ohms = cell (F12), calculated value.
ω = Resonant frequency in radians = cell (F10), calculated value = $2\pi f$ (f is the resonant frequency of the primary).
M = Mutual inductance of primary and secondary winding = cell (B9), enter calculated value in μH using the CH_4A.xls file, MUTUAL INDUCTANCE worksheet (1) or measured value using Figure 4-16. See Section 4.9. Converted to H in cell (C9) using a 1e-6 multiplier.
R_S = Resistance of secondary circuit in ohms = cell (B12), enter value.

We can infer from equations (4.41), (4.42) and (4.47):

$$C_{ref} = \frac{(\omega M)^2}{C_S} \tag{4.48}$$

Where: C_{ref} = Reflected self-capacitance of secondary circuit through the mutual inductance, into the primary circuit in ohms = cell (F11), calculated value.
ω = Resonant frequency in radians = cell (F10), calculated value = $2\pi f$ (f is the resonant frequency of the primary).
M = Mutual inductance of primary and secondary winding = cell (B9), enter calculated value in μH using the CH_4A.xls file, MUTUAL INDUCTANCE worksheet (1) or measured value using Figure 4-16. See Section 4.9. Converted to H in cell (C9) using a 1e-6 multiplier.
C_S = Capacitance of secondary circuit in farads = cell (B11), enter value in pF. Converted to farads in cell (C11) using a 1e-12 multiplier.

4.14 Hysteresis Curve in Air Core Resonant Transformers

The hysteresis curve for the spark gap coil is shown in Figure 4-26. The first two magnetizing cycles are displayed in the curve. Note the curve is very narrow indicating the air core has high permeability and easily transfers the primary energy to the secondary (less the leakage). The calculated curve was developed in the CH_4A.xls file, B vs. H worksheet (9). (See App. B.)

4.15 Measuring the Resonance of a Coil

To measure the resonant frequency of a coil, inject a sine wave from a signal generator into one end (bottom) of the winding in series with a 1.0-kΩ, $\frac{1}{4}$ W resistor and observe the frequency response across the resistor with an oscilloscope, as shown in Figure 4-27. Connect the ground probe of the oscilloscope to the ground (black) plug on the signal generator. If a function generator is used select the sine function. A current probe clamped around the wire feeding the base of the coil can also be used instead of the resistor. Do not measure the voltage waveform at the top of the coil (opposite end of waveform source) with the scope probe as it will top load the winding with its 50 pF capacitance and change the resonant frequency of the coil. As the function generator's output frequency is swept up and down, the coil's resonant frequency can clearly be seen by the waveform peak amplitude on the oscilloscope. Many thanks are extended to Tom Vales for this elegant technique.

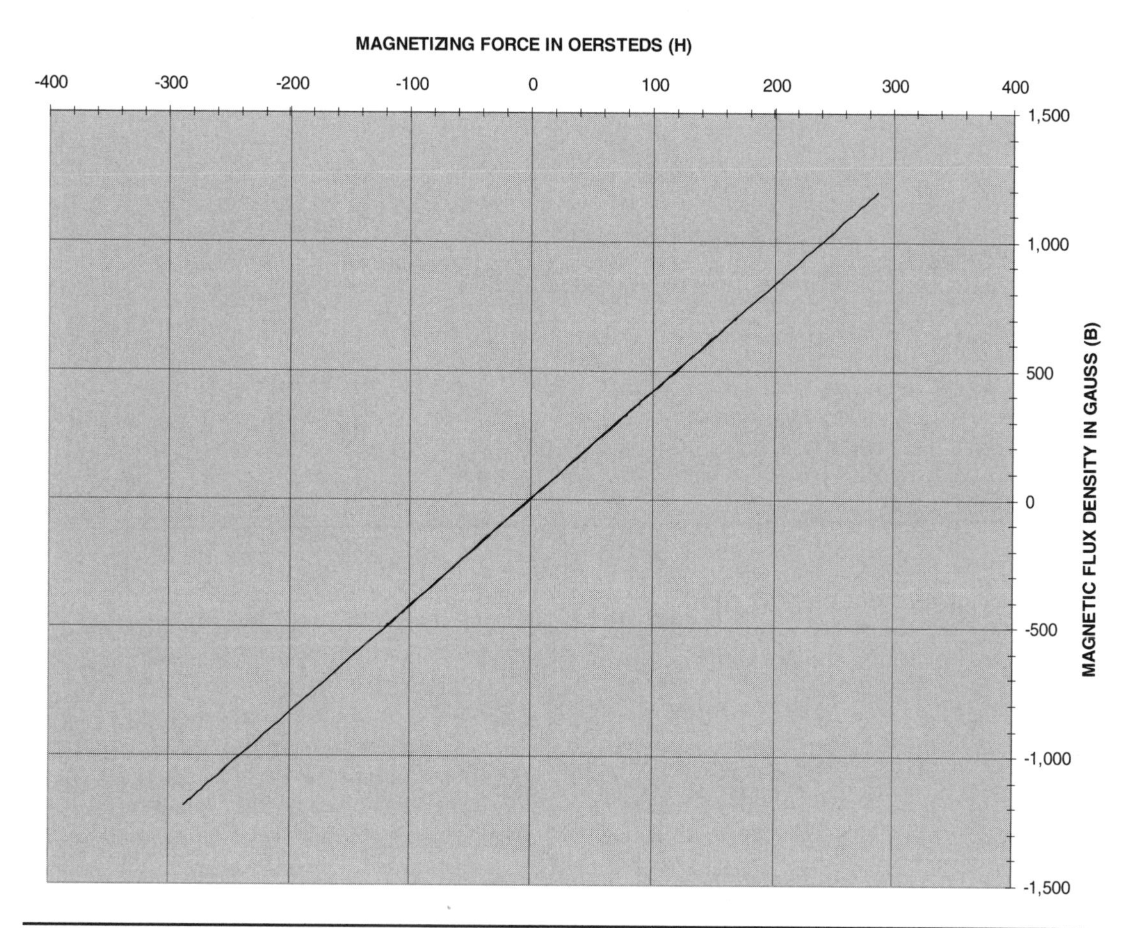

FIGURE 4-26 Hysteresis curve of spark gap Tesla coil.

Two parallel LEDs can also indicate the resonant frequency and harmonics when connected as shown in Figure 4-27. Many thanks to Richard Hull of the Tesla Coil Builders of Richmond (TCBOR) for this idea. The LEDs will illuminate as the signal generator's frequency approaches the resonant frequency. The LED intensity will change with the series current as shown in Figure 3-1. However, it may be difficult to discern the difference between the LED brightness at the resonant frequency with the harmonics above and sub-harmonics below this frequency. Tesla used a similar technique with incandescent lamps during his Colorado Springs experiments. You may also want a visual indication of this resonant frequency to verify the signal generator's frequency (dial) setting. An oscilloscope will of course display the waveforms so the frequency can be determined. An alternative to the scope is the newer Digital Multi-Meters (DMM) with a built-in frequency counter. These are sold for less than

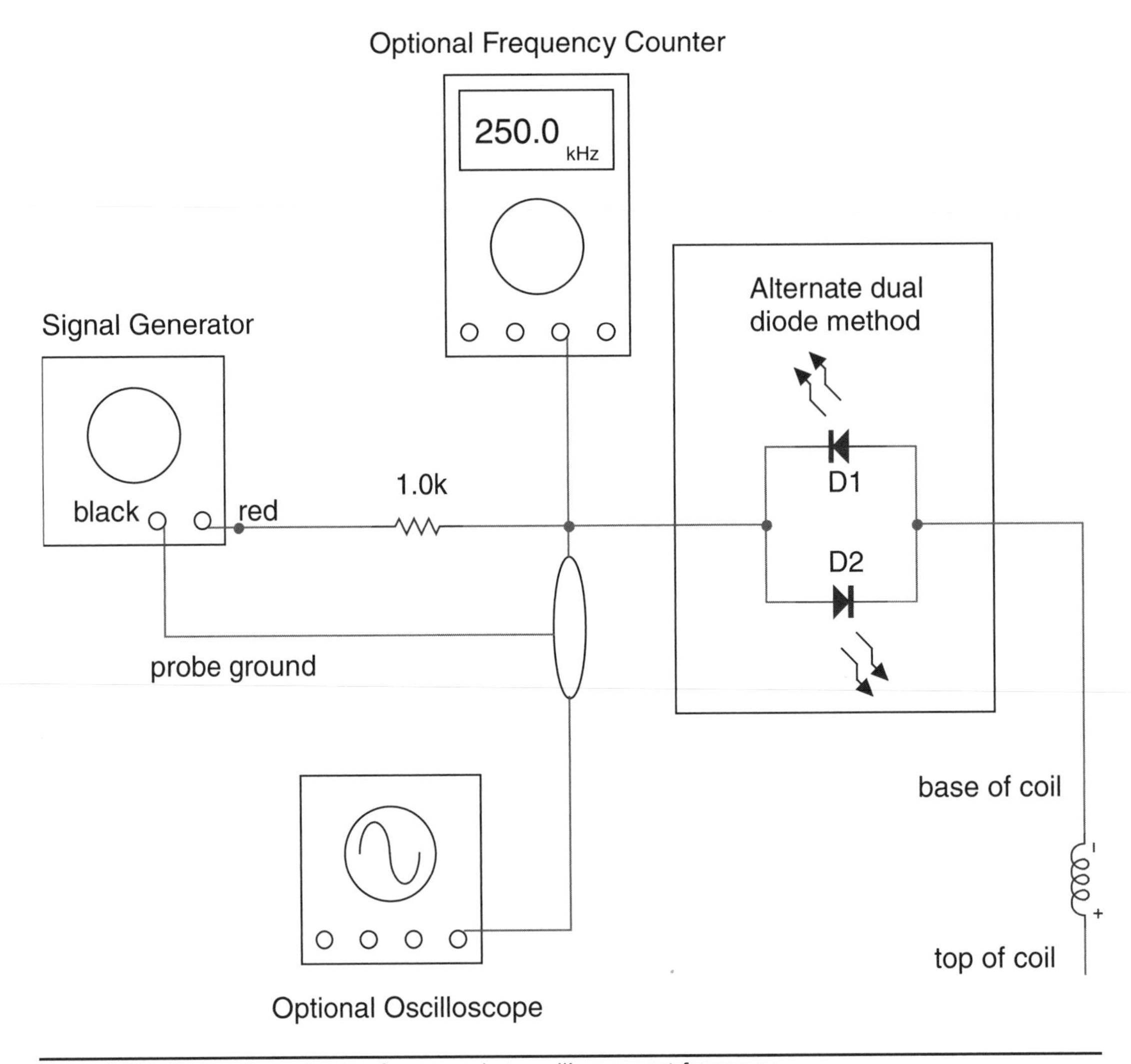

FIGURE 4-27 Test equipment setup for measuring a coil's resonant frequency.

$100. At your next Hamfest you may even find a $5 function generator, $20 to $50 oscilloscope and a $20 frequency counter.

References

1. Terman, F.E. *Radio Engineer's Handbook*. McGraw-Hill: 1943. Section 3: Circuit Theory, pp. 135–172.
2. *Reference Data for Radio Engineers*. H.P. Westman, Editor. Federal Telephone and Radio Corporation (International Telephone and Telegraph Corporation), American Book. Fourth Ed: 1956, p. 35.
3. *The VNR Concise Encyclopedia of Mathematics*. Van Nostrand Reinhold Company: 1977, pp. 184–191.
4. Magnetics, Inc. Ferrite Core Catalog, FC-601 9E, 1997.

5. MWS industries wire catalog for magnet wire, magnet wire insulation guide, pp. 2–3.
6. National Electrical Manufacturer's Association (NEMA) Manual Standard Magnet Wire Table For Heavy Build (K2), Thermal Class 200, Polyester (amide) (imide) Overcoated With Polyamideimide (MW 35-C), p. 109.
7. MWS industries wire catalog for magnet wire.
8. *The VNR Concise Encyclopedia of Mathematics*. Van Nostrand Reinhold Company: 1977, p. 192.
9. Unitrode Power Supply Design Seminar SEM-900, p. M2-4, pp. M8-1 thru M8-10.
10. P.L. Dowell. *Effects of Eddy Currents in Transformer Windings*. Proceedings IEE (UK), Vol. 113, No. 8, August 1966.
11. Kevan O'Meara. *Proximity Losses in AC Magnetic Devices*. PCIM, Dec. 1996, pp. 52–57.
12. Wheeler formula found in: H. Pender, W.A. Del Mar. *Electrical Engineer's Handbook*. Third Ed. John Wiley & Sons, Inc.: 1936. Air Core Inductors pp. 4–17 to 4–18. This reference cites the following source: Proc. I.R.E., Vol. 16, p. 1398. October 1928.
13. Terman, F.E. *Radio Engineer's Handbook*. McGraw-Hill: 1943. Section 2: Circuit Elements, pp. 71–73.
14. R.W. Landee, D.C. Davis and A.P. Albrecht. *Electronic Designer's Handbook*. McGraw-Hill: 1957, pp. 1–18 to 1–22.
15. *Electrical Engineering*. American Technical Society: 1929. Vol. VIII, pp. 94–99.
16. U.S. Department of Commerce, National Bureau of Standards, *Radio Instruments and Measurements, Circular 74*. U.S. Government Printing Office. Edition of March 10, 1924-reprinted Jan. 1, 1937, with certain type corrections and omissions.

5

CHAPTER

Capacitors

To get the true vibration we shall want at least 8 turns in the primary with present transformer to keep the capacity in the primary within the limits given by the output of transformer. p. 61

...the capacity should, as stated before, be best adapted to the generator which supplies the energy. This consideration is however, of great importance only when the oscillator is a large machine and the object is to utilize the energy supplied from the source in the most economical manner. This is the case particularly when the oscillator is designed to take up the entire output of the generator, as may be in the present instance. But generally, when the oscillator is on a supply circuit distributing light and power the choice of capacity is unrestricted by such considerations. p. 67

Nikola Tesla. Colorado Springs Notes: 1899-1900, pp. 61, 67.

Tesla discusses optimizing the capacitor value for the step-up transformer used in the primary circuit.

Actually, there was only one mile of secondary wire but owing to the large capacity (distributed) in the secondary the vibration was much slower than should be inferred from the length of wire.

Nikola Tesla. Colorado Springs Notes: 1899-1900, p. 59.

Tesla observes the interwinding capacitance in the secondary winding.

Thus with the sphere the capacity in the vibrating secondary system was increased... p. 232

The experiments seemed to demonstrate clearly that the augmentation of the capacity as the ball was elevated was in a simple proportion to the height, for that the middle position the value found was very nearly the arithmetic mean of the values in the extreme positions. p. 239

Nikola Tesla. Colorado Springs Notes: 1899-1900, pp. 232, 239.

Tesla observes the decreasing resonant frequency when adding a terminal capacitance to the interwinding capacitance in the secondary winding or elevating it above the secondary.

When two conductors are in close enough proximity to each other to establish an electric field in the insulation (dielectric) between them, a capacitor is formed. The capacitance (C) is a measure of the capacitor's ability to store electrons (capacity) and is directly proportional to the product of the conductor area (A) and dielectric constant (k) of the insulation and inversely proportional to the separation between the conductors (d) or $C = ([A \times k]/d)$. When a voltage differential is applied to the two conductors the dielectric stores a charge in an electric field. There is no actual current flow through the capacitor except a small leakage current typically in the μA to pA range. A voltage differential applied to the two plates will establish an electric field in the dielectric; however, the dielectric will not allow electrons to pass through it. A charge accumulates between the dielectric and the conductor with the most negative potential. The electric field (charge) will remain in this state until the applied voltage changes or until the charge leaks off, which can take a long time in large energy-storage capacitors. When an external circuit provides a path for current flow the capacitor will act much like a battery and the charge will flow out of the capacitor—through the circuit—and stack up on the opposite side of the dielectric, reversing the electric field polarity. Some of the charge will be lost in this action, proportional to the external circuit impedance. In the oscillating tank circuit found in a spark gap coil this will continue until the charge is dissipated (damped).

5.1 Capacitor Applications in Tesla Coils

There are many characteristics to consider when selecting capacitors for use in Tesla coil systems. The cost of new commercial high-voltage capacitors or the availability of these capacitors on the surplus market can present a design challenge for even the experienced coiler. Examples of commercial capacitors that can be used in Tesla coils are shown in Figure 5-1.

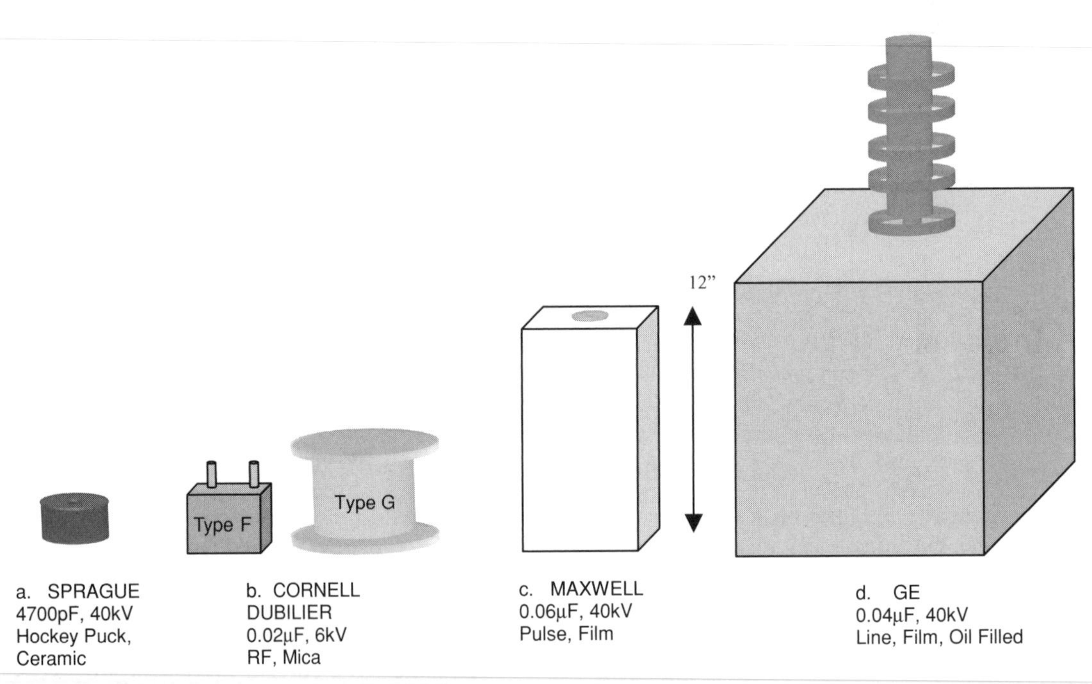

FIGURE 5-1 A sampling of high-voltage capacitors.

5.2 Increasing Capacitance or Dielectric Strength

Several capacitors can be arranged in series, parallel, or series–parallel to increase the capacitance value or working voltage (dielectric strength). Examples for increasing both are shown in Figure 5-2. These arrangements can be continued in unlimited series–parallel strings to obtain a very high-capacitance network or a higher working voltage.

Increasing Capacitance Value

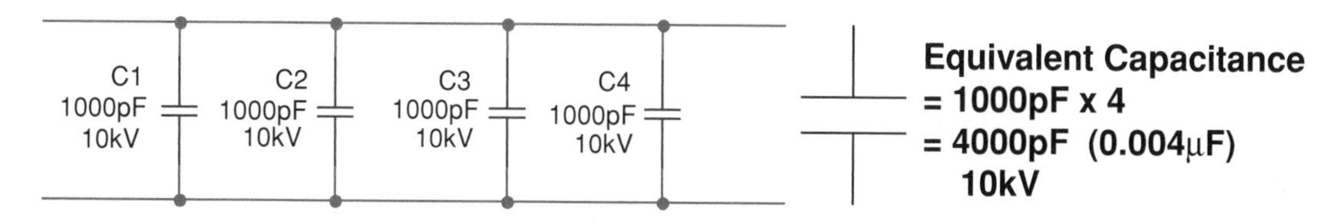

Increasing Dielectric Strength

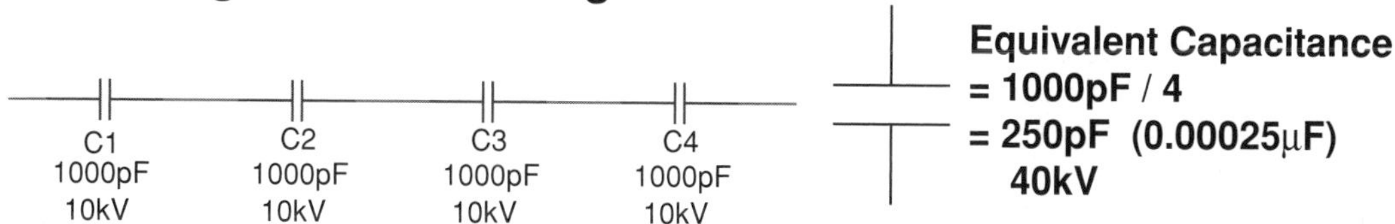

Increasing Capacitance Value And Dielectric Strength

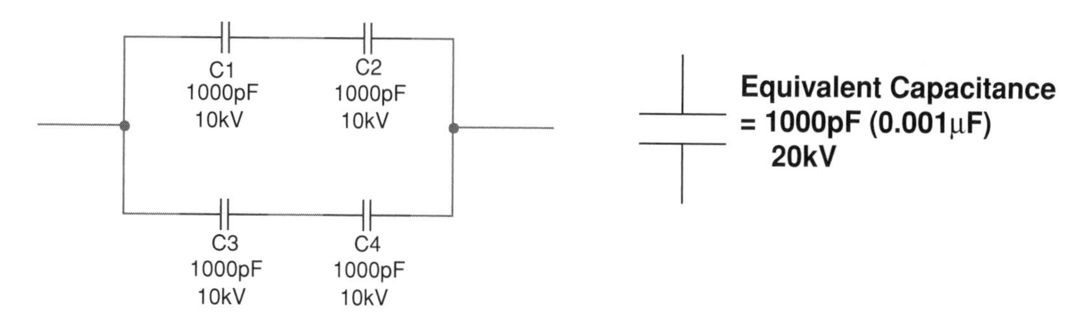

FIGURE 5-2 Circuit arrangement for increasing capacitance and dielectric strength.

Placing capacitors in series reduces the total capacitance to a value less than the smallest capacitor producing an equivalent capacitance of:

$$Ct = \frac{1}{\left(\frac{1}{C1} + \frac{1}{C2} + \frac{1}{C3} + \frac{1}{C4} + \frac{1}{Cn...}\right)} \tag{5.1}$$

Where: Ct = Total equivalent capacitance in farads.
$C1$ thru $Cn\ldots$ = Value of each series capacitor in farads.

The working voltage increases with each series capacitor for a total working voltage of:

$$\text{WVDC}_T = \text{WVDC}_{C1} + \text{WVDC}_{C2} + \text{WVDC}_{C3} + \text{WVDC}_{C4} + \text{WVDC}_{Cn...} \tag{5.2}$$

Where: WVDC_T = Total working voltage of series circuit.
$\text{WVDC}_{C1\,\text{thru}\,Cn}\ldots$ = DC working voltage of each series capacitor.

Placing capacitors in parallel increases the total capacitance producing an equivalent capacitance of:

$$Ct = C1 + C2 + C3 + C4 + Cn... \tag{5.3}$$

Where: Ct = Total equivalent capacitance in farads.
$C1$ thru $Cn\ldots$ = Value of each series capacitor in farads.

The total working voltage of the parallel circuit is equal to the lowest working voltage value. In other words the capacitor with the lowest working voltage will fail first if a higher voltage is applied to the parallel circuit.

In a series–parallel arrangement apply the series equations (5.1) and (5.2) to the series elements and the parallel equation (5.3) to the parallel elements.

Commercial high-voltage capacitors are generally made using series–parallel networks of lower voltage capacitors much like that shown in Figure 5-2. The principal difference between commercial capacitors and carefully assembled homemade capacitors is the dielectric. All dielectric materials have microscopic imperfections in their structure. These imperfections result in breakdown between the plates through the dielectric and the capacitor fails. For this reason commercial manufacturers generally use several layers of very thin (thousandths of an inch) dielectric sheeting between the capacitor plates, which are also very thin. The assembly techniques used with hundreds of layers of dielectric and plate sheeting a few thousandths of an inch thick could not easily be duplicated in homemade capacitors. The dielectric in commercial capacitors overlaps the capacitor plates uniformly to tolerances not obtainable in the homemade version. In the internal network of a commercial capacitor where the plates are connected together and to the external terminals, care is taken to minimize the resulting equivalent series resistance (ESR) and equivalent series inductance (ESL). This results in higher peak and rms current ratings, which cannot be duplicated in the homemade version. I have seen many homemade capacitors that performed well in a Tesla coil but the expense of the materials and time consumed in their construction may not have been an economic or efficient alternative to the surplus commercial capacitor.

5.3 Capacitor Limitations

High-voltage capacitors possess many differences to their low-voltage counterparts. Table 5-1 lists the limitations in high-voltage capacitors that can generally be found on the surplus market. Low-voltage capacitors are limited by the following characteristics:

- Rated working voltage in VDC. The applied DC voltage must be less than or equal to the rated working voltage. If a sinusoidal AC voltage is applied to the capacitor the WVDC must be greater than twice the rms voltage × 1.414 or greater than twice the peak voltage. The peak voltage in a sine wave is 1.414 times greater than its rms value. A sine wave has a positive and negative alternation, each reaching this peak value. When the sine wave is alternating at a high frequency these alternations occur almost instantaneously. The rated WVDC is conservative as manufacturers include a safety margin (typically 50%) meaning the WVDC rated limit can actually withstand a 50% higher applied voltage.
- Equivalent Series Resistance (ESR). Although no actual current passes through the dielectric, eddy current and hysteresis losses are produced from oscillating currents, which generate heat. The ESR is the resistance formed in the plates, interconnections, and any additional internal loss mechanisms. The term ESR is therefore used to describe the current handling and power dissipation characteristics of a capacitor. An alternative form of this limitation is to specify a rated rms current limit. If not exceeded the capacitor safely dissipates the $I^2 \times$ ESR power.

High-voltage capacitors have many additional limitations placed on them, which are specific to the type of capacitor. Often the rated voltage and current of the capacitor is very conservative but there is no general rule followed by manufacturers. Consulting the manufacturer datasheet (if available) may provide additional details not covered in the following subsections. Pulse operation is rough on capacitors and further limits the operating characteristics. A worksheet was constructed to evaluate the applied stresses for each type of high-voltage capacitor. General limitations for each type of high-voltage capacitor were identified using available manufacturer data and are considered applicable to similar type capacitors. The calculations for determining the total capacitance and dielectric strength of series–parallel networks detailed in Section 5.2 are contained in each worksheet.

5.3.1 Film/Paper-Foil Oil Filled Capacitor Limitations

High-voltage oil-filled capacitors such as shown in Figure 5-1d use a film or paper dielectric that is vacuum impregnated with an oil (e.g., mineral oil) to increase the dielectric strength and fill imperfections in the material. Power and power factor correction capacitors used in distribution systems are of this type. Identified in reference (4) are the failure mechanisms:

- Voltage stress and voltage reversal: Dielectric stress is proportional to dv/dt, or how much voltage is applied and how fast it is reversed. When the applied voltage is reversed, accumulated charge in the dielectric (space charge) adds to the voltage stress. Leakage currents break down polymer film at a microscopic level.

Capacitor Type	AC Rated Limit	Nominal ESR	Repetitive Peak Current Limit	Rated Life Span	Rated Power Dissipation Limit	Nominal ESL	Off-the-Shelf Capacitance Range
Ceramic Disc (Sprague), Class I KT series. Fig. 5-1a.	Applied frequencies above 50 kHz reduce AC voltage rating by a factor of: $\dfrac{V_{AC}\text{rated}}{\left(\dfrac{f(\text{kHz})}{50\,\text{kHz}}\right)^2}$	$\dfrac{100}{C(\text{pF}) \bullet f(\text{MHz})}$	≤100 pF = 10 kV/μs >100 pF = 5 kV/μs	Equation 5.7	< 250 mW or core < 25°C rise from case	2-34 nH, 50 n–0.8 μH	300 pF to 4,700 pF, 10 kV to 40 kV
Ceramic Disc (Sprague), Class II, III DK series. Fig. 5-1a.		$\dfrac{1}{C(\mu\text{F})} \bullet f(\text{kHz})$	<100 pF = 5 kV/μs 100 pF – 1 nF = 2 kV/μs 1 nF–10 nF = 1 kV/μs >10,000 pF = 500 V/μs	Equation 5.7	< 250 mW or core < 25°C rise from case	1-34 nH, 27 n–0.8 μH	300 pF to 9,300 pF, 15 kV to 40 kV
Mica, RF (Cornell-Dubilier), type F1(271)-F3(273), G1(291)-G4(294), G5. Fig. 5-1b.	Applied peak AC voltage < rated voltage	DF<0.05% @1 kHz = 79577 • C(pF)$^{-1}$ @1 MHz = 79.577 • C(pF)$^{-1}$	< 100 kV/μs	Unlimited pulse operation < 100 kV/μs Guaranteed 10 years	Current limited to (rms) marked on case	No mfg. Spec.	47 pF to 0.1 μF, 1.5 kV to 30 kV
Pulse (Maxwell), MDE Series. Fig. 5-1c.	Applied peak AC voltage < rated voltage	Not Specified	<25 A rms @ 25°C (CW) with <20% voltage reversal	1 × 10^8 to 1 × 10^9 shots = 55.6 to 556 hours @ 500 BPS	Current limited < 25 A rms	20 nH to 90 nH	7 nF–2.0 μF, 5 kV–50 kV
DM Series			<13 A rms to <60 A rms	Not Specified		No mfg. Spec.	25 μF–750 μF, 1 kV–3 kV
C Series			<50 kA peak to <260 kA peak	3 × 10^3 to 2 × 10^5 shots = 0.1 to 6.7 minutes @ 500 BPS		35 nH to 100 nH	1.3 μF–830 μF, 10 kV-100 kV
Metal Case			<5 A rms to <25 A rms	1 × 10^4 to 5 × 10^7 shots = 0.06 to 27.8 hours @ 500 BPS		15 nH to 170 nH	2.0 μF–185 μF, 2 kV–7.5 kV
Power (Montena) Film/Paper-Foil, Oil Filled. Fig. 5-1d.	Applied AC voltage (rms) < rated voltage (rms or DC)	mΩ for <1 μF @ 100 kHz tanδ <0.003 @ 50 Hz, 20°C	<90 A rms derated for applied frequency and ambient temperature	>180,000 hours	Limited to temp, freq derated current value	<650 nH	0.05 μF–4,000 μF, 400 V–50 kV

TABLE 5-1 Limitations for common high-voltage capacitors.

- Corona: When the medium surrounding or between plates in a capacitor becomes ionized, a current flows. This current deteriorates the polymer film and will lead to failure. Corona involves the emission of light (UV), which accelerates breakdown of the polymer. Polyethylene and Teflon are more vulnerable to corona than others (mica). In clear dielectrics (unfilled polyethylene), UV light generated by charge movement has been observed to deteriorate the dielectric.
- Moisture: Moisture absorption accelerates failure. Polycarbonate and polysulfone films absorb moisture better than others (polypropylene and polystyrene).
- Thermal stress: Elevated temperatures breakdown the dielectric.
- Film defects: Surface and interior imperfections, and conductive impurities in the film can cause failure. Oil impregnation will fill the imperfections, offering additional dielectric strength. Many thin layers of film or paper can be used instead of a single thicker layer to reduce film defect failure.

NOTE: *Avoid purchasing capacitors with polychlorinated bi-phenyl (PCB). The PCB presents an environmental hazard and becomes costly when trying to dispose of properly. Oil-filled capacitors containing PCB are not always easy to identify; however, those labeled NON-PCB have been tested and found to contain no PCB.*

CAUTION: *If an oil-filled capacitor short circuits in an operating primary circuit a resulting plasma arc is hot enough to ignite the oil resulting in a fire or the case can explode. Always wear appropriate protection or shield yourself and observers from the capacitor. If using an oil-filled capacitor ensure a fire extinguisher approved for class C (electrical) fires is available.*

From reference (11) is found application and derating data to characterize the electrical stresses and limitations shown in Table 5-1; this is considered equivalent to other manufacturer's film/paper-foil oil filled capacitors of similar design. To evaluate oil-filled film-foil capacitor ratings and applied electrical stresses open the CH_5.xls file, FILM-FOIL OIL CAP worksheet (8). (See App. B.) The stresses are evaluated as follows:

A data plate is usually mounted on this type of capacitor; however, the rated current may not be included. It is often necessary to estimate the rated current for the value of capacitance using whatever manufacturer's data are available. If the manufacturer specifies a current rating enter the rms value in cell (B14). An estimated rms current rating for the selected capacitance value in cell (B5) is provided in cell (E16) using the curve fit formula shown in Chart 1 developed from the data in Table 1 of rows 30 to 79. This estimation is usually low as the voltage rating of the capacitors used in the data was only 7.6 kV. The rated current is at a specified frequency, usually the standard line frequency of 60 Hz (US). This current must be derated for both applied frequency and temperature.

The manufacturer's temperature derating data are contained in Table 2 columns (H) through (J), rows 82 to 91. The data are displayed below in Figure 5-3 and in Chart 2 of the worksheet in columns (A) through (D), rows 82 to 123. Also shown in the figure is the trendline used to extend the derating multiplier to 0°C. The second-order polynomial curve fit formula shown in the figure is used to calculate the temperature derating multiplier in cell (E14) at the applied temperature entered in cell (B25).

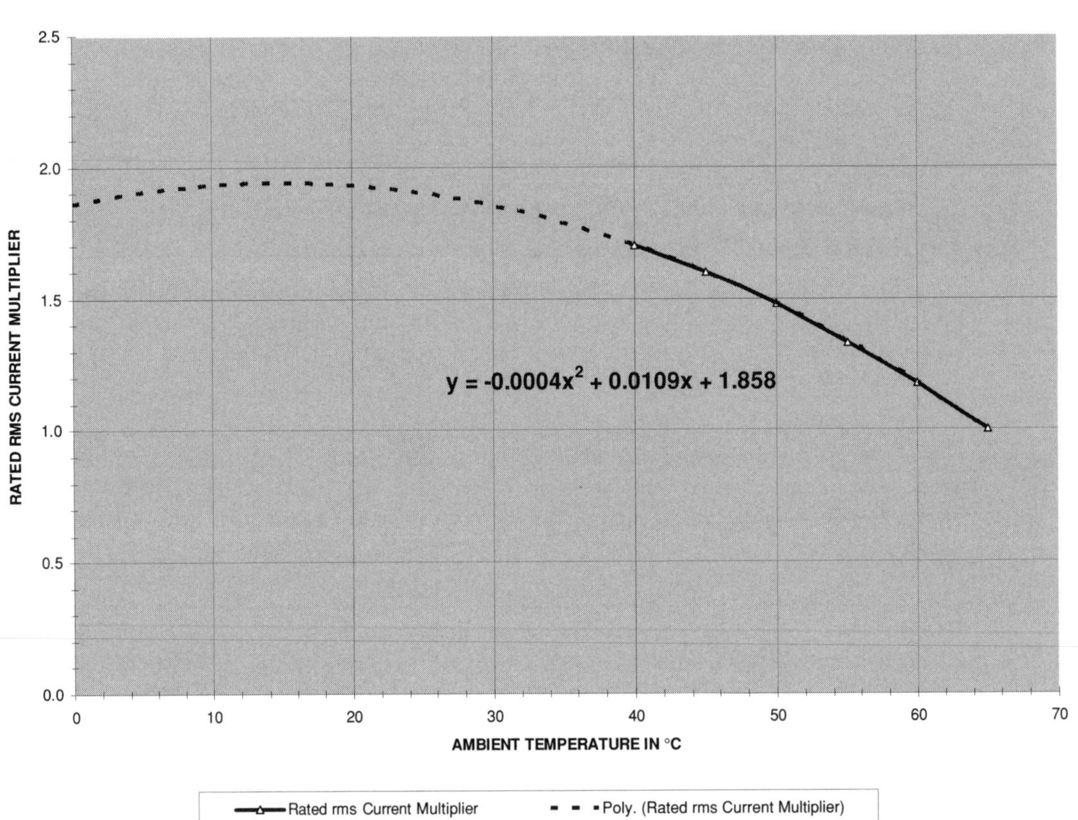

FIGURE 5-3 Current derating for ambient temperature in high-voltage film/paper-foil oil capacitors.

The manufacturer's frequency derating data are contained in Table 3 columns (H) through (J), rows 127 to 139. The data are displayed below in Figure 5-4 and in Chart 3 of the worksheet in columns (A) through (D), rows 127 to 168. Also shown in the figure is the trendline used to extend the derating multiplier to 100 kHz. The power curve fit formula shown in the figure is used to calculate the frequency derating multiplier in cell (E15) at the applied frequency entered in cell (B4).

The rated rms current entered in cell (B14) is derated using the temperature multiplier in cell (E14) and frequency multiplier in cell (E15) to the maximum derated current in cell (E17). This value must be greater than the applied rms primary current calculated in Chapter 2.

This type of capacitor is usually rated by a loss factor ($\tan\delta$). The loss factor is rated by the manufacturer at a maximum value of 0.003 @ 50 Hz for any capacitance value. Both the loss factor and dissipation factor (DF) increase as the applied frequency increases. Table 4 columns (H) through (J), rows 171 to 183, of the worksheet contain the manufacturer's rated loss factor over a 30-Hz to 1,000-Hz frequency range. The data are displayed below in Figure 5-5 and in Chart 4 of the worksheet in columns (A) through (E), rows 171 to 212. Also shown in the figure

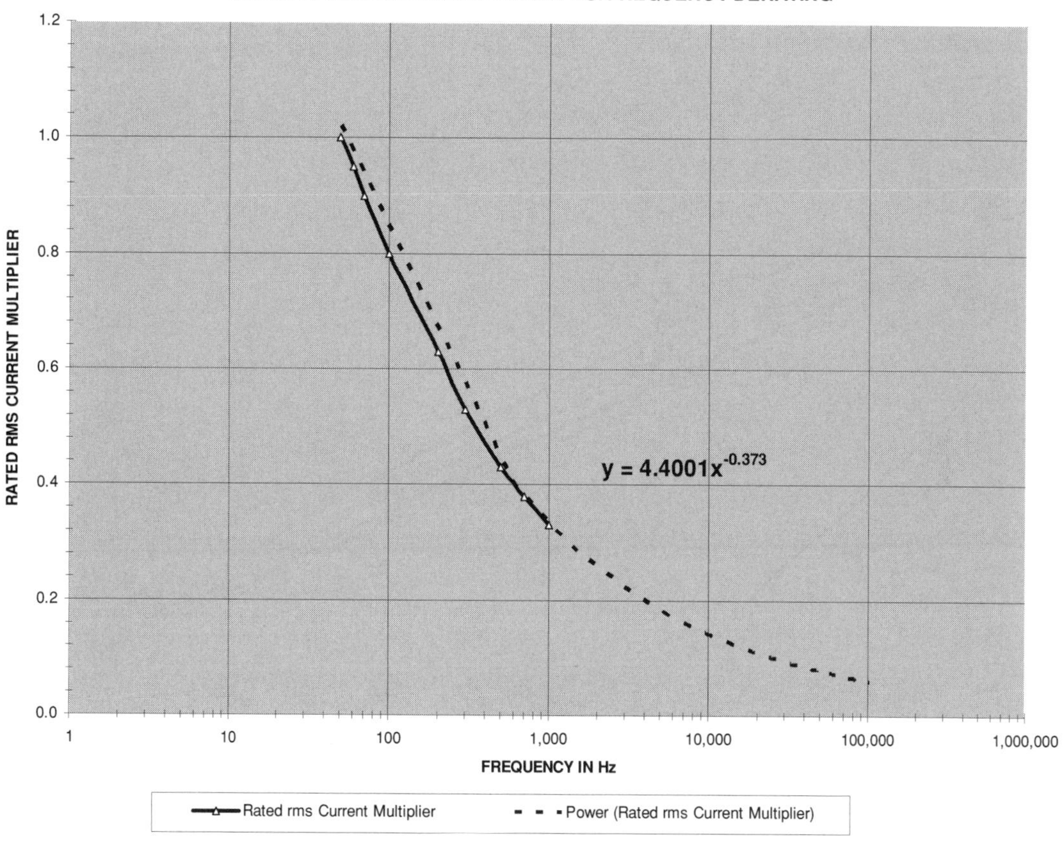

FIGURE 5-4 Current derating for applied frequency in high-voltage film/paper-foil oil capacitors.

is the trendline used to extend the loss factor to 100 kHz. The power curve fit formula shown in the figure is used to calculate the loss factor in cell (E7) at the applied frequency entered in cell (B4). The loss factor in cell (B7) is multiplied by a value of 100 to obtain the more familiar dissipation factor in cell (E8). For polypropylene and paper dielectric the loss factor exhibits negligible change over the operating temperature range of −35°C to 75°C.

The ratio of the dissipation factor (power lost in capacitor) to the reactance (power stored in capacitor) provides the equivalent series resistance value of the capacitor:

$$\mathrm{ESR} = \frac{\mathrm{DF}}{X_C} = \frac{\mathrm{DF}}{2\pi fC} \tag{5.4}$$

Where: ESR = Equivalent series resistance in ohms at applied frequency = cell (E9) calculated value.

DF = Dissipation Factor in percent (figure of merit) = cell (E8), calculated value. DF = $\tan\delta \times 100$. $\tan\delta$ is calculated in cell (E7) using curve fit formula in Figure 5-5.

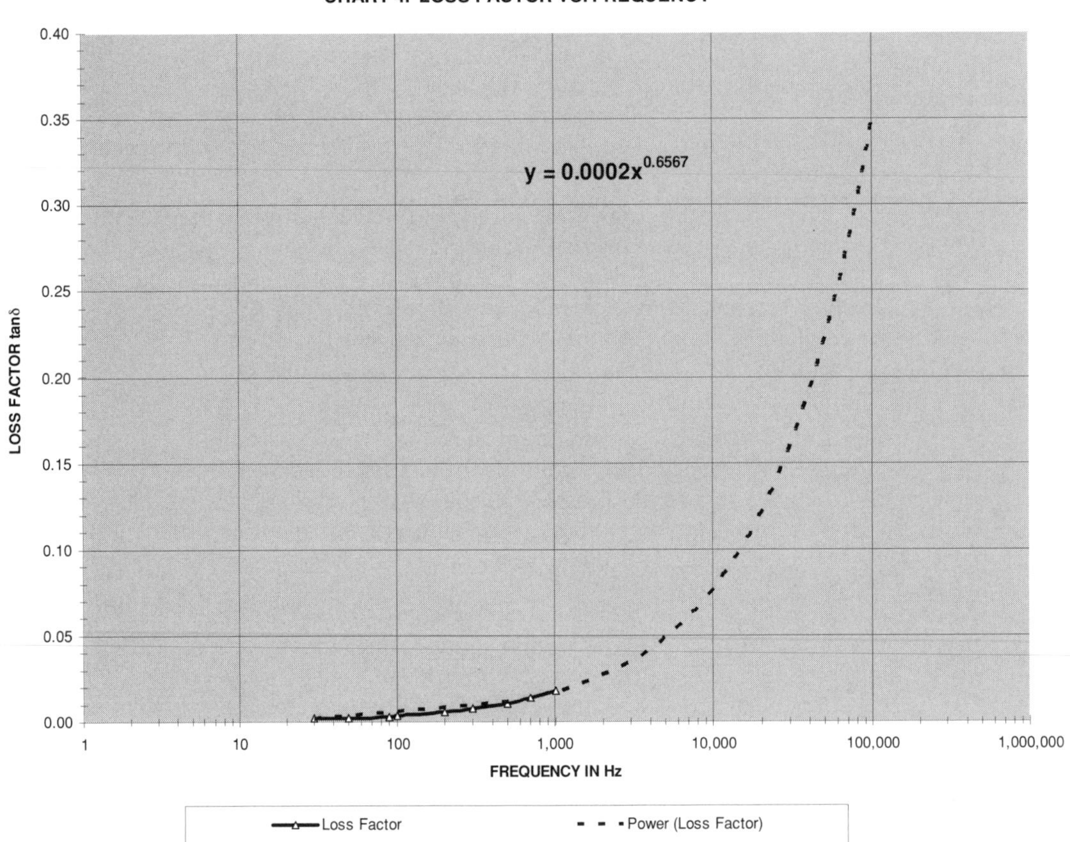

FIGURE 5-5 Typical loss factor vs. frequency in high-voltage film/paper-foil oil capacitors.

C = Total capacitance of network in farads = cell (B5) enter value in μF.
f = Applied frequency in Hz = cell (B4) enter value in kHz.

NOTE: *The total capacitance for the series–parallel network is calculated from values entered in cells (B5) to (B7) using the equations in Section 5.2. As this total capacitance is affected so is the calculated ESR in the worksheet.*

The power dissipated (Pd) in the capacitor's ESR is calculated in cell (E12) from the applied rms current entered in cell (B12): $Pd = \text{Irms}^2 \times \text{ESR}$.

Table 5 columns (H) through (J) and rows 215 to 225 of the worksheet contain the manufacturer's rated life expectancy in hours for the applied-to-rated voltage ratio and hot spot temperatures of 55°C (blue trace) and 75°C (black trace). The data are displayed in Figure 5-6 and in Chart 5 of the worksheet in columns (A) through (E), rows 215 to 256. Also shown in the figure is the trendline used to extend the life expectancy through an applied-to-rated voltage

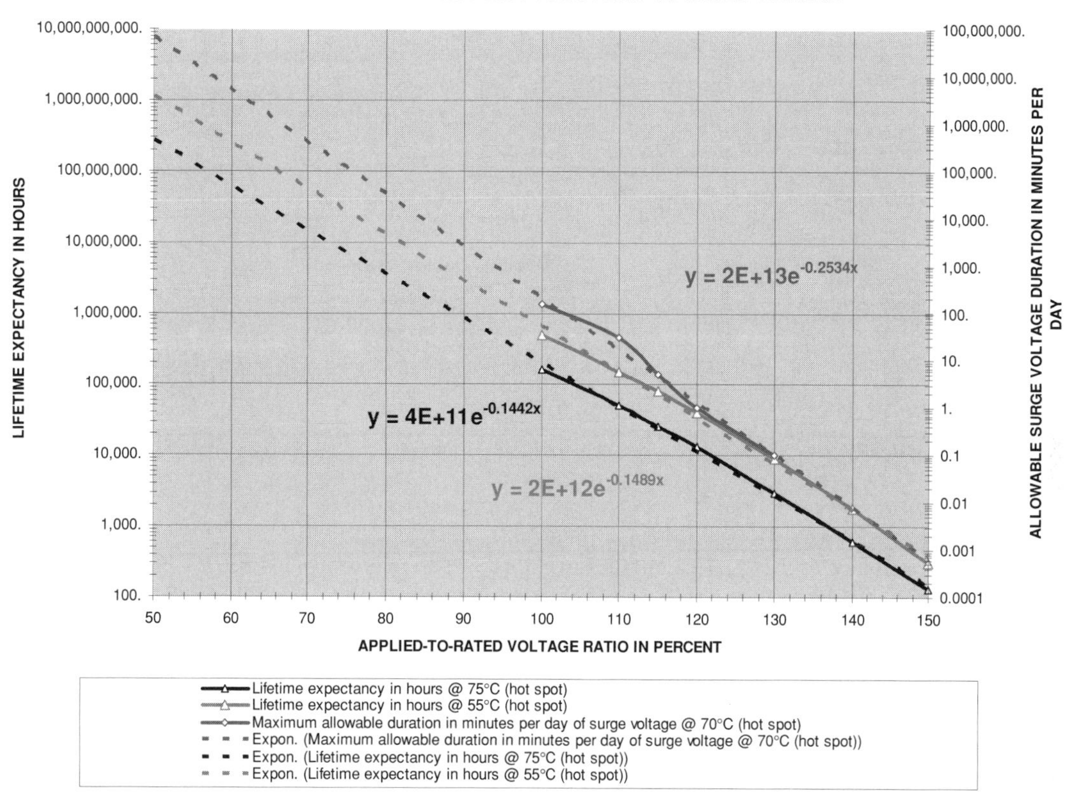

FIGURE 5-6 Life expectancy vs. applied-to-rated voltage in high-voltage film/paper-foil oil capacitors.

ratio of 50%. The exponential curve fit formula shown in the figure for a hot spot temperature of 75°C is used to calculate the life expectancy in hours in cell (E25) at the applied-to-rated voltage ratio calculated in cells (B21) and (B20) respectively. Also shown in the figure is the exponential curve fit formula used to calculate the maximum allowable surge voltage duration in minutes per day in cell (E24) at the applied-to-rated voltage ratio and for a conservative hot spot temperature of @ 70°C. Due to variations in manufacturing process (tolerance) the highest internal temperature rise from I^2R heating (rms current and ESR) is at an arbitrary location within the capacitor known as the hot spot.

The rated voltage of these capacitors is a continuous rating. The rms voltage applied to the capacitor can exceed the rated rms (DC) voltage on the case for allowable periods of time (surge). The allowable surge voltage duration in minutes per day for the applied-to-rated voltage ratio in percent is shown in Figure 5-6.

As shown in the example calculations in Figure 5-7 the tank capacitance of 0.043 μF entered in cell (B5) is chosen for use in a primary circuit resonant to 100 kHz entered in cell (B4). A (1) entered in cells (B6) and (B7) indicates no additional series–parallel arrangement for a total

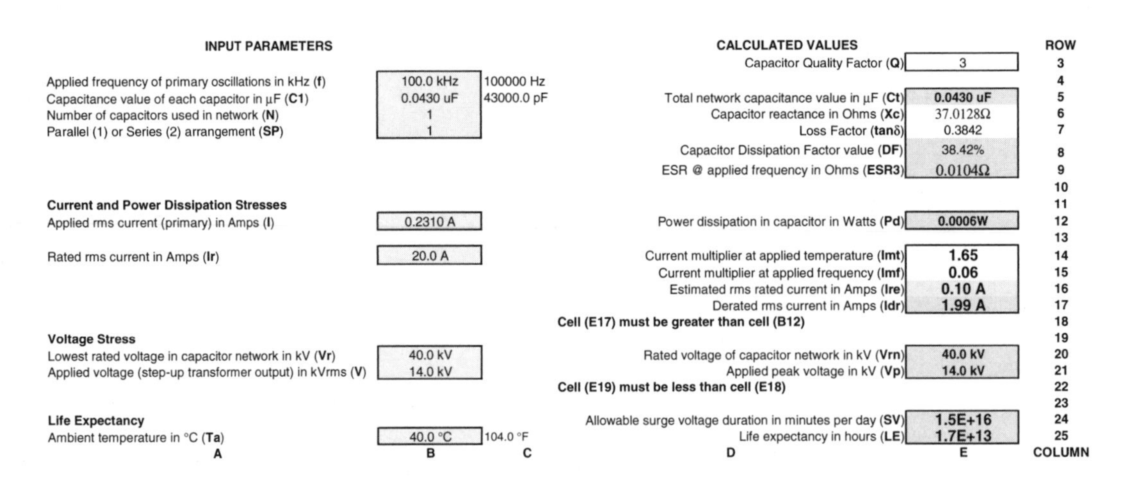

INPUT PARAMETERS			CALCULATED VALUES		ROW
			Capacitor Quality Factor (Q)	3	3
Applied frequency of primary oscillations in kHz (f)	100.0 kHz	100000 Hz			4
Capacitance value of each capacitor in µF (C1)	0.0430 uF	43000.0 pF	Total network capacitance value in µF (Ct)	0.0430 uF	5
Number of capacitors used in network (N)	1		Capacitor reactance in Ohms (Xc)	37.0128Ω	6
Parallel (1) or Series (2) arrangement (SP)	1		Loss Factor (tanδ)	0.3842	7
			Capacitor Dissipation Factor value (DF)	38.42%	8
			ESR @ applied frequency in Ohms (ESR3)	0.0104Ω	9
					10
Current and Power Dissipation Stresses					11
Applied rms current (primary) in Amps (I)	0.2310 A		Power dissipation in capacitor in Watts (Pd)	0.0006W	12
					13
Rated rms current in Amps (Ir)	20.0 A		Current multiplier at applied temperature (Imt)	1.65	14
			Current multiplier at applied frequency (Imf)	0.06	15
			Estimated rms rated current in Amps (Ire)	0.10 A	16
			Derated rms current in Amps (Idr)	1.99 A	17
			Cell (E17) must be greater than cell (B12)		18
Voltage Stress					19
Lowest rated voltage in capacitor network in kV (Vr)	40.0 kV		Rated voltage of capacitor network in kV (Vrn)	40.0 kV	20
Applied voltage (step-up transformer output) in kVrms (V)	14.0 kV		Applied peak voltage in kV (Vp)	14.0 kV	21
			Cell (E19) must be less than cell (E18)		22
					23
Life Expectancy			Allowable surge voltage duration in minutes per day (SV)	1.5E+16	24
Ambient temperature in °C (Ta)	40.0 °C	104.0 °F	Life expectancy in hours (LE)	1.7E+13	25
A	B	C	D	E	COLUMN

FIGURE 5-7 Electrical stress and life expectancy calculation worksheet for high-voltage film/paper-foil oil capacitors.

tank capacitance of 0.043 μF calculated in cell (E5). The maximum loss factor (tanδ) of 0.384 at 100 kHz is calculated in cell (E7) with a calculated dissipation factor (DF) of 38.4% in cell (E8), reactance of 37 Ω in cell (E6) and ESR of 0.0104 Ω in cell (E9). The rated rms current of 20 A entered in cell (B14) is derated to 1.99 A in cell (E17) using the applied temperature multiplier of 1.55 in cell (E14) at the 40°C entered into cell (B25), and the applied frequency multiplier of 0.06 in cell (E15) at the 100 kHz entered into cell (B4). The applied rms current entered in cell (B12) produces power dissipation in the ESR calculated in cell (E9), which is the 0.0006 W shown in cell (E12). The calculated lifespan of 1.7×10^{13} hours in cell (E25) and allowable surge voltage duration in cell (E24) is dependent on the applied-to-rated voltage ratio. The applied voltage of 14 kV rms entered into cell (B21) and the rated voltage of 40 kV rms entered into cell (B20) specifies the applied-to-rated voltage ratio. There are no electrical overstresses for this application and the reduced applied-to-rated voltage ratio results in a long-lasting capacitor.

5.3.2 Ceramic Capacitor Limitations

High-voltage ceramic capacitors generally use a deposited metallic layer to construct the plates on both sides of a ceramic disc. This type of plate construction does not enable high rms operating currents. The equivalent series resistance (ESR) is generally higher when compared to other high-voltage capacitors, which results in higher power dissipation with more internal heating. These capacitors tend to overheat as a result of the power dissipation and the capacitors often fail. Identified in reference (4) is the principal failure mechanism:

- Mechanical: Because ceramic is more brittle than other dielectrics, assembly, handling, soldering, and thermal cycling can produce mechanical stresses and failure. The dimensions of the capacitor can be changed to decrease mechanical stress.

To evaluate ceramic capacitor ratings and applied electrical stresses open the CH_5.xls file, CERAMIC CAP worksheet (9). (See App. B.) Reference (2) contains the manufacturer's data used to construct the worksheet and is considered representative for capacitors of this type. The stresses are evaluated as follows:

The nominal capacitance and rated DC voltage are usually printed on the case of these capacitors. An example of a ceramic capacitor (hockey puck) is shown in Figure 5-1a.

The total ESR of the network is calculated in cell (E8) using the formula provided in Table 5-1 for Class I or Class II and III types. The calculated ESR for a single capacitor is multiplied by the number of capacitors entered in cell (B8) for a series network. The calculated ESR for a single capacitor is divided by the number of capacitors entered in cell (B8) for a parallel network. The resulting power dissipation in the ESR is:

$$\text{PD} = \text{Irms}^2 \bullet \text{ESR} \tag{5.5}$$

Where: PD = Power dissipation of capacitor ESR in watts = cell (E16), calculated value. Converted to mW using a 1e3 multiplier.

Irms = RMS current in tank circuit in amps = cell (B16), enter value. Methodology for calculating the primary rms current value is detailed in Chapter 2. Enter the rated rms current value for a current-limited transformer if the transformer limits the current below the calculated value in Chapter 2.

ESR = Equivalent series resistance value in ohms = calculated value in cell (E8). The total series–parallel network ESR is calculated from values entered in cells (B7) to (B9) as follows:

$$\text{ESR} = \text{ESR1} \bullet N \text{ for series network or ESR} = \frac{\text{ESR1}}{N} \text{ for parallel network}$$

Where: ESR1 = equivalent series resistance of each capacitor in network $= \frac{100}{C(\text{pF}) \bullet f(\text{MHz})}$ for Class I or $\frac{1}{C(\mu\text{F}) \bullet f(\text{kHz})}$ for Class II, III.

N = Number of capacitors in network = cell (B8), enter value.

NOTE: *enter 1 for parallel or 2 for series arrangement in cell (B9).*

There is no rms current limit specified for the ceramic capacitor; however, there is a power dissipation limitation of 250 mW specified for small case sizes. This dissipation limit will be conservative for larger case sizes. The internal thermal rise or hot spot temperature of the core (TC) is also limited to 25°C rise from case temperature. Since the thermal impedance (θ) is unknown the temperature limit is of little use and will generally remain within limits if the 250-mW dissipation limit is not exceeded. For the applied rms current (I) entered in cell (B16) the internal power dissipation (Pd) in mW for the capacitor network is calculated in cell (E16) from: $\text{Pd} = \text{I}^2 \times \text{ESR}$. If the manufacturer's data include a thermal resistance rating the temperature rise in the core can be estimated using the calculated power dissipation in the ESR: $\text{TC} = (\theta \times \text{PD}) + \text{ambient (case) temperature}$.

The manufacturer specified self-resonant frequency is less than 50 MHz for small diameters and 10 MHz for large diameter capacitors. The equivalent series inductance (ESL) can be estimated from this specification:

$$ESL = \frac{1}{4\pi^2(fsr^2Ct)} \tag{5.6}$$

Where: ESL = Equivalent Series Inductance of capacitor network in henries = calculated value in cell (E11). Converted to nH using a 1e9 multiplier.
fsr = Manufacturer specified self-resonant frequency of 50 MHz for small diameters (enter 1 in cell B11) and 10 MHz for large diameter capacitors (enter 2 in cell B11).
Ct = Total capacitance of network = cell (E6), calculated value in pF. Converted to μF in cell (C6) using a 1e-6 multiplier. Calculates total series–parallel network capacitance from values entered in cells (B7) to (B9) using equations in Section 5.2.

As the applied frequency of these capacitors increases above 50 kHz the rated DC voltage limit must be derated to a lower value. The formula for determining the AC rated limit in these capacitors is provided in Table 5-1 for Class I, II, and III types. The derated AC voltage limit is calculated in cell (E20).

When repetitive peak currents as found in primary oscillations are applied to the capacitor the rate of change or *dv/dt* must be limited. The repetitive peak current limits are provided in Table 5-1 for each type. The rate of change is calculated:

$$\frac{\Delta v}{\Delta t} = \frac{Vp}{\frac{\left(\frac{1}{f}\right)}{1e\text{-}6}} \tag{5.7}$$

Where: $\Delta v/\Delta t$ = dv/dt = applied rate of change in volts per μsec (V/μs) = calculated value in cell (E14).
f = Applied frequency of operation in Hz = cell (B6), enter value in kHz. Converted to Hz in cell (C6) using a 1e3 multiplier.
Vp = Applied peak voltage in kV = cell (E21), calculated value. Peak applied voltage is rms voltage output of step-up transformer entered in cell (B21) × 1.414.

Reference (4) is used to estimate the life of ceramic capacitors at the applied voltage and temperature using the manufacturer's rated voltage and temperature limits:

$$LC = \left(\frac{Vr}{Vp}\right)^3 \bullet \left(\frac{TR}{TC}\right)^8 \tag{5.8}$$

Where: *LC* = Lifespan multiplier at applied temperature and voltage = cell (E24), calculated value.
Vr = Rated voltage of capacitor network = cell (E20), calculated value.
Vp = Applied peak voltage in kV = cell (E21), calculated value. Peak applied voltage is rms voltage output of step-up transformer entered in cell (B21) × 1.414.
TR = Rated temperature limit in °C = cell (B24), enter value.
TC = Estimated core temperature of capacitor (hot spot) in °C = cell (B25), enter value.

The life expectancy multiplier for ceramic capacitors is calculated in cell (E24) using equation (5.8). The multiplier indicates how many times longer the capacitor will last over the rated lifetime for the applied conditions of temperature and voltage. The accuracy of the calculated multipliers is highly dependent upon the estimated core temperature (TC) of the capacitor entered in cell (B25).

As an example, four 4,700 pF, 40-kV Sprague DK (Class III) ceramic capacitors are selected for use with a 15,000 V neon sign transformer. This results in a total circuit capacitance of 4 × 4,700 pF = 0.018 μF calculated in cell (E6) and a peak applied voltage of 15 kVrms × 1.414 = 21.2 kV calculated in cell (E21). The primary resonant frequency of 250 kHz has a time period (dt) of 1/250 kHz = 4.0 μs calculated in cell (E13). The applied *dv/dt* of 21.2 kV/4.0 μs = 5,303 V/μs shown in cell (E14) of the worksheet. The peak current limit in Table 5-1 for 4,700-pF ceramic disc capacitors of this type is 1 kV/μs. The manufacturer's limit has been exceeded by a factor of >5. Does this mean the capacitor will fail? When it only lasts for two minutes of coil firing time don't be too surprised. Let's examine the failure in detail.

Each 4,700-pF capacitor has an ESR = 1/(0.0047 μF × 250 kHz) = 1/(0.0047 × 250) = 0.851 Ω. The caps were wired in parallel to obtain a higher capacitance, so the total ESR is 0.851 Ω/4 = 0.213 Ω calculated in cell (E8). The tank circuit RMS current is applied to this ESR value. The RMS value of current for AC and pulse waveforms can be calculated using the methodology in Section 10.5 or in Chapter 2. For a Tesla coil the RMS value of primary current is equal to the RMS output current of the step-up transformer. The RMS current output for the example neon sign transformer is internally limited to 30 mA. The power dissipation of the capacitor's ESR is 30 mA^2 × 0.213 Ω = 0.2 mW calculated in cell (E16). The only power dissipation rating available from the manufacturer is for a much smaller ceramic disc. A conservative 250 mW allowable power dissipation limit is therefore selected. The 0.2 mW is below the conservative 250 mW allowable power dissipation limit, so we are safe on this limitation.

The rated DC voltage limit must be derated for applied AC voltages above 50 kHz. This is calculated in cell (E20) using the formula in Table 5-1. The 40-kV DC voltage limit is derated to an AC voltage limit of 1.6 kV at the applied resonant frequency of 250 kHz. The applied peak voltage of 21.2 kV exceeds the derated AC voltage limit of 1.6 kV by a factor of 13 and the capacitor network will undoubtedly fail. The calculated life expectancy multiplier of 0.0015 in cell (E24) also indicates the capacitor life at the applied voltage is reduced to 1/667 the nominal life span. The calculations are summarized in Figure 5-8.

Ceramic capacitors are a poor choice for tank capacitors in a Tesla coil unless a series arrangement is used to keep the applied peak voltage within the derated AC voltage limit. Also, the current-handling capability is often exceeded with series arrangements. The series arrangement will result in a very low capacitance value if each series element does not contain several parallel capacitors. A satisfactory ceramic capacitor network using 4,700-pF values would require 14 parallel elements, each element comprised of 4 capacitors in series. Each of these series elements would reduce the four 4,700-pF capacitors to 336 pF, requiring 14 series elements in parallel to bring the total capacitance back to a usable 0.018 μF. This network requires a total of 56 capacitors of 4,700 pF to eliminate premature failure.

5.3.3 Mica Capacitor Limitations

High-voltage mica capacitors are typically used in radio frequency (RF) oscillating circuits and are a good choice for tank capacitors, especially in high rms current vacuum tube coil

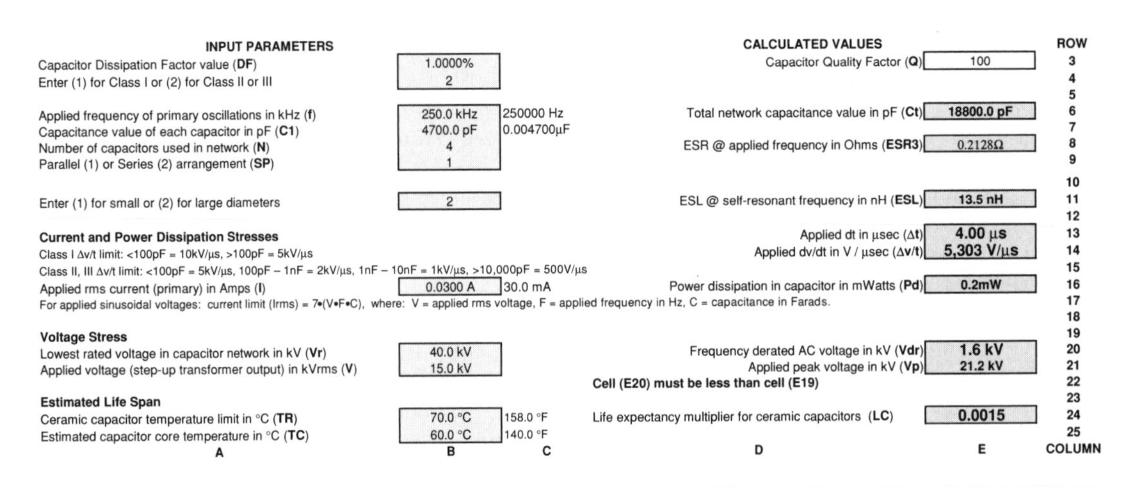

FIGURE 5-8 Electrical stress and life expectancy calculation worksheet for high-voltage ceramic capacitors.

applications. Two typical mica capacitor types (F and G) are illustrated in Figure 5-1b. A data plate is typically mounted on the case specifying the maximum rated peak voltage and rms current that can be applied to the capacitor and the frequency corresponding to the rated values (typically 1 MHz). From reference (8) is found the manufacturer specifications for high-voltage mica capacitors and is considered equivalent to other manufacturer's mica capacitors of similar design. To evaluate the electrical stresses in high-voltage mica capacitors open the CH_5.xls file, MICA CAP worksheet (10). (See App. B.) The stresses are evaluated as follows:

The rated rms current marked on the case must be derated to the applied frequency. As long as the derated rms current limit at the applied frequency is not exceeded the capacitor will be able to safely dissipate the power developed in its equivalent series resistance (ESR). Table 3 columns (H) through (Q) and rows 106 to 113 of the worksheet contain the manufacturer rated current limit for several capacitors over a 100-kHz to 3-MHz frequency range. The data are displayed below in Figure 5-9 and in Chart 3 of the worksheet in columns (A) through (D), rows 106 to 145. As seen in the figure the rated current at 100 kHz is approximately one half the rated current at 1 MHz, which is our area of interest. From 1 MHz to 3 MHz no comprehensive relationship exists for all capacitors; however, this is outside our area of interest. For the 100 kHz to 1 MHz range the current doubles over a frequency decade or a doubling per decade relationship. The rms current limit at a specific frequency over the 100-Hz to 1-MHz frequency range can therefore be estimated:

$$Imax = Irated \bullet 2^{\left(\frac{f-If}{10}\right)} \tag{5.9}$$

Where: $Imax$ = Derated rms current limit in amps at applied frequency = cell (E13), calculated value.

$Irated$ = Rated rms current limit in amps at specified frequency (If) = cell (B13), enter value.

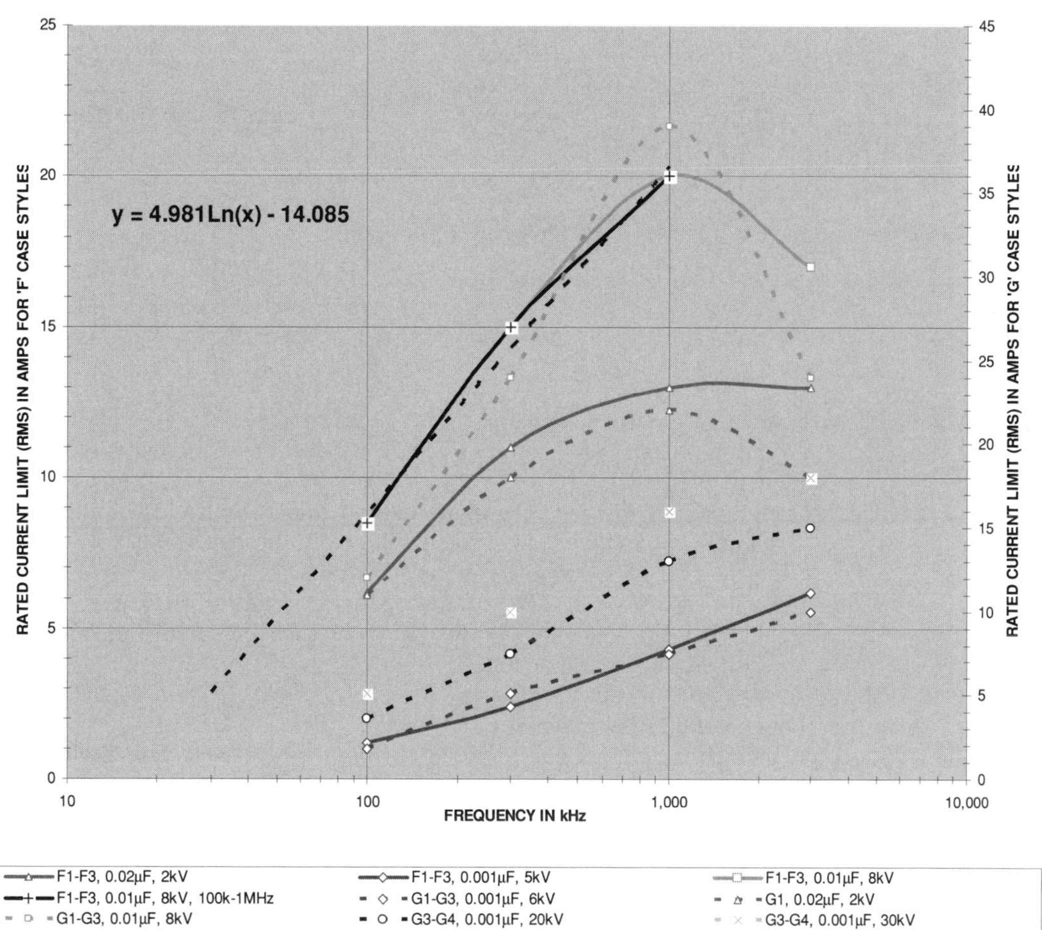

FIGURE 5-9 Typical RMS current limit for applied frequency in high-voltage mica capacitors.

f = Applied frequency of operation in Hz = cell (B4), enter value in kHz. Converted to Hz in cell (C4) using a 1e3 multiplier.

If = Specified frequency at manufacturer's rated rms current limit in Hz = cell (B14), enter value in kHz. Converted to Hz in cell (C14) using a 1e3 multiplier.

The logarithmic curve fit formula shown in Figure 5-9 is used to calculate the derated rms current limit in cell (E14) at the applied frequency entered in cell (B4).

As this is only an estimation, a more accurate method is to interpret the maximum current from the manufacturer data sheets on the Cornell-Dubilier website: http://www.cornell-dubilier.com/catalogs/4.025-4.026.pdf and http://www.cornell-dubilier.com/catalogs/4.027-4.029.pdf. If the datasheets are not obtainable the estimation is still quite close and the resulting calculations are considered to be conservative.

The manufacturer specifies the Dissipation Factor (DF) to be less than 0.05%. The ESR can be calculated from this maximum DF:

$$\mathrm{ESR} = \frac{\mathrm{Pd}}{\mathrm{Ps}} = \frac{\mathrm{DF}}{X_C} = \frac{\mathrm{DF}}{2\pi fC} \tag{5.10}$$

$$Q = \frac{1}{\mathrm{DF}} = co\tan\delta \quad \text{and} \quad \mathrm{DF} = \frac{1}{Q} = \tan\delta \bullet 100$$

Where: ESR = Equivalent series resistance in ohms at 1 kHz for >1,000 pF = cell (E6), or ESR in ohms at 1 MHz for ≤1,000 pF = cell (E7), calculated value in Table 1, column (I).
DF = Manufacturer's specified Dissipation Factor in percent (figure of merit) = cell (B3), enter value. For Cornell-Dubilier mica caps DF < 0.05%.
C = Capacitance in pF = Table 1, column (H) values.
f = 1 kHz for capacitance >1,000 pF, 1 MHz for capacitance ≤1,000 pF.
Q = Quality factor of capacitor (figure of merit) = cell (E3), calculated value.
Pd = Power dissipated in capacitor ESR.
Ps = Power stored in capacitor.
$\tan\delta$ = Tangent of the loss angle (δ).

The term $\tan\delta$ is defined in references (9) and (10) as the tangent of the loss angle (δ) in a reactive component. In an ideal capacitor the current will lead the voltage by 90°. In a real capacitor the current will lead the voltage not by 90° but some angle less than 90° due to the distributed resistive effects. These effects are lumped together as an ESR and are comprised of: friction caused from electric field interaction with the dielectric (absorption) which decreases as frequency increases, and resistive losses from the plates, leads, and connections which increase with frequency. The angle the current leads voltage in a real capacitor is Ø, and its compliment is δ, the loss angle. The loss angle is the component that interacts with the ESR to produce power loss and heat. Thus the lower the ESR or δ, the lower the stress on the capacitor.

Table 1, columns (H) through (J), rows 19 to 47, calculate the ESR values in column (J) for the capacitance values in column (H) using equation (5.10). The Table 1 data are shown in Figure 5-10 and in Chart 1 of the worksheet in columns (A) through (D), rows 18 to 61. The power curve fit equations shown in the figure calculate the ESR at 1 kHz (black trace) in cell (E6) and 1 MHz (red trace) in cell (E7) for the total network capacitance calculated in cell (E5). A large [X] indicates the ESR at the applied frequency and selected capacitance value. A linear interpolation can be performed for any frequency in between the 1 kHz and 1 MHz range entered in cell (B4) using the formula:

$$\mathrm{ESR3} = \left(\frac{\mathrm{ESR1} - \mathrm{ESR2}}{1\,\mathrm{kHz} - 1\,\mathrm{MHz}}\right) \bullet (f - 1\,\mathrm{MHz}) + \mathrm{ESR\,1} \tag{5.11}$$

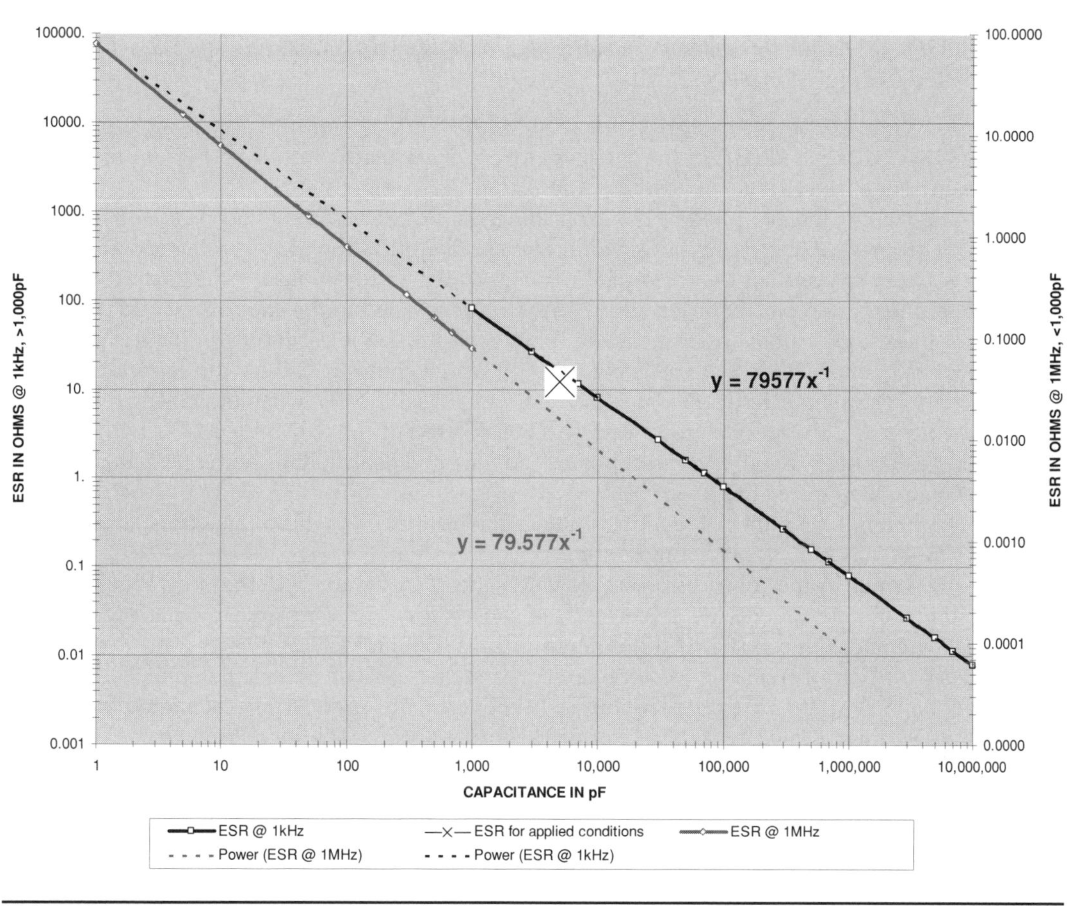

FIGURE 5-10 Typical ESR frequency response in high-voltage mica capacitors.

Where: ESR3 = Equivalent series resistance in ohms at applied frequency = cell (E8), calculated value.

ESR1 = Equivalent series resistance in ohms at 1 kHz = cell (E6), calculated value using power curve fit formula: $79577 \bullet Ct^{-1}$

ESR2 = Equivalent series resistance in ohms at 1 MHz = cell (E7), calculated value using power curve fit formula: $79.577 \bullet Ct^{-1}$

Ct = Total capacitance of network = cell (E5), calculated value in μF. Converted to pF in cell (F5) using a 1e6 multiplier. Calculates total series–parallel network capacitance from values entered in cells (B5) to (B7) using equations in Section 5.2.

f = Applied frequency of operation in Hz = cell (B4), enter value in kHz. Converted to Hz in cell (C4) using a 1e3 multiplier.

NOTE: *The total capacitance for the series–parallel network is calculated from values entered in cells (B5) to (B7) using the equations in Section 5.2. As this total capacitance is affected so is the calculated ESR in the worksheet.*

The power dissipated (Pd) in the capacitor's ESR is calculated in cell (E11) from the applied rms current in cell (B11): $Pd = \text{Irms}^2 \times \text{ESR}$.

The peak voltage applied to the capacitor must not exceed the rated peak voltage on the case. The mica capacitors are designed to withstand application of AC voltages with rms values equal to or less than the rated peak voltage on the case *at frequencies less than 100 Hz*. The line frequency is typically 60 Hz and the rms value applied to the capacitor could be as high as the peak rated value; however, as the spark gap ionizes the oscillating voltage is at a much higher frequency (100 kHz to 500 kHz typical), which could overstress the capacitor causing permanent damage. The applied peak voltage (Vp) in cell (E18) is calculated using the applied rms voltage (V, output of step-up transformer) entered in cell (B18): $Vp = V \times 1.414$. The applied peak voltage in cell (E18) must not exceed the rated peak voltage of the capacitor network (Vrn) calculated in cell (E17).

An unlimited number of repetitive pulses can be applied to these capacitors; therefore the life span requires no application derating.

As shown in the tube coil example calculations in Figure 5-11 the individual tank capacitance of 0.01 μF entered in cell (B5) is chosen for use in a primary circuit resonant to 250 kHz entered in cell (B4). A (2) entered in cells (B6) and (B7) indicates a network of two 0.01-μF capacitors in series for a total tank capacitance of 0.005 μF calculated in cell (E5). The ESR of 15.915 Ω at 1 kHz calculated in cell (E6) and 0.016 Ω at 1 MHz calculated in cell (E7) is linearly interpolated to 11.953 Ω at 250 kHz in cell (E8). The rated rms current of 4 A @ 1 MHz entered in cells (B13) and (B14) is derated to 2.4 A @ 250 kHz in cell (E13) using the doubling per decade from equation (5.9). The applied rms current entered in cell (B11) produces power dissipation in the calculated ESR in cell (E8) of 27 W shown in cell (E11). The applied voltage of 2.5 kV rms entered into cell (B18) results in a peak voltage stress of 3.5 kV

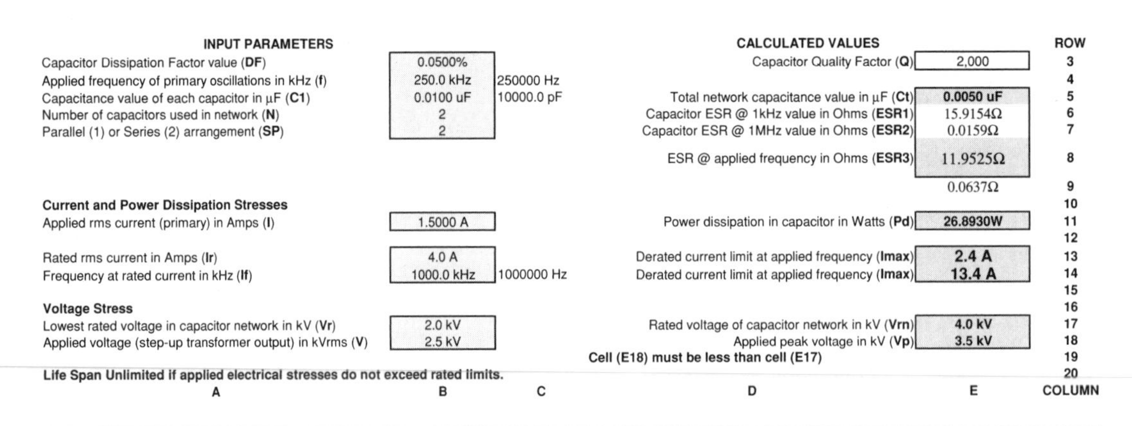

INPUT PARAMETERS			CALCULATED VALUES		ROW
Capacitor Dissipation Factor value (DF)	0.0500%		Capacitor Quality Factor (Q)	2,000	3
Applied frequency of primary oscillations in kHz (f)	250.0 kHz	250000 Hz			4
Capacitance value of each capacitor in µF (C1)	0.0100 uF	10000.0 pF	Total network capacitance value in µF (Ct)	0.0050 uF	5
Number of capacitors used in network (N)	2		Capacitor ESR @ 1kHz value in Ohms (ESR1)	15.9154Ω	6
Parallel (1) or Series (2) arrangement (SP)	2		Capacitor ESR @ 1MHz value in Ohms (ESR2)	0.0159Ω	7
			ESR @ applied frequency in Ohms (ESR3)	11.9525Ω	8
				0.0637Ω	9
Current and Power Dissipation Stresses					10
Applied rms current (primary) in Amps (I)	1.5000 A		Power dissipation in capacitor in Watts (Pd)	26.8930W	11
					12
Rated rms current in Amps (Ir)	4.0 A		Derated current limit at applied frequency (Imax)	2.4 A	13
Frequency at rated current in kHz (If)	1000.0 kHz	1000000 Hz	Derated current limit at applied frequency (Imax)	13.4 A	14
					15
Voltage Stress					16
Lowest rated voltage in capacitor network in kV (Vr)	2.0 kV		Rated voltage of capacitor network in kV (Vrn)	4.0 kV	17
Applied voltage (step-up transformer output) in kVrms (V)	2.5 kV		Applied peak voltage in kV (Vp)	3.5 kV	18
			Cell (E18) must be less than cell (E17)		19
Life Span Unlimited if applied electrical stresses do not exceed rated limits.					20
A	B	C	D	E	COLUMN

FIGURE 5-11 Electrical stress and life expectancy calculation worksheet for high-voltage mica capacitors.

in cell (E18). The lowest rated capacitor voltage of 2 kV rms entered into cell (B17) results in a rated voltage for the capacitor network of 4 kV in cell (E17), which is greater than the applied peak voltage of 3.5 kV. With no electrical overstresses a relatively unlimited lifespan can be expected.

5.3.4 Plastic Pulse Capacitor Limitations

Reference (3) was used to characterize the electrical stress and life span characteristics of plastic pulse capacitors and is considered applicable to any similar type capacitors. Unlike energy-storage capacitors that are designed for DC voltages, charge quickly, and store a charge for long time periods, the pulse capacitor is designed to work with AC voltages (pulsed), charge much more slowly (typically 60 seconds), and do not hold the charge as long as energy-storage capacitors (typically 30 seconds). To evaluate plastic pulse capacitor ratings and applied electrical stresses open the CH_5.xls file, PLASTIC PULSE CAP worksheet (11). (See App. B.) Consult the manufacturer's ratings if they are available or Table 5-1 for general limitations.

As the primary capacitance, step-up transformer output voltage, or primary resonant frequency is increased, the larger the peak current becomes. The manufacturer often limits the applied peak current to a rated value. To calculate the applied peak current in the capacitor:

$$Ip = C\frac{dv}{dt} \tag{5.12}$$

Where: Ip = di = Applied peak current in capacitor when used in a pulsed circuit = cell (E9), calculated value.

dv = Applied peak voltage to the capacitor = cell (E7), calculated value = RMS applied voltage from value entered in cell (B12) × 1.414 for sinusoidal waveform peak value.

dt = Time period of tank oscillations = cell (E6), calculated value = 1/resonant frequency in Hz from cell (B13), enter value in kHz. Converted to Hz using a 1e3 multiplier.

C = Total capacitance value of network in farads = cell (E4), calculated value. Converted to μF using a 1e-6 multiplier. Calculates total series–parallel network capacitance from values entered in cells (B4) to (B6) using equations in Section 5.2.

When AC or pulsed waveforms are applied to capacitors another factor should be evaluated. High-voltage capacitors are constructed using many stacked plates and although the plate interconnection resistances are in parallel with each other they can accumulate into a substantial equivalent series resistance (ESR). Mica, film/paper-foil, and pulse capacitors generally use metal foils for the plates with soldered/welded connections as opposed to the metallic deposits used in ceramic capacitors. As the ratio of height-to-diameter of the capacitor increases, so does the rms current limit (e.g., a taller capacitor has a higher current handling capability). Considering the pulse capacitor's ESR and internal power dissipation, pulse capacitor operation is generally limited only by the rated rms current. The ESR is nice to know for evaluating the effects of circuit parasitics, but not essential in this case. As long as the rated rms current is not exceeded the capacitor can safely dissipate any power developed in the ESR. The ESR in pulse capacitors is typically very low allowing high rms current ratings as shown in Table 5-1.

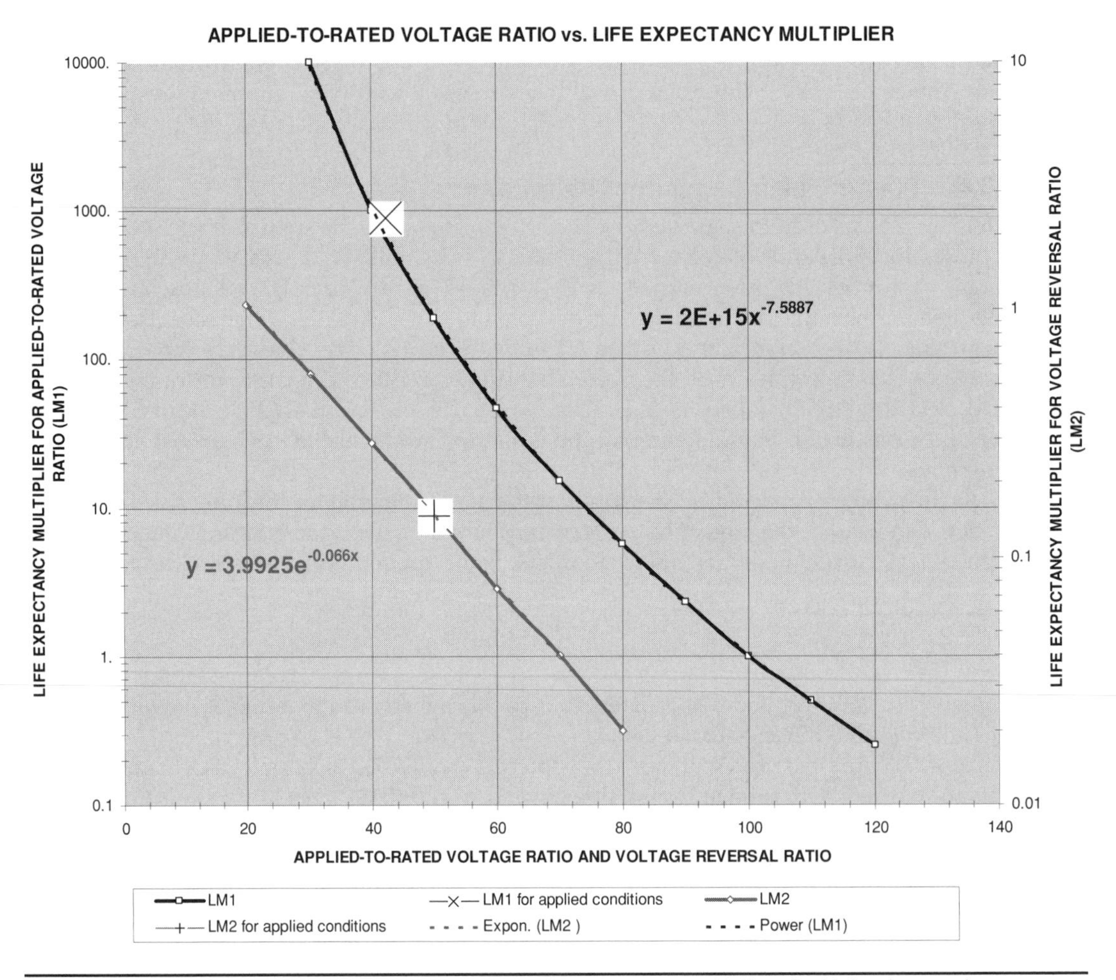

FIGURE 5-12 Curve fit formulae derivation of Maxwell life expectancy multipliers for high-voltage pulse capacitors.

To prolong the life of the capacitor, decrease the voltage applied to it (lower step-up transformer output voltage), or use a capacitor with a higher voltage rating. Application data from reference (3) are shown in Table 5-2. In the example worksheet calculations shown in Figure 5-13 an MDE series capacitance of 0.06 μF entered in cell (B4) with a rated life expectancy of 1×10^8 charge/discharge cycles (from Table 5-1) entered in cell (B9) is used in a Tesla coil design with a rotary spark gap break rate of 500 BPS entered in cell (B15). A (1) entered in cells (B5) and (B6) indicates no additional series–parallel arrangement for a total tank capacitance of 0.06 μF calculated in cell (E4). The applied charge/discharge cycle rate for a tank capacitor in a Tesla coil is equal to the break per second (BPS) rate of the spark gap, therefore the rated life expectancy $= 1 \times 10^8 \div 500$ BPS $= 200{,}000$ seconds, or 55.6 hours calculated in cell (E13). The capacitor voltage rating is 40 kV and the rotary spark gap was adjusted to ionize

Applied Voltage – to – Rated Voltage Ratio vs. Life Expectancy Multiplier		Voltage Reversal Ratio (%) vs. Life Expectancy Multiplier	
Applied Voltage – to – Rated Voltage Ratio (%)	Life Expectancy Multiplier	Voltage Reversal Ratio (%)	Life Expectancy Multiplier
30	10,000	20	1.05
40	1,000	30	0.55
50	190	40	0.29
60	47	50	0.15
70	15	60	0.075
80	5.7	70	0.04
90	2.3	80	0.02
100	1		
110	0.5		
120	0.25		

TABLE 5-2 Maxwell life expectancy multipliers for high-voltage capacitors.

at 171.4 Vrms × 70:1 step-up = 12 kVrms entered in cell (B12). The peak voltage is calculated in cell (E7) using the rms value in cell (B12) × 1.414 = 16,968 V. The applied-to-rated voltage ratio of 16.97 kV/40 kV = 0.42 or 42% is calculated in cell (E12). Table 1, columns (H) through (K), rows 19 to 32 was constructed to reproduce the manufacturer's life data in the graph in rows 18 to 60 and shown in Figure 5-12. Table 5-2 and Table 1 of the worksheet list a multiplier of 1,000 for a 40% ratio. Figure 5-12 shows the power curve fit formula used to calculate the life expectancy multiplier of 891 in cell (E14) for the applied-to-rated voltage ratio in cell (E12). This means our 55.6 hours of run time at the rated voltage of 40 kV rms will be extended 891 times longer or 49,540 hours when the applied voltage is reduced to 12 kV rms.

But don't get excited yet, we have not yet applied the manufacturer's voltage reversal vs. life expectancy multiplier, which decreases the life expectancy when applying a sinusoidal 50% duty cycle waveform. The 50% voltage reversal ratio entered in cell (B14) is typical for the symmetrical damped sinusoidal primary oscillations. Table 5-2 lists a multiplier of 0.15 for a 50% voltage reversal ratio. Figure 5-12 shows the exponential curve fit formula used to calculate the voltage reversal ratio multiplier in cell (E15). As shown the calculated multiplier is 0.147 for a 50% voltage reversal ratio. The 55.6 hours of run time extended to 49,540 hours with an applied-to-rated voltage ratio of 42% will be reduced by a factor of 0.147 for the applied voltage reversal ratio. The calculated life expectancy in cell (E16) is 7,289 hours for the applied conditions. The 55.6 hours has been extended to 7,289 hours by running the capacitor at a low applied-to-rated voltage ratio.

This means a new capacitor (≈ \$700) can be expected to last for about 7,289 hours, when another \$700 would be required for a new replacement. If you purchase from surplus stock and Hamfests (< \$75) like I do, there is no telling how long it has previously been run or at what

INPUT PARAMETERS			ELECTRICAL STRESS CALCULATIONS			ROW
Rated Parameters						3
Capacitance value in μF (**C1**)	0.0600 uF		Total capacitance value in μF (**C**)	**0.0600 uF**	C = C1 * N or C1 / N	4
Number of capacitors used in network (**N**)	1		Instantaneous peak energy in Joules (**E**)	17.27 joules	E = C * dv^2	5
Parallel (1) or Series (2) arrangement (**SP**)	1		Period of oscillations (**dt**)	4.00 usec	dt = 1 / Fo	6
			Peak voltage applied to capacitor (**dv**)	16968 V	dv = V * 1.414	7
Capacitor rated voltage in kV (**VR**)	40.0 kV		Applied dv/dt in V / μsec (**Δv/t**)	**4,242 V/μs**	Δv/t = dv / (dt / 1.0μsec)	8
Capacitor life span in shots (**LS**)	1.0E+08		Peak current applied to capacitor (**Ip**)	**254.52 A**	Ip = C * (dv / dt)	9
						10
Applied Parameters			**LIFE EXPECTANCY CALCULATIONS**			11
RMS step-up voltage applied to tank in kV (**V**)	12.00 KV		Applied-to-Rated voltage ratio (**AR**)	42.4	AR = dv / VR or dv / (VR*4)	12
Resonant frequency (**Fo**)	250.000KHz		Life expectancy of pulse capacitor without applied multipliers in hours (**LE**)	55.6	LE = (LS / BPS) / 3600	13
Voltage reversal ratio (**VRR**)	50		Life expectancy mulitplier for applied-to-rated voltage ratio (**LM1**)	891.0	LM1 = 2E+15*AR^-7.5887	14
Spark gap breaks per second (**BPS**)	500		Life expectancy mulitplier for voltage reversal (**LM2**)	0.147	LM2 = 3.9925*EXP(-0.066*VRR)	15
			Life expectancy of pulse capacitor with applied multipliers in hours (**LP**)	**7289.3**	LP = LE * LM1 * LM2	16
A	B	C	D	E	F	COLUMN

FIGURE 5-13 Electrical stress and life expectancy calculation worksheet for high-voltage pulse capacitors.

applied voltage and pulse characteristic. Other than the unknown initial condition of a used capacitor the calculations can be used with some degree of accuracy if no other information on your capacitor can be obtained from the manufacturer. Used capacitors are cost effective, even if their history is unknown.

Let's look at the effect of adjusting the spark gap to ionize at a line voltage of 220 Vrms instead of the 171 Vrms in the previous example. The 220 Vrms × 70:1 step-up = 15,400 Vrms × 1.414 = 21,776 V peak. This gives us an applied -to- rated voltage ratio of 21.78 kV/40 kV = 0.54 or 54%. The applied-to-rated voltage ratio multiplier decreases from 891 to 134 and the voltage reversal ratio multiplier remains at 0.147. This means our 55.6 hours of run time will only be extended to 1,098 hours with the higher applied voltage. If the step-up transformer is changed to a 120:1 ratio and the spark gap is set to ionize at a line voltage of 225 Vrms the peak voltage of 38.2 kV produces an applied-to-rated voltage ratio multiplier of only 1.9 with the voltage reversal ratio multiplier remaining at 0.147. Our life expectancy is now only 15.5 hours. Increasing the applied voltage to a value near the rated maximum will severely limit the life of the pulse capacitor in a Tesla coil.

Section 5.3 detailed the common types of high-voltage commercial capacitors. The applied-to-rated parameters can be evaluated to ensure the capacitor will not be overstressed, which decreases the design lifespan. Not all high-voltage capacitor manufacturers were included but most surplus capacitors will be from the types covered. When commercial capacitors are used in the design, obtain the manufacturer's application data to ensure the applied voltage and current do not overstress the part. Occasionally the surplus capacitor will have no data plate or markings and not even the manufacturer can be identified. In these cases you will just have to try it in the circuit using the lowest applied voltage as the spark gap ionization permits. The spark gap spacing can be decreased so the gap will ionize at a lower peak voltage. The secondary voltage will be lower but the unknown capacitor can often be safely operated without failure. If the capacitor can be identified and the manufacturer's application data are different than that covered in Section 5.3 use the methodology covered in the section and the application data to ensure the rated voltage and current are not exceeded. The lifespan of any high-voltage capacitor can be increased as the applied-to-rated voltage ratio is decreased.

Although commercial capacitors will last longer than homemade capacitors and use superior materials their cost and surplus availability may prove difficult. The following sections provide calculations and data for making your own capacitors.

5.4 Dielectric Constants

Table 5-3 details dielectric material characteristics found in reference (1). The data includes measured characteristics at selected frequency decades and can be used to speculate how the material will react to higher frequencies. Note the data does not seem to support the concern for insulating degradation with humidity and moisture in PVC materials when compared with other insulting materials available to coilers. The dielectric constant at the selected frequency decades can be used in the capacitance calculations for additional accuracy. All dielectric constants are relative to air, which has a standard value of 1.0.

5.5 Leyden Jar Capacitors

Reference (7) provides the formula for calculating a cylindrical capacitance. When a cylinder is used to construct a capacitor (Leyden jar) as shown in the CH_5.xls file, LEYDEN JAR worksheet (2) (see App. B), the capacitance is:

$$C(\mu\text{F}) = \frac{(0.224\,k\text{OD})\,(H + 0.25\,\text{OD})}{T \bullet 1 \times 10^{6}} \tag{5.13}$$

Where: C = Capacitance of cylinder in μfarads = cell (C14), calculated value.
k = Dielectric constant of cylinder (insulator) = cell (C11), enter value, see Table 5-3 for table of dielectric constants or worksheet (1).
H = Height of cylinder in inches = cell (C7), enter value.
OD = Outside diameter of cylinder in inches = cell (C3), enter value.
T = Thickness of cylinder wall in inches = cell (C6), enter value.

The maximum allowable voltage that can be applied to the capacitor before breakdown of the dielectric is:

$$V = kS \bullet T \tag{5.14}$$

Where: V = Maximum voltage applied to capacitor before breakdown = cell (C16), calculated value. Converted to kV using a 1e-3 multiplier.
kS = Dielectric strength of cylinder (insulator) in volts/mil = cell (C12), enter value, see worksheet (1) for table of dielectric constants.
T = Thickness of cylinder wall in inches = cell (C6), enter value. 1 inch = 1000 mils.

From reference (6) is found the formula for calculating the energy stored in the capacitor:

$$E = \frac{CV^{2}}{2} \tag{5.15}$$

Parameters @ 25°C	Dielectric Constant (Relative to Air = 1)					Dissipation Factor (DF)					Dielectric Strength in volts/mil	DC Resistivity in ohms-cm	Moisture Absorption in Percent
Material	60 Hz	1 kHz	1 MHz	100 MHz	1 GHz	60 Hz	1 kHz	1 MHz	100 MHz	1 GHz			
Ceramics													
Ceramic (NPOT96)		29.5	29.5	29.5	29.5		0.00049	0.00016	0.0002		>75	$> 1 \times 10^{14}$	<0.1
Ceramic (N750T96)		83.4	83.4	83.4	83.4		0.00045	0.00022	0.00046		>75	$> 1 \times 10^{14}$	<0.1
Ceramic (Coors Al-200)		8.83	8.8	8.8	8.79		0.00057	0.00033	0.0003		>75	$> 1 \times 10^{14}$	<0.1
Porcelain (wet process)	6.5	6.24	5.87	5.8		0.03	0.018	0.009	0.0135		>75	$> 1 \times 10^{14}$	<0.1
Porcelain (dry process)	5.5	5.36	5.08	5.04		0.03	0.014	0.0075	0.0078		>75	$> 1 \times 10^{14}$	<0.1
Glasses													
Corning 0120 (soda-potash-lead silicate)	6.76	6.7	6.65	6.65	6.64	0.005	0.003	0.0012	0.0018	0.0041		1×10^{10}@250°C	
Corning 7040 (soda-potash-borosilicate)	4.85	4.82	4.73	4.68	4.67	0.0055	0.0034	0.0019	0.0027	0.0044		1×10^{10}@250°C	
Corning 7060 (PYREX)		4.97	4.84	4.84	4.82		0.0055	0.0036	0.003	0.0054		1×10^{10}@250°C	
Quartz (silicon dioxide)	3.78	3.78	3.78	3.78	3.78	0.0009	0.00075	0.0001	0.0002	0.00006	410	$> 1 \times 10^{19}$	
Plastics													
Bakelite BM120 (Phenol-formaldehyde)	4.9	4.74	4.36	3.95	3.7	0.08	0.022	0.028	0.038	0.0438	300	1×10^{11}	<0.6
Bakelite BM250 (w/66% asbestos fiber, preformed and preheated)		22	5.3	5	5		0.37	0.125			<300	$< 1 \times 10^{11}$	<0.6
Bakelite BT-48-206 (100% Phenol-formaldehyde)	8.6	7.15	5.4	4.4	3.64	0.15	0.082	0.06	0.077	0.052	277		0.42
Catalin 200 base (Phenol-formaldehyde)	8.8	8.2	7		4.89	0.05	0.029	0.05		0.108	200		

TABLE 5-3 Dielectric constants for some materials used in Tesla coil construction.

Parameters @ 25°C	Dielectric Constant (Relative to Air = 1)					Dissipation Factor (DF)					Dielectric Strength in volts/mil	DC Resistivity in ohms-cm	Moisture Absorption in Percent
Material	60 Hz	1 kHz	1 MHz	100 MHz	1 GHz	60 Hz	1 kHz	1 MHz	100 MHz	1 GHz			
Formica XX (Phenol-formaldehyde & 50% paper laminate)	5.25	5.15	4.6	4.04	3.57	0.25	0.0165	0.034	0.057	0.06			
Formvar E (Polyvinyl formal)	3.2	3.12	2.92	2.8	2.76	0.003	0.01	0.019	0.013	0.0113	860	$> 5 \times 10^{16}$	1.3
Hysol 6020 (Epoxy resin)		3.9	3.54	3.29	3.01		0.0113	0.0272	0.0299	0.0274			
Hysol 6030 (Epoxy resin, flexible potting)		6.15	4.74	3.61	3.2		0.048	0.084	0.09	0.038			
Melmac resin 592 (Melaminel-formaldehyde, mineral filler)	8	6.25	5.2	4.7	4.67	0.08	0.047	0.0347	0.036	0.041	450	3×10^{13}	0.1
Micarta 254 (Cresylic acid-formaldehyde, 50% α cellulose)	5.45	4.95	4.51	3.85	3.43	0.098	0.033	0.036	0.055	0.051	1020	3×10^{13}	1.2
Nylon 610 (Polyhexamethylene-adipamide)	3.7	3.5	3.14	3	2.84	0.018	0.0186	0.0218	0.02	0.0117	400	8×10^{14}	1.5
Plexiglass (Polymethyl methacrylate)	3.45	3.12	2.76		2.6	0.064	0.0465	0.014		0.0057	990	5×10^{16}	0.3–0.6
Polystyrene	2.56	2.56	2.56	2.55	2.55	<0.00005	<0.00005	0.00007	<0.0001	0.00033	500–700	1×10^{18}	0.05
Styrofoam 103.7 (Foamed Polystyrene, 0.25% filler)	1.03	1.03	1.03		1.03	<0.0002	<0.0001	<0.0002		0.0001			Low
Teflon (Polytetraflouroethylene)	2.1	2.1	2.1	2.1	2.1	<0.0005	<0.0003	<0.0002	<0.0005	<0.00015	1 k-2 k	1×10^{17}	0
Vinylite QYNA (100% Polyvinyl-chloride PVC)	3.2	3.1	2.88	2.85	2.84	0.0115	0.0185	0.016	0.0081	0.0055	400	1×10^{14}	0.05–0.15

TABLE 5-3 Dielectric constants for some materials used in Tesla coil construction. (*continued*)

Parameters @ 25°C	Dielectric Constant (Relative to Air = 1)					Dissipation Factor (DF)					Dielectric Strength in volts/mil	DC Resistivity in ohms-cm	Moisture Absorption in Percent
Material	60 Hz	1 kHz	1 MHz	100 MHz	1 GHz	60 Hz	1 kHz	1 MHz	100 MHz	1 GHz			
Plasticell (Expanded Polyvinyl-chloride PVC)		1.04	1.04	1.04	1.04		0.0011	0.001	0.001	0.0055			
Geon 2046 (59% PVC, 30% dioctyl phosphate, 6% stabilizer, 5% filler)	7.5	6.1	3.55	3	2.89	0.08	0.11	0.089	0.03	0.0116	400	8×10^{14}	0.5
Organic Liquids													
Aviation gasoline (100 Octane)			1.94	1.94	1.92				0.0001	0.0014			
Jet fuel (JP-3)			2.08	2.08	2.04			0.0001		0.0055			
Bayol-D (77.6% paraffins, 22.4% naphthenes)	2.06	2.06	2.06	2.06	2.06	0.0001	<0.0001	<0.0003	0.0005	0.0013	300		Slight
Ethyl alcohol			24.5	23.7	6.5				0.09	0.062			
Ethylene glycol			41	41	12			0.03	0.045				
Pyranol 1467 (Chlorinated benzenes, diphenyls)	4.4	4.4	4.4	4.08	2.84		0.0003	0.0025	0.13	0.12	300		
Pyranol 1478 (Isomeric trichlobenzenes)	4.55	4.53	4.53	4.5	3.8		0.02	0.0014	0.0002	0.014			
Silicon Fluid SF96-1000		2.73	2.73	2.73	2.71		<0.00001	<0.0001		0.0106			
Transil oil 10C (Aliphatic, aromatic hydrocarbons)	2.22	2.22	2.22	2.2	2.18	0.001	<0.00001	<0.0005	0.0048	0.0028	300		
Waxes													
Beeswax, yellow	2.76	2.66	2.53	2.45	2.39		0.014	0.0092	0.009	0.0075			
Parafin wax, 132° ASTM (aliphatic saturated hydrocarbons)	2.25	2.25	2.25	2.25	2.25	<0.0002	<0.0002	<0.0002	<0.0002	<0.0003	1060	5×10^{16}	Very low

TABLE 5-3 Dielectric constants for some materials used in Tesla coil construction. *(continued)*

Parameters @ 25°C	Dielectric Constant (Relative to Air = 1)					Dissipation Factor (DF)					Dielectric Strength in volts/mil	DC Resistivity in ohms-cm	Moisture Absorption in Percent
Material	60 Hz	1 kHz	1 MHz	100 MHz	1 GHz	60 Hz	1 kHz	1 MHz	100 MHz	1 GHz			
Rubbers													
GR-1 (butyl rubber, copolymer of 98-99% isobutylene, 1-2% isoprene)	2.39	2.38	2.35	2.35	2.35	0.0034	0.0035	0.001	0.001	0.0009			
Hevea rubber, vulcanized (100 pts pale crepe, 6 pts sulfer)	2.94	2.94	2.74	2.42	2.36	0.005	0.0024	0.0446	0.018	0.0047			
Neoprene compound (38% GN)	6.7	6.6	6.26	4.5	4	0.018	0.011	0.038	0.09	0.034	300	8×10^{12}	Nil
SE-972 (silicone rubber compound)		3.35	3.2	3.16	3.13		0.0067	0.003	0.0032	0.0097			
Miscellaneous													
Douglas Fir	2.05	2	1.93	1.88	1.82	0.004	0.008	0.026	0.033	0.027			
Douglas Fir, plywood	2.1	2.1	1.9			0.012	0.0105	0.023					
Amber (fossil resin)	2.7	2.7	2.65		2.6	0.001	0.0018	0.0056		0.009	2300	Very high	
Mycalex 400 (Mica, glass)		7.45	7.39				0.0019	0.0013					
Mycalex K10 (Mica, glass, titanium dioxide)		9.3	9				0.0125	0.0026		0.004			<0.5
Paper, Royalgrey	3.3	3.29	2.99	2.77	2.7	0.01	0.0077	0.038	0.066	0.056			
Water, distilled			78.2	78	76.7			0.04	0.005	0.157	1×10^6		
Ice, from distilled water			4.15	3.45	3.2				0.12	0.035			
Snow, freshly fallen		3.33	1.2	1.2	1.2		0.492	0.0215		0.00029			

Dissipation Factor (DF) is a dimensionless ratio of power dissipated-to-power stored in the dielectric to the applied AC. The AC is dissipated in any dielectric resistance or distributed resistance. It is stored in the reactive component (XL or XC). The smaller the DF the less power it will dissipate, more power it will store, and is a better dielectric than a material with a higher DF. If the DF is less than 0.1 the DF is considered to be equivalent to the Power Factor (PF) for the same material. The DF may peak at a frequency and fall as the frequency is further increased. This is a resonance effect of some materials (e.g., polarized). As the temperature is increased above 25°C the resonant frequency will increase in these polar materials and decrease at colder temperatures. DF in non-polarized materials will essentially remain the same over frequency. Some materials exhibit an ionic or conduction loss at the lower power distribution frequencies (e.g., 50 to 1000 Hz), which varies inversely with frequency. As the temperature is increased above 25°C these losses increase which increases the power dissipated and the DF.

TABLE 5-3 Dielectric constants for some materials used in Tesla coil construction. *(continued)*

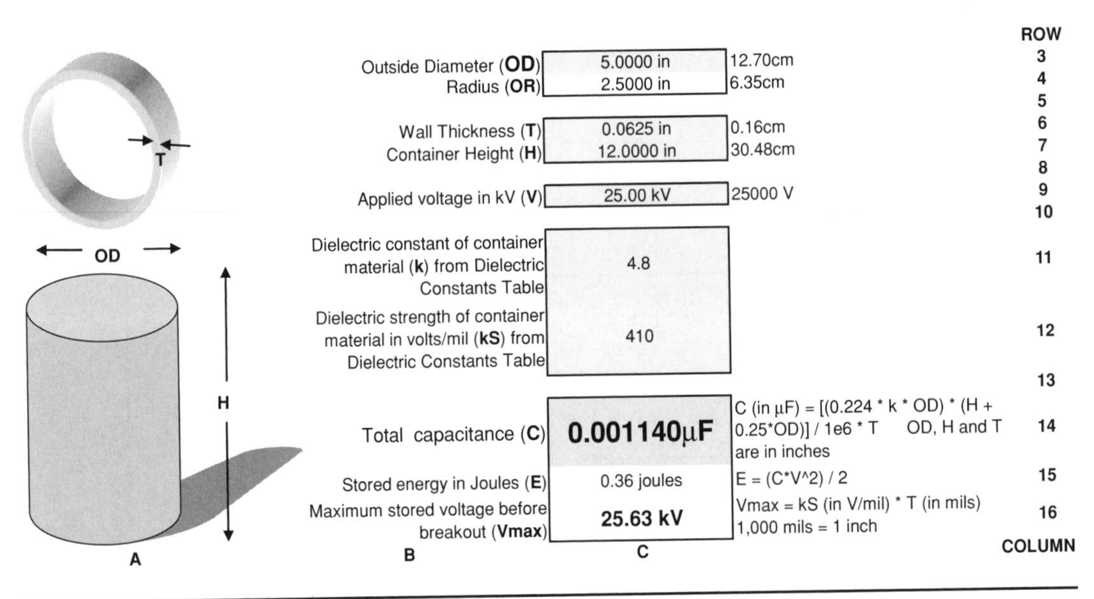

FIGURE 5-14 Leyden jar capacitor worksheet.

Where: E = Energy stored in capacitor in joules (1 joule = 1 watt for one second) = cell (C15), calculated value.

C = Capacitance in farads = calculated value in cell (C14) from equation (5.9). Converted to farads using a 1e-6 multiplier.

V = Voltage applied to capacitor = cell (C9), enter value in kV. Converted to volts in cell (D9) using a 1e3 multiplier.

Using the dynamics of the worksheet the capacitor's outside diameter, height, k, and thickness can be changed to find the optimum design for a desired capacitance value before construction work is started.

The example calculations shown in Figure 5-14 are for a glass cylinder of 5″ diameter, 12″ high and 1/16″ thick. The dielectric strength of the container material is 410 V/mil (glass) and the dielectric constant is 4.8 for the type of glass chosen (typically ranges between 4.0 and 7.0). This produces 0.0011 μF of capacitance with a working voltage of 25.6 kV. The peak voltage applied to this capacitor should only be 25.6 kV/2 = 12.8 kV. Safe design dictates a 50% safety margin when determining the working voltage, in other words apply no more than one half the voltage shown in cell (C16) to the capacitor.

5.6 Plate Capacitors

Two types of plate capacitor construction can be used. A film or other dielectric material can be placed between square or rectangular plates, forming a conventional plate capacitor. The other type uses air or an oil dielectric material placed between the rectangular stationary plates and semicircular rotating vanes of an air–vane capacitor. Reference (6) provides the formula for calculating parallel plate capacitance. For either type shown in the CH_5.xls file, PLATE worksheet (3) (see App. B), the capacitance is:

$$C(\mu\text{F}) = \frac{(0.2244\,\text{kA})(N-1)}{T \bullet 1 \times 10^6} \tag{5.16}$$

Where: C = Capacitance in μfarads = cell (C16), calculated value.
k = Dielectric constant of insulator material = cell (C13), enter value, see Table 5-3 for table of dielectric constants or worksheet (1).
N = Number of plates used in capacitor = cell (C10), enter value. The formula considers the capacitor as a number of series capacitors which is one less than the number of plates used. For air–vane capacitors enter the number of rotating vane or stationary plates, whichever is greater.
T = Thickness of dielectric material or distance between plates in inches = cell (C8), enter value.
A = Area of plate in square inches = calculated value in cell (C7). For rectangular plates:

$$A = L \bullet W$$

Where: L = Length of plate in inches = cell (C6), enter value.
W = Width of plate in inches = cell (C5), enter value.
For air–vane types the area is approximately that of the semicircular vanes. The area of a circle is $\pi \times (D/2)^2$. The area of the vane is one half the area of a circle:

$$A = \frac{\pi\left(\frac{D}{2}\right)^2}{2}$$

Where: D = Diameter of semicircular vane in inches = cell (C4), enter value.

NOTE: *Enter a (1) into cell (C3) to perform air–vane capacitor calculations or a (2) for rectangular plate capacitor calculations.*

Using the dynamics of the worksheet the capacitor's plate area, number of plates, k, and thickness can be changed to find the optimum design for a desired capacitance value before construction work is started. The methodology shown in equations (5.14) and (5.15) also apply to this worksheet. Remember to divide the working voltage calculation in cell (C18) by two for a 50% safety factor.

Air–vane capacitors are generally built for use with an air dielectric. They provide a variable capacitance between some minimum value (typically 25–100 pF) when the rotating vanes are positioned outside the stationary plates, to a maximum value (typically 150–300 pF) when the rotating vanes are positioned inside the stationary plates. The maximum value can be determined by calculation.

Figure 5-15 shows the calculations for an air–vane capacitor used in a tube coil. A surplus RF tuning capacitor was obtained with a separation of 1/2″ between the rotating vanes and stationary plates using 12 rotating vanes of 6″ diameter (13 stationary plates). The maximum

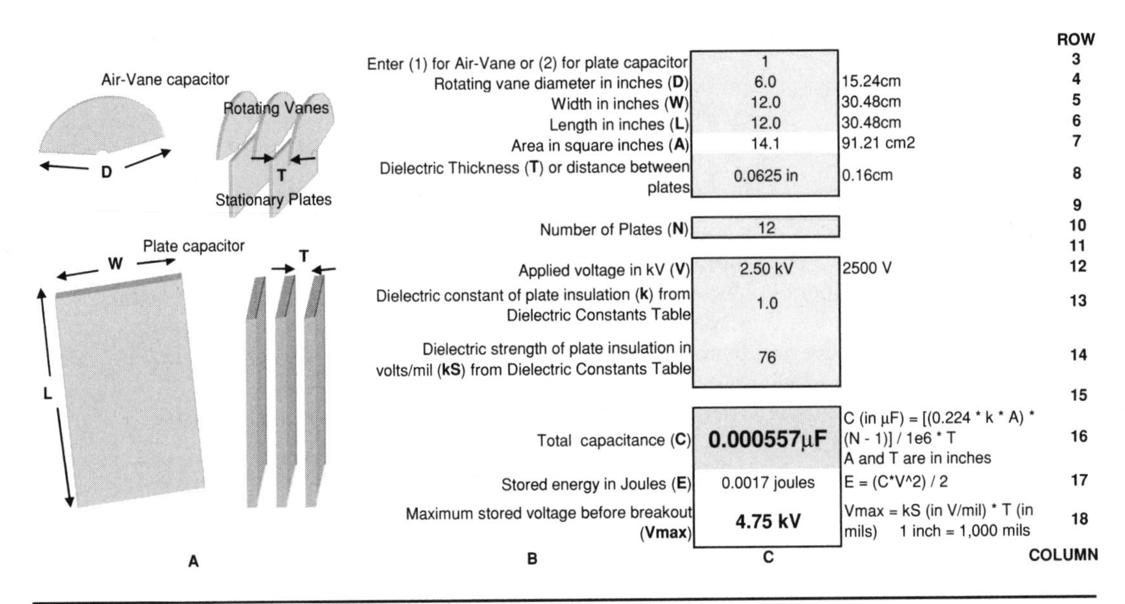

FIGURE 5-15 Plate capacitor worksheet.

capacitance obtainable was only 70 pF with a working voltage of 38 kV. Since the working voltage was very conservative for use with a tube coil the plate separation was decreased from 1/2″ to 1/16″. The dielectric constant for air is 1.0 and the dielectric strength is approximately 76.2 kV/inch, therefore 76.2 kV/inch = 76 V/mil. This provided a maximum capacitance of 557 pF and a working voltage of 4.75 kV.

To increase the working voltage and the capacitance even further the air–vane capacitor can be placed into a leak-proof container and filled with mineral oil or suitable oil dielectric. The oil replaces the air dielectric increasing the dielectric constant to 2.0 and the dielectric strength to approximately 300 V/mil. This provides a maximum capacitance of 1,115 pF (0.0011 μF) and a working voltage of 18.75 kV.

The minimum capacitance also increases in the same proportion as the maximum capacitance. The example began with a minimum measured capacitance of 25 pF and a maximum capacitance of 70 pF. When the plate separation was decreased from 1/2″ to 1/16″ the maximum capacitance increased to 557 pF for air dielectric (557 pF/70 pF = 8 times greater) and 1,115 pF using oil dielectric (1,115 pF/70 pF = 16 times greater). This ratio will also increase the minimum capacitance by a factor of 8 for air dielectric (25 pF $\times$ 8 = 200 pF) and a factor of 16 for oil dielectric (25 pF $\times$ 16 = 400 pF). This is not objectionable when the tuning range is examined. With no modification there was only a 25 pF to 70 pF tuning range provided. This will produce negligible tuning effects in the primary of a tube coil where the tank capacitance used is 0.002 μF. By decreasing the plate separation from 1/2″ to 1/16″ a 200 pF to 557 pF tuning range is obtained which produces noticeable tuning effects in the tube coil. By placing the air–vane capacitor in a container filled with mineral oil a 400 pF to 1,115 pF (0.0004 μF to 0.001115 μF) tuning range is obtained, which produces noticeable tuning effects. In this manner more than one surplus air–vane capacitor can be added together in parallel and the plate separations modified to produce very large variable capacitors.

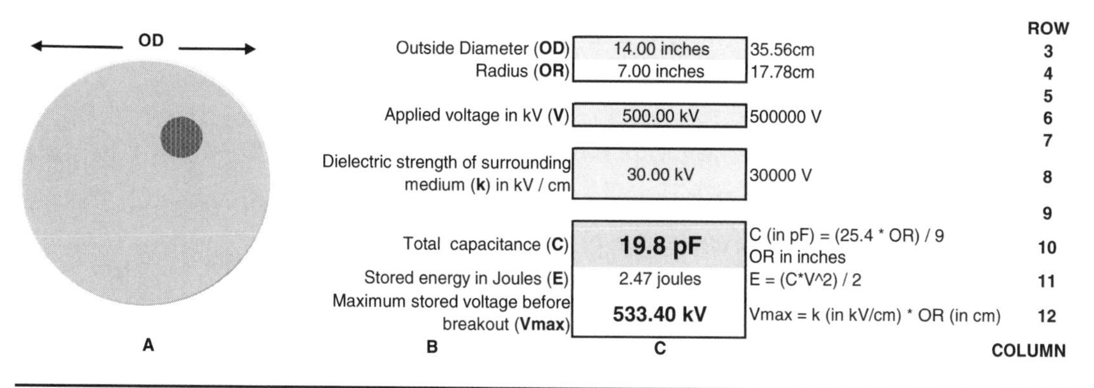

FIGURE 5-16 Spherical capacitor worksheet.

5.7 Spherical Capacitors

Reference (6) provides the formula for calculating spherical capacitance. When a sphere is used to construct a capacitor as shown in the CH_5.xls file, SPHERICAL worksheet (4) (see App. B), the capacitance is:

$$C(\text{pF}) = 1.412D \quad \text{or} \quad C(\text{pF}) = \frac{25.4R}{9} \tag{5.17}$$

Where: C = Capacitance of sphere in pF = cell (C10), calculated value.
R = Radius of sphere in inches = calculated value in cell (C4) from 1/2 the outside diameter entered in cell (C3).

Using the dynamics of the worksheet the capacitor's outside diameter can be changed to find the optimum design for a desired capacitance value before construction work is started. The methodology shown in equations (5.14) and (5.15) also apply to this worksheet.

The example calculations shown in Figure 5-16 indicate a 14″ diameter sphere produces 19.8 pF for the secondary terminal capacitance of a spark gap coil. The sphere will store 533 kV under nominal conditions before a spark breaks out. The air dielectric breaks down with less than the 30 kV/cm used in the calculations when considering the applied voltage is at a high frequency and the proximity-to-ground plane effect detailed in Chapter 6, therefore the secondary voltage does not need to reach 533 kV for the spark to breakout. Nominal conditions are approximately 20°C, 760 torr, 50% relative humidity and no proximity-to-ground effect, which are explained in Section 5.9.

5.8 Toroidal Capacitors

When a toroid is used to construct a capacitor as shown in the CH_5.xls file, TOROID worksheet (5) (see App. B), the capacitance is calculated from reference (12) as:

For d/D ratios <0.25: $C(\text{pF}) = \dfrac{1.8(D-d)}{\ln\left(\frac{8D-d}{d}\right)}$ For d/D ratios >0.25: $C(\text{pF}) = 0.37\text{D} + 0.23\text{d}$

$$\text{Or the more traditional: } C(\text{pF}) = 1.4\left(1.2781 - \frac{\text{CS}}{\text{OD}}\right) \bullet \sqrt{\pi\text{CS}(\text{OD}-\text{CS})} \quad (5.18)$$

Where: C = Capacitance of toroid in pF = cell (C14), calculated value.
CS = Cross-sectional diameter (*d* or *d*2) of tube in inches = cell (C6), enter value.
OD = Outside diameter (D or d1) of toroid in inches = cell (C3), enter value.
d/*D* ratio = cell (C12), calculated value from CS/OD ratio.

NOTE: *D and d must be in meters. The dimensions in inches in column C are converted to cm in column D and to meters using a 0.01 multiplier. An accuracy of 1–2% for practical d/D ratios was derived from data in the reference and shown below in Figure 5-17.*

Using the dynamics of the worksheet the capacitor's cross-sectional diameter (*d* or *d*2) and toroid diameter (*D* or *d*1) can be changed to find the optimum design for a desired capacitance value. The methodology shown in equations (5.14) and (5.15) also apply to this worksheet.

The example calculations shown in Figure 5-18 indicate a 24″ diameter toroid with a 3″ diameter cross-section produces 23.9 pF for the secondary terminal capacitance of a spark gap coil.

5.9 Stray Capacitance and Proximity-to-Ground Effect in Terminal Capacitances

There is a parasitic or stray capacitance that exists between the toroid or spherical terminal capacitance and the ground plane beneath the coil. The terminal capacitance forms one plate of a parallel capacitor and the ground plane forms the other plate with the air between forming the dielectric. To estimate the stray capacitance developed from the terminal capacitance-to-ground plane, open the STRAY CAPACITANCE Worksheet (6) in the CH_5.xls file. (See App. B.)

The worksheet as shown in Figure 5-19 will calculate stray capacitance values using the parallel plate methodology covered in Section 5.6. The dielectric in stray capacitances is air therefore the dielectric constant (*k*) in cell (C15) is fixed at 1.0 for all cases. The number of plates in cell (C14) is also a fixed value of two since there is only the terminal capacitance and the ground plane to consider. For the toroid or sphere terminal diameter (Da) entered in cell (C3) and distance between this terminal and the ground plane (D) entered in cell (C12), the calculated stray capacitance from terminal capacitor-to-ground plane is displayed in cell (C18). This is used in the Spice model in Chapter 8 to simulate a spark gap coil's operation. There may be other stray capacitances if you are using a lab with metal walls and roof, or large metal objects are in the vicinity of the operating coil. Change the distance value (D) in cell (C12) to estimate these values.

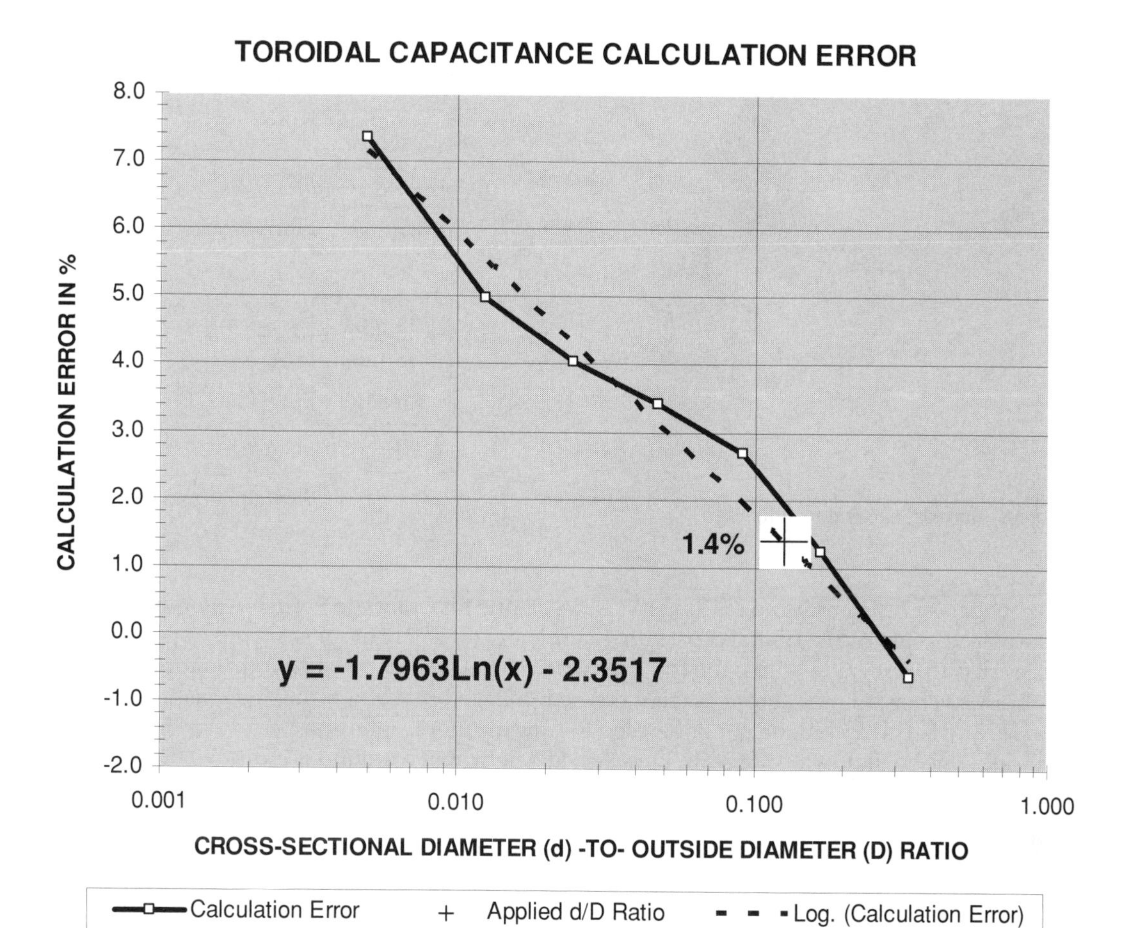

FIGURE 5-17 Toroidal capacitance calculation error.

The proximity to the ground plane will affect the breakdown threshold of the terminal capacitance. The proximity-to-ground effect is characterized in Chapter 6. Figure 5-20 shows the stray capacitance of the terminal-to-ground plane and proximity-to-ground effect for a 24″ diameter toroid. The terminal height above ground (D) is nominally 48″ entered in cell (C12) and the spark gap coil produces a spark length (S) of 30″ entered in cell (C11). The X-axis of the graph in Figure 5-20 displays the calculated stray capacitance and proximity-to-ground effects over a terminal elevation above ground (D) range of 12″ to 540″. Also displayed are the curve fit formulae to calculate the breakout voltage adjusted for the proximity-to-ground effect shown in cell (C21). The stray capacitance for the selected height above ground is indicated by a large [+] in the figure.

As shown in Figure 5-20 the stray capacitance (red trace) decreases as the terminal is elevated above ground. This parasitic element affects the terminal capacitance value, changing the resonant frequency of the secondary. To minimize this effect elevate the terminal

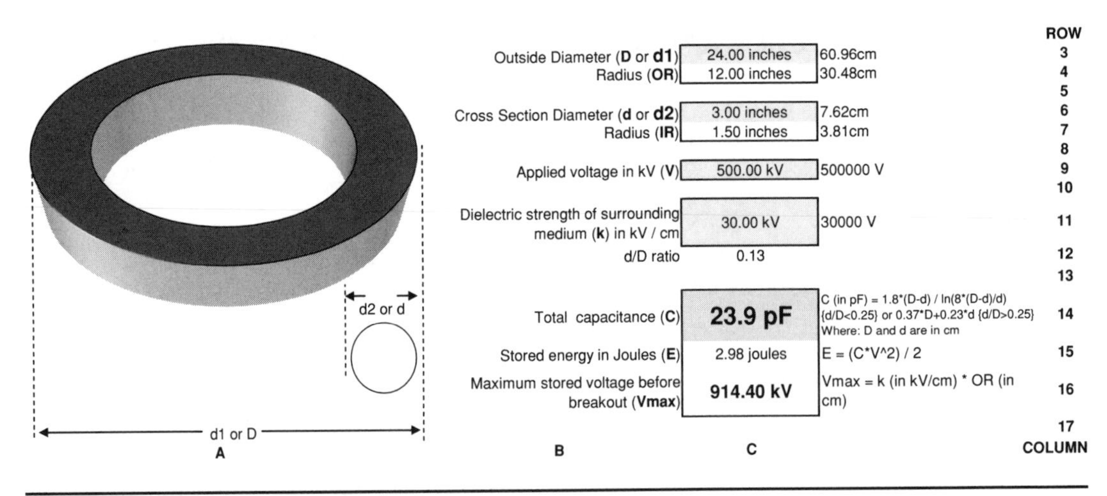

FIGURE 5-18 Toroidal capacitor worksheet.

capacitance. Tesla produced his longest sparks during his Colorado Springs experiments when he placed a spherical terminal capacitance atop the extra coil in his magnifying transmitter and elevated the sphere to adjust the coil's resonant response and the voltage breakdown threshold. Also shown in the figure is the increased voltage breakdown threshold as the terminal is elevated. This technique could be used to tune a coil for optimum performance and spark length. The breakdown voltage for the selected height above ground is indicated by a large [X] in the figure.

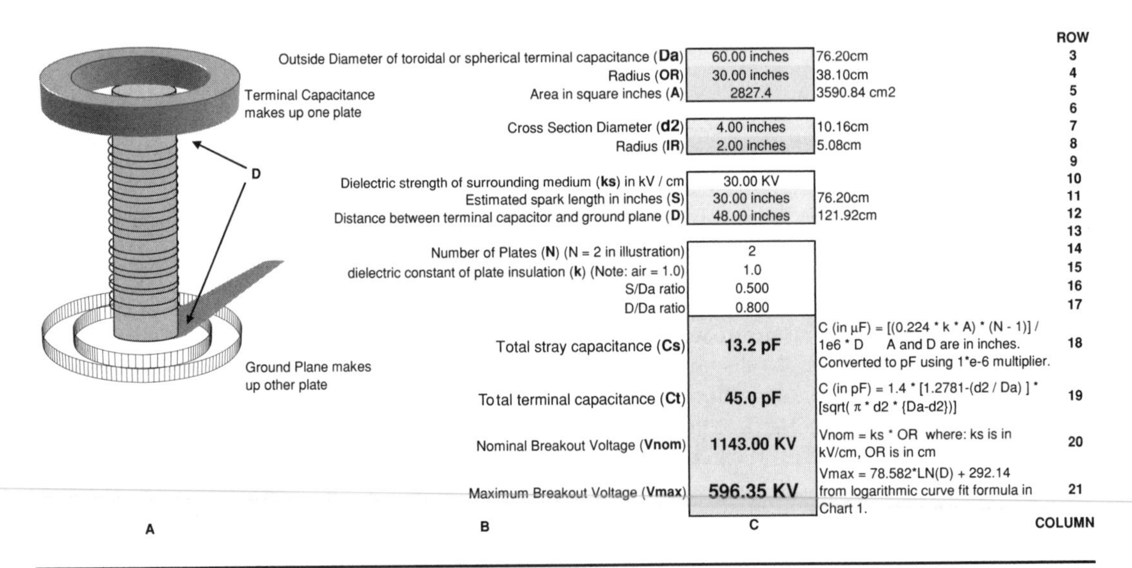

FIGURE 5-19 Stray terminal-to-ground capacitance worksheet.

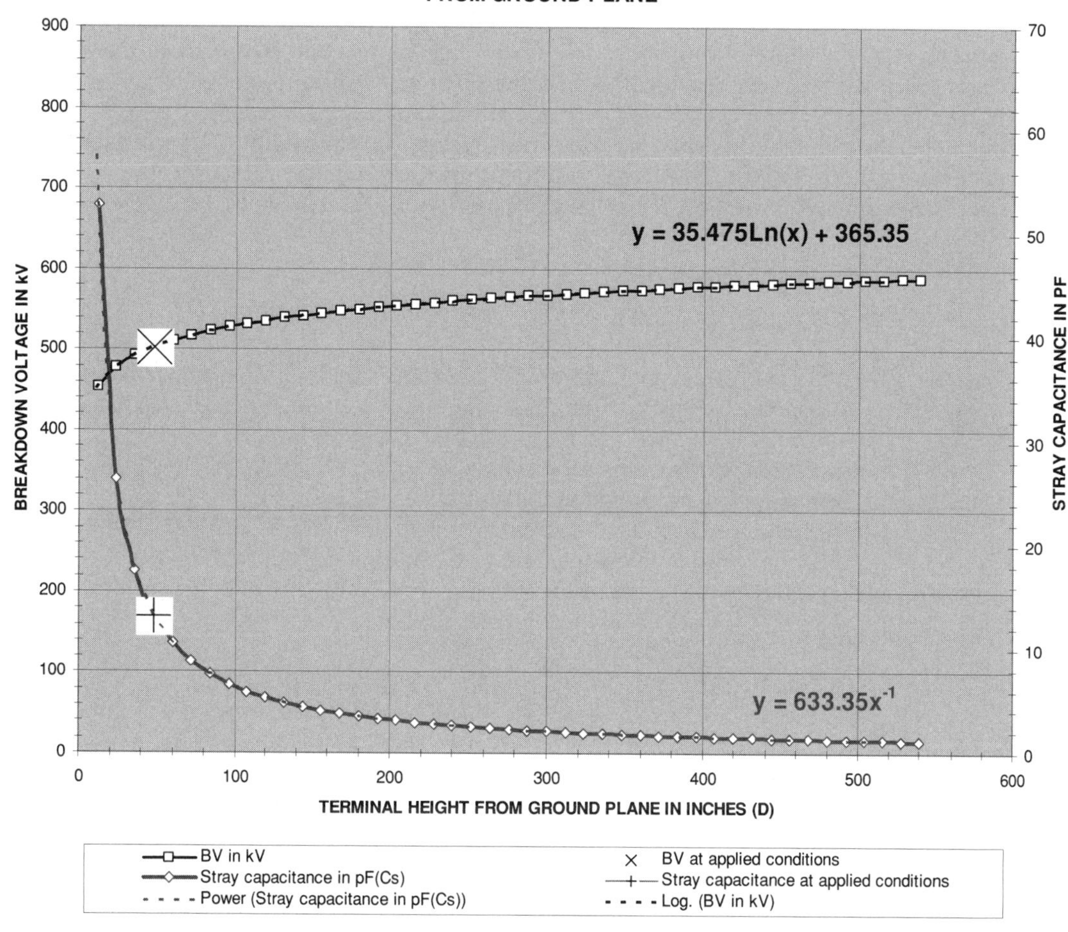

FIGURE 5-20 Stray terminal-to-ground capacitance and proximity-to-ground effect characteristic for a 24″ diameter toroid.

5.10 Effects of Terminal Capacitance on Resonant Frequency

When a terminal capacitance is added to the top of a grounded series resonant LCR circuit, such as that used in the secondary winding of a Tesla coil, the resonant frequency changes. This is similar to top loading a radio antenna with capacitance. When a terminal capacitance is added to the self-capacitance of the secondary winding they effectively combine as two parallel capacitances. As the two capacitances are added together the increased secondary capacitance will shift the resonant frequency to a lower value. As more terminal capacitance is added the lower the resonant frequency of the secondary becomes. This generally requires retuning in the primary to compensate for the shift in secondary resonant frequency.

To determine the effects of the terminal capacitance on the resonant frequency open the CH_5.xls file, CT vs fo worksheet (12). (See App. B.) A table of calculations was constructed to determine whether the terminal capacitance and self-capacitance of the secondary winding form an effective series or parallel capacitance. Table 1, rows 3 to 26 calculates the effective capacitance for a parallel arrangement in column (C) and resonant frequency in column (D) using equation (5.20) for the corresponding terminal capacitance value in column (B). Table 2 calculates the effective capacitance for a series arrangement in column (F) and resonant frequency in column (G) for the corresponding terminal capacitance value in column (E). Frequency measurements indicated that the two capacitances combine in a parallel arrangement with an effective capacitance of:

$$Ceff = Ct + Cs \tag{5.19}$$

Where: $Ceff$ = Effective value of the terminal capacitance and self-capacitance of the secondary winding in farads = column (C) and cell (D35), calculated value.

Cs = Self-capacitance of the secondary winding in farads = cell (D30), enter value in pF. Converted to farads using a 1e-12 multiplier.

Ct = Terminal capacitance in farads = value in column (B) and cell (D34) in pF. Converted to farads using a 1e-12 multiplier.

The resonant frequency of the secondary winding with the added terminal capacitance is:

$$fso = \frac{1}{2\pi\sqrt{Ls(Ct + Cs)}} \tag{5.20}$$

Where: fso = Resonant frequency of secondary with terminal capacitance in Hz = column (D) and cell (D36), calculated value. Converted to kHz using a 1e-3 multiplier.

Ls = Inductance of the secondary winding in henries = cell (D29), enter value in mH. Converted to henries using a 1e-3 multiplier.

Cs = Self-capacitance of the secondary winding in farads = cell (D30), enter value in pF. Converted to farads using a 1e-12 multiplier.

Ct = Terminal capacitance in farads = value in column (B) and cell (D34) in pF. Converted to farads using a 1e-12 multiplier.

Table 3, columns (R) through (V), rows 3 to 54, was constructed to determine the Archimedes spiral primary turn required to tune the primary oscillations to the resonant frequency of the secondary with the added terminal capacitance. Rows 58 to 109 perform the same calculation for a helically wound primary. The tank capacitance is entered into cell (D39) and the primary winding characteristics are entered in cells (D44) to (D50) as detailed in Chaper 2. The primary oscillating frequency for the selected characteristics entered in cells (D39) and (D44) to (D50) is calculated in column (R) and cell (D51) using equation (2.23). Column (S) uses a conditional statement to limit the primary frequency range below the resonant secondary frequency. This is seen in the graph in column (A) and in Figure 5-21 as a vertical line for the lower number of primary turns. Column (U) reverses the calculations in column (S) at the corresponding primary turn number in column (T) for use in a lookup function calculating the Archemedes

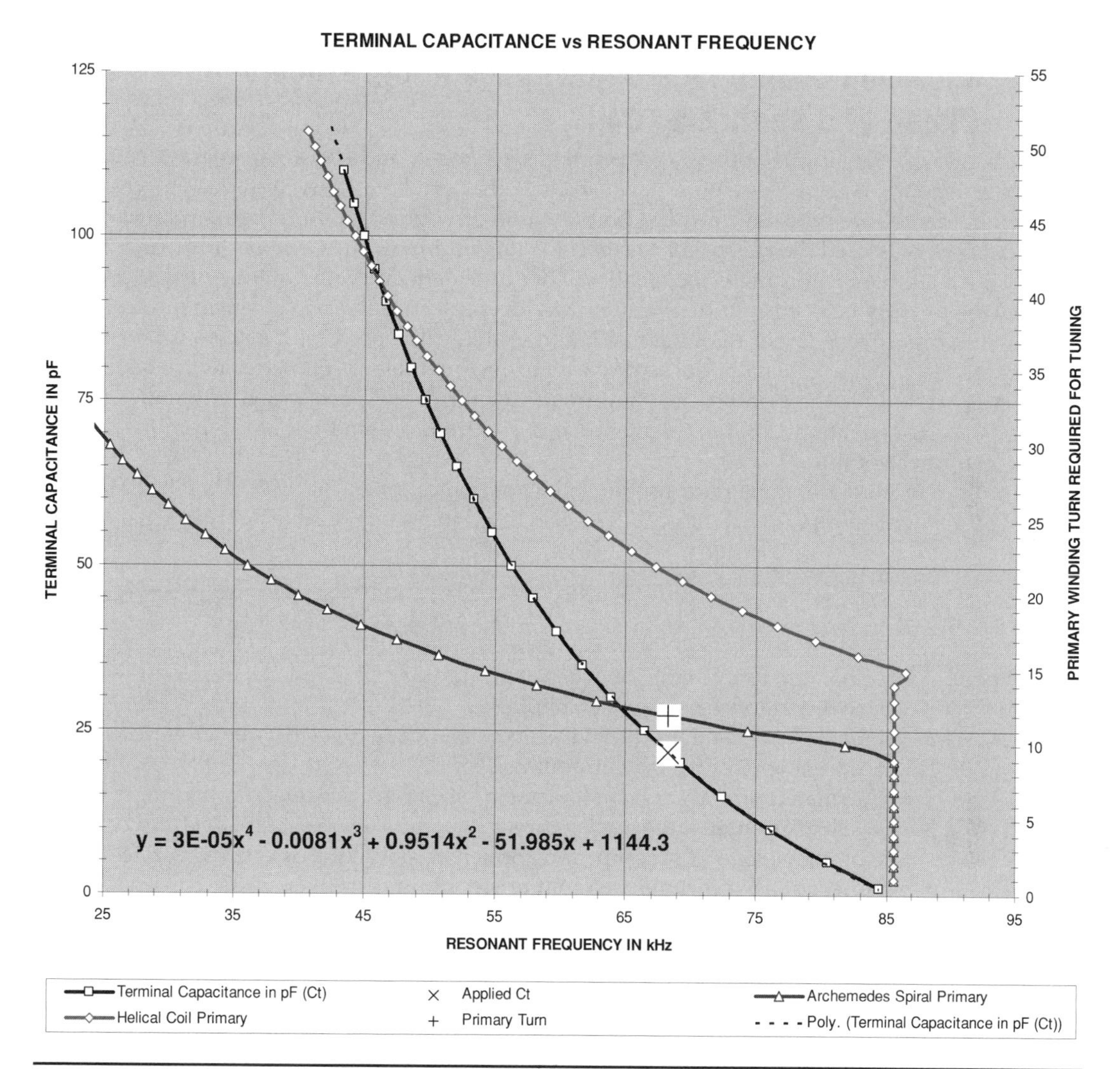

FIGURE 5-21 Effects of terminal capacitance on resonant frequency of secondary winding.

spiral primary turn required to tune in cell (D40) and helically wound primary turn in cell (D41). Column (S) is displayed in the graph and Figure 5-21 as a blue trace for an Archemedes spiral primary and a red trace for a helically wound primary. The black trace indicates the resonant frequency of the secondary at the corresponding value of terminal capacitance calculated in column (D). A large [X] marks the applied terminal capacitance value entered in cell (D34) and [+] marks the primary turn required to tune the primary oscillations to the resonant frequency of the secondary. The calculated results shown in the figure apply to the coil design detailed in Chapter 2.

5.11 Maximum Usable Tank Capacitance in the Primary Circuit of a Spark Gap Coil

The step-up transformer turns ratio, primary decrement, and spark gap characteristics will determine the largest value of tank capacitance that can be effectively used. The calculated maximum usable capacitance value is really a design optimization. When a current-limited transformer is used the rms primary current is limited to the rated output of the transformer. Using a tank capacitor that is too small will not utilize the full rated output current and lowers the primary peak current. This also means less peak current and voltage in the secondary. A tank capacitance larger than the maximum usable value used with a non–current-limited transformer will not increase the secondary voltage or enhance coil performance but may be necessary to tune the primary resonant frequency for the selected primary winding characteristics. See equation (2.31) to calculate the rms current drawn by the coil from the step-up transformer's output.

To determine the maximum usable tank capacitance, open the CH_5.xls file, CMAX VS STEP-UP worksheet (7). (See App. B.) The calculated maximum usable tank capacitance is:

$$C = dt\frac{I}{dv} = \frac{1}{\text{BPS}} \bullet \frac{I}{V} \tag{5.21}$$

Where: C = Maximum usable tank capacitance in farads = cell (F7), calculated value. Converted to μF using a 1e-6 multiplier.

BPS = Breaks Per Second produced by rotary spark gap = cell (B7), enter value. See Chapter 2 or 6 for determining BPS.

I = Calculated primary current (output of step-up transformer) in rms amps = cell (B4), enter value.

V = Output voltage of step-up transformer = Measured or estimated line voltage with coil running (input of step-up transformer) in kV rms entered in cell (B5) multiplied by the transformer turns ratio entered in cell (B6).

Table 1, columns (I) through (M), rows 18 to 96, calculates the primary power level, line current, and maximum usable capacitance for a non–current-limited transformer application. The primary rms current (output of step-up transformer) is increased from 1 mA to 2 A, in 10-mA increments, shown in column (I) of the worksheet. Column (J) shows the calculated line current (step-up transformer input) at the corresponding primary current in column (I) for the transformer turns ratio entered in cell (B6). For the line voltage (input of step-up transformer with coil running) entered in cell (B5) the calculated transformer output power in kW is shown in column (L) for the transformer turns ratio entered in cell (B6). Column (M) shows the calculated maximum usable tank capacitance for the calculated output power values in column (L). A graph was constructed to display the maximum usable tank capacitance and line current for the selected BPS entered in cell (B7) and transformer turns ratio in cell (B6). Two graphs are provided in the worksheet to display the Table 1 calculations. Chart 1 in rows 19 to 55 displays the power levels through 70 kW and the Chart in rows 57 to 99 display the power levels through 5 kW. The maximum usable tank capacitance and line current at selected power levels for a BPS of 460, 0.161 primary decrement, 70:1 transformer ratio and a measured line voltage

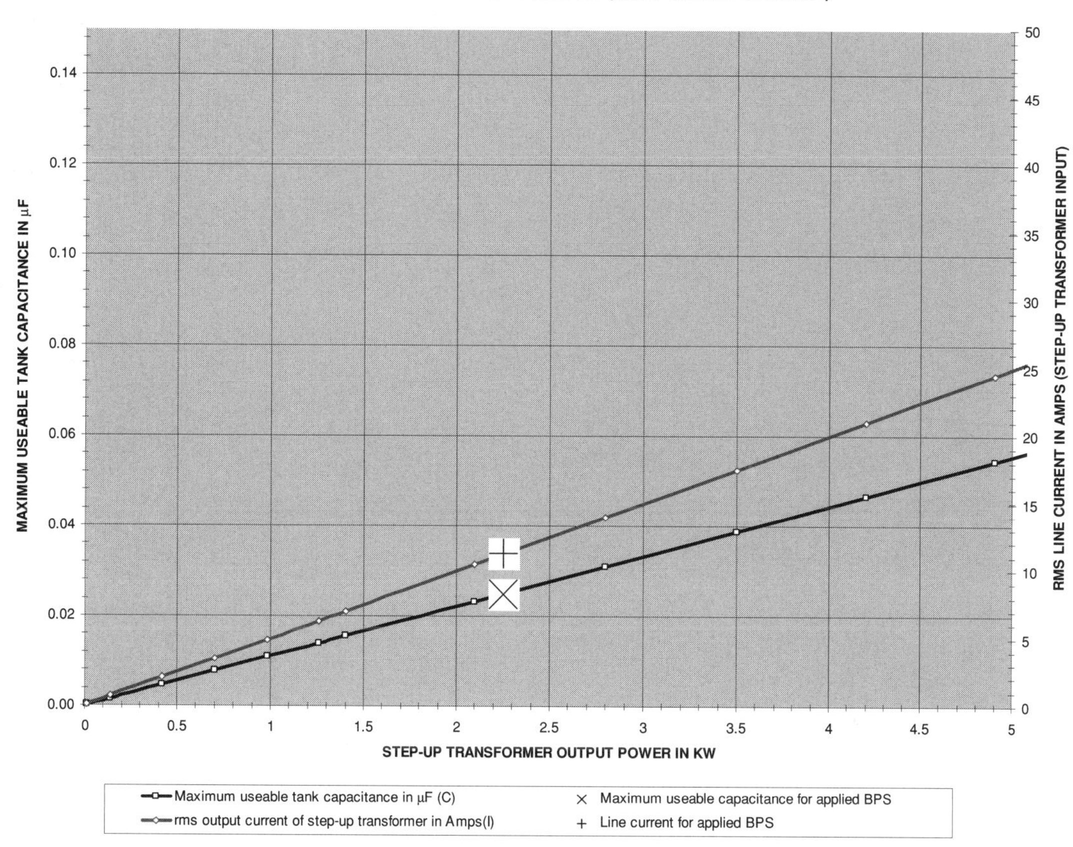

FIGURE 5-22 Maximum usable capacitance and line current for selected BPS and non–current-limited step-up transformer output.

of 200 V (14 kV output) is shown in Figure 5-22. A large [X] marks the calculated capacitance in cell (F7) and [+] marks the calculated line current in cell (F5) at the transformer output power in cell (F6).

For comparison, columns (V) through (AA) rows 18 to 49 in Table 3 were constructed for a medium spark gap coil using a 70:1 (16.1 kV output with 230-V input) non–current-limited transformer with rotary spark gap BPS of 250, 500, 750 and 1,000, which is shown in Figure 5-23.

Of course the higher BPS rates assume the rotary gap will quench between firings. Rates of 1,000 BPS and higher are attainable, but should not be presumed until the coil is actually built and tested. The BPS will be limited to the spark gap ionization and quenching characteristics, therefore the maximum usable capacitance is also limited. Section 6.11 provides details and calculations to determine the maximum rotary spark gap BPS. Higher break rates are

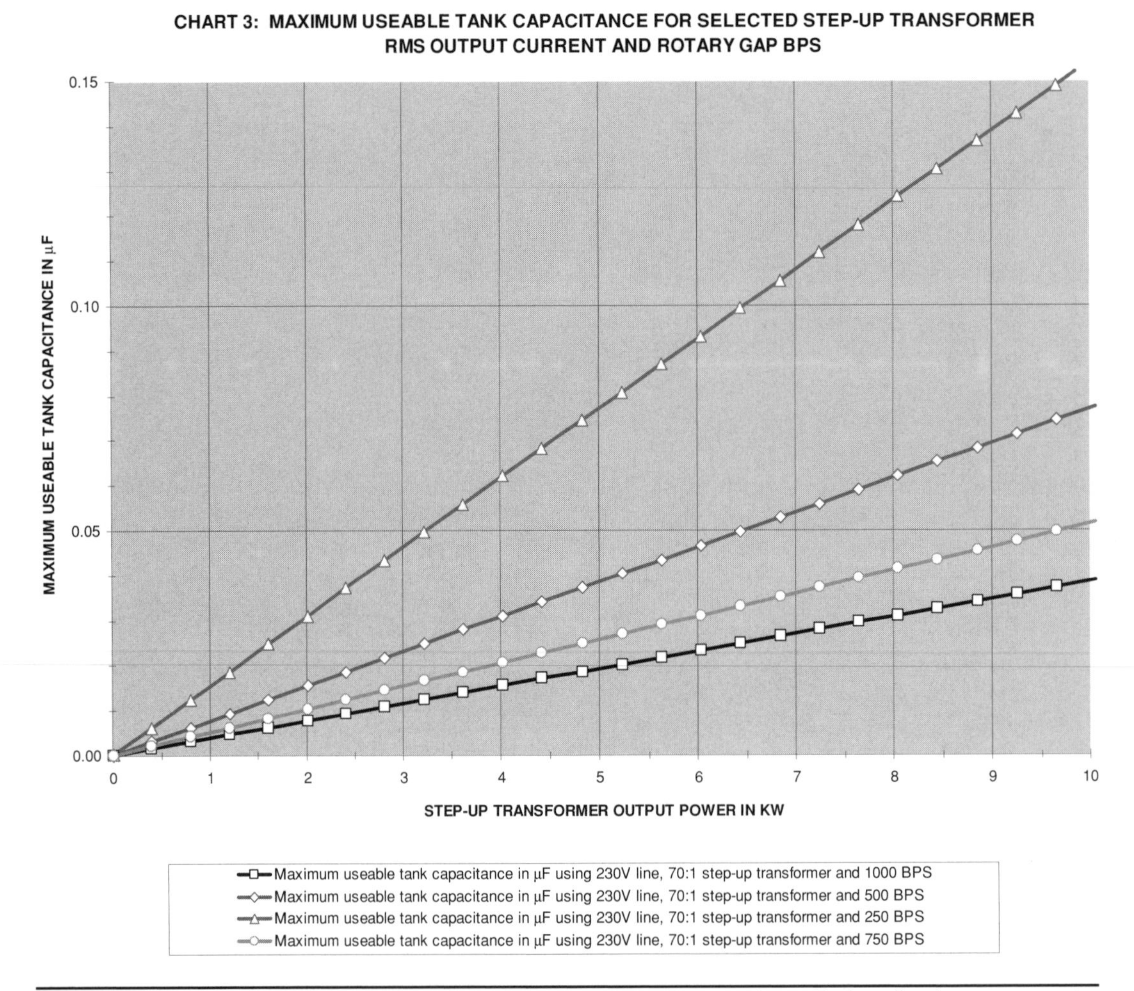

FIGURE 5-23 Maximum usable capacitance vs. power level for medium spark gap Tesla coil using rotary spark gap.

obtainable only with high primary decrements (few primary turns). Remember that higher break rates require more line current and only brighten the secondary spark which does not significantly increase the peak voltage or length of the spark. The rotary gap can be designed for 1,000 BPS and a speed control provision (a simple autotransformer for an AC motor) will allow a variable range of about 700 to 1,000 BPS. If an even number of rotary electrodes were selected in the rotary gap design, removing every other electrode on the disk will reduce the break rate to a range of 300 to 500 BPS. This would be a good design for experimenting with the BPS. When AC motors are used in a rotary gap they tend to stop running when the applied voltage decreases below about 1/3 of the nominal line voltage, which prevents a variable speed range of 0 to the maximum calculated BPS. DC motors will not have this limitation.

Any value of tank capacitance can be used with a non–current-limited transformer. The only limitation that may be encountered is the rms line current drawn by the operating coil (load). This is dependent on the turns ratio of the transformer and the rms current drawn by the coil. When using step-up transformers made for line voltages of 120 V the rms current drawn by the load can be higher than your equipment allows. For example, I am running a coil at 240 BPS that will demand 1.1 kW from the line. I want to use a 120-V line service with a 15-A circuit breaker that is installed in the house wiring and a 120:1 non–current-limited potential transformer. The calculated rms output of the secondary using equation (2.31) is 0.23 A. The spark gap is adjusted to break down at a line voltage of 40 V. The input current of the step-up transformer (line) would be 0.23 A × 120:1 = 27.6 A. The high line current requires a large variable autotransformer and is too high for the 15 A service. By using a 60:1 step-up transformer and adjusting the spark gap separation if necessary the line current drawn by the coil can be lowered to 0.23 A × 60:1 = 13.8 A. The reduced line current can utilize a smaller variable autotransformer and is not too high for the 15-A service.

Table 2, columns (I) through (M), rows 102 to 123, calculates the primary power level, line current, and maximum usable capacitance for current-limited transformer applications. The step-up transformer rated rms current limit entered in cell (B11) is increased from 10% to 100% in 5% increments in column (I) of the worksheet. Column (J) shows the calculated line current (step-up transformer input) at the corresponding primary current in column (I) for the transformer turns ratio calculated in cell (F11). The turns ratio is calculated from the rated transformer output voltage in kV entered in cell (B12) and rated line voltage entered in cell (B13). The output voltage of the step-up transformer is calculated in cell (F12) from the calculated turns ratio and the measured or estimated line voltage entered in cell (B14). The calculated transformer output power in kW is shown in column (L) at the corresponding output current in column (I) and calculated output voltage in cell (F12). This assumes the spark gap is adjusted to break down near the maximum rated output voltage of the transformer. Column (M) shows the calculated maximum usable tank capacitance for the calculated output power values in column (L). Chart 2 in rows 103 to 145 was constructed to display the maximum usable tank capacitance and line current for the selected BPS entered in cell (B15). The maximum usable tank capacitance and line current at selected power levels for a fixed gap BPS of 120, transformer output rating of 12 kV @ 0.03 A, transformer input voltage rating of 115 V and a measured line voltage of 100 V (10.4 kV output) is shown in Figure 5-24. A large [X] marks the calculated capacitance in cell (F14).

In order to run a coil using a fixed gap at optimum power levels the tank capacitance must be carefully selected since the BPS is not adjustable. The maximum usable capacitance values for a small spark gap coil, using a 15 kV, 12 kV, or 9 kV current-limited neon sign transformer up to 150 mA output current, and a fixed gap with a BPS of 120, are calculated in Table 3, columns (O) through (U) rows 18 to 49 and shown in Figure 5-25. Figure 5-25 is of general use in designing a small Tesla coil using a NST and simpler fixed gap. The data shown in the figure assumes the line current drawn by the coil is at the current-limited output of the step-up transformer and the spark gap is adjusted to provide breakdown at the full line voltage (120 V). See Section 9.4 for further details on determining the maximum usable capacitance when using NSTs.

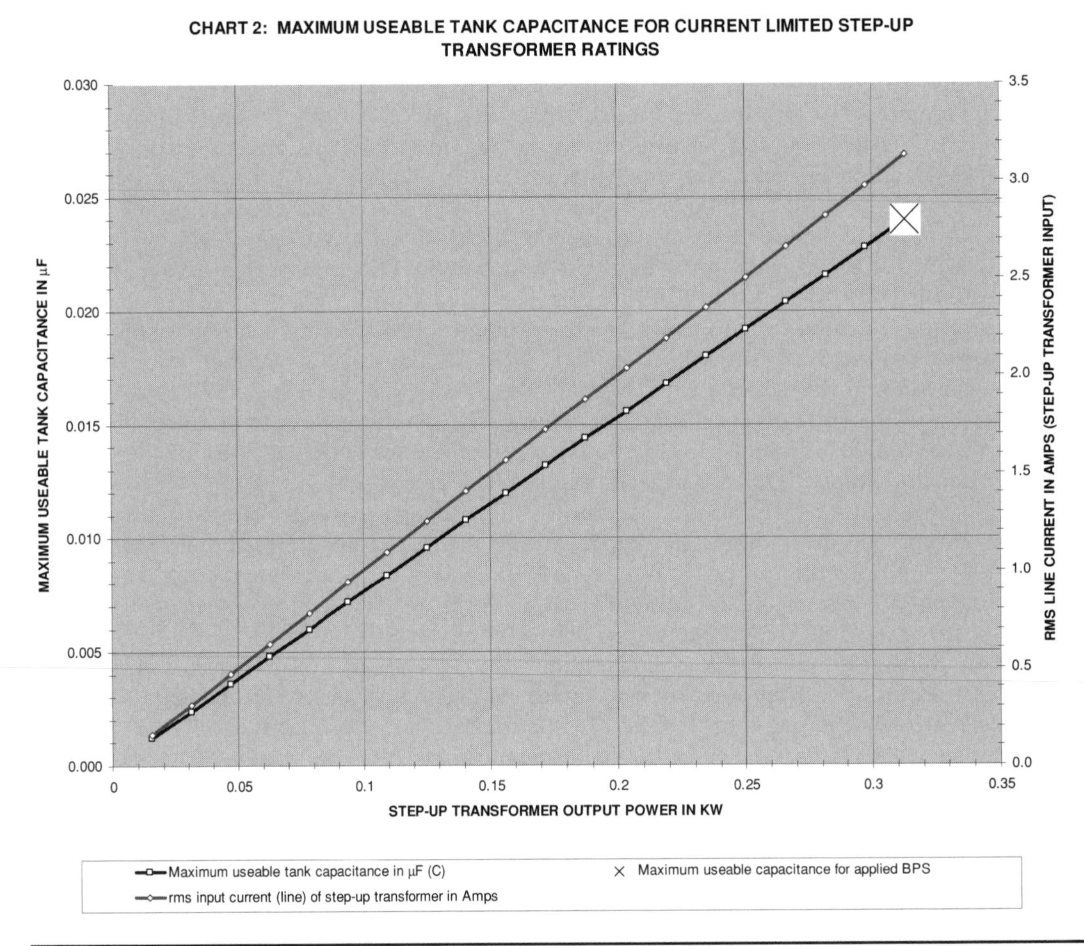

FIGURE 5-24 Maximum usable capacitance and line current for 120 BPS and current-limited step-up transformer output.

5.12 Discharging and Storing High-Voltage Capacitors

After a coil run don't forget to discharge the tank capacitor before touching anything in the primary circuit and **ENSURE THAT THE SUPPLY LINE IS DISCONNECTED!!!** The capacitor has a path of discharge through the primary winding in the tank and the secondary winding of the step-up transformer when connected as shown in Figure 2-1. If you have seen a can crusher or other dramatic demonstration of how many joules of energy these capacitors can release you will not want to take any chances on whether there is any stored charge. Work safely not quickly. Discharge the tank capacitor as shown in Figure 5-26. Connect the discharge wand to both plates of the capacitor at the same time and hold it there for a count of ten. Lift the wand slightly from one of the capacitor plates and reconnect observing for a spark or corona. If there is none the capacitor should be safely discharged. If the supply line

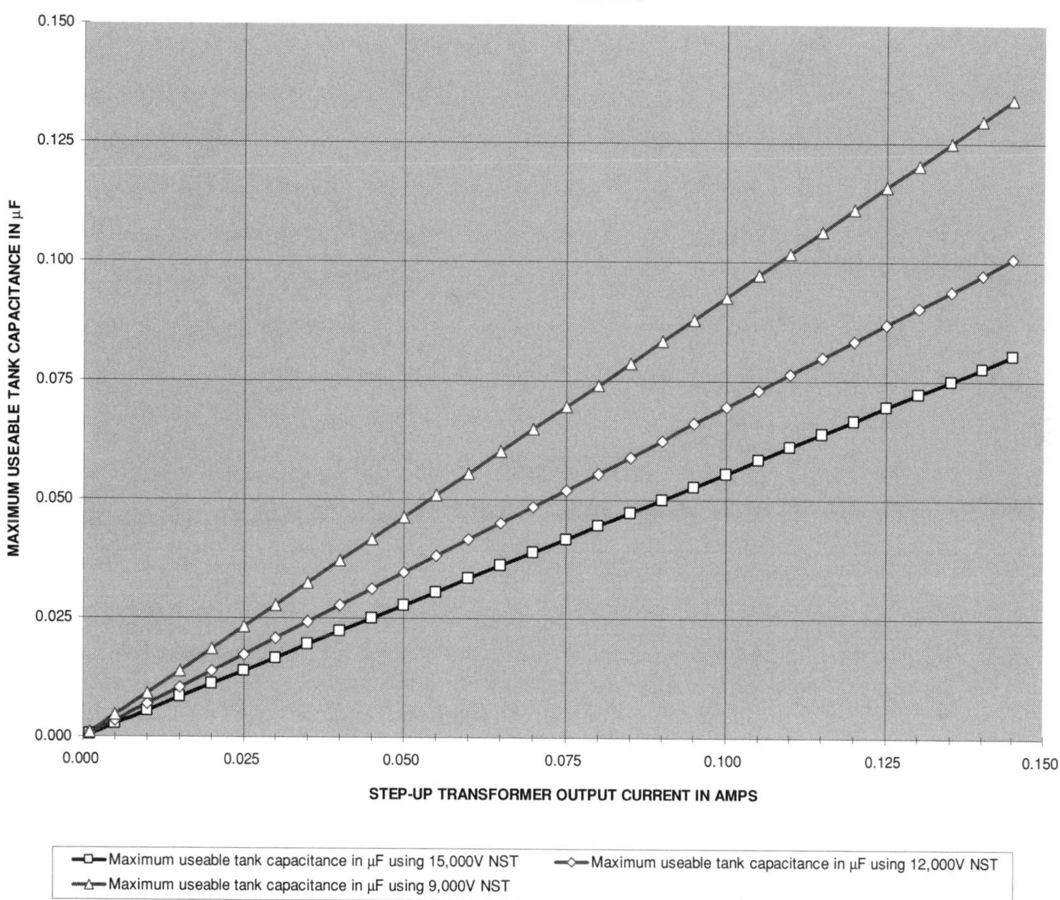

FIGURE 5-25 Maximum usable capacitance vs. power level for small spark gap Tesla coil using neon sign transformer and 120 BPS.

is still connected and/or your control scheme leaves the line connected to the high-voltage step-up transformer and you try to discharge the tank capacitor, you will short circuit the capacitor and produce a large explosion.

Construct a discharge wand from appropriate insulator and conductor materials. Use spherical or rounded conductor ends to maintain more uniform electrical fields. This will help direct and control the discharge. One of each type end is shown in Figure 5-26. If you use points on the ends and there is a good charge built up you are likely to generate a small explosion instead of a discharge. A copper toilet bowl float with a typical 4″ diameter makes a good inexpensive sphere end. The insulator can be made from G-10, plexiglass, or appropriate material with a high dielectric strength. A suitable one was made from a surplus high-voltage oscilloscope probe (1000:1). These are made to safely isolate the user from a source of up to

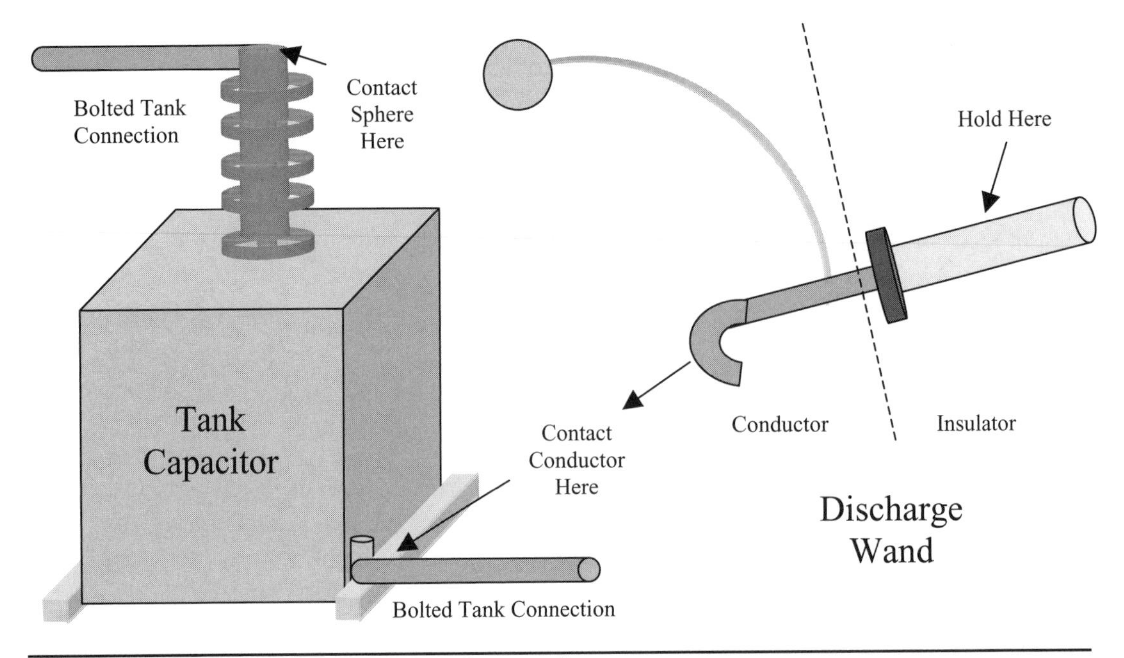

FIGURE 5-26 Discharging the tank capacitor after operation.

45 kV. Disassemble the probe and anchor a threaded brass rod into the end of the insulator handle that will be used as the conductor end with adhesive or tap receiving threads for the brass rod. Using threaded fasteners connect two spheres, rounded ends, or a combination of either to the threaded brass rod in the handle. Copper tubing provides a rigid conductor to support the sphere end as shown in the figure. Do not use solder on any of these connections unless they are also crimped, bolted, or otherwise fastened.

When storing high-voltage capacitors ensure that both plates are shorted together with wire to prevent a static charge build up. They will charge just sitting on the shelf if the plates are left unconnected. The shorting wire will keep both plates at the same electrical potential, preventing any charge accumulation.

The energy contained in a charged capacitor is calculated using equation (5.15). In a typical Tesla coil a primary capacitance of 0.043 μF and applied peak voltage of 20 kV can produce a discharge equivalent to $(0.043\ \mu\text{F} \times 20\ \text{kV}^2)/2 = 8.6$ joules. This may not seem like much but the energy contained in 8.6 joules discharged over a time period of 10 ns (see Section 1.5) can produce the same destructive effects as 8.6 Joules/10 ns = 860 mW or 1,153,000 horsepower. Think of what over a million stampeding horses could do to the tiny nerve tissues in your nervous system. Remember that 1 second is an awfully long time for electricity with most electrical activity occurring in fractions of a second. You should have a healthy respect for the energy in a capacitor.

References

1. *Reference Data For Radio Engineers*. Editor: H.P. Westman, Federal Telephone and Radio Corporation (International Telephone and Telegraph Corporation), American Book, Fourth Ed: 1956, pp. 62–71.

The reference cites contributions from the following sources: *Tables of Dielectric Materials,* vols I–IV, prepared by the Laboratory for Insulation Research of the Massachusetts Institute of Technology, Cambridge, MA. Jan 1953; and *Dielectric Materials and Applications,* A.R. von Hipple, editor. John Wiley & Sons, Inc. NYC: 1954.

2. Sprague Ceramic Capacitor Catalog. Application Notes, p. 6.
3. Pulse capacitor datasheets and application notes found on Maxwell Technologies website: http://www.maxwell.com.
4. Capacitor life formulae found on Voyager Net website: http://www.execpc.com and Illinois Capacitor, Inc. website http://www.illiniscapacitor.com/techcenter/lifecalculators.asp.
5. Ceramite Corporation Capacitor Catalog. Application Notes, pp. 32–33.
6. Terman F.E. *Radio Engineer's Handbook*. McGraw-Hill: 1943. Section 2: Circuit Elements, pp. 112–129.
7. *Handbook of Chemistry and Physics*. Editor: Charles D. Hodgeman, 30th Ed: 1947. Chemical Rubber Publishing Co., pp. 2247–2249.
8. Cornell-Dubilier Capacitor Catalog.
9. AVX Corporation Application Note: S-RTC2.5M895-R. *Equivalent Series Resistance of Tantalum Capacitors*.
10. Kemet Electronics Corporation Application Note: F-2856E 1/96. *What is a Capacitor*?
11. Montena Components: Power Capacitor (Condar) datasheet. Found on Montena website: http://www.montena.com.
12. Dr. Gary L. Johnson. *Solid State Tesla Coil*, Chapter 2—Ideal Capacitors. December 27, 2001 found on website: http://www.eece.ksu.edu/~gjohnson/tcchap2.pdf. Dr. Johnson cites references used in his work as: Moon, Parry and Domina Eberle Spencer, *Field Theory for Engineers*, D. Van Nostrand Company, Princeton, New Jersey, 1961; and Schoessow, Michael, *TCBA News*, Vol. 6, No. 2, April/May/June 1987, pp. 12–15.

6

CHAPTER

Spark Gaps

The scheme of connections illustrated in Fig. 1 [tank capacitance in shunt and spark gap in series with the step-up transformer output] has the disadvantage that the primary discharge current passes through the break hence, the resistance of the latter being large, the oscillations are quickly damped and there is besides a large current through the break which makes good operation of the latter difficult.

Nikola Tesla. Colorado Springs Notes: 1899-1900, p. 56.

Tesla observes spark gap operation.

There is more than one spark gap application in a Tesla coil. The primary tank circuit must be isolated from the line (open circuit) and a closed circuit formed for the tank current oscillations. The spark gap determines the number of times a second the damped primary oscillations are created. It is often cited as the most critical element in a Tesla coil having the greatest effect on the coil's performance. A safety gap is typically used to protect the step-up transformer and tank capacitor from damaging transients in the oscillating primary circuit. The high-voltage output of the secondary winding creates a very large spark and in simple terms is merely a very large spark gap. All of these spark gap applications will be discussed in this chapter.

6.1 Breakdown Voltage of an Air Gap

Just how much voltage does it take to ionize an air gap at a specific air temperature, pressure, and relative humidity? To answer this question requires the voltage difference across the air gap to be defined as a uniform or non-uniform field. Figure 6-1 shows ideal uniform and non-uniform air gap applications in a Tesla coil.

When the secondary winding of the resonant transformer is not terminated in its own dedicated ground or when the output of the step-up transformer is center tapped, the field in the spark gap will become non-uniform. A uniform field has approximately equal electric field distribution on each of the two gap ends. By definition a fairly uniform field can be produced using spherical or rod end gaps of the Rogowski type (resemble bullet-shaped ends). A non-uniform field has unequal electric field distribution on each of the two gap ends and includes any voltage differential to ground.

To calculate the breakdown voltage of an air gap open the CH_6.xls file, HVEF 1984 worksheet (1). (See App. B.) The worksheet was constructed from data found in reference (14). This reference was the most comprehensive source found and produced the most useful

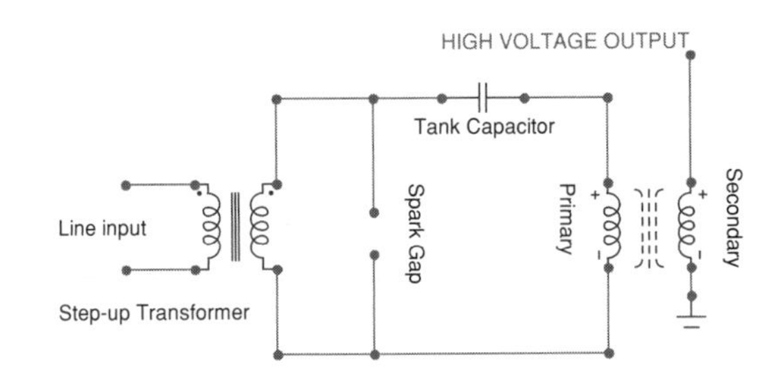

Spark Gap Producing a Uniform Field

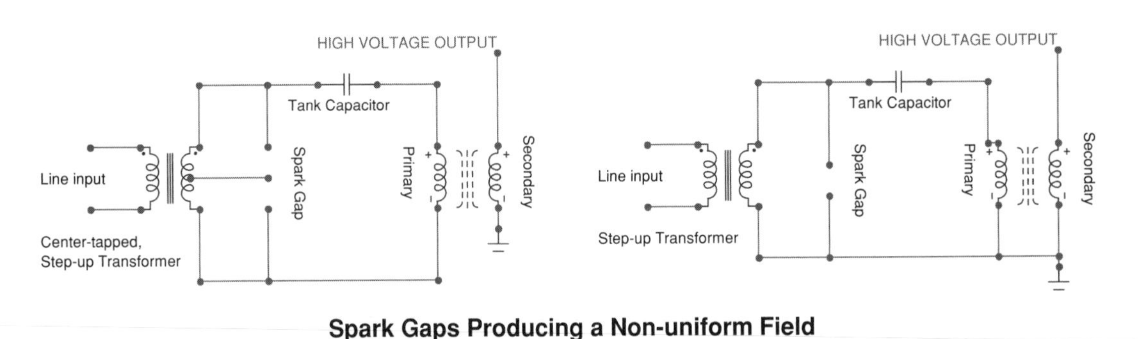

Spark Gaps Producing a Non-uniform Field

FIGURE 6-1 Ideal uniform and non-uniform air gaps found in Tesla coil construction.

calculations. Graphs of the breakdown voltage for different gap distances at nominal laboratory conditions (room temperature and sea level pressure) were generated for each gap type. The reference data also provided adjustment factors for temperature, air pressure, humidity, and proximity-to-ground effects, which were added to the worksheet calculations. The worksheet entry and calculation cells are shown in Figure 6-11.

6.2 Breakdown Voltage of an Air Gap with Spherical Ends in a Non-Uniform Field

The breakdown voltages shown in Table 2, columns (B) to (L), rows 97 to 149 and displayed on chart 2A, columns (M) to (Z), rows 95 to 137 are from data in reference (14). The data were generated using spherical gap ends and are valid for applied DC and AC voltages, negative lightning impulses, and negative switching impulses. The fourth-order polynomial curve fit formula in Chart 2A is used to calculate the breakdown voltage for a specific gap distance *greater than two inches* in cell (B22). The breakdown voltage is also adjusted for the applied air temperature, pressure, and relative humidity using the multiplier shown in cell (B20). Chart 2A is shown in Figure 6-4 at nominal conditions. A large [X] appears in all the graphs to mark the calculated value at the applied conditions entered in cells (B6) through (B9). The accuracy of the data shown in Chart 2A is ±3% for gap spacing that is less than one half the sphere diameter of the gap ends. As the gap spacing exceeds one half the sphere diameter the accuracy decreases to ±5% and can be seen to fall off the main curve in Chart 2A. Dust particles in the air were attributed as the cause for the decreasing accuracy at longer gap spacing. The breakdown voltage adjusted for the applied air temperature, pressure, and relative humidity is calculated:

$$BV = k \bullet BVn \tag{6.1}$$

Where: BV = Breakdown voltage of gap distance *greater than two inches*, adjusted for the applied air temperature, pressure, and relative humidity = calculated value in cell (B22), or interpreted value from chart 2A (Figure 6-3), or corresponding value in Table 2, columns (B) to (L), rows 97 to 149.

k = Breakdown voltage multiplier to adjust breakdown voltage at controlled laboratory conditions, for applied conditions of air temperature, pressure, and relative humidity = cell (B20).

BVn = Breakdown Voltage at controlled laboratory conditions.

NOTE: *Set cell (B6) to a value of 20°C, cell (B7) to a value of 760 torr, and cell (B8) to a value of 50% relative humidity to return the values in Table 2 and chart 2A, to nominal laboratory conditions.*

The breakdown voltage multiplier (k) is:

$$k = (-0.1061\text{RAD}^2 + 1.1102\text{RAD} - 0.004) + \text{kHL} \tag{6.2}$$

Where: k = Breakdown voltage multiplier to adjust breakdown voltage of gap distance *greater than two inches*, at controlled laboratory conditions, for applied conditions of air temperature, pressure, and relative humidity = calculated value in cell (B20). The second-order polynomial equation was developed from the curve fit formula shown in Figure 6-2.

RAD = Relative Air Density for the applied air temperature and pressure:

$$RAD = \frac{AP}{Pn} \bullet \frac{Tn}{Ta}$$

Where: RAD = Calculated value in cell (B15).

AP = Applied Air Pressure in torr = cell (B7), enter value.

Pn = Nominal air pressure in torr = 760.

Tn = Nominal air temperature in °C = 20.

Ta = Applied air temperature in °C = cell (B6), enter value. Converted to °F in cell (C6).

kHL = Breakdown voltage multiplier contribution for the applied air temperature and humidity:

$$kHL = 0.0035(AH - An)$$

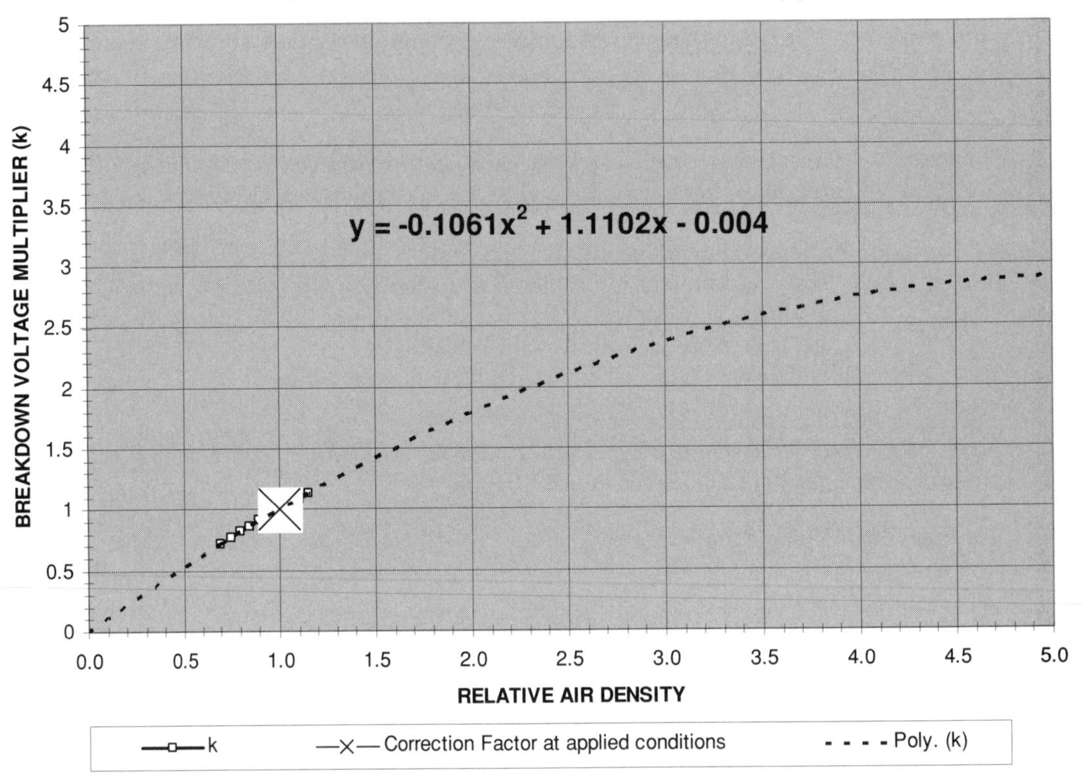

FIGURE 6-2 Derived k multiplier to adjust breakdown voltage in Figures 6-4 thru 6-8 for air temperature and pressure.

Where: kHL = Calculated value in cell (B19).
An = Absolute Humidity in g/m^3 at a nominal 50% Relative Humidity = 9.0.
AH = Applied Absolute Humidity in g/m^3:

$$AH = VP \bullet \frac{288.8}{273 + Ta}$$

Where: AH = Calculated value in cell (B18).
Ta = Applied air temperature in °C = cell (B6), enter value. Converted to °F in cell (C6).
VP = Vapor Pressure in torr (mmHg or g/m^2):

$$VP = RH \bullet SVP$$

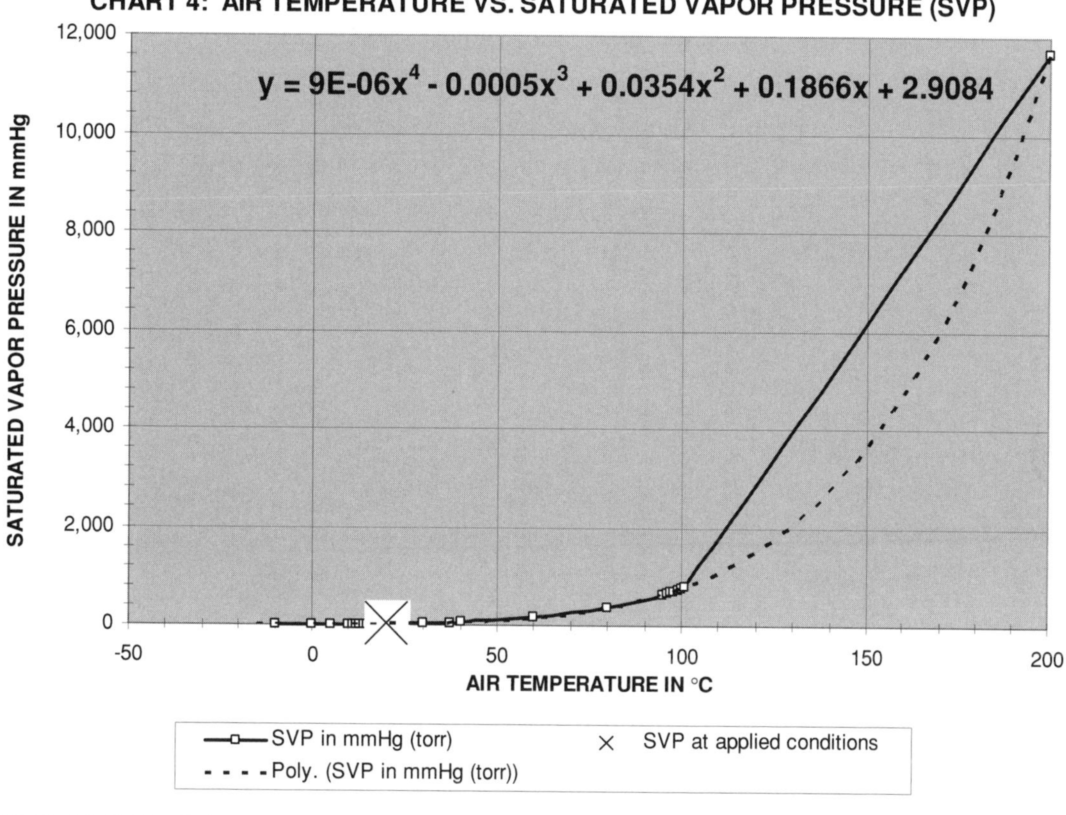

FIGURE 6-3 Derived kH multiplier to adjust breakdown voltage in Figures 6-4 thru 6-8 for humidity.

Where: VP = Calculated value in cell (B17).
RH = Relative Humidity in % = cell (B8), enter value.
SVP = Saturated Vapor Pressure in torr at applied air temperature entered in cell (B6) = calculated value in cell (B16).

NOTE: *The curve fit formula shown in Figure 6-3 was developed from reference (9) and is used to calculate SVP in cell (B16). The data used to generate Chart 4 shown in the figure can be seen in the worksheet in rows 197–228.*

The fourth-order polynomial curve fit formula in Chart 2A (Figure 6-4) was modified to calculate the breakdown voltage for a specific gap distance *equal to or less than two inches*. The second-order polynomial curve fit formula shown in the green trace in Chart 2B (Figure 6-5) is used to calculate the breakdown voltage in cell (B28) using the same methodology in equations (6.1) and (6.2). The only difference is the calculated breakdown voltage multiplier (k) in equation (6.2) is shown in cell (B26) and uses kHS, shown in cell (B25) instead of kHL in

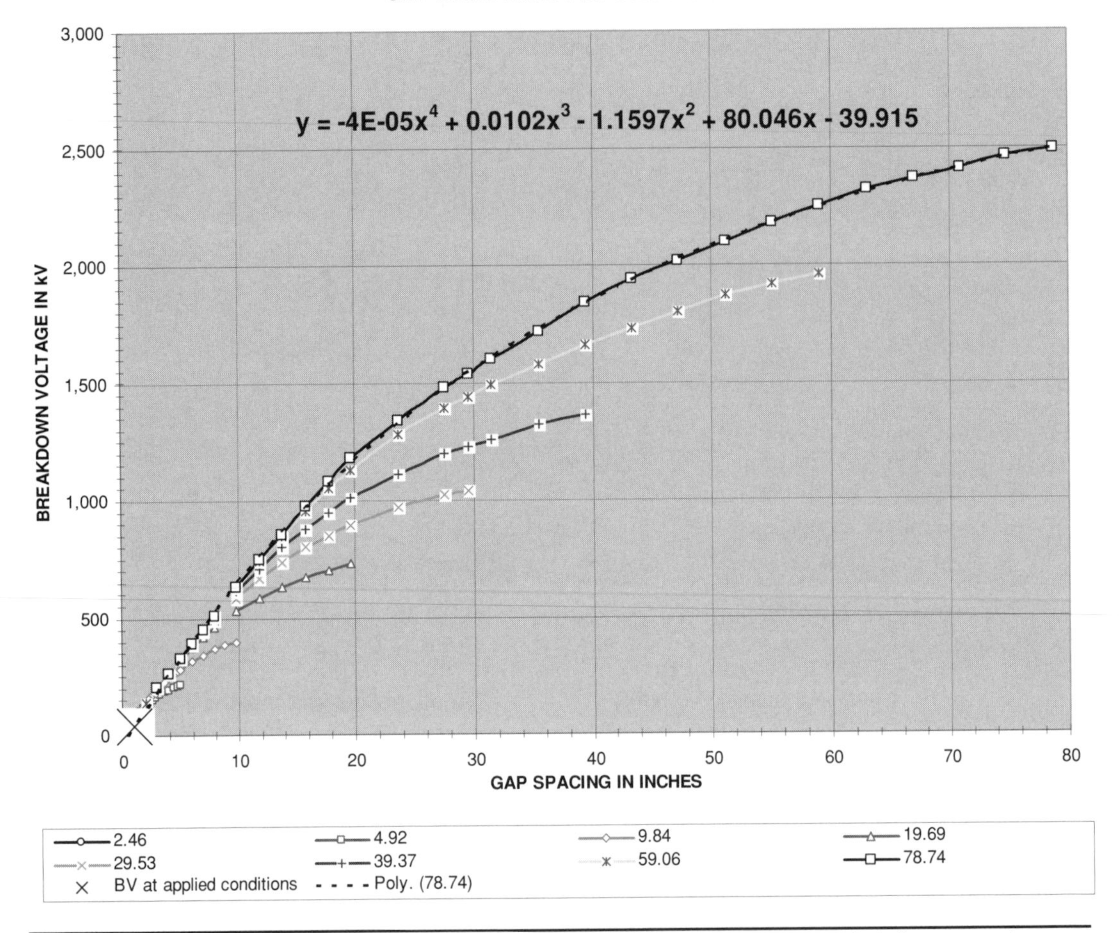

FIGURE 6-4 Breakdown voltages at nominal conditions for spherical end gaps of specified diameter and separation greater than two inches (non-uniform field).

cell (B19). The multiplier kHS is calculated:

$$kHS = 0.002(AH - An)$$

Where: kHS = Breakdown voltage multiplier contribution for the applied air temperature and humidity.
An = Absolute Humidity in g/m^3 at a nominal 50% Relative Humidity = 9.0.
AH = Applied Absolute Humidity in g/m^3.

The curve fit formula used to calculate the breakdown voltage for gap distance *greater than two inches* is also shown for comparison in Chart 2B as the black trace but is inaccurate for

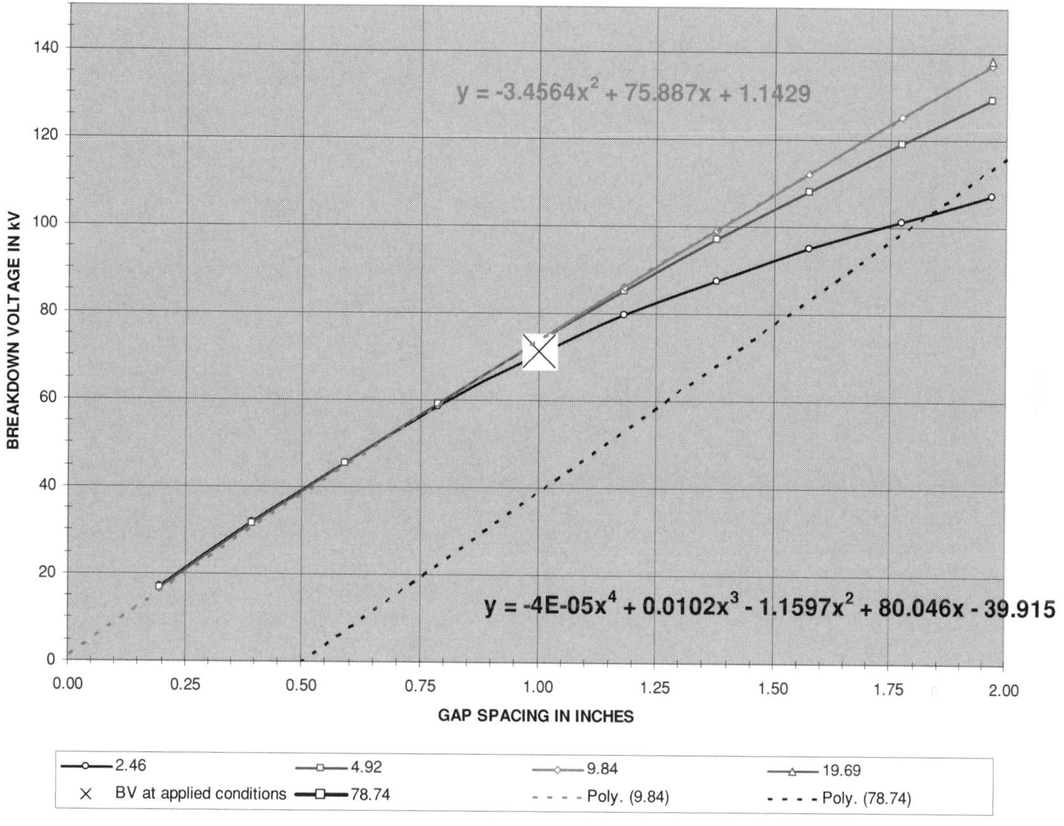

Figure 6-5 Breakdown voltages at nominal conditions for spherical end gaps of specified diameter and separation less than 2 inches (non-uniform field).

small gap distances. The breakdown voltage is also adjusted for the applied air temperature, pressure, and relative humidity using the multiplier shown in cell (B26). Chart 2B is shown in Figure 6-5 at nominal conditions.

The breakdown voltages in Table 1, columns (B) to (*k*), rows 41 to 91, were generated from the reference using spherical gap ends and are valid for positive lightning impulses, and positive switching impulses. The fourth-order polynomial curve fit formula in Chart 1A (Figure 6-6) is used to calculate the breakdown voltage in cell (B21) for a specific gap distance *greater than two inches*. The power curve fit formula shown in the green trace in Chart 1B (Figure 6-7) is used to calculate the breakdown voltage in cell (B27) for a specific gap distance *equal to or less than two inches*. Both calculations use the same methodology in equations (6.1) and (6.2). The breakdown voltage is also adjusted for the applied air temperature, pressure, and relative humidity using the multipliers shown in

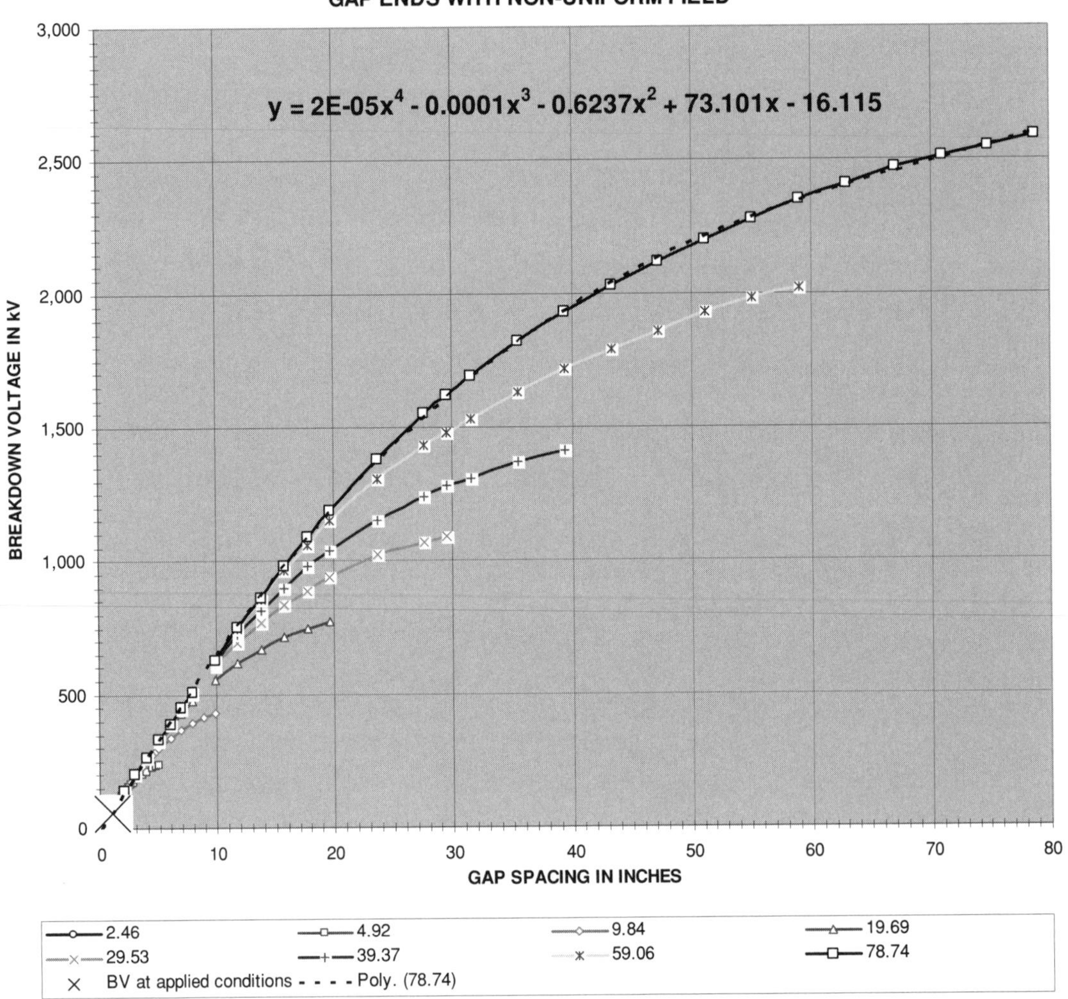

FIGURE 6-6 Breakdown voltages at nominal conditions for spherical end gaps of specified diameter and separation greater than two inches (non-uniform field) for positive lightning and switch impulses.

cells (B20) and (B26). Charts 1A and 1B are shown in Figures 6-6 and 6-7 at nominal conditions.

Note the breakdown voltage marked by the large [X] shown in Figures 6-5 (Chart 2B, calculated value in cell B28) and 6-7 (Chart 1B, calculated value in cell B27) for a 1″ gap at nominal conditions. In both cases the breakdown voltage is approximately 72 kV. Therefore the breakdown voltage required to ionize an air gap with spherical ends in a non-uniform field is defined as 72 kV/inch.

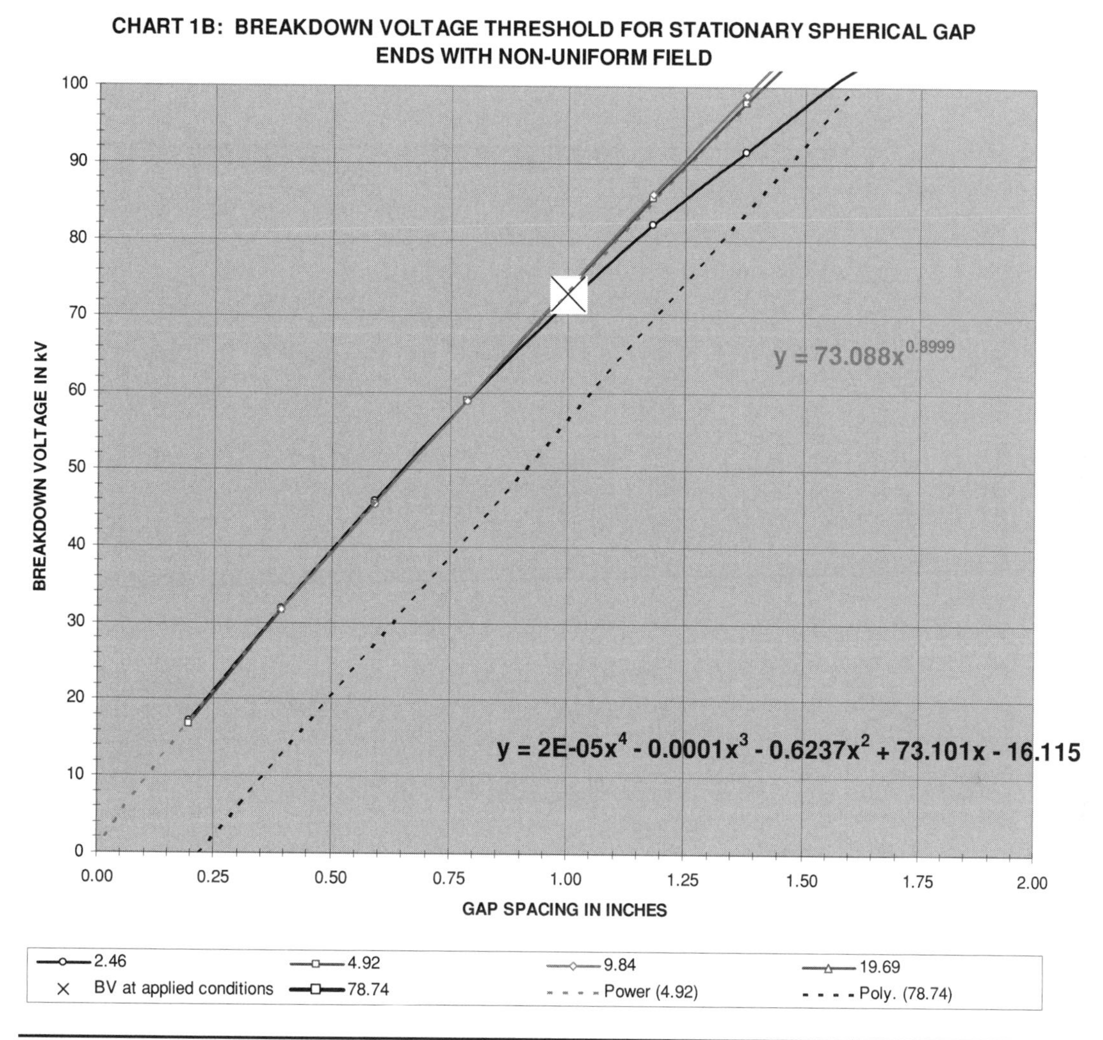

FIGURE 6-7 Breakdown voltages at nominal conditions for spherical end gaps of specified diameter and separation less than two inches (non-uniform field) for positive lightning and switch impulses.

6.3 Breakdown Voltage of an Air Gap with Rod Ends in a Non-Uniform Field

When the gap uses rod ends that are typically flat the electric field formed at the gap ends is non-uniform. Reference (14) provides equation (6.3) to calculate the breakdown voltage of an air gap using rod ends. Note there are two versions of the formula. The $\sqrt[4]{\ }$ function is equivalent to its inverse function. The inverse function of the root is the power: $1/4 = 0.25$.

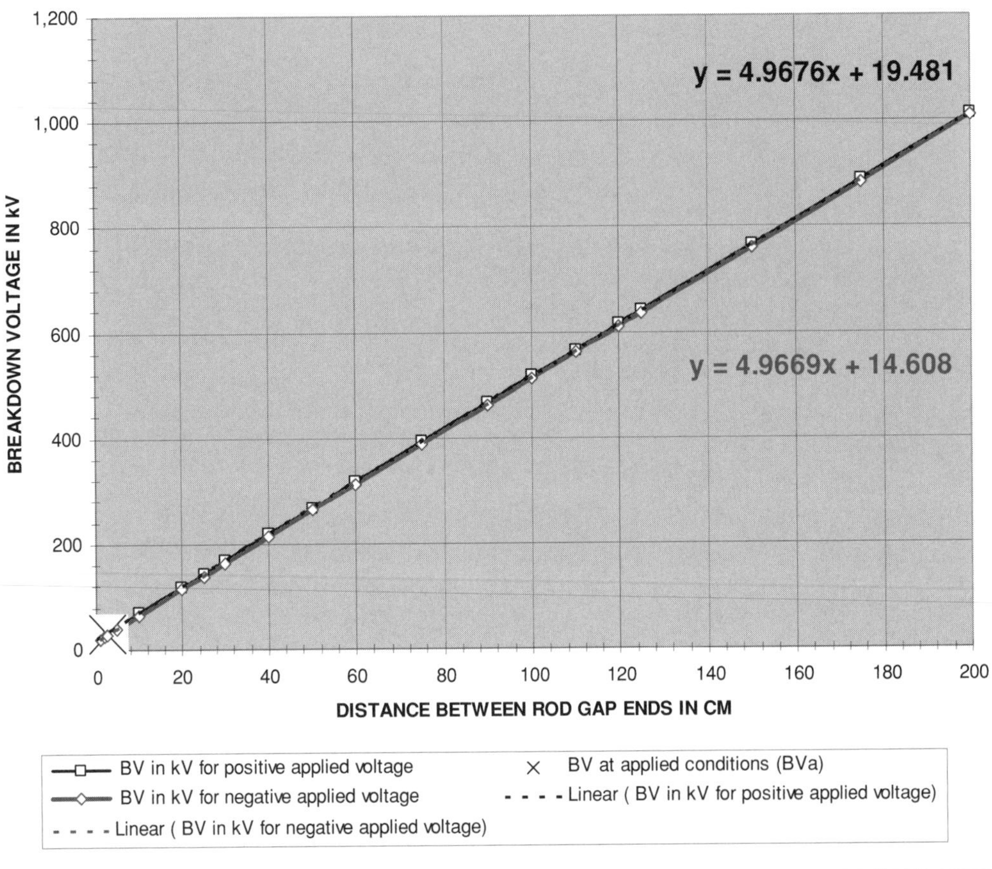

FIGURE 6-8 Breakdown voltages at nominal conditions for rod end gaps of specified separation and applied voltage transient characteristics.

Excel cannot perform root functions higher than $\sqrt[2]{}$ (inverse is $x^{0.5}$) but can perform the inverse $x^{0.25}$ function. Both equations are accurate to ±2% for relative humidity between 20 to 100% and breakdown voltages up to 1,300 kV. The $\sqrt[4]{}$ function was used in Mathcad 7.0 to generate a table of breakdown voltages (Table 6, rows 273 to 312, columns B and C) and displayed on chart 6 (rows 273 to 312, columns F to AB). Chart 6 is shown in Figure 6-8 and in detail for gap lengths less than 4 inches in Figure 6-9. A large [X] appears in the graph to mark the calculated value at the applied conditions entered in cells (B6) through (B9). The breakdown voltage is also adjusted for the applied air temperature, pressure, and relative humidity using the multiplier shown in cell (B20) for gap separations greater than 2 inches and cell (B26) for gap separations 2 inches and below. Chart 6 is shown in Figures 6-8 and 6-9 at nominal conditions. A large [X]

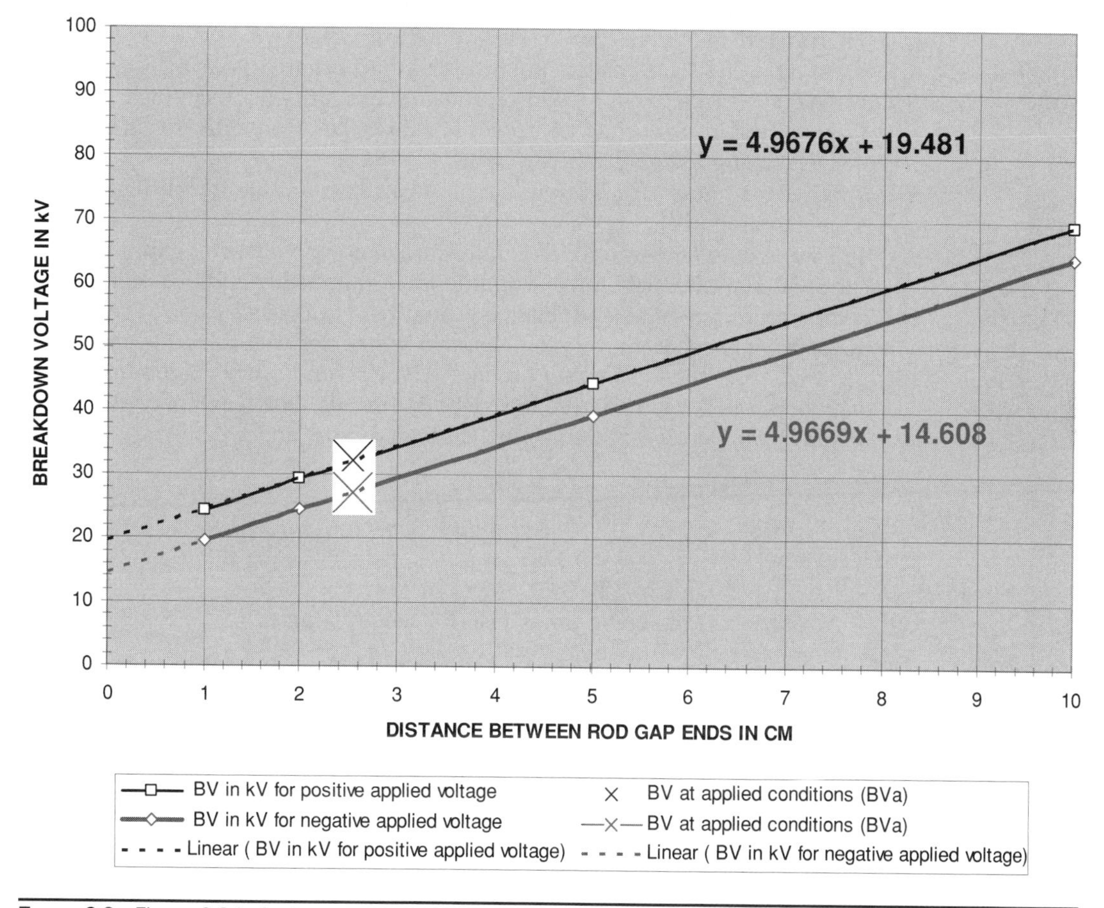

FIGURE 6-9 Figure 6-8 in further detail for gap separation <4 inches.

appears in all the graphs to mark the calculated value at the applied conditions entered in cells (B6) through (B9). The breakdown voltage adjusted for the applied air temperature, pressure, and relative humidity is calculated:

$$BV = (RAD(A + BS) \bullet \sqrt[4]{0.051(AH + 8.65)})k$$

or: (6.3)

$$BV = (RAD(A + BS) \bullet [0.051(AH + 8.65)]^{0.25})k$$

Where: BV = Breakdown voltage of gap distance adjusted for the applied air temperature, pressure, and relative humidity = calculated value in cell (B38) for positive applied voltage or cell (B39) for negative applied voltage, or interpreted value from chart 6 (Figures 6-8 and 6-9).
A = constant = 20 kV for positive polarity, 15 kV for negative polarity.
B = constant = 5.1 kV/cm for positive or negative polarity (2.01 kV/inch).
S = Gap spacing in cm = cell (B9), enter value in inches. Converted to cm in cell (C9) using 2.54 multiplier.
AH = Applied Absolute Humidity in g/m^3 = calculated value in cell (B18) from equation (6.2).
k = Breakdown voltage multiplier to adjust breakdown voltage of gap distance at controlled laboratory conditions, for applied conditions of air temperature, pressure, and relative humidity = calculated value in cell (B20) for gap separation *greater than two inches*, and cell (B26) for gap separation *equal to or less than two inches*. The second-order polynomial equation was developed from the curve fit formula shown in Figure 6-2.
RAD = Relative Air Density for the applied air temperature and pressure:

$$RAD = \frac{AP}{Pn} \bullet \frac{Tn}{Ta}$$

Where: RAD = Calculated value in cell (B15).
AP = Applied Air Pressure in torr = cell (B7), enter value.
Pn = Nominal air pressure in torr = 760.
Tn = Nominal air temperature in °C = 20.
Ta = Applied air temperature in °C = cell (B6), enter value. Converted to °F in cell (C6).

Note the breakdown voltage marked by the large [X] shown in Figure 6-9 (Chart 6, calculated value in cells B38 and B39) for a 1″ (2.54 cm) gap at nominal conditions. In both cases the breakdown voltage is less than one half the 72 kV/inch required in spherical gaps. The 5 kV difference between the breakdown voltage with a positive voltage applied and the lower breakdown voltage with a negative voltage applied can also be seen. These differences will become important when setting the initial separation for a desired breakdown voltage in a gap.

6.4 Breakdown Voltage of an Air Gap in a Uniform Field

When the electric field formed at the gap ends is uniform the breakdown characteristics will follow Paschen's Law which states the sparking potential of a gas in a uniform field depends only on the product of the gas pressure and the electrode separation. Paschen's Law can be found in reference (14), pages 354–361 and shown in equation (6.4) to calculate the breakdown voltage of an air gap in a uniform field. The breakdown voltage is also adjusted for the applied air temperature, pressure, and relative humidity using the multiplier shown in cell (B20) for gap separations greater than 2 inches and cell (B26) for gap separations 2 inches and below. The breakdown voltage adjusted for the applied air temperature, pressure, and relative humidity is calculated:

$$BV = (6.72\sqrt{APbS} + 24.36APbS) \bullet k \tag{6.4}$$

Where: BV = Breakdown voltage of gap distance adjusted for the applied air temperature, pressure, and relative humidity = calculated value in cell (B35).

S = Gap spacing in cm = cell (B9), enter value in inches. Converted to cm in cell (C9) using 2.54 multiplier.

k = Breakdown voltage multiplier to adjust breakdown voltage of gap distance at controlled laboratory conditions, for applied conditions of air temperature, pressure, and relative humidity = calculated value in cell (B20) for gap separation *greater than two inches*, and cell (B26) for gap separation *equal to or less than two inches*. The second-order polynomial equation was developed from the curve fit formula shown in Figure 6-2.

APb = Air pressure in bar:

$$APb = AP \bullet 0.001333$$

Where: APb = Calculated value in cell (B34).

AP = Applied Air Pressure in torr = cell (B7), enter value. Nominal: 760 torr = 1.013 bar.

Note the calculated breakdown voltage for a 1″ gap at nominal conditions is 73.5 kV. This agrees with the generally accepted rule in high voltage of 75 kV/inch required to break down an air gap. It is assumed this general rule applies to a uniform electric field. Spherical gaps produce an approximately uniform field as shown by the 72 kV/inch concluded in Section 6.2 and the 73.5 kV/inch from Pashen's Law. In conclusion, under nominal conditions an air gap exhibits the following characteristics:

- The breakdown voltage for an air gap with a uniform field distribution is 73.5 kV/inch from Pashen's Law.
- The breakdown voltage for an air gap using spherical ends with a non-uniform field distribution is 72 kV/inch from Section 6.2.
- The breakdown voltage for an air gap, using rod ends with a non-uniform field distribution, is 27 kV/inch for an applied negative voltage and 32 kV/inch for an applied positive voltage from Section 6.3. When the rod ends are rounded or of the Rogowski shape the field becomes more uniform and the breakdown voltage moves closer to that characterized in Section 6.2.

6.5 Breakdown Voltage and the Proximity-to-Ground Effect

The proximity of the gap to the ground plane will affect the breakdown threshold of the gap. A graph of measured effects was developed from data in reference (14). The test conditions

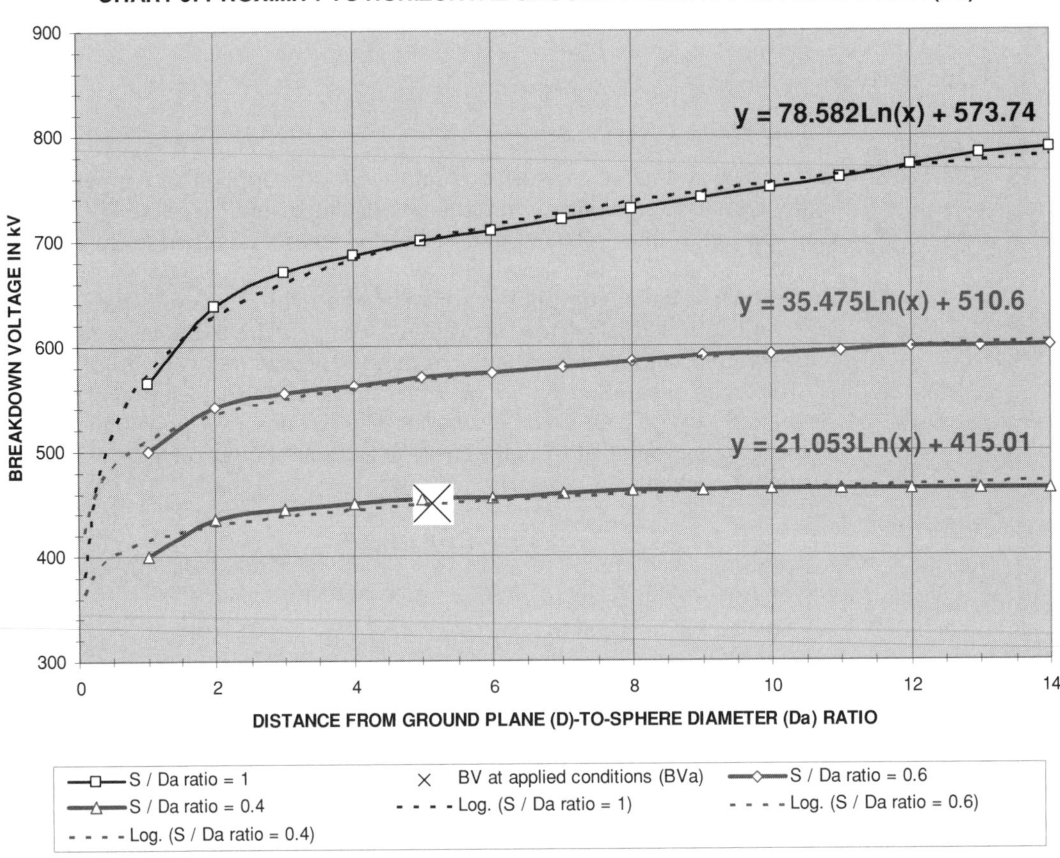

FIGURE 6-10 Proximity-to-ground effect characteristics.

used to generate the data were a negative impulse voltage of 1.5 μs to 40 μs duration applied to a 50-cm (20 inch) diameter spherical gap.

The gap separation (S)–to–sphere diameter (Da) dimensions were changed to develop the three ratios (0.4, 0.6, and 1.0) displayed in Chart 5 and shown in Figure 6-10. The calculated S-to-Da ratio is shown in cell (B253) for the S value entered in cell (B9) and the Da value entered in cell (B12). With the proximity-to-ground plane effect isolated the nominal breakdown voltage (BVn) for an S-to-Da ratio of 0.4 is 460 kV, for an S-to-Da ratio of 0.6 is 585 kV, and for an S-to-Da ratio of 1.0 is 730 kV.

The distance from the spherical gap to the ground plane (D) was also changed to produce the D–to–sphere diameter (Da) ratio on the X-axis of the chart. The calculated D-to-Da ratio is shown in cell (B252) for the D value entered in cell (B11) and the Da value entered in cell (B12). The Y-axis is the corresponding breakdown voltage measured for each of the three S-to-Da ratios. Chart 5 (rows 233 to 268) is shown in Figure 6-10 at nominal conditions.

The three logarithmic curve fit formulae shown in Figure 6-10 were used with a conditional statement to calculate the breakdown voltage (BVa) in cell (B251) for the applied D-to-Da and S-to-Da ratios. The applied breakdown voltage (BVa)–to–nominal breakdown voltage (BVn) ratio calculated in cell (B31) can be used as a breakdown voltage multiplier (kGP) to adjust the applied breakdown voltages calculated in equations (6.2) through (6.4) to account for the proximity-to-ground plane effect. The multiplier is not used in the worksheet calculations and must be performed manually after the applied conditions are entered in cells (B6) through (B12).

In conclusion, the proximity-to-ground plane effect exhibits the following characteristics:

- The terminal capacitor diameter is inversely proportional to the breakdown voltage. When the proximity-to-ground plane effect is not considered or is isolated in laboratory conditions the larger the diameter of the spherical or toroidal capacitance the higher the breakdown voltage threshold becomes. However, when the proximity–to–ground plane effect is considered for a fixed height above ground the larger the sphere diameter (Da) the lower the breakdown voltage threshold becomes. Da is comparable to the high-voltage terminal of the Tesla coil when a toroid or spherical terminal capacitance is used. For a predetermined terminal height from ground (D), as the terminal diameter (Da) is increased, the lower the applied S/Da and D/Da ratios become. This decreases the breakdown voltage threshold of the terminal by decreasing the proximity–to–ground plane effect multiplier (kGP) by as much as 25% as the terminal diameter equals the terminal height from ground.
- In a Tesla coil this does not necessarily mean that the spark will be 25% longer. As the terminal is lowered toward the ground plane the voltage rise on the terminal will break out at a lower voltage than if the terminal is elevated. To achieve longer sparks it is necessary to generate a higher voltage in the secondary, elevate the terminal higher above the ground plane (D) or increase the terminal diameter (Da). When the terminal diameter is increased the terminal capacitance increases which decreases the resonant frequency of the secondary. Don't forget to include these effects in your design calculations. Design compromises can optimize this parameter for your laboratory space.

6.6 Additional Breakdown Voltage Tables

Similar high-voltage breakdown threshold data were developed using several other sources for comparison. CH_6.xls file, SPHERICAL GAP ENDS worksheet (2) was developed from reference (4) for spherical gap ends. (See App. B.) By entering the air temperature into cell (B6) and air pressure in cell (B7) the air density (d) is calculated in cell (B12). Graph 3 (rows 121–158) displays the second-order polynomial curve fit formula used to calculate the air density multiplier (kP) in cell (B13). Entering the relative humidity in cell (B8) calculates the vapor pressure (VP) in cell (B15). The SVP parameter in the calculation is the same saturated vapor pressure shown in Figure 6-3 and equation (6.2). Graph 4 (rows 171–209) displays the second-order polynomial curve fit formula used to calculate the air temperature and humidity multiplier (kH) in cell (B16). The air density multiplier (kP) in cell (B13) is combined with the air temperature and humidity multiplier (kH) in cell (B16) for a combined effects correction factor (k) in cell (B17).

Table 1 in columns (B) to (L), rows 25 to 52, displays the measured breakdown voltage for selected sphere diameters and gap separation in inches for a non-uniform electric field. One gap end was grounded and an AC impulse was applied to the gap. The combined effects correction factor (k) in cell (B17) is applied to the measured values in Table 1. Chart 1A displays the measured values in Table 1 and the second-order polynomial equation used to calculate the breakdown voltage of a non-uniform field for large gap spacing in cell (B18) for the gap spacing (S) entered in cell (B9). Chart 1B shows the measured values in Table 1 and the power equation used to calculate the breakdown voltage of a non-uniform field for small gap spacing in cell (B22) for the gap spacing (S) entered in cell (B9).

Table 2 in columns (B) to (L), rows 75 to 104, displays the measured breakdown voltage for selected sphere diameters and gap separation in inches for a uniform electric field. The combined effects correction factor (k) in cell (B17) is applied to the measured values in Table 2. Chart 2A shown in Figure 6-12 displays the measured values in Table 2 and the second-order polynomial equation used to calculate the breakdown voltage of a uniform field for large gap spacing in cell (B19) for the gap spacing (S) entered in cell (B9). Chart 2B in Figure 6-13 shows the measured values in Table 2 and the power curve fit equation used to calculate the breakdown voltage of a uniform field for small gap spacing in cell (B23) for the gap spacing (S) entered in cell (B9). A large [X] marks the calculated result in cell (B19) on Chart 2A and cell (B23) on Chart 2B.

The measurements used in Tables 1 and 2 of the worksheet were recorded at the nominal conditions of 20°C, 760 torr, and 80% relative humidity. For gap separation less than one half the sphere diameter the accuracy is ±3% increasing to ±5% as the gap separation becomes greater than one half the sphere diameter. This can be seen in the charts as the measurements fall off the main curve indicated by the trendline. A one-inch air gap at nominal conditions with spherical gap ends requires a breakdown voltage of 72.5 kV/inch in a uniform field (Figure 6-13) and 71.8 kV/inch in a non-uniform field using this reference.

CH_6.xls file, ROD GAP ENDS worksheet (3) was also developed from reference (4) for rod gap ends. (See App. B.) Worksheet 3 uses the same methodology as worksheet 2 for a gap using rod ends instead of spherical terminals. The multiplier (k) was also derived in the same manner. A variety of voltage characteristics were applied to the gap including DC and AC impulses of both polarities with 0.2 μs and 0.02 μs durations. Table 1 in columns (B) to (I), rows 22 to 42, displays the measured breakdown voltage for selected applied voltage characteristics and gap separation in inches for a non-uniform electric field. The combined effects correction factor (k) in cell (B17) is applied to the measured values in Table 1. Chart 1A shown in Figure 6-14 displays the measured values in Table 1 and the second-order polynomial equation used to calculate the breakdown voltage for an applied AC impulse in cell (B18) for the gap spacing (S) entered in cell (B9).

Chart 1A also displays the sixth-order polynomial equation used to calculate the breakdown voltage for an applied DC voltage in cell (B19) for the gap spacing (S) entered in cell (B9). The sixth-order polynomial equation produces excessive error with gap separations greater than 15 inches therefore Chart 1B shown in Figure 6-15 provides more detail for the first 5 inches of gap separation. A large [X] marks the result in cell (B18) and a large [+] marks the result in cell (B19) on both charts.

A	B	C–K	ROW
For all calculations:	**ENTER VALUES**		
Air Temperature in °C (**Ta**)	20.0 °C	68.0 °F	6
Air Pressure in torr (**AP**)	760		7
Relative Humidity in Percent (**RH**)	50%		8
Distance between gap ends in inches (**S**)	1.000 inches	2.54cm	9
For proximity of ground plane effect calculation:			10
Distance between gap end and ground plane in inches (**D**)	72.0 inches	182.88cm	11
Diameter of spherical gap end in inches (**Da**)	14.0 inches	35.56cm	12
			13
For larger diameter spherical gap ends and spacings:	**CALCULATED VALUES**		14
Relative Air Density (**RAD**)	1.00	RAD = (AP / 760)*(20 / Ta) where: 760 is nominal air pressure in torr, 20 is nominal air temp in °C	15
Saturated Vapor Pressure (**SVP**)	18.2	SVP = 0.000009*Ta^4 - 0.0005*Ta^3 + 0.0354*Ta^2 + 0.1866*Ta + 2.9084 from curve fit formula in Chart 4	16
Vapor Pressure in torr [mmHg or g/m^2] (**VP**)	9.1	VP = RH * SVP @ T_A	17
Absolute Humidity in g/m^3 (**AH**)	9.0	AH = VP * (288.8 / [273+Ta])	18
Humidity effect multiplier for large diameters and spacings (**kHL**)	0.000	kHL = (AH - 9) * 0.0035 where 9.0 is AH at 50% RH	19
Temperature-Pressure-Humidity Correction Factor (**k**)	**1.00**	k = (-0.1061*RAD^2 + 1.1102*RAD - 0.004) + kHL from curve fit formula in Chart 3	20
Breakdown Voltage for Chart 1A (**BV1A**)	**56.4 kV**	BV1A = (2E-05*S^4 - 0.0001*S^3 - 0.6238*S^2 + 73.109*S - 16.116)*k from curve fit formula in Chart 1A for nominal conditions	21
Breakdown Voltage for Chart 2A (**BV2A**)	**39.0 kV**	BV2A = (-4E-05*S^4 + 0.0102*S^3 - 1.1598*S^2 + 80.055*S - 39.92)*k from curve fit formula in Chart 2A for nominal conditions	22
			23
For smaller diameter spherical gap ends and spacings:			24
Humidity effect multiplier for small diameters and spacings (**kHS**)	0.000	kHS = (AH - 9) * 0.002 where 9.0 is AH at 50% RH	25
Temperature-Pressure-Humidity Correction Factor (**k**)	**1.00**	k = (-0.1061*RAD^2 + 1.1102*RAD - 0.004) + kHS from curve fit formula in Chart 3	26
Breakdown Voltage for Chart 1B (**BV1B**)	**73.1 kV**	BV1B = (73.088*S^0.8999)*k from curve fit formula in Chart 1B for nominal conditions	27
Breakdown Voltage for Chart 2B (**BV2B**)	**71.3 kV**	BV2B = (-3.4564*S^2 + 75.887*S + 1.1429)*k from curve fit formula in Chart 2B for nominal conditions	28
			29
Additional Correction Factors:			30
Proximity of parallel ground plane effect multiplier (**kGP**)	0.977	From data in rows 226 to 261	31
			32
For uniform fields:			33
Air pressure in bar (**APb**)	1.013	APb = AP * 0.001333 760 torr = 1.013 bar	34
Breakdown Voltage for uniform fields (**BVU**)	**73.5 kV**	BVU = 6.72*sqrt(APb*S) + 24.36*(APb*S)*k from Paschen's Law: E. Kuffel and W.S. Zaengl.	35
		High Voltage Engineering: Fundamentals. Pergamon Press, Oxford: 1984. Pp.354-361.	36
For rod gap ends:			37
Breakdown Voltage for rod gap ends and (+) voltage applied (**BVR+**)	**32.1 kV**	BVR+ = (RAD*(20+5.1*S)*(0.051*[AH+8.65])^0.25)*k or (4.9676*S + 19.481)*k from curve fit formula in Chart 6 for nominal conditions	38
Breakdown Voltage for rod gap ends and (-) voltage applied (**BVR-**)	**27.2 kV**	BVR- = (RAD*(15+5.1*S)*(0.051*[AH+8.65])^0.25)*k or (4.9669*S + 14.608)*k from curve fit formula in Chart 6 for nominal conditions	39
A	B	C D E F G H I J K	COLUMN

FIGURE 6-11 Breakdown voltage worksheet.

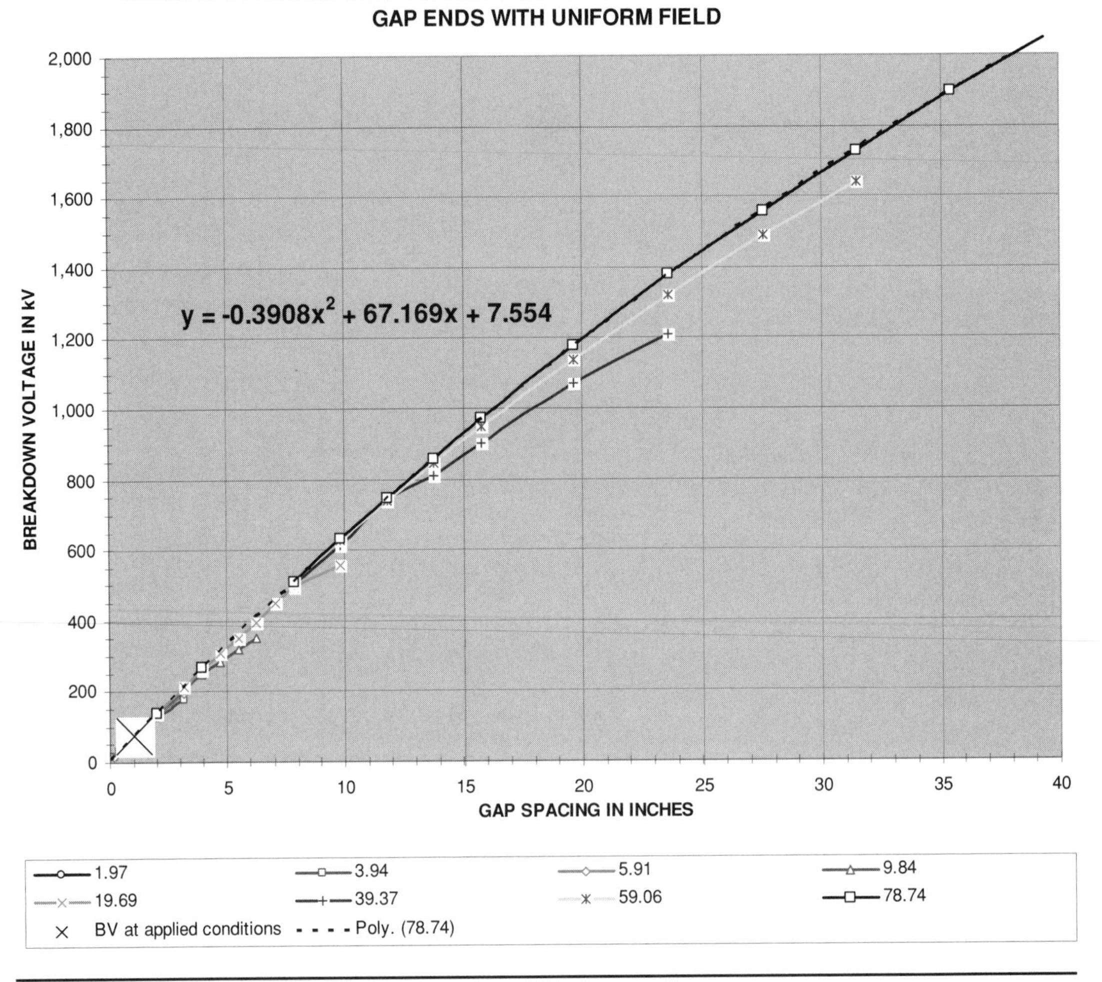

FIGURE 6-12 Breakdown voltage for gap separation >2″ (reference 4).

The measurements used in Table 1 of the worksheet were recorded at the nominal conditions of 27°C, 760 torr (1,013 mbar), and 55% relative humidity (15.5 torr water vapor pressure). Note under nominal conditions the multiplier is 0.98 resulting from the elevated air temperature (i.e., the nominal air temperature of 20°C from previous measured data is a nominal 27°C for this measured data). The accuracy of the measurements is specified as ±8%. A one-inch air gap at nominal conditions with rod gap ends requires a breakdown voltage of 33.5 kV/inch with an applied AC impulse and 32.7 kV/inch (Figure 6-15) with an applied DC voltage using this reference. The calculations developed from reference (4) agree quite well with those from reference (14).

Another breakdown voltage chart was constructed in the CH_6.xls file, AIEE 1947 worksheet (4), using reference (5) for selected sphere diameters and needle points. (See App. B.) By

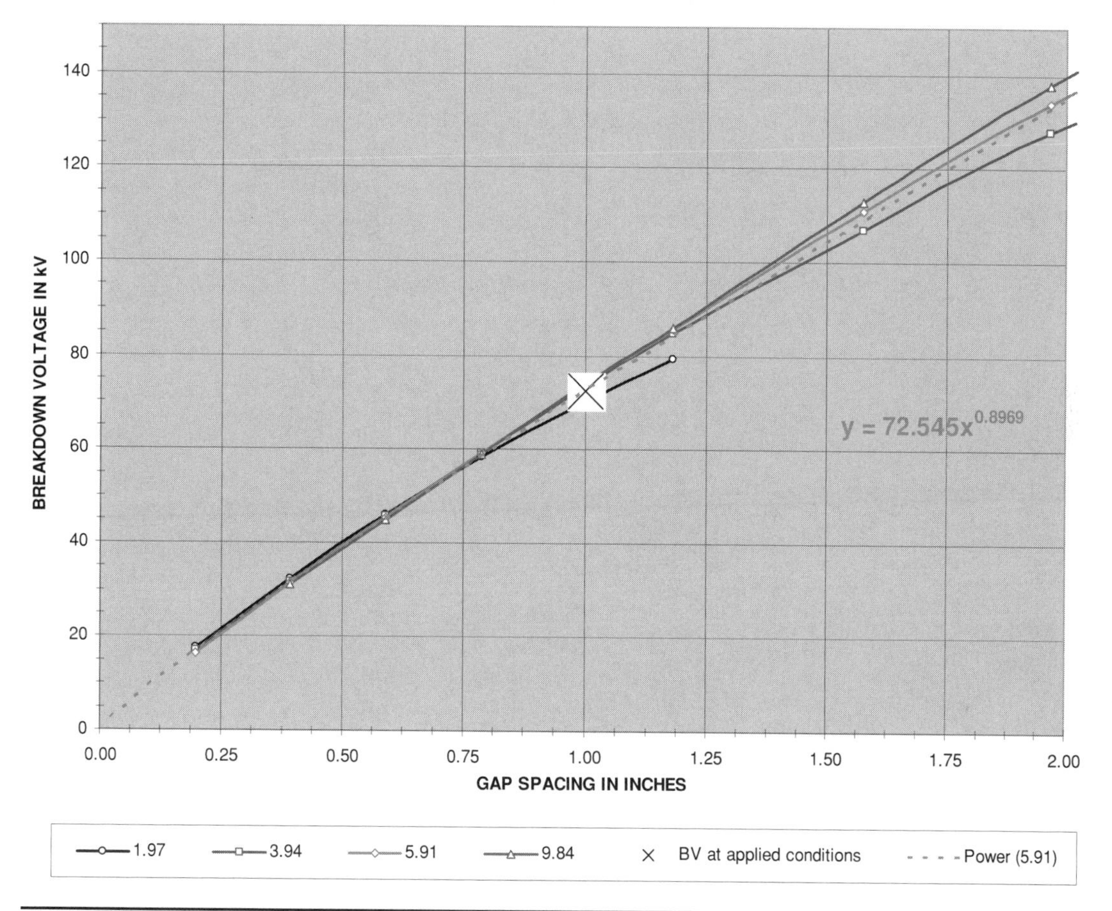

FIGURE 6-13 Breakdown voltage for gap separation <2″ (reference 4).

entering the air temperature into cell (B6) the air temperature correction factor (kT) is calculated in cell (B11). Graph 2A (rows 66–110) displays the linear curve fit formula used to calculate the air temperature correction factor (kT) in cell (B11). Entering the air pressure in cell (B7) the air pressure correction factor (kP) is calculated in cell (B12). Graph 2B (rows 66–110) displays the power curve fit formula used to calculate the air pressure correction factor (kP) in cell (B12). There was no correction provided for humidity. The air temperature multiplier (kT) in cell (B11) is combined with the air pressure multiplier (kP) in cell (B12) for a combined effects correction factor (k) in cell (B13).

Table 1 in columns (B) to (I), rows 18 to 135, displays the measured breakdown voltage for selected sphere diameters and gap separation in inches for an applied DC voltage with a uniform electric field. The combined effects correction factor (k) in cell (B13) is applied to the measured values in Table 1. Chart 1A shown in Figure 6-16 displays the measured values in

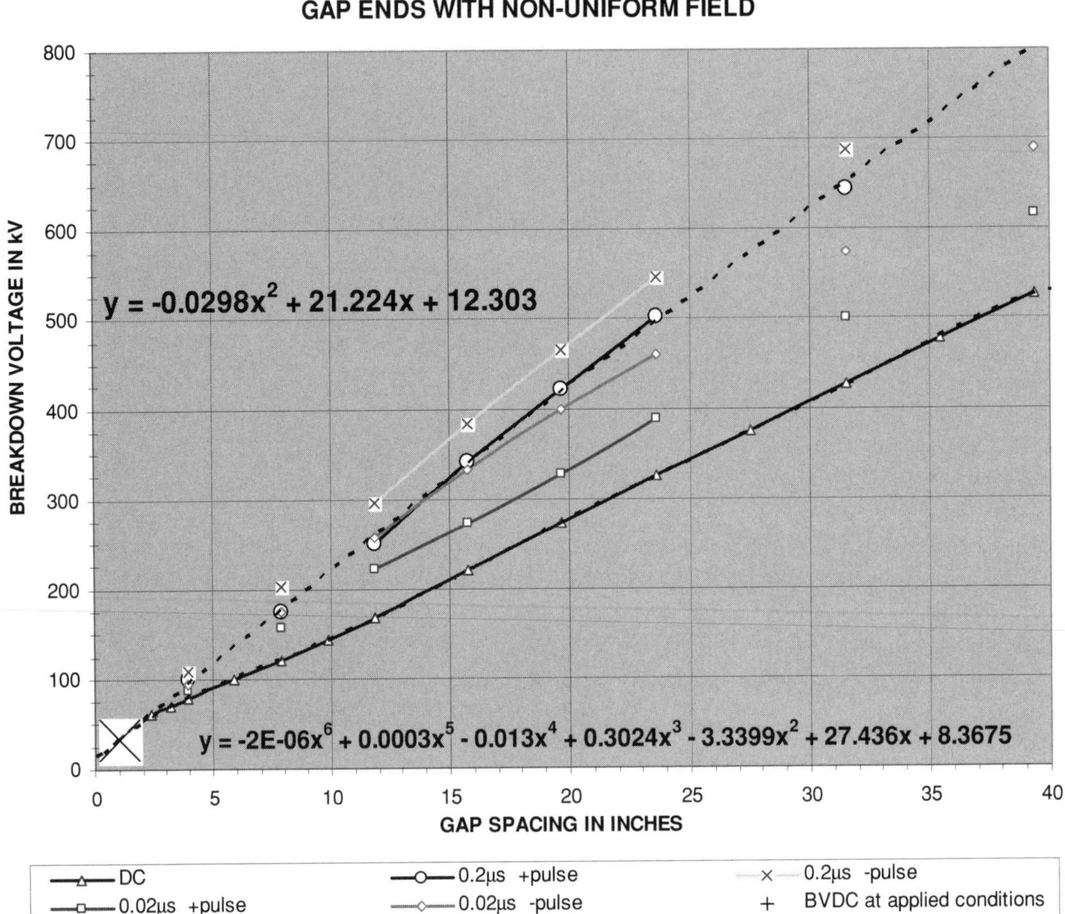

FIGURE 6-14 Breakdown voltage for rod gap ends (reference 4).

Table 1 with the third-order polynomial equation used to calculate the breakdown voltage for the uniform field found in spherical gap ends in cell (B14) and sixth-order polynomial equation used to calculate the breakdown voltage for the non-uniform field found in the needle point gap ends in cell (B15) for the gap spacing (S) entered in cell (B8). Chart 1B shown in Figure 6-17 displays more detail than chart 1A for the first 5 inches of gap separation. A large [X] marks the calculated result in cell (B14) and a large [+] marks the calculated result in cell (B15) on Charts 1A and 1B.

The measurements used in Table 1 of the worksheet were recorded at the nominal conditions of 25°C, and 760 torr. No calculation accuracy was specified in the reference. A one-inch air gap at nominal conditions with spherical gap ends requires a breakdown voltage of

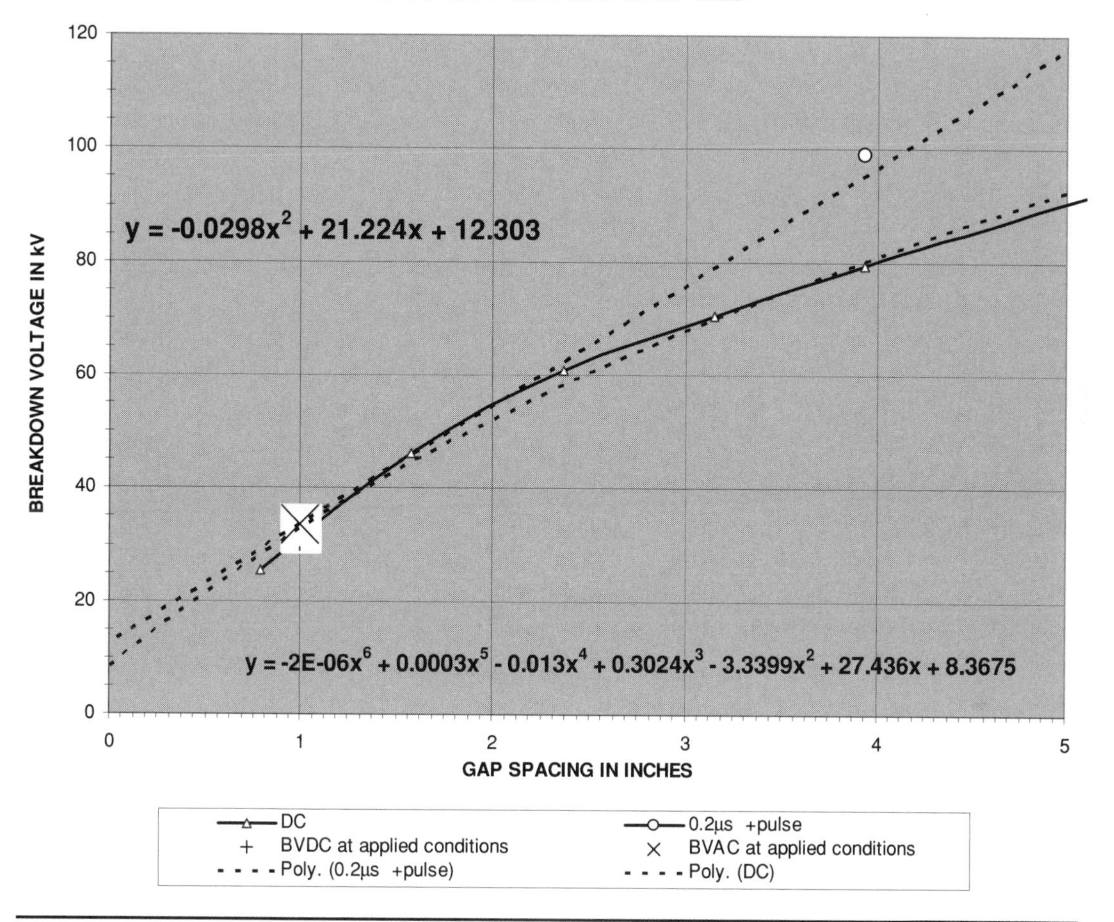

FIGURE 6-15 Additional detail of Figure 6-14.

73.3 kV/inch in a uniform field and 28.0 kV/inch (Figure 6-17) with needle point gap ends (non-uniform field) using this reference. Although the data are only for an applied DC voltage with no humidity correction the data from reference (5) agree quite well with those from references (4) and (14).

Another breakdown voltage chart was constructed in the CH_6.xls file, WIRELESS TELEGRAPHY 1915 worksheet (5), using reference (6) for selected sphere diameters and shallow bowl gaps. (See App. B.) By entering the air pressure into cell (B7) the air pressure correction factor (kP) is calculated in cell (B11). Graph 2 (rows 67–104) was constructed from the Table 2 values in rows 107 to 124 and displays the power curve fit formula used to calculate the air pressure correction factor (kP) in cell (B11). The only correction factor provided in the reference was for air pressure (kP); therefore the combined effects correction factor (*k*) in cell (B12) is equal to kP in cell (B11).

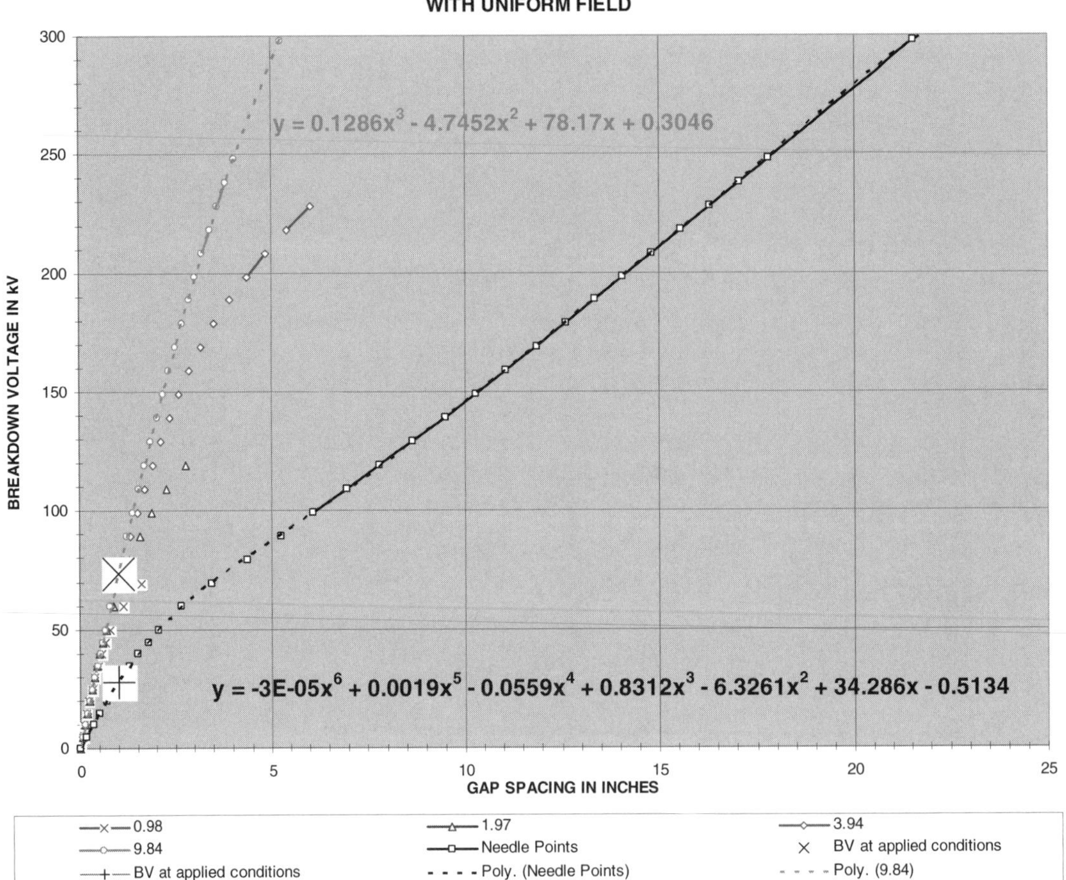

FIGURE 6-16 Breakdown voltage for spherical gap ends (reference 5).

Table 1 in columns (B) to (I) rows 16 to 64, displays the measured breakdown voltage for selected sphere diameters and gap separation in inches for an applied DC voltage with a uniform electric field. The combined effects correction factor (*k*) in cell (B12) is applied to the measured values in Table 1. Chart 1A shown in Figure 6-18 displays the measured values in Table 1 with the power equation used to calculate the breakdown voltage for the uniform field found in spherical gap ends in cell (B14) and the power equation used to calculate the breakdown voltage for the shallow bowl gap ends in cell (B13) for the gap spacing (S) entered in cell (B8). A large [X] marks the calculated result in cell (B13) on Chart 1A.

The measurements used in Table 1 of the worksheet were recorded at the nominal conditions of 18°C, and 745 torr. No calculation accuracy was specified in the reference. A one-inch air gap at nominal conditions with spherical gap ends requires a breakdown voltage of 70.0 kV/inch in a uniform field and 66.9 kV/inch with shallow bowl gap ends (less uniform

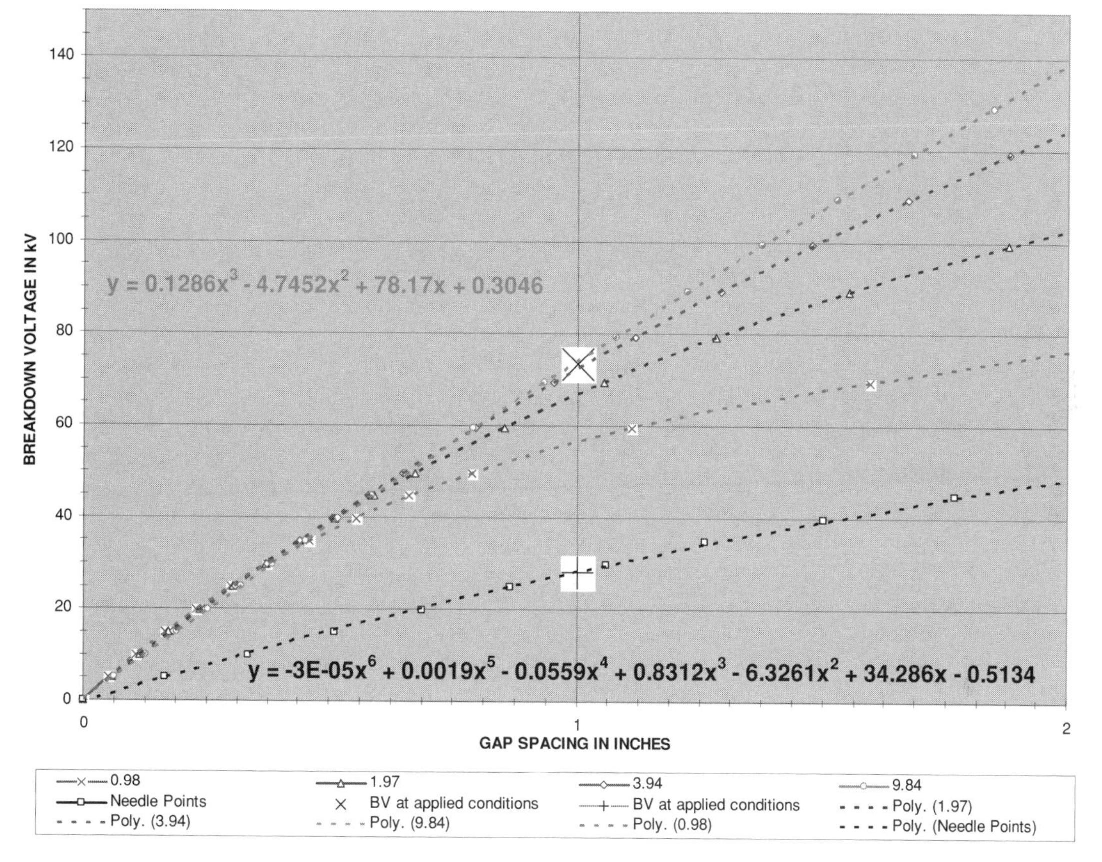

FIGURE 6-17 Additional detail of Figure 6-16.

field) using this reference. Although the data are only for an applied DC voltage with no temperature or humidity correction the data from reference (6) agree quite well with those from references (4), (5) and (14). The shallow bowl end data may prove useful when constructing an air gap with shallow bowls.

6.7 General Relationships of Applied Environmental Conditions to Breakdown Voltage in Air Gaps

Refer to the CH_6.xls file, HVEF 1984 worksheet (1) and reference (14). By varying the applied environmental conditions of air temperature in cell (B6), air pressure in cell (B7), relative humidity in cell (B8), distance between gap ends in inches in cell (B9), distance between gap end

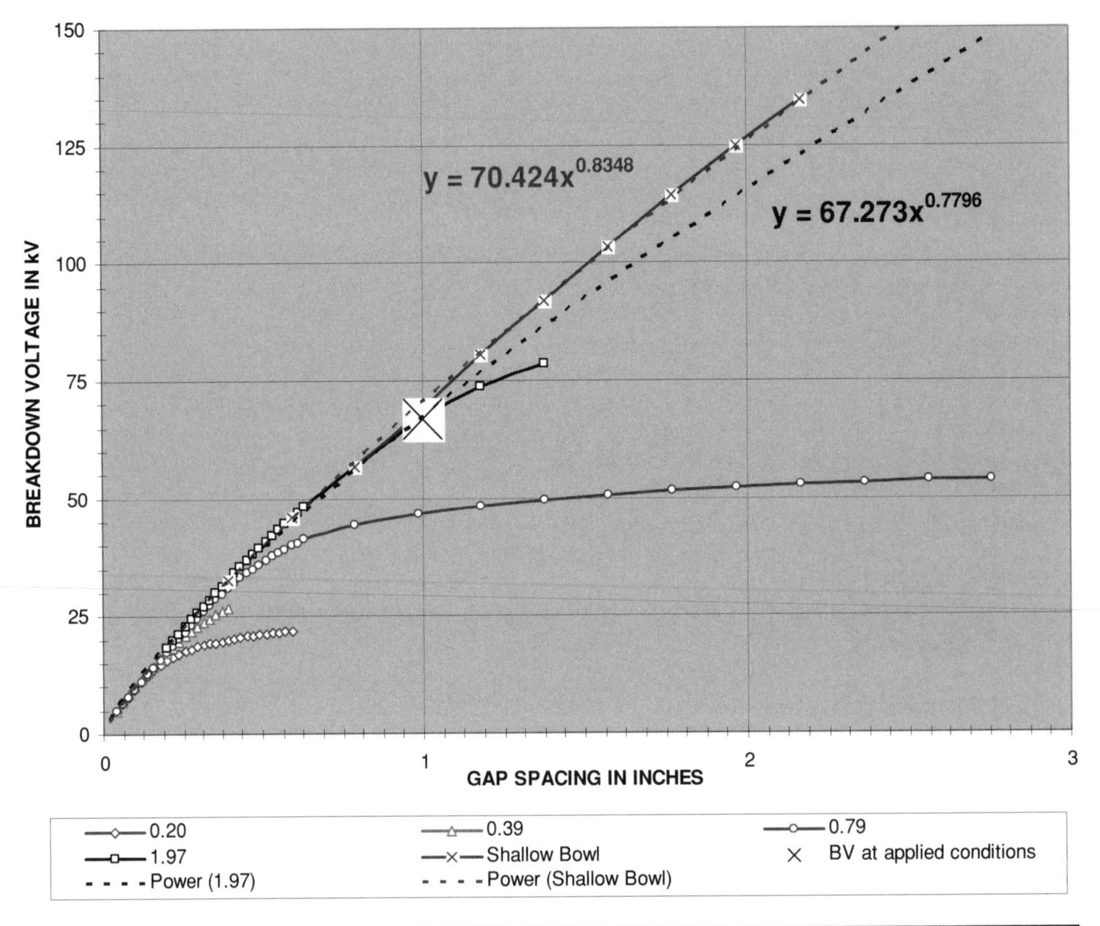

FIGURE 6-18 Breakdown voltage for spherical gap ends (reference 6).

and ground plane in inches in cell (B11), and diameter of the spherical end gap in inches in cell (B12), the following general relationships in an air gap will be seen:

- The effects of air pressure are directly proportional. As air pressure decreases (toward vacuum) the voltage required to break down an air gap decreases. A trick in the aerospace industry is to seal high-voltage transmitting equipment from the outside air and pressurize the air inside to about 5 PSI. As an aircraft increases altitude the air thins out approaching vacuum conditions and radio transmitter equipment will tend to arc over as the air between high-voltage points thins out and looses its insulating ability. When the equipment is sealed and pressurized the insulating ability of the air inside is increased and minimally affected by changes in outside air density at different altitudes.

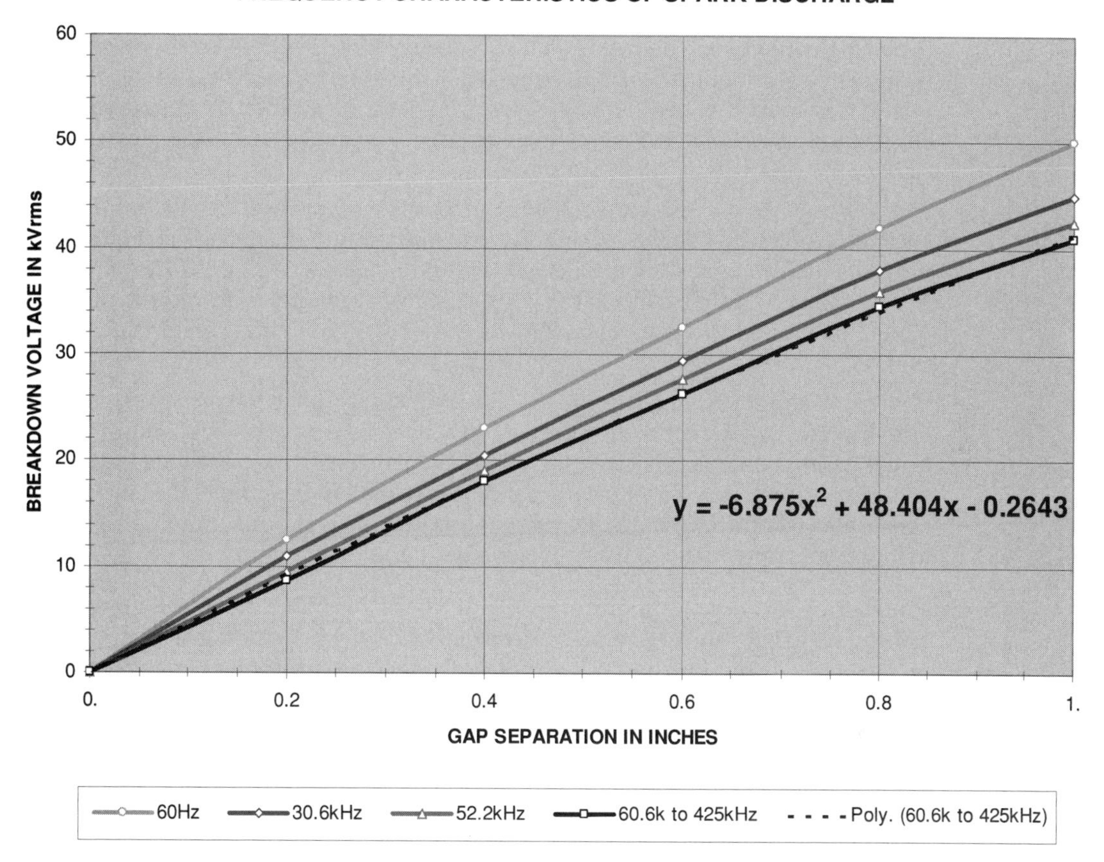

Figure 6-19 Frequency effects on breakdown voltage.

- The effects of temperature are inversely proportional. As temperature increases the voltage required to break down an air gap decreases. You can observe this in an operating Tesla coil when using a director (metal protrusion used to direct the spark breakout from a specific point in the toroid or sphere) on the terminal capacitance. As the spark breaks out it superheats a channel of air and since the spark is directed to breakout from the same point it continues to heat the air channel and grow in length with each pulse from the coil. The heated air and the director will generally give you a longer, hotter spark than letting the coil run without a director.
- The effects of humidity are directly proportional. As humidity decreases the voltage required to break down an air gap decreases. This can be observed during the fall and winter as the drier air increases the number of electrostatic shocks generated from walking on carpet.
- The effects of frequency are inversely proportional. Reference (10) provides the following rule: 30 kV for each cm of gap length (75 kV for each inch) is required for breakdown

at 25°C and 760 mm Hg up to 300 MHz, therefore the effects characterized are valid to 300 MHz. However, page 185 of reference (20) provides the data in Figure 6-19 showing the breakdown voltage decreases as the frequency increases. The breakdown voltage is decreased by about 20% as the frequency is increased from 60 Hz to over 60 kHz in a 1″ gap, which is our range of interest in Tesla coils. To account for frequency effects in the 60-kHz to 425-kHz range decrease the breakdown voltage by as much as 20%.

- The proximity-to-ground plane effect is inversely proportional. As the terminal capacitance is elevated above ground the proximity-to-ground plane effect multiplier increases and the breakdown voltage threshold is increased. See Figure 6-10. However, as already stated the breakdown voltage is decreased with the proximity-to-ground plane effect by as much as 25% for the typical geometries found in most Tesla coils.
- With small gap separation (less than 1.0 inch) the breakdown voltage can be reduced by as much as 20% by irradiating the air gap between the electrodes with ultraviolet or X-rays. Using a rod material such as thoriated tungsten will also reduce the breakdown voltage in a small gap. The thorium (a radioactive material) increases the ionization of the air between the electrodes. I would not utilize this feature as it presents a hazard to the coiler. I only mention it as it is sometimes used in industry.

The data in Figures 6-5 and 6-7 support the general rule of 75 kV/inch to break down an air gap. The dependent factors are the gap end geometry and whether a uniform electric field is produced. There are always exceptions to general relationships and high-voltage breakdown in liquids, solids, and gases are not completely understood or quantified as stated in reference (11). My generalizations include:

- If a rotary spark gap is used a non-uniform field is produced and the 75 kV/inch rule will not apply. Refer to Section 6.11 for calculation methodology for breakdown voltage threshold in operating spark gaps.
- The safety gap in the primary circuit can produce a uniform or non-uniform field depending on the type of gap end geometry. When a non-spherical gap end is used and the secondary winding of the resonant transformer is not terminated in its own dedicated ground or when the output of the step-up transformer is center tapped (see Figure 6-1), the field is non-uniform and the 75 kV/inch rule will not apply. The 32 kV/inch or 27 kV/inch shown in Figure 6-9 should be used instead. Refer to Section 6.11 for calculation methodology for breakdown voltage threshold when a flat rod end gap is used.
- The spherical or toroidal terminal capacitance in the secondary circuit produces a non-uniform field when the spark is attracted to ground or dissipated in air, and the 75 kV/inch rule will not apply. There is also the proximity-to-ground effect. If the spark is attracted through air between two spheres a more uniform field can be expected. However, the terminal diameter, spark length, and terminal separation from ground will influence the 75 kV/inch rule as supported in Section 6.5. To evaluate the secondary terminal voltage after breakdown use 10 kV/inch as detailed in Section 6.13.

Atmospheric Pressure Measurement	Space (Vacuum)	Below Nominal	At Nominal (Baseline)	Above Nominal
Altitude	>100 miles above S.L.	18,000 feet above S.L.	Sea level (S.L.)	33 feet ocean depth
Atmospheres	0	0.5	1.0	2.0
Pounds per Square Inch (PSI)	0	7.35	14.7	29.4
inches Hg	0	14.96	**29.92**	59.84
mm Hg (torr)	0	380	**760**	1,520
bar	0	0.505	1.01	2.02
millibar	0	506.6	1,013.2	2026.4
Kilopascal (kPa)	0	50.65	101.3	202.6
cm H_2O	0	494	988	1,976

TABLE 6-1 Air pressure at selected conditions converted to terms commonly used in gap breakdown voltage measurements.

The term torr is simply inches of mercury converted to mm, or Hg (inches) × 25.4 = Hg (mm).

- The data presented are useful in design to approximate the behavior of a high voltage present in a gap. It also serves as a guideline for initial adjustments. Always include an adjustment provision in your spark gap design so that it can be fine tuned once the coil is built and running (meaning adjust it with the power off between run observations).

Seemingly unrelated terms are used in voltage breakdown data, which are confusing. Using reference (8), Table 6-1 is constructed for cross-referencing air pressure terms at both higher than normal and lower than normal (toward vacuum) pressure. Table 6-2 shows the conversion

	kPa	mmHg	psi	cm H_2O	bar	torr	atm
1 mmHg =	0.133	1	0.0193	1.33	0.0013	1	0.0013
1 torr =	0.133	1	0.0193	1.332	0.0013	1	0.0013
1 bar =	100	750	14.5	1000	1	750	1
1 atm =	100	750	14.5	1000	1	750	1
1 psi =	6.9	51.72	1	69	0.069	51.72	0.069
1 cm H_2O =	0.1	0.736	0.0145	1	0.001	0.736	0.001

TABLE 6-2 Conversion multipliers for pressure units in Table 6-1.

multiplier for these pressure units and was constructed from data found in reference (13). The CH_6.xls file, PRESSURE CONVERSIONS worksheet (6) performs conversions of the pressure units shown in Table 6-2. (See App. B.)

6.8 Rotary Spark Gaps

To design a rotary spark gap open the BPS CALCULATOR worksheet (1) in the CH_6B.xls file. (See App. B.) The four steps listed below will calculate the Breaks Per Second (BPS) for the selected motor speed, rotor diameter, and number of electrodes used in the rotor.
Step 1. Select a rotor diameter for a given motor speed:

$$Vr = C\frac{RPM}{60} \tag{6.5}$$

Where: Vr = Rotor rotational velocity in inches per second = cell (H5), calculated value.
RPM = Rotational speed of motor in Revolutions Per Minute = cell (C5), enter value.
C = Rotor circumference in inches = cell (H4), calculated value.

$$C = \pi\, Dr \tag{6.6}$$

Where: Dr = Rotor diameter in inches = cell (C4), enter value.

Step 2. Determine the electrode dwell time:

$$T_{ON} = \frac{De}{(Vr\, Ne)} \tag{6.7}$$

Where: T_{ON} = Electrode dwell time in seconds (total time that rotating electrodes are in electrical contact with stationary electrodes) = cell (H7), calculated value.
De = Diameter of rotating electrodes in inches = cell (C7), enter value.
Ne = Number of rotating electrodes mounted in rotor = cell (C6), enter value.
Vr = Rotor rotational velocity in inches per second = cell (H5), calculated value.

NOTE: *$Vr \times Ne$ is also the electrode velocity (Ve) shown in cell (H6).*

It is assumed that the high voltage on the electrodes can begin to ionize the gap once they reach position 1 as shown in Figure 6-20. Depending on the ionization characteristics (see Section 6.11 for number of primary oscillations and ionization time) the gap can remain ionized

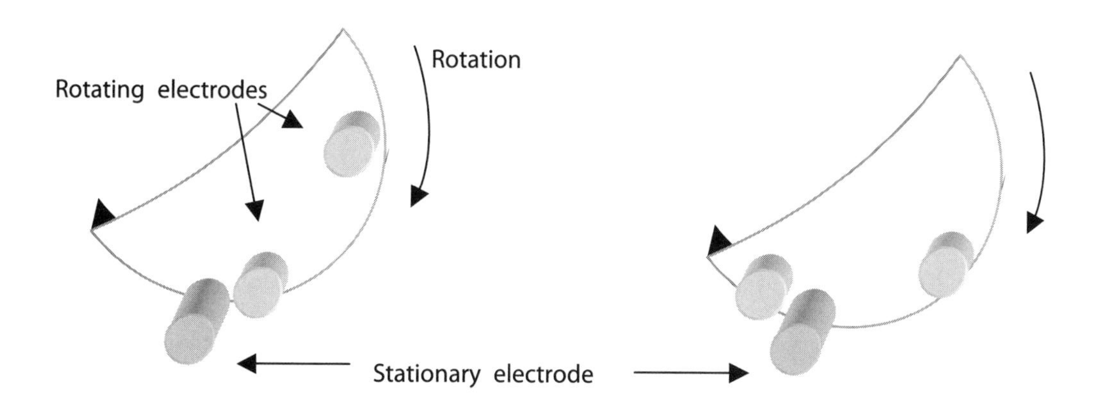

FIGURE 6-20 Rotor and stator positions at gap ionization and deionization (dwell time).

until it reaches position 2. The maximum estimated dwell time for the gap is the time required for the electrode to travel through its arc from position 1 to position 2 or twice the electrode diameter. Once the arc is ionized it is difficult to extinguish (deionize) until the tank capacitor has released enough energy through the damped primary oscillations to reach the minimum ionization current threshold.

I have experimented with low decrements having an ionization time of 350 μs and rotary spark gap electrode dwell time of 53 μs. The gap ionization characteristics were observed using welding goggles indicating the gap remained ionized long after the electrodes passed position 2 until the minimum ionization current threshold was reached. The longer the arc formed in the gap, the more energy was being drawn from the line and consumed in the arc as heat. None of the energy dissipated in the arc as heat can be transferred to the secondary. To accommodate a low decrement having a long ionization time, it may be necessary to use larger diameter electrodes for optimum performance. A more efficient design will use a higher decrement with the required ionization time calculated in Section 6.11 coinciding with the calculated spark gap dwell time. The experiment was repeated using less primary turns for a higher decrement and calculated ionization time of about 70 μs. The primary was retuned to resonance using a larger capacitance. The observed arc did not ionize past position 2 in Figure 6-20 and the coil used much less energy to produce the same spark length.

Step 3. Determine the electrode quench time:

$$T_{OFF} = \frac{\left(\frac{C}{Ne}\right)}{Vr} - T_{ON} \tag{6.8}$$

Where: T_{OFF} = Electrode quench time in seconds (total time that rotating electrodes are not in electrical contact with stationary electrodes) = cell (H9), calculated value.

T_{ON} = Electrode dwell time in seconds (total time that rotating electrodes are in electrical contact with stationary electrodes) = cell (H7), calculated value from equation (6.7).

C = Rotor circumference in inches = cell (H4), calculated value from equation (6.6).

Ne = Number of rotating electrodes mounted in rotor = cell (C6), enter value.

Vr = Rotor rotational velocity in inches per second = cell (H5), calculated value.

Step 4. Determine the Breaks Per Second of the rotary gap:

$$BPS = \frac{1}{(T_{ON} + T_{OFF})} Ne$$

or: (6.9)

$$BPS = \frac{Vr}{C} Ne$$

Where: BPS = Breaks Per Second produced by rotary spark gap (frequency of charge-to-discharge of tank circuit) = cell (H10), calculated value.

T_{OFF} = Electrode quench time in seconds (total time that rotating electrodes are not in electrical contact with [between] stationary electrodes) = cell (H9), calculated value from equation (6.8).

T_{ON} = Electrode dwell time in seconds (total time that rotating electrodes are in electrical contact with stationary electrodes) = cell (H7), calculated value from equation (6.7).

C = Rotor circumference in inches = cell (H4), calculated value from equation (6.6).

Ne = Number of rotating electrodes mounted in rotor = cell (C6), enter value.

Vr = Rotor rotational velocity in inches per second = cell (H5), calculated value.

Using the dynamics of the worksheet the rotary gap's rotor diameter, number of electrodes, and motor speed can be changed to find the optimum design for a desired BPS value before construction work is done.

Let's not forget safety. The G force the electrodes exert in their rotational path is:

$$G = VeM \tag{6.10}$$

Where: G = Gravity force (centrifugal) exerted by rotating electrodes = cell (H13), calculated value.

Ve = Electrode velocity in inches per second = cell (H6), calculated value.

M = Weight of electrodes in grams (454 grams = 1.0 lb) = cell (C13), enter value.

The mechanical strength of the rotor is dependent on the type of material used and the thickness:

$$PSI = RPSI \bullet T \tag{6.11}$$

Where: PSI = Available strength of selected rotor material and thickness in Pounds per Square *Inch* = cell (H14), calculated value.

$RPSI$ = Rated strength of selected rotor material and thickness in Pounds per Square Inch = cell (C15), enter value.

T = Thickness of rotor in inches = cell (C14), enter value.

To convert the electrode force in G's to Newtons for comparison to the rotor's mechanical strength:

$$N = \frac{Ve}{12} \bullet \frac{M}{454} \tag{6.12}$$

Where: N = Force exerted by rotating electrodes in lb/ft = cell (H15), calculated value.

Ve = Electrode velocity in inches per second = cell (H6), calculated value.

M = Weight of electrodes in grams (454 grams = 1.0 lb) = cell (C13), enter value.

Comparing the calculated force in Newtons and the available mechanical strength of the rotor, a safety margin is derived:

$$S = \frac{PSI}{N} \tag{6.13}$$

Where: S = Safety margin or ratio of available strength to applied force = cell (H16), calculated value.

PSI = Available strength of selected rotor material and thickness in Pounds per Square Inch = cell (H14), calculated value from equation (6.11).

N = Force exerted by rotating electrodes in lb/ft = cell (H15), calculated value from equation (6.12).

Note a safety factor of less than 1.0 would indicate the force exerted by the electrodes is greater than the mechanical strength of the rotor material. For a safe design do not work with a safety factor of less than 10. This will allow a conservative margin for manufacturing tolerance and the estimations used in the calculations. The above calculations are performed for the rotary spark gap used in the coil designed in Chapter 2 and shown in Figure 6-21.

Once you have selected the motor speed, number of electrodes, and rotor diameter that produces the desired BPS, begin construction by selecting the rotor material. Rotary spark gaps, as well as complete Tesla coils, were once available from companies such as EICO, when commercial radio transmitters resembled Tesla coils and hobbyists turned exclusively to radio experimenting to occupy their time. You could buy a ready-to-run rotary gap,

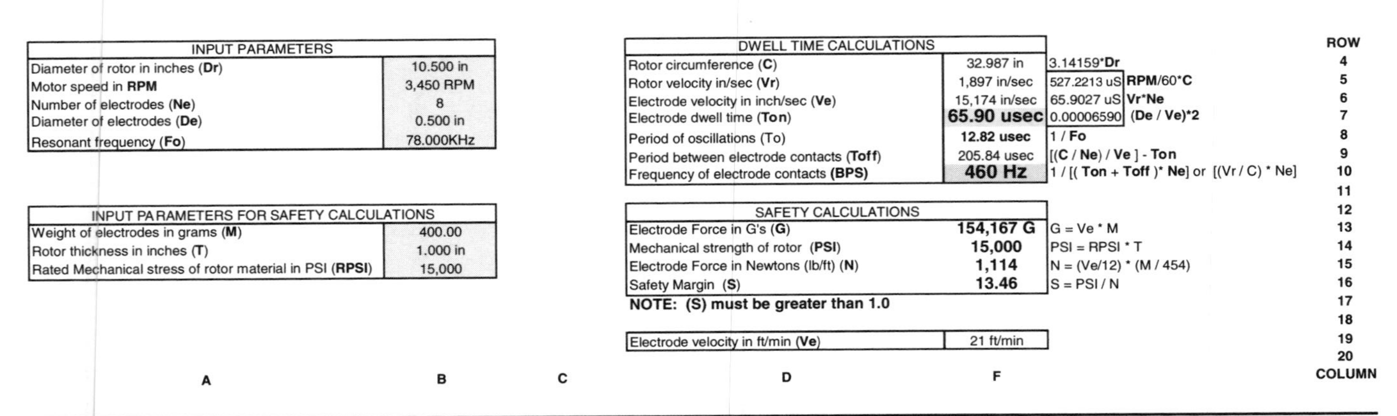

INPUT PARAMETERS	
Diameter of rotor in inches (**Dr**)	10.500 in
Motor speed in **RPM**	3,450 RPM
Number of electrodes (**Ne**)	8
Diameter of electrodes (**De**)	0.500 in
Resonant frequency (**Fo**)	78.000KHz

INPUT PARAMETERS FOR SAFETY CALCULATIONS	
Weight of electrodes in grams (**M**)	400.00
Rotor thickness in inches (**T**)	1.000 in
Rated Mechanical stress of rotor material in PSI (**RPSI**)	15,000

DWELL TIME CALCULATIONS				ROW
Rotor circumference (**C**)	32.987 in	3.14159***Dr**		4
Rotor velocity in/sec (**Vr**)	1,897 in/sec	527.2213 uS	**RPM**/60***C**	5
Electrode velocity in inch/sec (**Ve**)	15,174 in/sec	65.9027 uS	**Vr*****Ne**	6
Electrode dwell time (**Ton**)	**65.90 usec**	0.00006590	**(De / Ve)*2**	7
Period of oscillations (To)	**12.82 usec**	1 / **Fo**		8
Period between electrode contacts (**Toff**)	205.84 usec	[(**C** / **Ne**) / **Ve**] - **Ton**		9
Frequency of electrode contacts (**BPS**)	**460 Hz**	1 / [(**Ton** + **Toff**)* **Ne**] or [(Vr / C) * Ne]		10
				11

SAFETY CALCULATIONS			ROW
Electrode Force in G's (**G**)	**154,167 G**	G = Ve * M	13
Mechanical strength of rotor (**PSI**)	**15,000**	PSI = RPSI * T	14
Electrode Force in Newtons (lb/ft) (**N**)	**1,114**	N = (Ve/12) * (M / 454)	15
Safety Margin (**S**)	**13.46**	S = PSI / N	16

NOTE: (S) must be greater than 1.0

Electrode velocity in ft/min (**Ve**)	21 ft/min

FIGURE 6-21 Example rotary spark gap breaks per second worksheet and calculations.

made with a hard rubber rotor for your radio experiments, for a dollar! Those were the days.

My first rotary gap rotor was made from plexiglass and was used until cautioned by veteran coiler Ed Wingate that this approach was dangerous. After removing the electrodes the rotary disc was closely examined. With only several minutes of use the material supporting the electrodes had become discolored and crazed. The electrodes can heat to several hundred if not thousand degrees while the gap is ionized. The electrodes attempt to conduct the heat away from the ionized gap, but the thermally insulating rotor material surrounding the electrode prevents efficient conduction of heat and the temperature of the material supporting the electrodes will be quite high.

To illustrate the potential danger, take a 0.25″ thick piece of scrap plexiglass, heat a soldering iron (above 600°F to melt solder) and poke it into the plexiglass. It will go through easily. Now heat a butane, propane, or acetylene torch (hotter than 1500°F) and poke the flame into the plexiglass. This is closer to what the heated electrodes are doing to your rotor disc. The arc temperature is typically >6,000°C using tungsten electrodes. Now repeat with the G-10 material. I will bet you feel more comfortable with the G-10!

The materials list in Table 6-3 developed from references (1) and (2) show why the G-10 type laminate is the best material to use for the rotor. It has the highest strength and operating temperature of the thermoplastic materials. Compared to what was available in the old EICO catalog, modern materials are superior insulators and last much longer. As you sweep through the antiques at your next Hamfest the aging effects on some of these older materials can easily be seen.

Refer to Chapter 9 for rotary spark gap construction techniques.

6.9 Stationary Spark Gaps

The stationary spark gap is much easier to construct than the rotary gap, but will not be as effective in medium and large coils. Any type of end gap geometry can be used and experimented with. Because of the rotary gap's rotational forces and high speed, it is generally limited to rod end gap geometry. Stationary gaps can be placed in series with rotary gaps to produce different quench and ionization characteristics. A stationary gap is essentially two electrodes placed end-to-end with an air gap in between as shown in Figure 6-22. Additional gaps can be placed in series with one another, the more gaps the better the performance. In contrast to the safety gap, the stationary spark gap will closely follow the characteristics in Figure 6-28.

The design in Figure 6-23 works quite well and slightly increases the quenching ability by supplying a constant source of fresh, non-ionized, cooling airflow to the gap. The PVC form must be long enough to offer complete isolation of the fan wiring with the high voltage on the gap surfaces, usually requiring several inches of separation. Minimize the use of any parts made of ferrous materials including the cooling fan. Also include dedicated (separate) line filtering to the fan motor to reduce noise and high-voltage transients from entering your line service. To decrease the line power drawn by the coil or increase the secondary spark length, try adjusting the gap separation.

	THERMAL			MECHANICAL				ELECTRICAL				
Material Type	Continuous Service Temperature Rated Limit (°F)	Thermal Conductivity (10^{-5} cal-cm/sec-cm^2 - °C)	Coefficient of Thermal Expansion (10^{-5} in/in - °C)	Tensile Strength in Pounds per Square Inch (PSI) Lengthwise / Crosswise	Minimum Flexural Strength in PSI Lengthwise / Crosswise (1 / 8" stock)	Rockwell Hardness Scale	Minimum Bonding Strength (lb)	Dielectric Breakdown (kV) (1 / 16" stock)	Water Absorption (% in 24 hours 1 / 16" stock)	Insulation Resistance in MegΩ (35°C, 90%RH)	Dissipation Factor at 1MHz	Dielectric Constant at 1MHz
Laminates												
G-10 (glass fabric – epoxy laminate)	250	–	–	45,000/ 40,000	55,000/ 45,000	M111	2,000	45	0.25	200 k	0.025	5.2
G-9 (glass fabric – melamine laminate)	325	7.0	1.0	50,000/ 40,000	55,000/ 35,000	M120	1,700	60	0.80	10 k	0.017	7.2
G-7 (glass fabric – silicone laminate)	460	7.0	–	23,000/ 18,500	20,000/ 18,000	M100	650	32	0.55	2,500	0.003	4.2
G-3 (glass fabric – phenolic laminate)	290	–	1.8	23,000/ 20,000	40,000/ 30,000	M100	850	–	2.70	–	0.003	5.5
LE (Linen, fine weave cotton – phenolic laminate)	250	7.0	2.0	12,000/ 8,500	15,000/ 13,500	M105	1,600	40	1.95	–	0.055	5.8
CE (canvas, cotton fabric – phenolic laminate)	250	7.0	2.0	9,000/ 7,000	17,000/ 14,000	M105	1,800	35	2.20	–	0.055	5.8
P (paper – phenolic laminate)	250	7.0	2.0	12,400/ 9,500	12,000/ 10,500	M105	750	60	1.00	20 k	0.035	4.6
AA (asbestos fabric – phenolic laminate)	275	–	1.5	12,000/ 10,000	18,000/ 16,000	M103	1,800	–	3.00	–	–	–
LEXAN	280			8,900–		M70						
				9,000								
Acrylic (plexiglass)	185–214		3.6–3.9	10,500	16,500	M95		0.5/mil	0.3		0.02	2.3
NYLON 6/6 (polyamide)	190		4.4	12,000					1.5			
TEFLON (flouroplastics) PTFE (TFE)	250	1.7	5.5–8.4	3,350	No break	D50-65		0.5-0.6 /mil	<0.01		0.00005 @ 1 kHz	2.1 @ 1 kHz
PVC (polyvinyl chloride)	165			7,000				R118				

TABLE 6-3 Properties of thermoplastic materials.

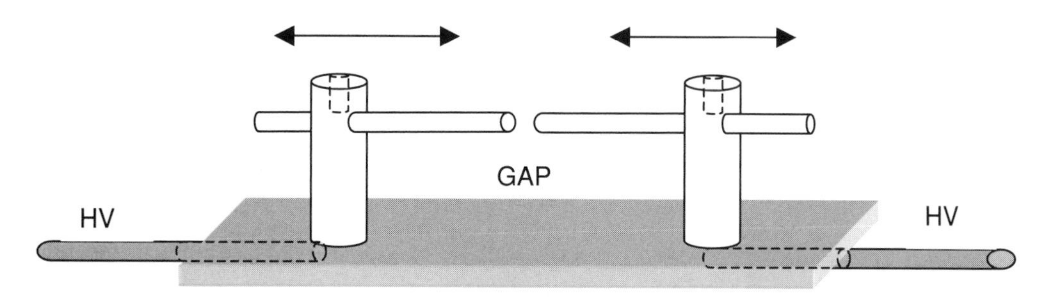

Figure 6-22 Details of stationary spark gap construction.

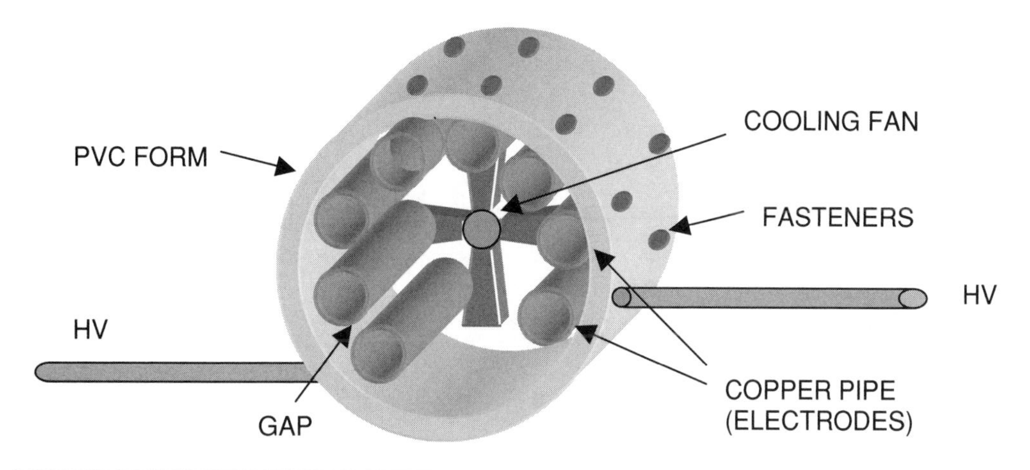

Figure 6-23 Details of stationary spark gap construction with cooling airflow to increase quenching ability.

6.10 Ionization and Deionization in a Spark Gap

The term quenching is often shrouded in fog and mystery. I will define quenching as the deionization of the spark gap. The ionization characteristics of a gap should be understood before examining deionization. The ionization characteristics are defined for a 50-cm tube filled with neon at 1 mm Hg in reference (19) and shown in Figure 6-24. Although the gas is neon at vacuum conditions, it is typical for all spark discharges. At room temperature, atmospheric pressure (760 mm Hg), and no voltage differential applied, air is an excellent insulator. When enough voltage is applied to a gap of air the molecules ionize, giving off electrons. An ion is an atom or molecule with a positive or negative electric charge and ionization is the production of ions in a gas. In our discussion of gaseous conductors and ionization of gases, air is the gas used in the gap at atmospheric pressure. This is known as a high-pressure arc, a low-pressure arc being toward vacuum (<200 mm Hg). When high voltage is applied to an air gap, electron movement in the gap produces a current flow. This operating point is shown in the left of Figure 6-24 as the breakdown threshold. As ionization continues (moving right) the Townsend region, corona region, and glow discharge region are surpassed by

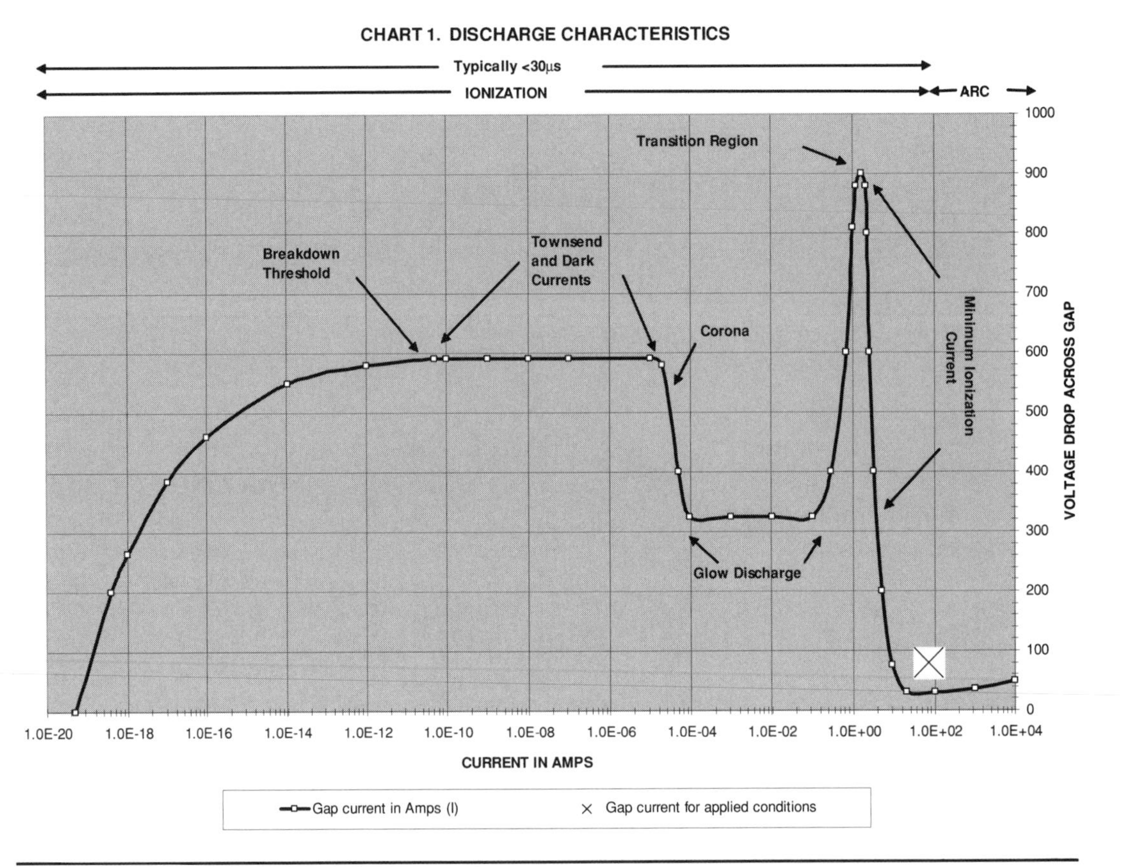

FIGURE 6-24 Discharge characteristics in a spark.

the increasing current flow. The transition region is typically reached within 30 μs of applying the breakdown voltage and an arc is established across the gap. In this arc state the air becomes an excellent conductor. The arc is actually plasma considered to be a fourth state of matter. As the current increases through the transition region to the arc region, the voltage drop across the ionized gap decreases rapidly. The voltage across the gap at arc formation is designated the initial voltage (Vfo) in reference (17) and will be retained throughout this discussion.

Reference (23) shows the same characteristic curve in Figure 6-24 for a gas discharge tube at low pressure. In this curve the end of the transition region where the arc forms is 0.1 A instead of the 1.0 A in Figure 6-24. The minimum ionization current will therefore be defined as anywhere in the range of 0.1 A to 1.0 A. Ambient air temperature, pressure, and humidity can shift the value outside this range; however, it will work quite well for our purposes.

Reference (22) provides the fundamental study of ionized gases. Pages 345 to 355 of reference (20) discuss arcs with applied AC waveforms. In frequencies below 500 Hz, as in 60-Hz power frequencies, the voltage applied across the gap crosses through 0 V to the opposite polarity in time periods longer than 1 ms. As the voltage approaches 0 V the arc is deionized and must be re-ignited as the arc looses heat rapidly. To reignite the arc the voltage must either increase quickly as in a higher frequency (high dv/dt, 100 V/< 40 μs) or a voltage approximate to the glow discharge region ($\approx$300 V) shown in Figure 6-24 is required to reignite the arc. This characteristic is advantageous to manufacturers of circuit protection equipment (e.g., fuses, circuit breakers, contactors). If an arc is established as the protection device opens the circuit, the arc will be extinguished within one alternation of the line frequency as it passes through 0 V. It is not advantageous when trying to maintain an arc in a high-frequency oscillator.

Thermal equilibrium is established in about 1 ms therefore the arc is maintained by thermal ionization in applied frequencies greater than 500 Hz, where electrical energy is converted to heat. The time period of 500 Hz $= 1/500$ Hz $= 2$ ms with the positive and negative alternations occurring over a 1-ms time period. Re-ignition is not required in the high-frequency oscillations of the primary of a Tesla coil as thermal ionization maintains the arc as the oscillations pass through 0 V to the opposite polarity. The time period for typical primary oscillations of 100 kHz $= 1/100$ kHz $= 10$ μs with each positive and negative alternation being 5 μs. The temperature of the arc at atmospheric pressure is quite high, typically reaching 3,777°C using copper electrodes and 5,877°C to 6,167°C for tungsten electrodes. These high temperatures sustain thermal ionization of the air in the gap. These high temperatures decrease as the pressure approaches vacuum.

While in the transition region the current is increasing while the voltage drop across the gap decreases, exhibiting the characteristics of a negative resistance. Once the arc is formed the ionization of the gas increases with temperature, lowering the resistance even further, making it an even better conductor.

Shown in Figure 6-25 are the initial voltage vs. arc current characteristics for several electrode materials. The graph shown in the figure was constructed using reference (20) in the CH_6.xls file, Ea VS. I worksheet (11). (See App. B.) The initial voltage for carbon, silver, copper, and iron electrodes in air at atmospheric pressure and selected arc currents was calculated in Table 1, columns (I) through (O) of the worksheet using the Ayrton equation from reference (21). The initial voltage for zinc and tungsten electrodes in air at atmospheric pressure and selected arc currents was plotted on the graph from data on page 294 of reference (20). Easily seen in the figure is the gradual increase in initial voltage as the arc current decreases from 3 A to 1 A. As the current decreases from 1 A the initial voltage increases rapidly indicating the arc is deionizing. This will become important when the minimum ionization current threshold is discussed. Also seen in the figure is the small variation in initial voltage regardless of the electrode material used, typically within 10 V of each other at any arc current.

The initial voltage is calculated using equation (6.14) from reference (17) as it also provides a solution for the arc resistance. Reference (20) cautions using Ohm's Law for calculating the gap resistance can be misleading. The Ayrton equation can also be used to calculate the initial voltage for carbon, silver, copper, and iron electrodes in the CH_6.xls file, Ea VS. I worksheet (11):

$$Vfo = a + bS + \frac{c + dS}{I} \tag{6.14}$$

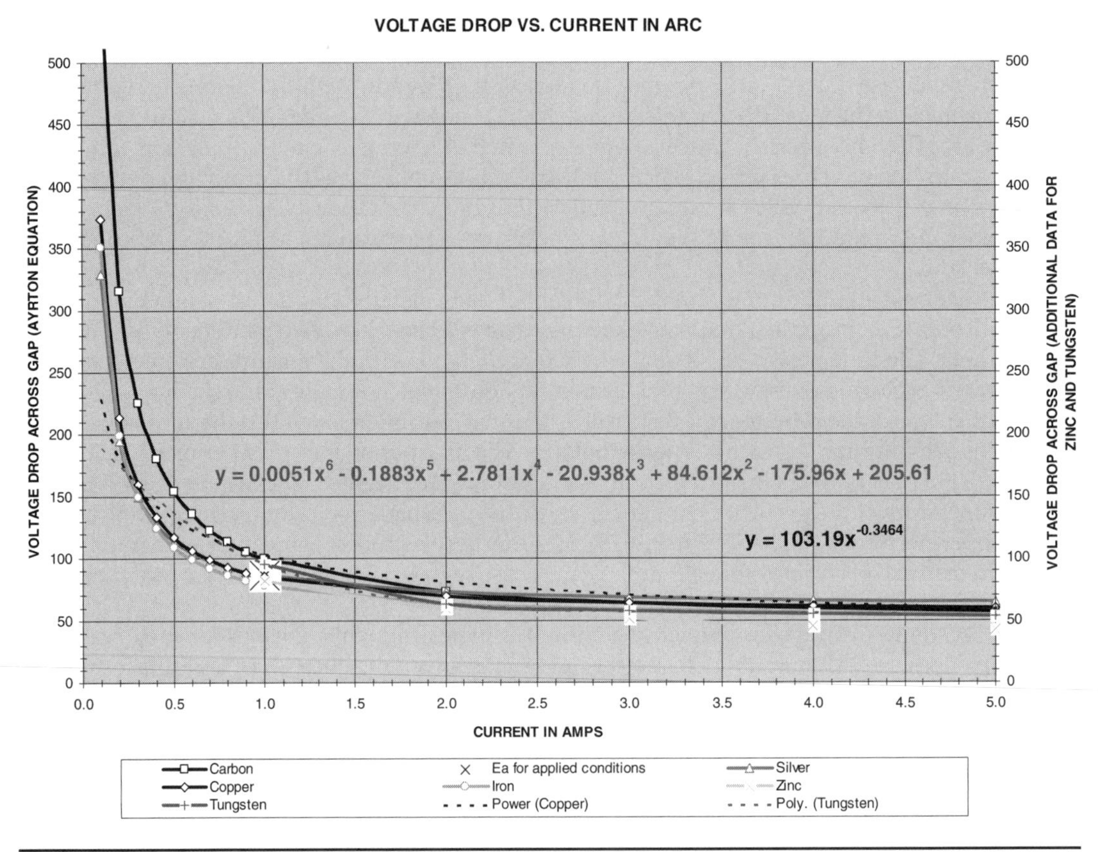

FIGURE 6-25 Initial voltage vs. arc current.

Where: *Vfo* = Ea = Initial voltage or voltage drop across the ionized gap = cell (F6) calculated value for selected electrode material entered into cell (B8). Also calculated in Table 1, columns (J) through (O) for the respective electrode material.

S = Spark gap spacing in mm = cell (B6) enter value in inches. Converted to mm in cell (C6) using S (inches) × 25.4.

I = Arc current in amps = column (I).

Coefficients a, b, c, and d are dependent on the electrode material shown below:

	Carbon	Silver	Copper	Iron
a	38.9	19.0	15.2	15.0
b	2.0	11.4	10.7	9.4
c	16.6	14.2	21.4	15.7
d	10.5	3.6	3.0	2.5

The calculated initial voltage in cell (F4) using reference (17) is retained in the worksheet for comparison. Either method produces calculated results typically within 10 V of each other.

In the deionized spark gap the step-up transformer applies a high voltage difference of potential to the electrodes. In a rotary spark gap the rotating electrodes come in proximity to the stationary electrodes reducing the gap spacing

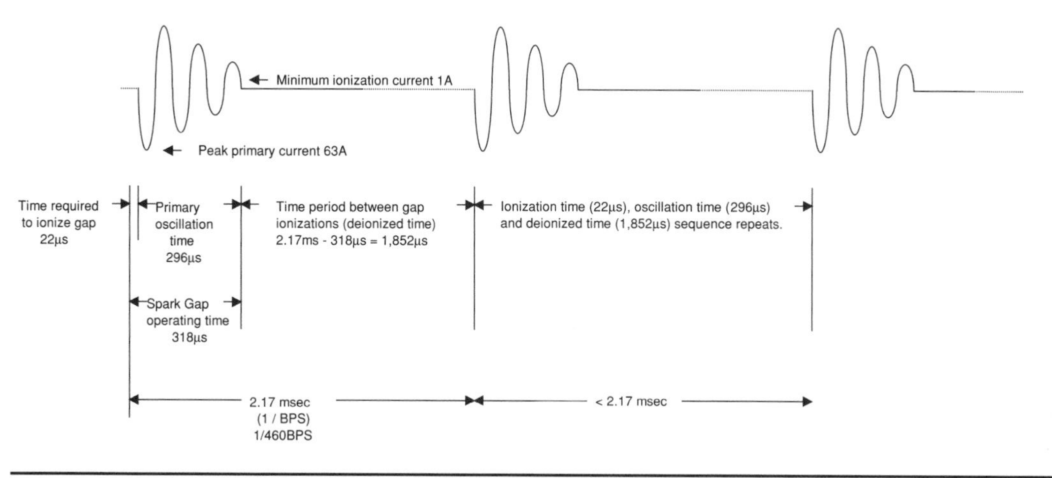

FIGURE 6-26 Ionization characteristics in a spark gap coil.

(see Figure 6-20). When the voltage differential across the gap is high enough the gap will fire (ionize). During ionization the charge in the tank capacitor and primary current are decreasing with each oscillation by the primary decrement value. Actually the capacitor charge is dissipated in the arc of the ionized gap (energy conversion to heat and light), which dampens the oscillating current much as a resistance would. The primary current flows through the series RLC circuit formed by the spark gap resistance, primary winding, and tank capacitance. The voltage drop across the ionized gap has been defined as the initial voltage (*Vfo*), which remains relatively constant while the gap is ionized. The high-frequency oscillations maintain the arc through thermal ionization. The resistance of an ionized gap is very low, the value dependent on the peak primary oscillating current that ionizes the gap, the gap separation, and electrode material. As the primary current is decreasing (damped) with each oscillation the minimum ionization current threshold is eventually reached. The gap will deionization (quench) when the peak oscillating current drops below the minimum ionization current threshold. The primary decrement and minimum ionization current threshold determine how many primary oscillations are produced each time the gap is ionized. The BPS determines how many times a second the gap is ionized. While the gap is ionized the primary and secondary are coupled and the primary is inducing a current in the secondary. Once the gap is quenched the primary becomes an open circuit and is no longer coupled to the secondary. This allows the secondary to oscillate independently with its stored energy, Q, and decrement.

The ionization characteristics in an operating spark gap are shown in Figure 6-26. The time periods shown were calculated in the worksheet shown in Figure 6-37 for the coil detailed in Chapter 2. The minimum ionization current threshold was determined to be 1 A in this application, which is supported by the data in Figure 6-25.

Circuit breaker and contactor manufacturers often use magnetic quenching to aid in extinguishing the high-voltage arcs that form across the contacts during fast switching and interrupt

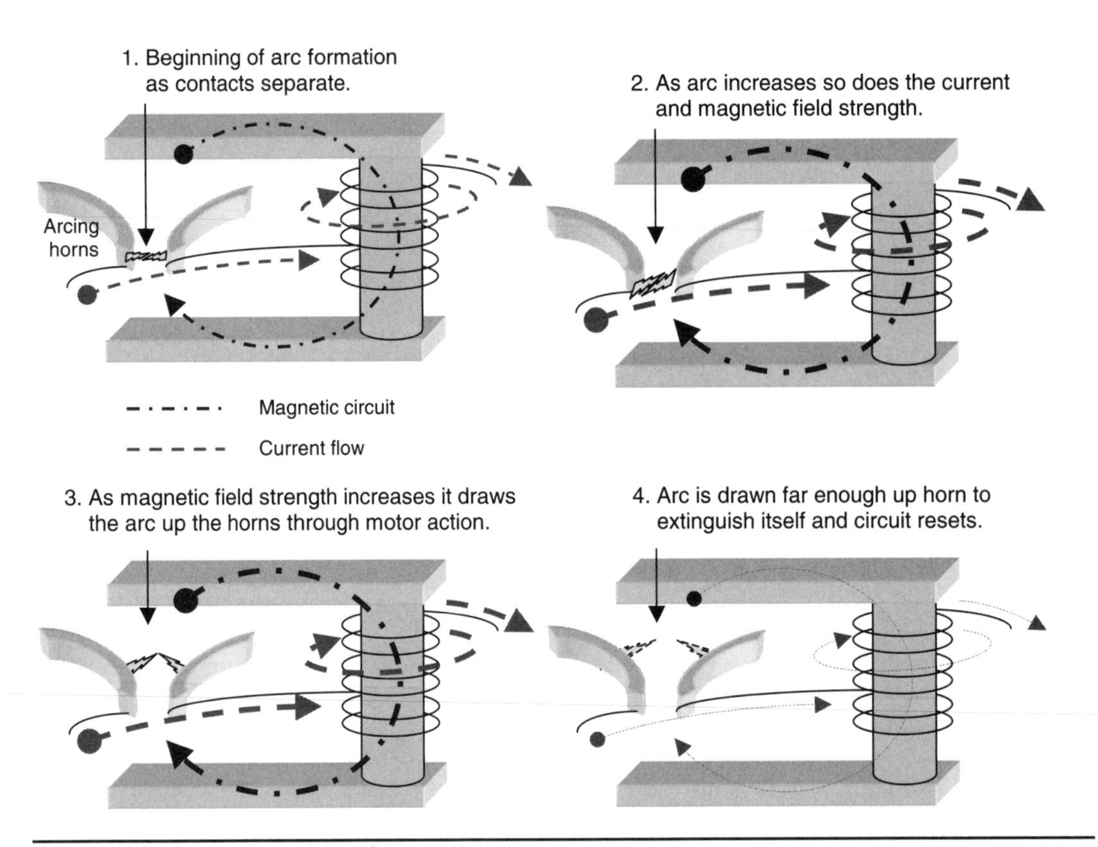

FIGURE 6-27 Magnetic quenching of an arc.

events. A description of magnetic quenching (blowout) is detailed in reference (15) and shown in Figure 6-27. When a switch or contact is quickly opened while maintaining a high load current a high-voltage transient results (dv/dt). An arc will form if the dv/dt is high enough to initiate ionization of the air between the contacts.

As the arc forms it draws more circuit current, which increases the magnetic field strength in the magnetic circuit formed by a solenoid, magnetic core, and pole pieces. The magnetic field produces a magnetic repulsive-attractive effect (motor action) on the arc and draws it up a Jacob's ladder type structure (arcing horns). The rising temperature also helps draw the arc up the horns. As the arc travels up the horns it extinguishes itself (blows out) before it produces damage. The arcing horns must be properly oriented in the magnetic field (as shown in the figure) to produce the Jacob's ladder effect on the arc. This technique could be adapted in a spark gap to improve quenching.

6.11 Spark Gap Operating Characteristics

To evaluate the operating characteristics of a spark gap open the CH_6A.xls file, IONIZATION TIME worksheet (1). (See App. B.) As most spark gap construction uses rod end geometry the

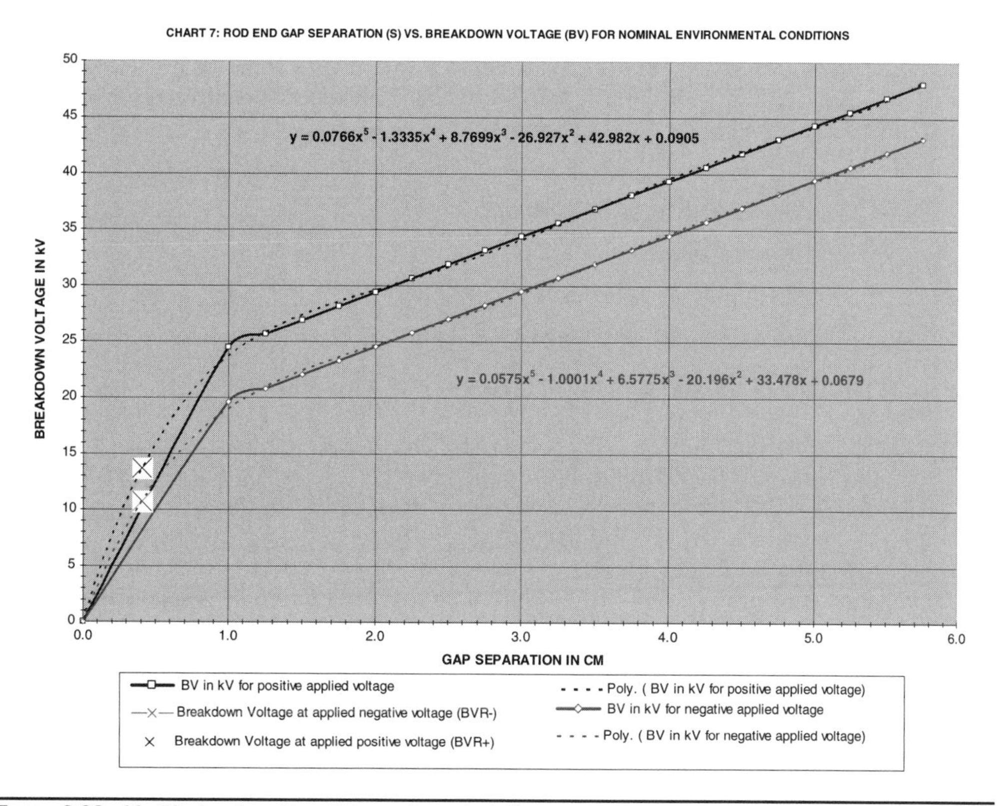

FIGURE 6-28 Modified rod end breakdown characteristics.

calculated operating characteristics will be limited to this type. When using a stationary spark gap it is limited to a break rate of 120 BPS. Even if multiple series gaps are used in a stationary gap it will ionize only once as the positive and negative line alternations reach their peaks (60 Hz × 2 = 120). Higher break rates can often pump more energy into the secondary. This generally increases the secondary current but not the voltage so the sparks will be brighter but not longer. An undesirable side effect is that the coil will draw more power from the line without producing a longer spark. To obtain a break rate higher than 120 BPS a rotary gap must be used. For this reason the worksheet calculations were developed to be used with a rotary spark gap as it can be designed to exhibit selected operating characteristics, whereas the stationary gap has a fixed 120 BPS.

The breakdown characteristics of a stationary gap with rod end geometry are detailed in Section 6.3. Equation (6.3) was used to calculate the voltage at the breakdown threshold; however, as shown in Figure 6-9 it is not comprehensive enough to determine the threshold for gap separations less than 1 cm. Equation (6.3) was used to construct the breakdown threshold voltages for the selected gap separations listed in Table 7 of the worksheet in rows 218 to 272. The table values were started at 0 V and 0 cm and graphed in Figure 6-28 (Chart 7 of the worksheet) to obtain the fifth-order polynomial equations shown in the figure. The curve fit formulae enable calculating the breakdown threshold for both the positive and negative line alternations at gap separations of less than 1 cm (0.4″), which is typical for rotary spark gaps. The calculated breakdown thresholds are shown in cell (E19) for the positive alternation (black trace) and cell (E20) for the negative alternation (red trace) and marked by a large [X] on the graph. The calculations require the following worksheet entries:

- Air temperature in °C into cell (B5).
- Air pressure in torr in cell (B6).
- Relative humidity in % in cell (B7).
- The gap separation in inches in cell (B8).

The ionization of the gap is not instantaneous and requires a portion of time. This formative time lag increases with gap length and field non-uniformity, but decreases with applied overvoltage. Reference (14), page 390 provides the time period required to ionize an air gap at 0% (0 multiplier) to 100% (1.0 multiplier) applied overvoltage for four gap separations (0.12, 0.24, 0.39, and 0.55″). These data were used to construct Table 4 in rows 89 to 131. The table was used to construct Chart 4A, and the axes reversed in Chart 4B providing the curve fit formulae shown in the charts. The extended trendlines shown in Chart 4B were used to predict the ionization time at 0% overvoltage. Chart 4 shown in Figure 6-29 displays the second-order polynomial equation developed to calculate the ionization time at 0% overvoltage for the selected gap separation entered in cell (B8). The line voltage supplying the tank circuit of a spark gap coil has a frequency of 60 Hz and the rate of voltage change can be considered negligible. The applied overvoltage is also considered to be negligible or 0%. A 60-Hz, 9.0-kVrms line voltage will only change about 6 V in 50 μs which is only 6 V/9.0 kV = 0.067%, not enough to be considered an overvoltage. A better example of an applied overvoltage is the faster rate of change in the primary tank oscillations; however, this does not apply to the line voltage. The large [X] in Figure 6-29 marks the 19-μs time period (lag time) required to ionize a 0.14″ air gap with an applied overvoltage of 0%. This will be typical of most spark gap Tesla coil applications. The curve fit formula shown in the figure is used to calculate the time period required to ionize the gap in cell (E7) for the selected overvoltage in percent entered in cell (B9) and selected gap separation entered in cell (B8). A conditional statement in cell (E7) uses the four curve fit formulae shown in Chart 4B when an overvoltage greater than 0% is entered in cell (B8).

A very small portion of the ionization time is required for streamer formation, which leads to the breakdown of the gap. Reference (14), page 373, provides the time required for streamer formation and used to construct Table 6 and Chart 6 in rows 175 to 215 of the worksheet. Chart 6 is shown in Figure 6-30, which displays the propagation velocity in 10^6 m/sec for selected gap spacing and applied voltages of 25 kV (black trace) and 38.7 kV (red trace). The curve fit formulae shown in the figure are used to calculate the propagation velocity for 25 kV in cell (E28) and 38.7 kV in cell (E27) at the gap spacing entered in cell (B8). The calculated propagation velocity is converted to a time period required for streamer formation in μs at 25 kV in cell (E30) and 38.7 kV in cell (E29). It can be seen the time period required for streamer formation in a small gap is typically less than 1.0 ns, which is considered negligible.

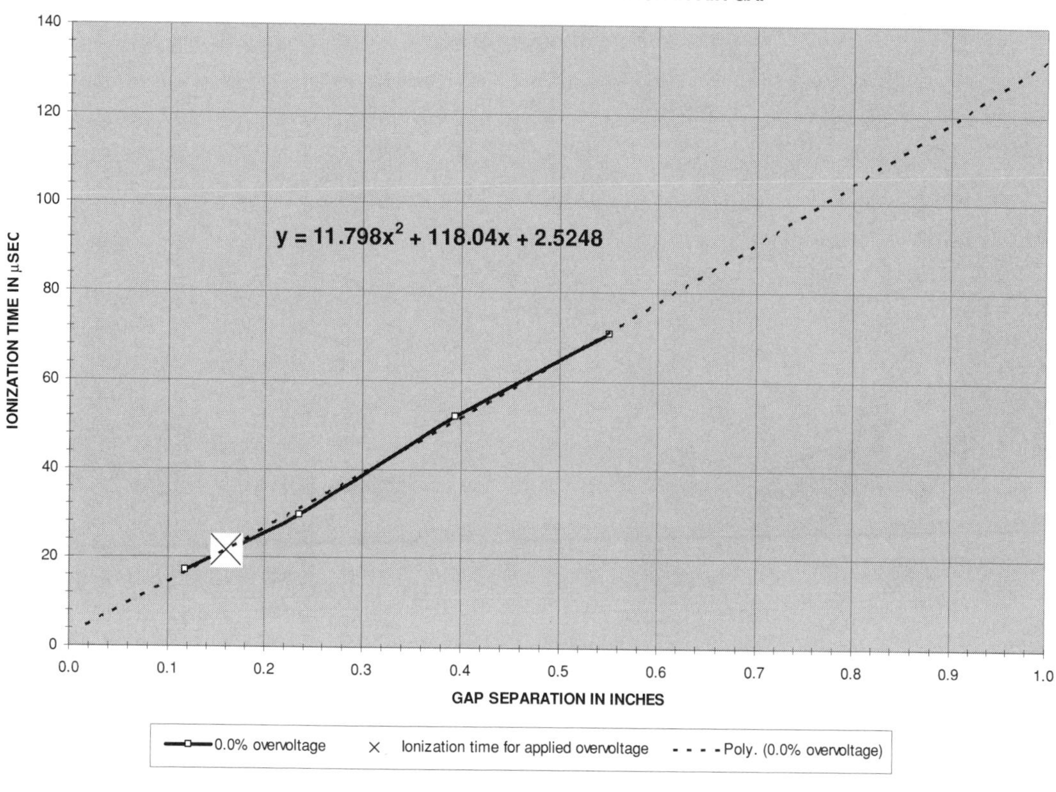

FIGURE 6-29 Ionization time of air gap with 0% applied overvoltage.

The ionization and break rate characteristics can be used to calculate and visually display the performance of the spark gap. Enter the remaining spark gap characteristics:

- Applied voltage (RMS output of step-up transformer) into cell (B10).
- Primary oscillation time in μs into cell (B11) from PRIMARY OSCILLATIONS worksheet (2).
- Applied line frequency in Hz into cell (B12).
- Phase shift in degrees in cell (B13).

The calculations and displayed data in Chart 1 require lookup functions in Excel, which are rendered inoperative unless the calculated data in the columns are arranged in a certain order (i.e., ascending or descending). For this reason the phase shift should be left at the –89.6° entered in cell (B13). If it is changed and the display in Chart 1 becomes erratic, return the phase shift to its –89.6° default. The calculation methodology for the primary and secondary impedance, current, and voltage is detailed in Chapter 2. The data in columns (AC)

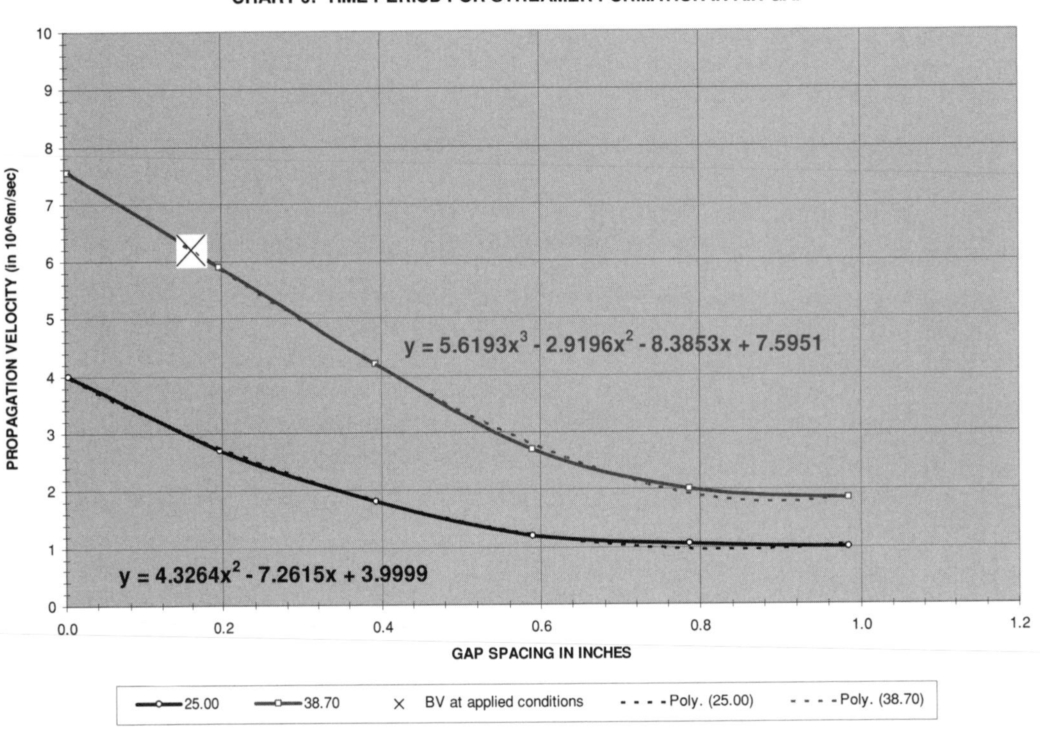

FIGURE 6-30 Streamer formation time of air gap at selected gap spacings.

to (AG) are used to calculate the line voltage, which forms the blue trace in Chart 1 shown in Figure 6-31.

Continue by entering the following parameters:

- The primary decrement into cell (B16).
- Tank capacitance in μF into cell (B17).
- Resonant frequency of the primary oscillations in kHz in cell (B18).
- The primary impedance at resonance into cell (B19).

The worksheet calculations shown in Table 2, columns (AL) to (AR) continue to Row 75, columns (AI) to (AK) and (AS) to (AT) continue to Row 800. The resulting primary tank voltage oscillations were selected for display in Chart 1 and seen as the red trace in Figure 6-31. Refer to both the red and blue traces in Figure 6-31. The breakdown voltage threshold for the positive line alternation is calculated in cell (E19) and for the negative alternation in cell (E20) from the curve fit formula shown in Figure 6-28. These breakdown voltages also determine the time period where gap ionization can occur. The ionization time for the positive line alternation shown in cell (E22) and negative alternation in cell (E23) uses a lookup function from Table 1 in the worksheet. The time period between the negative and positive line

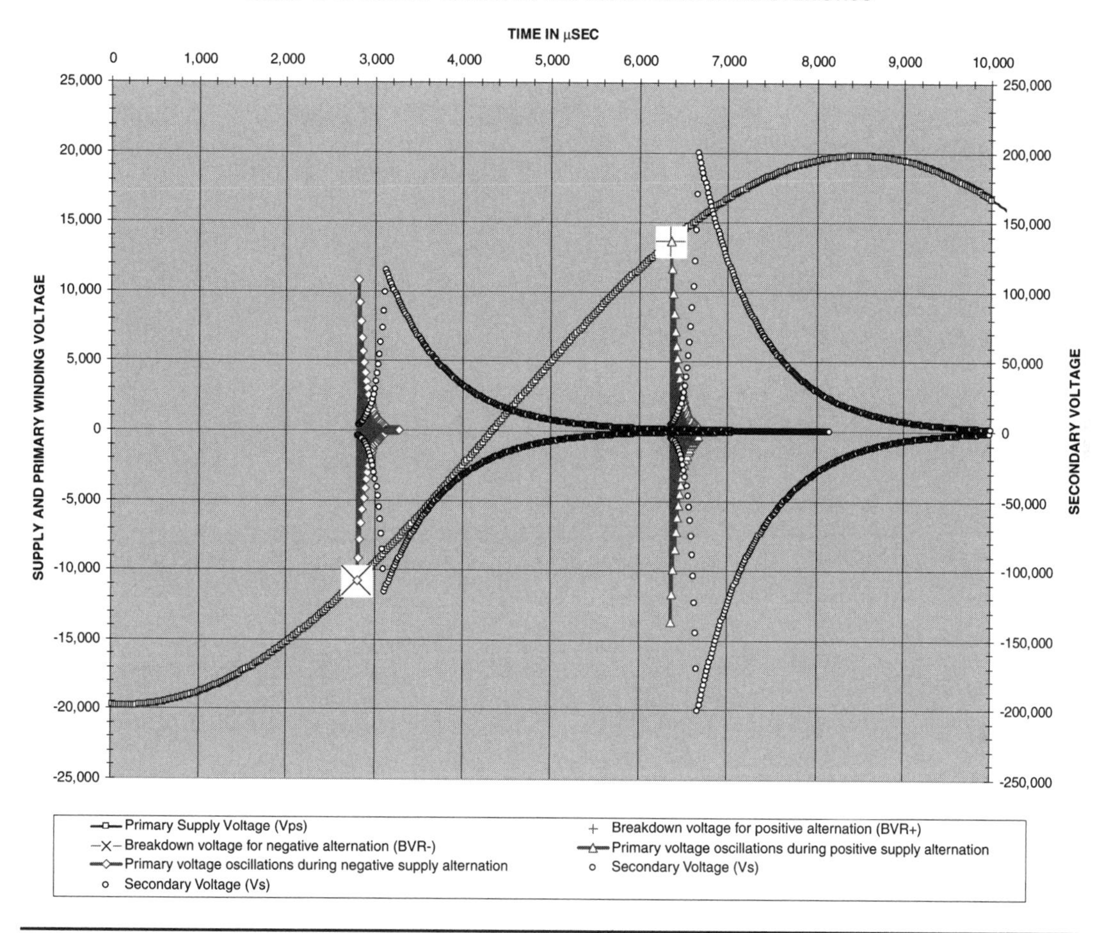

FIGURE 6-31 Calculated performance for selected gap operating characteristics.

alternation ionization thresholds determine the time the gap cannot be ionized due to insufficient line voltage to initiate breakdown in the gap. This time period is shown in cell (E24). The remaining time period from the voltage threshold, up through the peak line voltage and back down to the voltage threshold, is the calculated time period that ionization can occur in cell (E25).

The last parameters that need to be entered are:

- The secondary decrement in cell (B22).
- Secondary impedance at resonance in cell (B23).

The calculations shown in Table 3, columns (AY) to (BE) continue to Row 75, columns (AV) to (AX) and (BF) continue to Row 800. The resulting secondary voltage was selected for display in Chart 1 and seen as the black trace in Figure 6-31. Only the peaks of the secondary voltage are

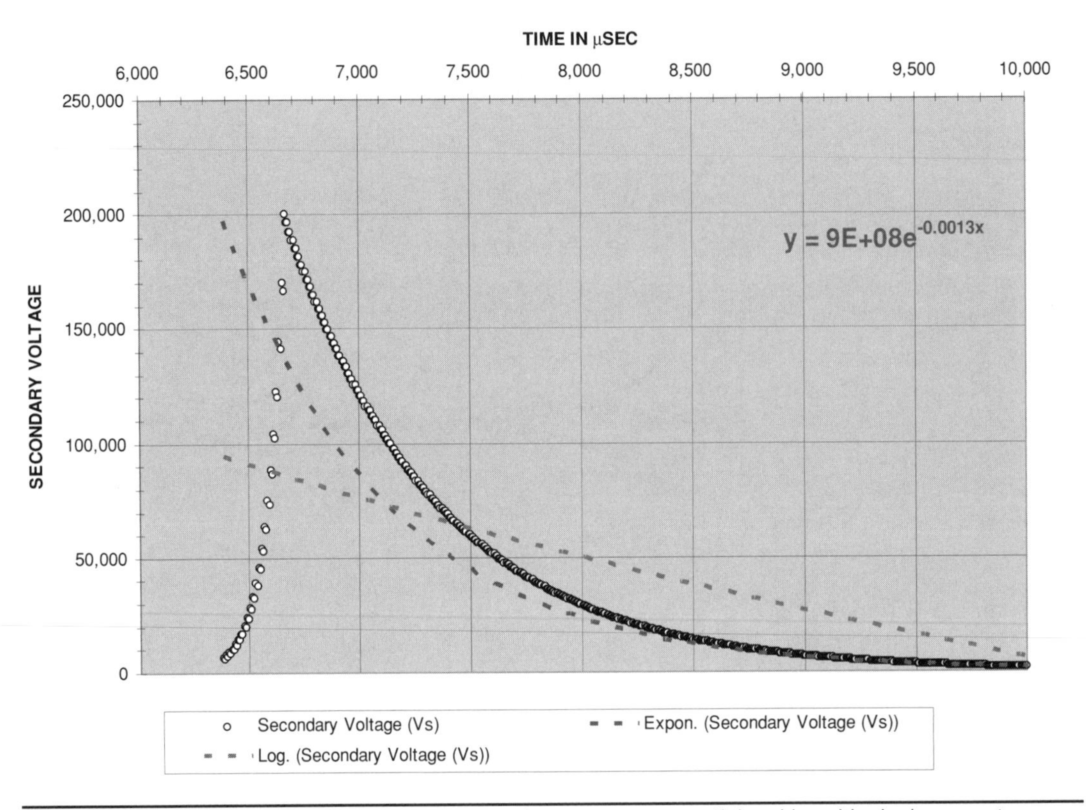

FIGURE 6-32 Comparison of calculated secondary voltage to an exponential and logarithmic decrement.

displayed forming a waveform envelope that prevents covering over the displayed primary waveform.

Once all the parameters are entered the high-voltage output of the step-up transformer at the line frequency of 60 Hz, primary tank voltage and secondary voltage are displayed in Chart 1. As the oscillation time in cell (B11) is adjusted the primary and secondary voltages will change. Adjust the oscillation time to that calculated in PRIMARY OSCILLATIONS worksheet (2). This will determine the time period required for the primary oscillations to diminish to their minimum ionization current. This time period is also shown in cell (E6).

The maximum break rate (BPS) for our selected rotary spark gap characteristics can now be determined. Added to the required primary oscillation time period in cell (E6) is the ionization time (formative lag time) required for the gap shown in cell (E7) from the curve fit formula in Figure 6-29. The time required for streamer formation in the gap is considered negligible but is added to the calculations anyway. The formation time period in cell (E30) for the slower propagating 25 kV was selected to represent this ionizing characteristic. When these time periods are added together the required primary oscillation time period is calculated in cell (E8). Some

Spark Gap Characteristics	ENTER VALUES	
Air Temperature in °C (**Ta**)	20.0 °C	68.0 °F
Air Pressure in torr (**AP**)	760	
Relative Humidity in Percent (**RH**)	50%	
Distance between gap ends in inches (**S**)	0.160 inches	0.41cm
Applied Overvoltage in percent (**Vo**)	0.0%	
Applied RMS voltage (**V**)	14.00 KV	19796.0 V
Enter primary oscillation time in μsec (**to**)	**317.50 usec**	0.000318 sec
Applied LINE frequency in Hz (**Lf**)	60 Hz	0.0167 sec
Enter Phase Shift of Synchronous Gap in Degrees (**PS**)	-89.6	
Primary Characteristics		
Primary decrement (δ**P**)	0.147	6.789
Tank capacitance in μF (**Cp**)	0.0430 uF	0.000000043
Resonant frequency of oscillations in kHz (**fo**)	74.38 KHz	74380Hz
Enter primary impedance @ resonance (**Zps**)	80.20 Ω	
Secondary Characteristics		
Secondary decrement (δ**S**)	0.01920	52.083
Enter secondary impedance @ resonance (**Zs**)	238.40 Ω	
Quench Characteristics		
Quench time in μs (**tQ**)	0.00 usec	0.000000 sec
Time resolution in μs (multiplier)	25.00 usec	0.000025 sec
A	B	C

	CALCULATED VALUES
Resonant oscillation time in μs (**t**)	13.4 μs
Required primary oscillation time in μs (**tP**)	317.5 μs
Required ionization time of spark gap in air in μs (**ti**)	21.7 μs
Required primary oscillation time in μs (**tPR**)	339.2 μs
Required quench time in μs (**tQ**)	0.0 μs
Minimum required operating time of spark gap in μs (**tO**)	**339.2 μs**
Maximum allowable breaks per second of spark gap (**BPS**)	**1692**
Relative Air Density (**RAD**)	1.00
Saturated Vapor Pressure (**SVP**)	18.2
Vapor Pressure in torr [mmHg or g/m^2] (**VP**)	9.1
Absolute Humidity in g/m^3 (**AH**)	9.0
Humidity effect multiplier for large diameters and spacings (**kHL**)	0.000
Temperature-Pressure-Humidity Correction Factor (**k**)	1.00
Breakdown Voltage at applied positive voltage (**BVR+**)	**13.665 kV**
Breakdown Voltage at applied negative voltage (**BVR-**)	**-10.753 kV**
Ionization Time in seconds for positive alternation (**tp**)	0.00635
Ionization Time in seconds for negative alternation (**tn**)	0.00280
Time period gap cannot ionize in seconds (**tni**)	0.00710
Time period gap can ionize in seconds (**tgi**)	0.00957
Propagation velocity of axial streamers at 38.7kVDC (**V1**)	6.20
Propagation velocity of axial streamers at 25.0kVDC (**V2**)	2.95
Formative time lag of axial streamers for 38.7kVDC (**tS1**)	0.0007 μs
Formative time lag of axial streamers for 25.0kVDC (**tS2**)	0.0014 μs
Applied **kV/m**	3444.88
Flashover time in μs (**tf**)	0.00000
D	E

FIGURE 6-33 Rotary spark gap performance worksheet.

time is required to deionize the gap and allow the secondary to decouple from the primary; however, it is very short and not considered essential to the calculations. The required quench time entered into cell (B26) allows a selectable time period for this characteristic. The required primary oscillation time period and quench time are added together in cell (E10) for a minimum required operating time for the gap.

The maximum allowable break rate can now be calculated:

$$BPS = \frac{1}{tO} \bullet \frac{tgi}{\left(\frac{1}{Lf}\right)} \tag{6.15}$$

Where: BPS = Maximum allowable breaks per second of rotary spark gap = cell (E11), calculated value.

tO = Minimum required operating time (ionization and quenching) of spark gap in μs = cell (E10), calculated value from the required primary oscillation time period (ionization) in cell (E8) and quench time in cell (E9).

tgi = Time period the line voltage can ionize the gap in μs = cell (E25), calculated value from Table 1, columns AC to AG.

Lf = Line frequency in Hz = cell (B12), enter value.

The calculated secondary voltage was also used to construct Figure 6-32. The secondary voltage is plotted by the black trace and an exponential trendline developed by the red trace. The exponential trendline closely follows the calculated secondary voltage, illustrating the exponential characteristic of the secondary decrement, whereas the logarithmic trendline shown in the blue trace does not approximate the voltage waveform. This should not be confused with the primary decrement, which is dependent upon the material used in the spark gap electrodes (reference 6) but independent from the secondary.

The worksheet calculations are shown in Figure 6-33. By changing the gap separation in cell (B8), ionization time in cell (B11), and the deionization time in cell (B26), the spark gap primary and secondary operating characteristics can be evaluated. The maximum BPS can also be determined for the characteristics entered.

Note the peak secondary voltage of 200 kV shown in Figure 6-31 is much less than that calculated in Chapter 2. The calculations in Chapter 2 assume the spark gap ionizes at the peak of the applied line voltage, which is the theoretical maximum possible. Note in Figure 6-31 the gap does not ionize at the line voltage peak therefore the secondary voltage will be less than the theoretical maximum. This worksheet was developed to illustrate the gap ionization characteristics and their correlation to the voltage developed in the secondary, and by itself does not represent all details of coil operation. When the PRIMARY OSCILLATIONS worksheet (2) is used to determine the total ionization time of the gap for the worksheets in Chapter 2, all operating characteristics are accounted for.

As shown in Figure 6-33 the 14 kV rms output of the step-up transformer will not ionize the 0.160″ gap until the positive alternation reaches 13.665 kV and the negative alternation reaches −10.753 kV. A 317.5-μs operating time produces the approximate 200-kV secondary voltage shown in Figure 6-31. This voltage increases when the gap is set to ionize at the line peak. The quench time was set to 0 μs to determine a maximum 1,692 BPS shown in Figure 6-33. This is slightly high as there is some small quench time. When the quench time is increased from 0 to 50 μs the maximum BPS decreases to 1,475. Higher quench times may be realized in the actual coil after it has been built and tested which will further decrease the BPS. In his Colorado Springs experiments Tesla was able to produce 4,200 BPS in a single turn primary winding. Early commercial spark gap transmitters used large motor–generator sets to produce very high BPS rates of several thousand per second; however, their purpose was to produce *undamped* oscillations. This was primarily used to increase the transmitted power in the antenna. As the BPS doubles, the power drawn from the line doubles, the relationship being linear. The secondary power increases but does not result in increased spark length, just increased brightness.

The ionization time is mostly dependent on the number of damped primary oscillations. To calculate the number of primary oscillations, open the CH_6A.xls file, PRIMARY OSCILLATIONS worksheet (2) shown in Figure 6-37. (See App. B.) Only the primary characteristics are needed for the calculations as detailed in Chapter 2. Enter the primary, spark gap, and tuning characteristics into the blue cells in column (B) rows 5 through 34. How many damped oscillations are produced in the primary depend on the following factors:

- The primary decrement (δP) determines how fast the primary oscillations will decay. In Chapter 2 are found the formulae for calculating δP and shown below:

$$\delta P = \frac{\pi}{Qp} \tag{6.16}$$

Where: δP = Primary decrement = calculated value in cell (F10).
Qp = Quality of primary circuit (figure of merit) = cell (F9), calculated value from:

$$Qp = \frac{\omega p L p}{Zpss}$$

Where: ωp = Frequency of primary oscillations in radians per second = cell (F6), calculated value. $\omega p = 2\pi p$ where: p is the resonant frequency of the primary oscillations calculated in cell (B36).
Lp = Calculated primary inductance for turns used in henries = cell (C35), calculated value.
$Zpss$ = AC impedance of primary circuit without reflected secondary at the resonant frequency in ohms = cell (F8) from:

$$Zpss = \sqrt{(Rp + Rg)^2 + \left(\omega p L p - \frac{1}{\omega p C p}\right)^2}$$

Where: Cp = Primary capacitance in farads = cell (B7), enter value in μF.
Rp = DC Resistance of primary winding in ohms = cell (B8), enter value.
Rg = Spark gap resistance for selected material characteristic.

- Whether the primary decrement and waveform is linear or exponential depends on the electrode material (e.g. copper, brass, aluminum, zinc, and silver are linear, magnesium is exponential) as cited in reference (6) pages 1 through 23.
  Entering a (1) into cell (B20) for a linear decrement characteristic (Rgl) uses the calculated resistance in cell (F13) from:

$$Rgl = \frac{6Vfo}{\pi I p} = \frac{6\,(264.16S + 42)}{\pi I p}$$

Entering a (2) into cell (B20) for an exponential decrement characteristic (Rge) uses the calculated resistance in cell (F14) from:

$$Rge = \frac{8Vfo}{\pi I p} = \frac{8\,(193.04S + 34)}{\pi I p}$$

Where: S = Spark gap spacing in inches = cell (B15), enter value.
Ip = Peak oscillating current in the primary circuit = cell (F12), calculated value from equation (6.17).
Vfo = Initial voltage of ionized gap = curve fit formulae from Figure 6-33 for selected gap spacing and electrode material.

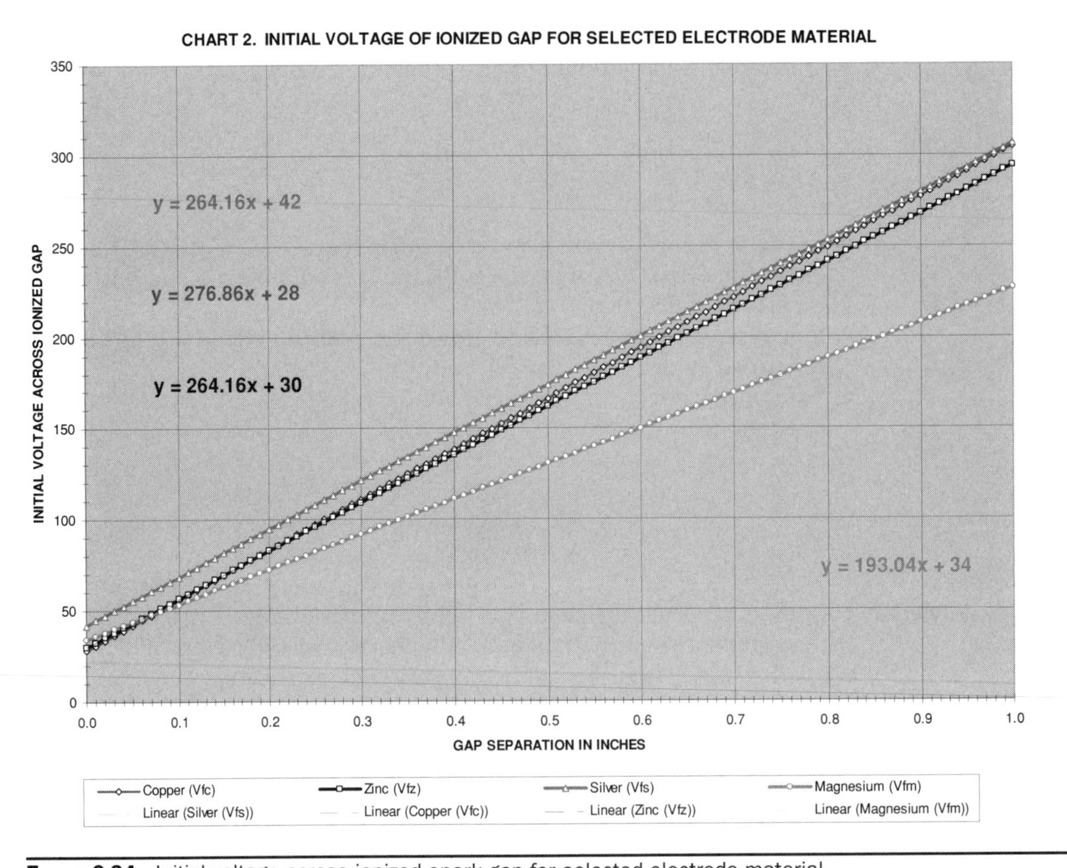

FIGURE 6-34 Initial voltage across ionized spark gap for selected electrode material.

The initial voltage amplitude of the ionized gap (Vfo) for selected gap separation and electrode materials is calculated in the CH_6.xls file, GAP RESISTANCE worksheet (8) using methodology in reference (17) for damped high-frequency currents:

$$Vfo = a_0 + a_1 \bullet S(mm)$$

Where: S = Spark gap spacing in inches = column (N). Converted to mm in column (O) using S (inches) × 25.4.
Coefficients a_0 and a_1 are dependent on the electrode material shown below:

	Magnesium	Zinc	Copper	Silver
a_0	34.0	30.0	28.0	42.0
a_1	7.6	10.4	10.9	10.4

Table 2 in rows 4 to 108 calculates the initial voltage (Vfo) in column (P) for magnesium, column (Q) for zinc, column (R) for copper, and column (S) for silver at the corresponding gap separation in mm in column (N). Chart 2, rows 63 to 103, columns (A) to (F), displays the calculations in Table 2 and shown in Figure 6-34.

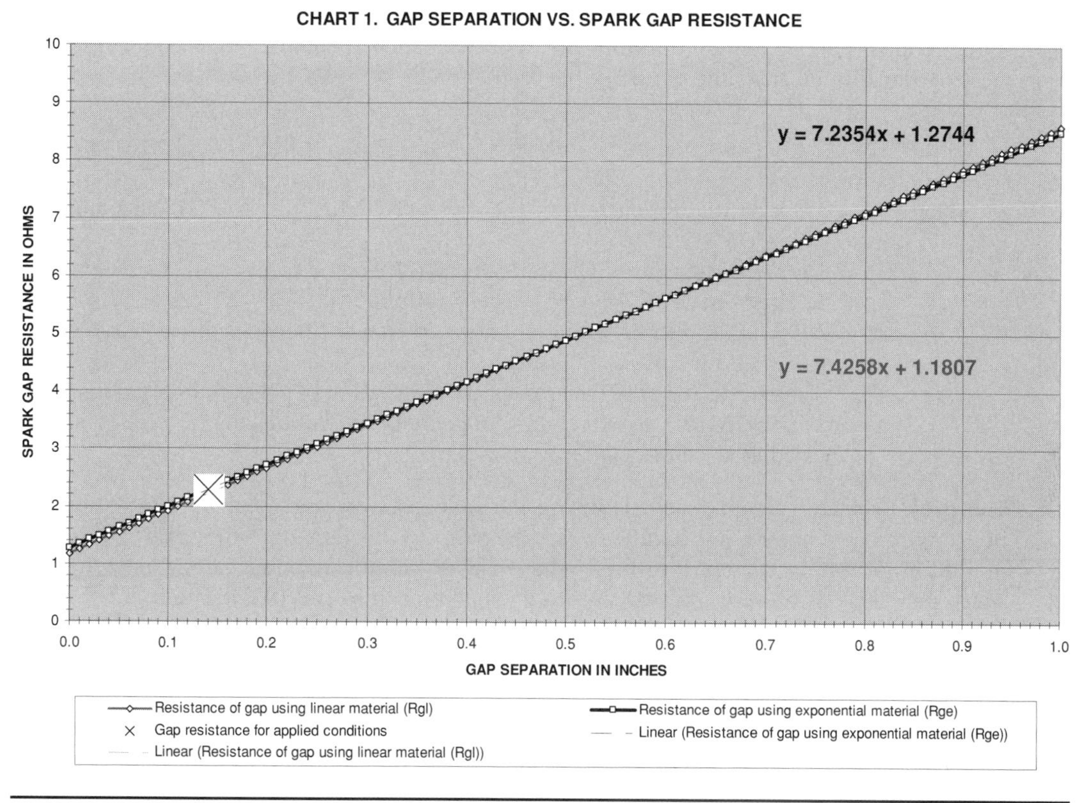

FIGURE 6-35 Spark gap resistance for selected primary current, gap separation, and electrode material.

The linear electrode materials exhibit little difference from each other and silver was considered representative for all linear electrode materials. The magnesium data are considered representative for all exponential materials. The linear curve fit formulae shown in Figure 6-34 are used to calculate Vfo for the applied gap spacing in all gap resistance formulae.

Table 1 in rows 4 to 108 calculates the gap resistance for linear electrode materials (Rgl) in column (*k*) and exponential electrode materials (Rge) in column (L) at the corresponding gap separation in mm in column (J). Chart 1, rows 9 to 48, columns (A) to (F), displays the calculations in Table 1 and shown in Figure 6-35. Both the linear and exponential materials exhibit a linear increase in gap resistance as the separation is increased. The difference in resistance between the two materials is typically a few mΩ at any gap separation. A large [X] marks the gap resistance at the selected separation entered in cell (B6), and initial peak current calculated in cell (F3) using equation (6.17) for the applied peak primary voltage in cell (B3), tank capacitance in cell (B4), and primary oscillating frequency in cell (B5).

- The primary current will be at its peak value during the first oscillation. This peak value is dependent on the value of tank capacitance, applied voltage from the step-up transformer, and time period of the oscillations:

$$Ip = Cp\frac{dv}{dt} \tag{6.17}$$

Where: Ip = Peak oscillating current in the primary circuit = cell (F12), calculated value.
Cp = Primary capacitance in farads = cell (B7), enter value in μF. Converted to farads in cell (C7) using a 1e-6 multiplier.
dv = Vp = Peak output voltage of step-up transformer in volts = cell (C9), calculated value.
dt = Time period of positive or negative alternation of primary oscillations = 1/(2fp), where fp = frequency of primary oscillations in Hz = cell (B36), calculated value.

The peak primary current reached during the first negative and positive alternation will decay with each succeeding oscillation by the primary decrement value until it reaches a minimum ionization current threshold where the arc cannot maintain ionization across the spark gap. This minimum ionization current is entered into cell (B21). The number of oscillations required for the peak current to decay to the minimum ionization current threshold is calculated in Table 1 of the worksheet, columns (*k*) through (CX), rows 2 to 27. Rows 2 to 14 calculate the decrement of the primary oscillating current for the selected primary characteristics entered into cells (B5) through (B34). A conditional statement in rows 16 to 27 calculates the number of primary oscillations required to decay to the minimum ionization current entered into cell (B21) for the selected decrements. Column (L) uses a conditional statement to count the number of primary oscillations required to decay to the minimum ionization current calculated in columns (M) through (CX). The total number of primary oscillations will change with the minimum ionization current threshold set in cell (B21).

For design comparison the calculated number of oscillations for decrement values 0.1 to 1.0 in 0.1 increments is shown in cells (F17) through (F26) and displayed in the graph shown in columns (A) to (D), rows (38) to (65) using the CH_6A.xls file, PRIMARY OSCILLATIONS worksheet (2) and shown in Figure 6-36. The calculated number of oscillations required to decay from the peak value to the minimum ionization current threshold for the applied decrement is shown in cell (F28). The time period of these primary oscillations is calculated in cell (F29). The aforementioned ionization characteristics are added to this oscillation time for a minimum required ionization time period calculated in cell (F34) and the maximum allowable BPS in cell (F35).

In the graph and in Figure 6-36 the large [X] shows the number of calculated oscillations using the logarithmic curve fit equation and the large [+] the calculated results using the sixth-order polynomial equation shown in the figure. Unfortunately the curve fits change somewhat with the primary characteristics entered into cells (B5) through (B34), therefore a comprehensive formula to calculate the number of oscillations cannot be developed. Several cells in the worksheet were used in an attempt to develop a comprehensive formula and were retained but are not addressed in this section. The calculations in cells (F39) and (F40) must be changed

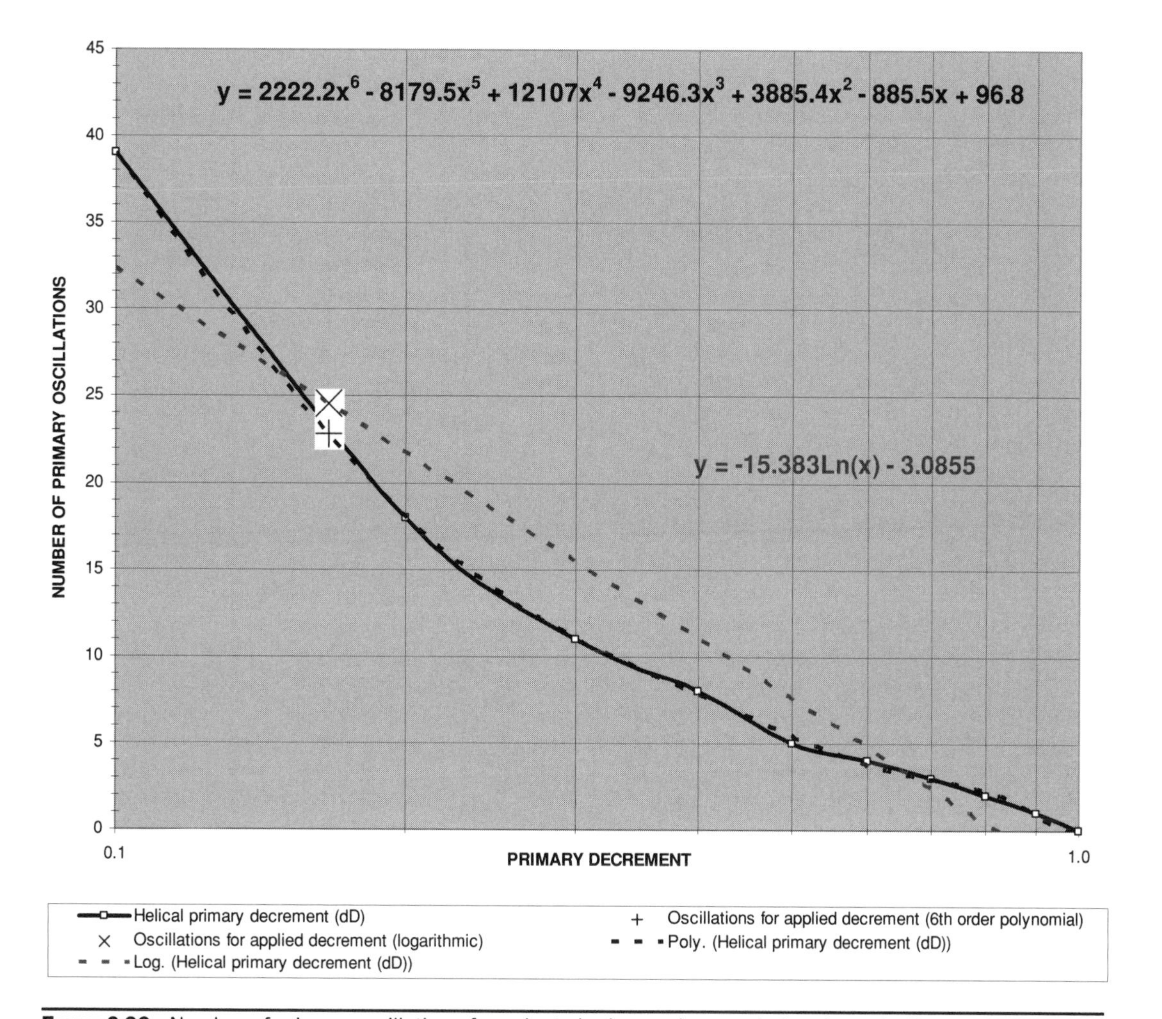

FIGURE 6-36 Number of primary oscillations for selected primary characteristics shown in Figure 6-37.

to match the curve fit formulae shown in the graph as the primary characteristics entered into cells (B5) through (B34) are changed, but can be left as they are since they do not affect the calculations at the applied decrement characteristics in cells (F28), (F29), (F34), and (F35). It is interesting, however, that a logarithmic curve (broken red trace) will approximate the number of oscillations produced in the primary for decrement values 0.1 to 1.0, indicating a pseudo-logarithmic primary decrement.

For comparison the methodology used in reference (24) to calculate the number of oscillations was examined. When the natural logarithm is used the number of oscillations can be

PRIMARY CHARACTERISTICS

A	B (ENTER HERE)	C
Measured or estimated (coil running) AC line voltage (**LV**)	200 V	70:1
Turns ratio of step-up transformer (**NT**)	**1:70**	
Tank capacitance in μF (**Cp**)	**0.0430μF**	0.0000000430000 F
Primary winding DC resistance in Ω (**Rp**)	0.0100 Ω	
Calculated applied output voltage of step-up transformer in kVrms (**Vp**)	**14.00 KV**	19796 V

SPARK GAP CHARACTERISTICS

A	B	C
Air Temperature in °C (**Ta**)	20.0 °C	68.0 °F
Air Pressure in torr (**AP**)	760	
Relative Humidity in Percent (**RH**)	50%	
Distance between gap ends in inches (**Sg**)	0.160 inches	0.41cm
Applied Overvoltage in percent (**Vo**)	0.0%	
Applied LINE frequency in Hz (**Lf**)	60 Hz	0.0167 sec
Enter Phase Shift of Synchronous Gap in Degrees (**PS**)	89.5	466.5311
Spark gap breaks per second (**BPS**)	460	0.002174 sec
Enter: (1 for linear) or (2 for exponential) gap material characteristics	2	
Minimum ionization current (**Imin**)	**1.0 A**	
Quench time in μs (**tQ**)	0.00 usec	0.000000 sec
Calculated breakdown voltage at applied positive alternation in kV (**BVp**)	**13.67 KV**	13665 V

Peak applied voltage (Vp) in cell (C9) must be greater than breakdown voltage (BVp) in cell (B22).

PRIMARY TUNING

A	B	C
Enter: (1 for Archemedes Spiral) or (2 for Helical) Wound Primary	1	
Enter Inside Diameter of Archemedes Spiral (**ID**) or Outside Diameter of Helical Primary in inches (**OD**)	18.0 in	45.72cm
Interwinding Distance in inches (**IWD**)	1.000 in	2.54cm
Enter Total Number of Turns in Primary Winding (**Ttp**)	13	
Enter Angle of Inclination in° if using Archemedes Sprial (θ)	50.0 °	
Enter Desired Primary Turn Number Used To Tune (**Tp**)	**11**	
Enter Tuning Capacitance in μF (**Cpt**)	0.00000 μF	0.0000000000000 F
Enter Tuning Inductance in μH (**Lpt**)	0.0 uH	0.0000000000000 H
Calculated Primary Inductance in μH (**Lp**)	**106.48 uH**	0.00010648 H
Primary resonant frequency (**fP**)	**74,378 Hz**	fp=1/[2*π*sqrt([Lp+Lpt]*Cp)]

PRIMARY CALCULATIONS

ROW	E	F
5	Find ωp	467331
6	Find resonant oscillation time period in seconds (**tp**)	1.34448E-05
7	Find **tan ϕ** of primary oscillations	0.000000
8	Resonant primary impedance with S.G., w/o reflected sec (**Zpss**)	2.6198 Ω
9	Find primary Quality factor (**Qps**) with spark gap	18.92
10	Find decrement factor of primary (δP)	**0.16602**
11	Maximum primary winding voltage (**Vpp**)	3271 V
12	Peak primary tank current in Amps (**Ip**)	**63.3 A**
13	Lineal spark gap resistance in Ohms (**Rgl**)	2.5419 Ω
14	Exponential spark gap resistance in Ohms (**Rge**)	2.6098 Ω
15		
16	Total number of primary oscillations greater than minimum ionization current (**nP**)	
17	Primary decrement (δD) 1.0	0
18	0.9	1
19	0.8	2
20	0.7	3
21	0.6	4
22	0.5	5
23	0.4	8
24	0.3	11
25	0.2	18
26	0.1	39
27	0	90
28	Primary oscillations for selected decrement characteristics (**nP**)	**22**
29	Calculated primary oscillation time period (**tP**)	**295.8 μs**
30		
31	Required ionization time of spark gap in air in μs (**ti**)	21.7 μs
32	Required primary oscillation time in μs (**tPR**)	317.5 μs
33	Required quench time in μs (**tQ**)	0.0 μs
34	Minimum required operating time of spark gap in μs (**tO**)	**317.5 μs**
35	Maximum allowable breaks per second of spark gap (**BPS**)	**1817**
36		
37		

Columns: A, B, C, D, E, F (COLUMN)

FIGURE 6-37 Number of primary oscillations (ionization time) worksheet.

approximately found using the formula from Bureau of Standards, Circular 74 (C74):

$$nP = Ln\frac{\left(\frac{Ip}{Im}\right)}{\delta P} \tag{6.18}$$

Where: nP = Number of primary oscillations for selected primary decrement.
Ln = Natural logarithm.
Ip = Peak primary current in amps
Im = Minimum ionization current for selected spark gap characteristics (see Figure 6-25).
δP = Primary decrement.

This assumes the decrement follows a logarithmic characteristic, which may not be the case. The approximation method used in C74 may not produce the same results as calculating each oscillation with the appropriate decrement for the gap material used (i.e., exponential for magnesium, linear for brass, aluminum, zinc, or silver).

The primary decrement itself was referred to as a logarithmic decrement as far back as 1915 as evidenced in reference (6). It was customary to use the natural logarithm of the ratio of initial amplitude to the amplitude of succeeding oscillations. Damped oscillations such as those found in the primary are illustrated in Figures 6-38 to 6-43 for selected decrement values. Do not confuse these with the pseudo-exponential characteristic found in the secondary.

To optimize the spark gap performance, do one or more of the following:

- Increasing the gap separation allows the gap to break down at a higher voltage threshold. This produces a higher primary voltage and peak current, which in turn produces a higher secondary voltage. Increasing the separation decreases the maximum BPS that can be obtained; however, this is usually not of concern.
- Increasing the electrode diameter will increase the dwell time of the rotary spark gap. This decreases the maximum BPS obtainable; however, as long as break rates of at least 500 BPS are still available in the design, efficient operation is obtainable. The dwell time should approximate the ionization time or the arc may still be ionized long after the rotating electrodes have passed the stationary electrodes. This is inefficient and excess energy is drawn from the line by the arc. If a specific dwell time is desired and the ionization time exceeds it, there is a way to compromise. The ionization time is predominately affected by the primary decrement (δP). As δP is increased the ionization time decreases. To increase δP simply use fewer primary turns and use a larger tank capacitance to retune the primary oscillations to the resonant frequency of the secondary. Thus decreasing the primary inductance (Lp) by using fewer turns decreases the primary Q, which increases the δP producing fewer primary oscillations (ionization time). If the stationary or rotating electrodes quickly burn away, requiring frequent replacement, the electrode diameter can be increased to enable better electrode cooling between ionization periods (see Section 9.4). The heat sink mass used for the stationary electrodes can be increased without increasing their diameter; however, the rotating electrodes have no heat sink. Larger electrodes provide a larger surface area increasing their current carrying capacity. The larger surface area also provides more efficient cooling through convection with the surrounding

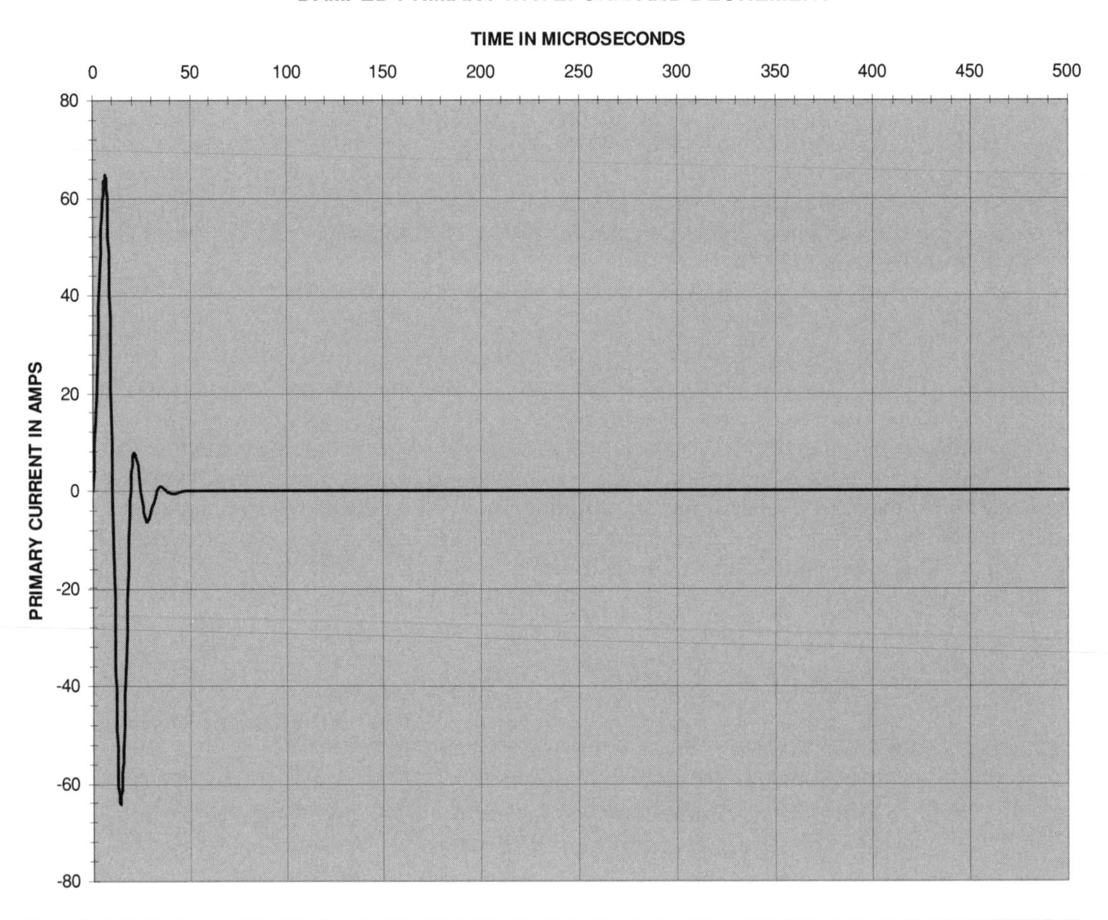

FIGURE 6-38 Damped oscillations for decrement value of 0.9.

airflow. Ensure the calculated safety ratio in equation (6.13) is not exceeded with the larger rotating mass. A simpler solution is to include additional series stationary gaps, which will help distribute the power dissipation in the series gaps while they are ionized.

- As the number of primary turns used to tune decreases, the selected tank capacitance should be increased to retune the primary oscillations to the resonant frequency of the secondary. The higher the capacitance (Cp) and faster the oscillating frequency (dt), the higher the peak current $Ip = Cp(\mathrm{dv}/\mathrm{dt})$. If the tank capacitance is not increased when decreasing the primary turns, the peak primary current increases from the higher primary oscillating frequency; however, the secondary voltage may not increase as the frequency of oscillations in the primary are detuned from resonance.

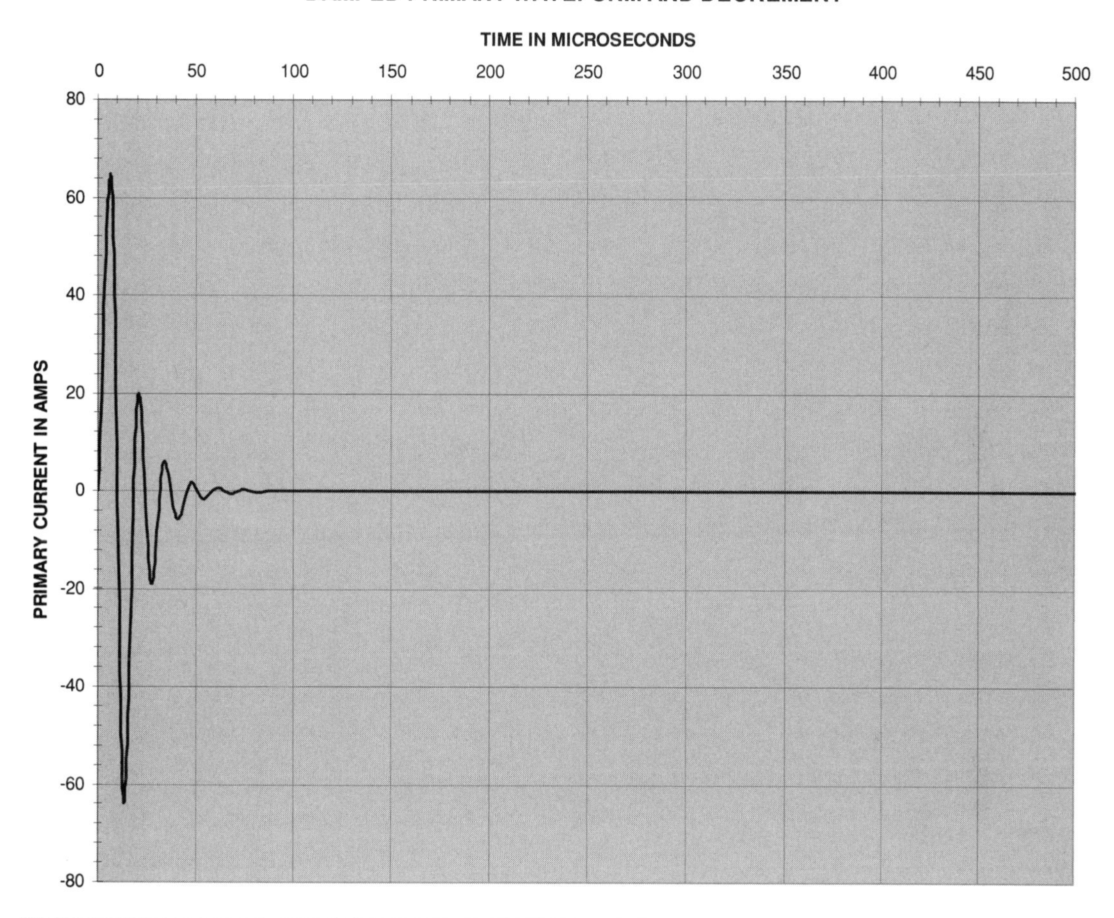

FIGURE 6-39 Damped oscillations for decrement value of 0.7.

- A higher BPS does not necessarily mean a higher secondary voltage. It does mean the coil will use more line power. The intensity of the secondary spark will increase with a higher BPS.
- The electrode material and gap spacing determine the frequency stability of the primary oscillations. Page 9 of reference (6) indicates that copper or silver electrodes with a separation less than 2 mm (0.08 inch) and small tank capacitance value (corresponding to a high primary inductance and low decrement) exhibit frequency variations up to ±10%. When tin, zinc, cadmium, or magnesium electrodes with a separation greater than 4 mm (0.157 inch) to 5 mm (0.197 inch) and larger tank capacitance value (corresponding to a low primary inductance and high decrement) were used, the frequency variations decreased to as low as ±1.0%. This was critical when rotary gaps were used

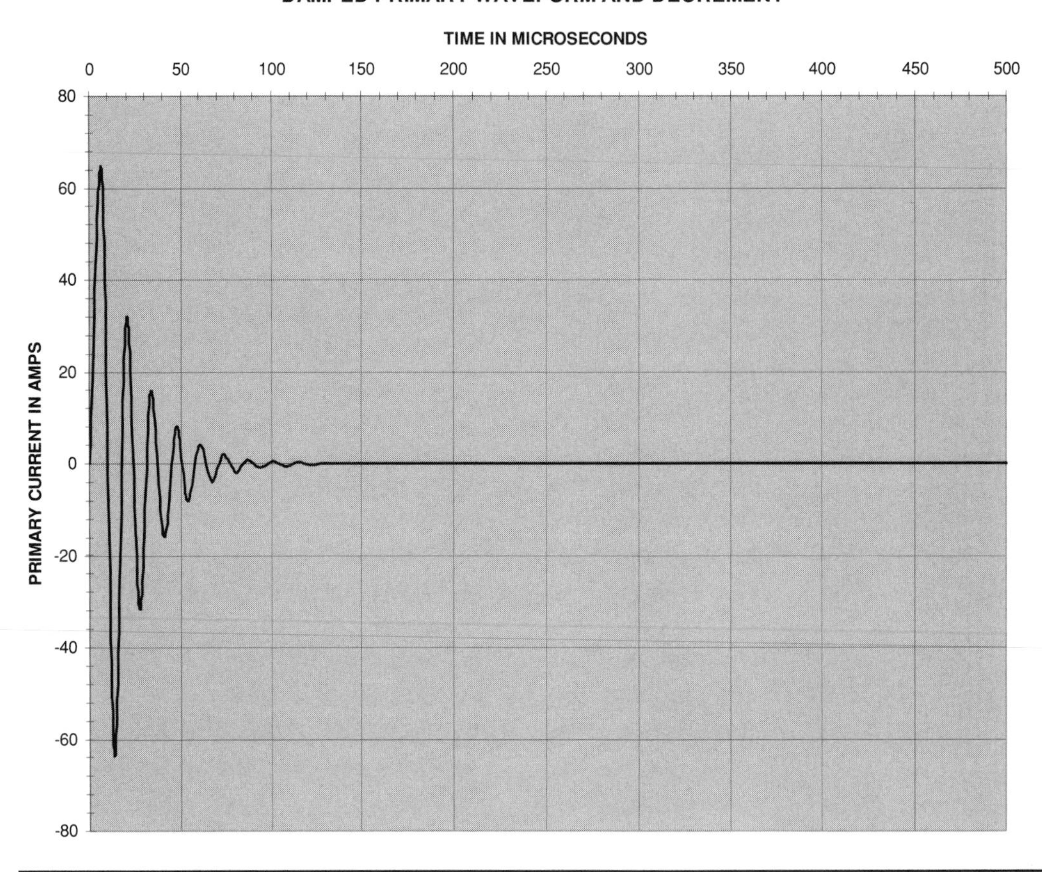

FIGURE 6-40 Damped oscillations for decrement value of 0.5.

in spark gap radio transmitters. It also indicates electrode material and separation distance is open to further experimentation.

I inadvertently discovered an interesting quenching characteristic when the speed control (variable autotransformer) to my rotary gap motor was turned down to a very low setting after moving the control cabinet. I estimate the BPS was running at less than 120 BPS, which is less than a fixed gap and difficult or not really obtainable in practice. When I tried to run the coil, no secondary spark was produced. My cabinet was producing a *noisy* 60-Hz buzz, and the cabinet's wattmeter indicated almost 20-kW output (the coil normally ran at 3 kW). I am sure this 20 kW was right because the 60-Hz buzz was the result of all the magnetic material (lamination strips) of the step-up and variable autotransformers vibrating from the excessive eddy currents and power levels. Those variable autotransformers and potential transformers can really take a pounding and still deliver. Finally looking at the rotary spark gap I saw a huge arc extending three quarters of the rotor circumference. Several adjustments to the system produced the same annoying result until I found the speed control turned down. Running up the

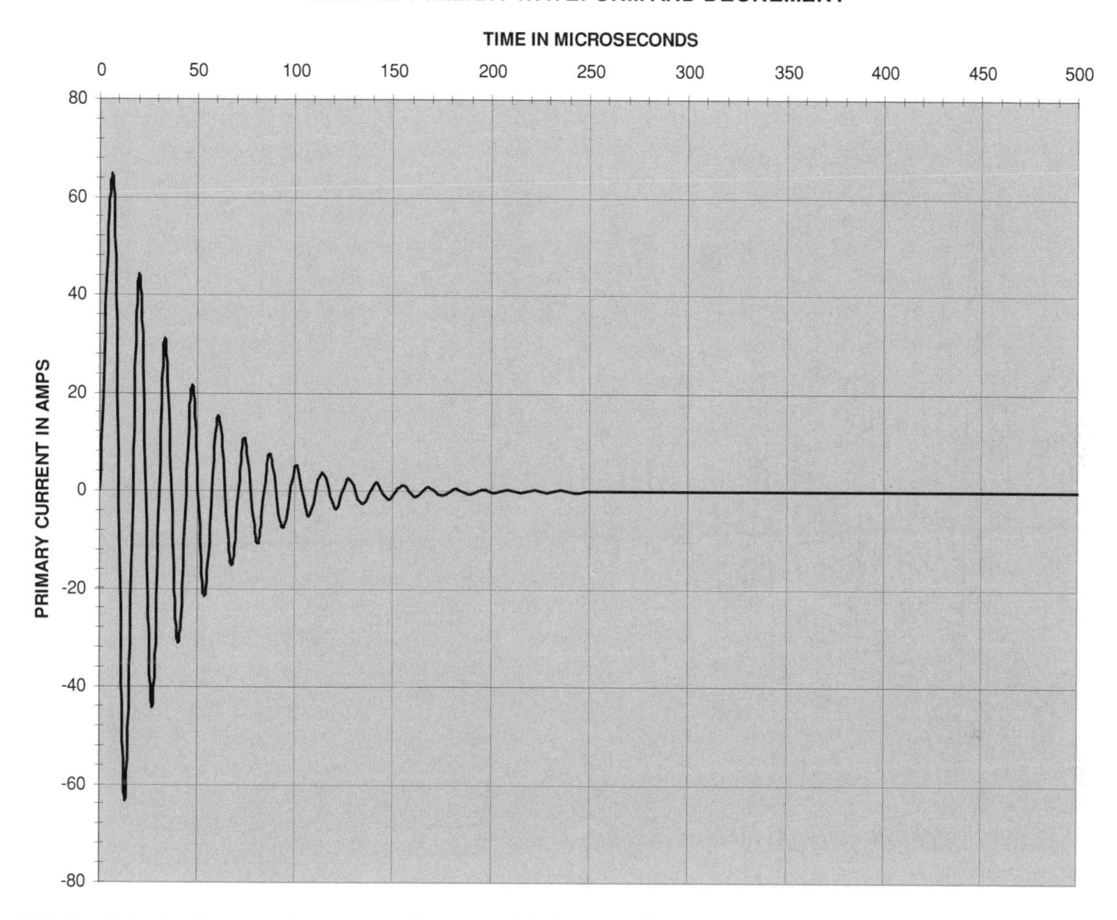

FIGURE 6-41 Damped oscillations for decrement value of 0.3.

speed toward full had the coil running smoothly. The 15″ arc formed along the circumference of the rotor was drawing 20 kW from the line and dissipating it. This was just an observation and is not recommended for use.

6.12 Safety Gaps

The safety gap is essentially a stationary spark gap, but should only fire (ionize) when an over-voltage transient is generated or induced in the primary tank. Large voltage transients created in detuned or improperly operating Tesla coils can damage the step-up transformer or tank capacitor, or transfer to the primary (line) side, creating EMI or dangerous voltage transients on the supply line. Refer to Figure 2-1. The distance between electrodes in the safety gap must be adjusted to break down when a predetermined voltage threshold is reached. Using Sections 6.2

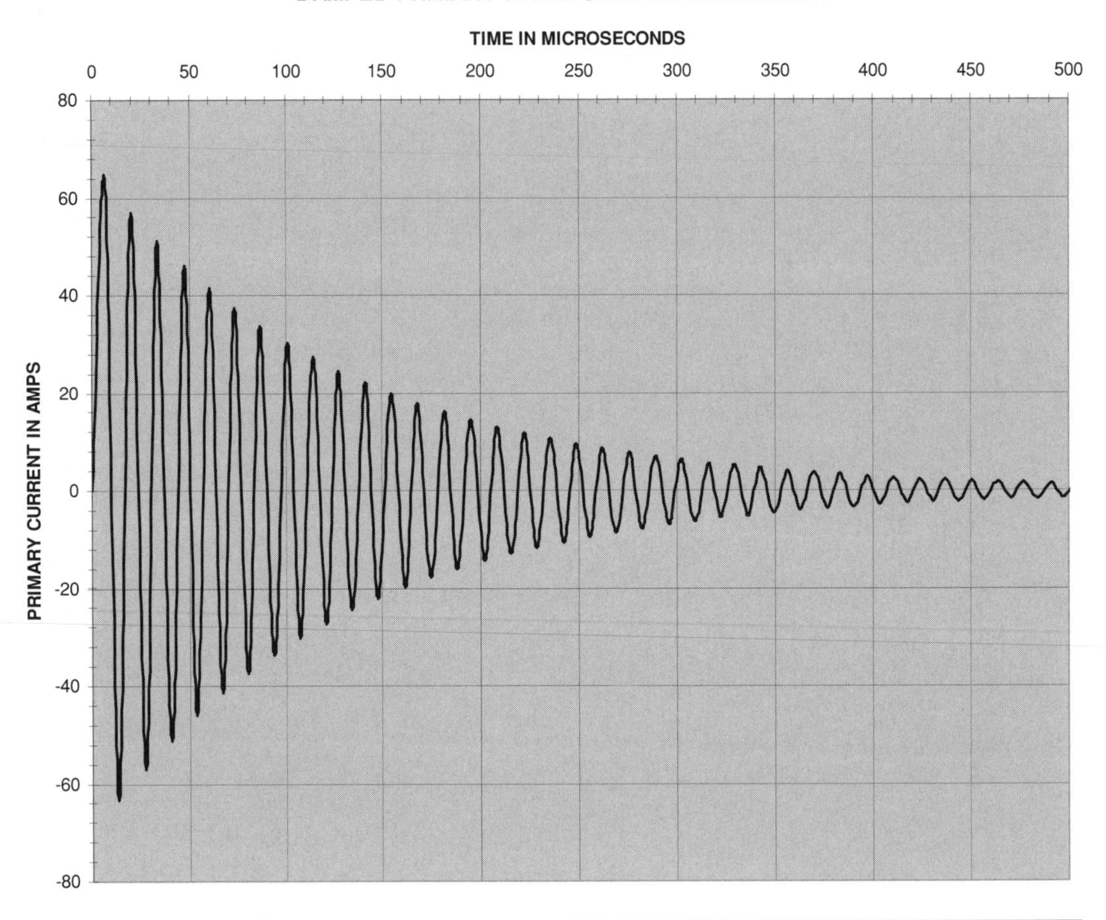

FIGURE 6-42 Damped oscillations for decrement value of 0.1.

through 6.4, determine the gap distance needed for the type of gap ends being used, and the operating temperature, humidity, and air pressure that will break down the gap at the peak applied voltage of the primary tank circuit. Add enough additional gap distance to keep it from arcing during normal operation.

Say we are using a 70:1 step-up transformer with a 230 V input. The output of the step-up transformer is 70 × 230 V = 16.1 kV. The peak voltage is 16.1 kV × 1.414 = 22.8 kV. We want the safety gap to fire above the peak 22.8 kV. Follow these steps to adjust the gap:

1. Using a safety gap such as shown in Figure 6-44 and the appropriate non-uniform spherical gap end geometry data from Figure 6-5, the gap spacing at 25 kV would need to be about 0.3 inch for an initial setting. For comparison, if the spherical ends were removed and the rod gap end geometry data from Figure 6-9 used, the gap spacing at 25 kV would need to be about 1 inch for an initial setting.

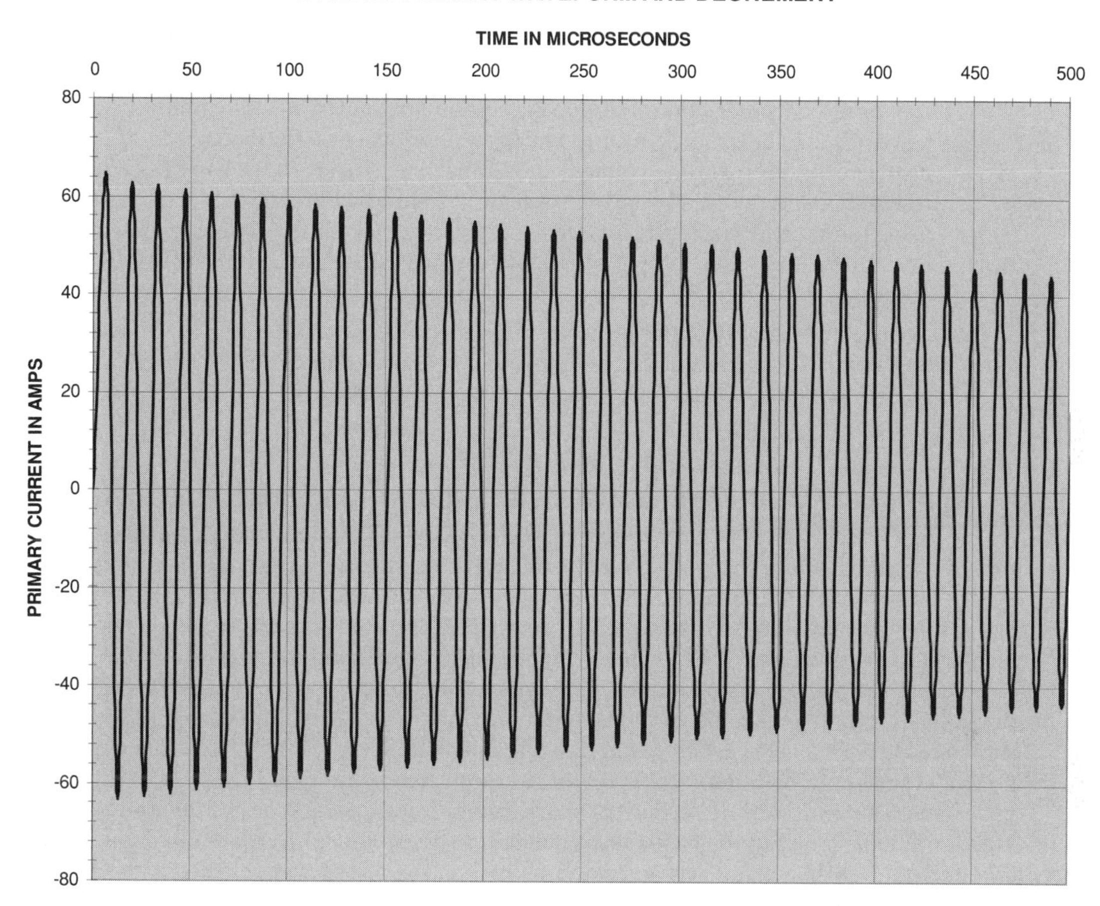

FIGURE 6-43 Damped oscillations for decrement value of 0.01.

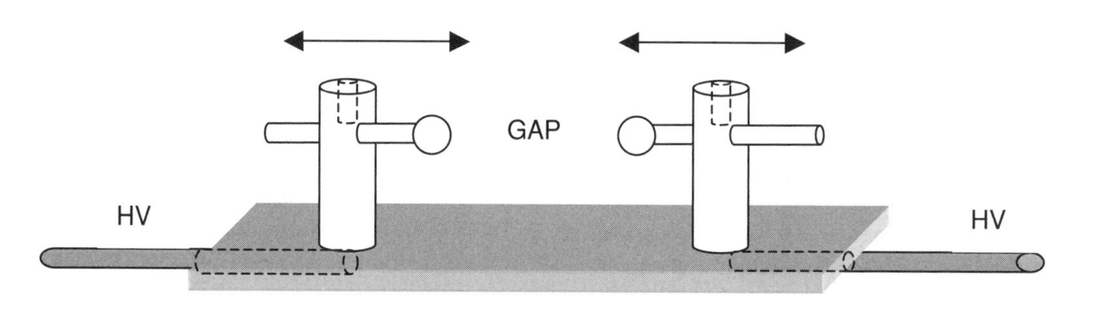

FIGURE 6-44 Details of safety gap construction.

2. With the primary circuit of the Tesla coil disconnected (spark gap, tank capacitor, and primary winding) and the safety gap connected to the step-up transformer, slowly apply the full line voltage to the step-up transformer using the variable autotransformer (voltage control). Again, you want the safety gap to fire at a level above the step-up transformer's output. If the gap fires before you reach maximum the gap distance needs to be increased. *Disconnect power* and increase the distance in 1/16″increments, reapplying the line voltage until the gap stops firing. If the gap does not fire the coil will operate normally and the voltage threshold where the gap fires is estimated using the appropriate figure in Sections 6.2 through 6.4 for the gap end geometry used. Remember, the wider the gap separation the higher the voltage transients developed in the primary circuit.
3. Connect the primary of the Tesla coil to the step-up transformer and check overall coil operation. The gap may fire at a slightly lower setting with the load connected. Further adjustment may be necessary to achieve the desired results.

6.13 Spark Length in High-Voltage Terminal of an Operating Tesla Coil

As the high voltage on the terminal capacitance exceeds the breakdown value a spark erupts from the terminal. Once the ionized air channel is formed the air is continuously superheated by the discharge. The plasma in the discharge is very hot. Thermal ionization lowers the voltage required to maintain the discharge and thermal equilibrium is nearly reached in operating break rates approaching 500 BPS. The voltage needed to maintain the spark once it is ionized is much less than that characterized in Sections 6.2 through 6.4, which apply to a single discharge event. Reference (7) contains measured data from operating Tesla coils. The graph shown in Figure 6-45 was generated in the CH_6B.xls file, VS vs. SPARK LENGTH worksheet (3) from the data in the reference. The broken trace extends the linear characteristic of these measurements to project the secondary spark length for a 2,000-kV output. Figure 6-46 provides additional detail of the measured data and the linear curve fit. As seen in the figure the curve fit and 10 kV/inch is a very good approximation supported by published data. Once the spark reaches about 10 inches it takes about 10 kV to produce each additional inch of spark or 10 kV/inch. The linear curve fit formula generated by this extrapolation of the measured data is:

$$SL = 0.1058Vs - 1.2027 \tag{6.19}$$

Where: SL = Length of high-voltage discharge in inches = cell (D19), calculated value.
Vs = Peak voltage in the secondary winding = cell (D17), enter value in kV.

The terminal can be made large enough that the high voltage generated in the secondary cannot overcome the insulating threshold of the terminal capacitance and a spark never breaks out. This detunes the coil and standing waves are produced in the secondary. Unless your intention is generating a large electromagnetic pulse (EMP) and electromagnetic interference (EMI), I would not recommend this approach. See Chapter 5 for details on terminal capacitance construction and calculating the high-voltage threshold where the spark will break out.

When the spark does not travel a straight line between discharge terminals as shown in Figure 6-47 the length is increased. To calculate the additional length when the spark does not

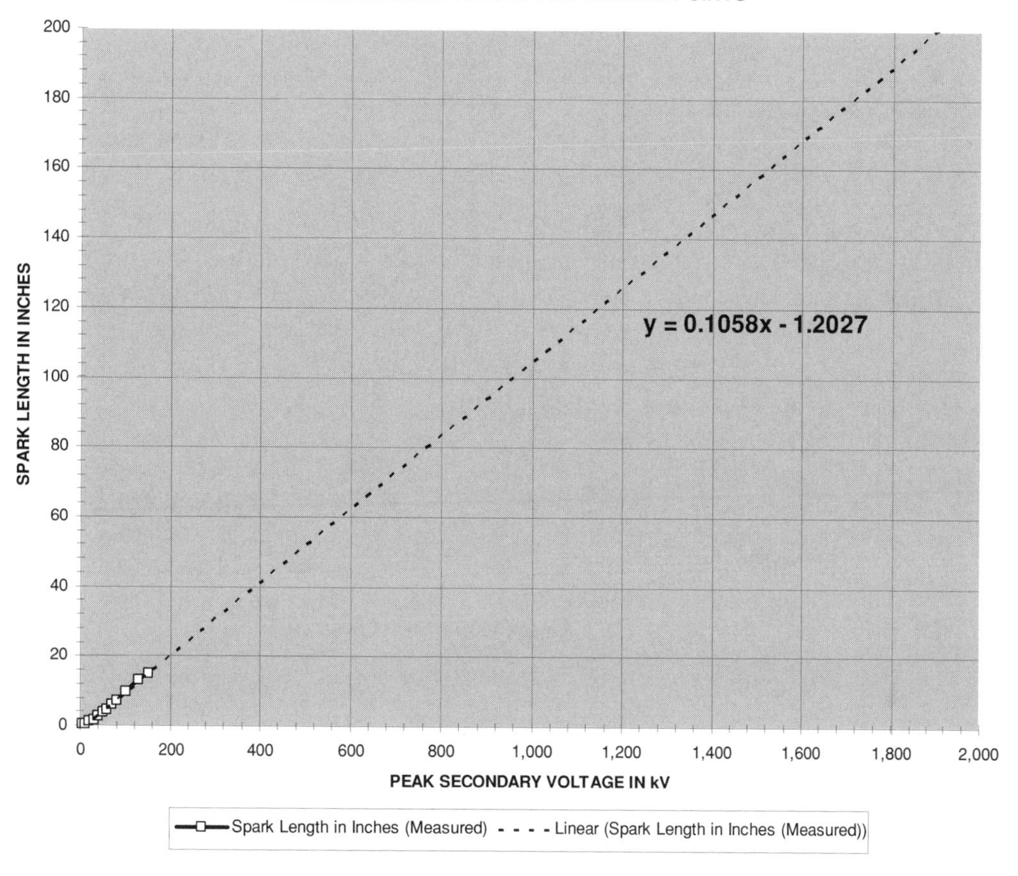

FIGURE 6-45 Spark length in inches produced by peak secondary voltage.

follow a straight path, open the CH_6A.xls file, ARC LENGTH worksheet (4). From reference (16) is found the formula for calculating the arc length:

$$SL = \frac{1}{2}\sqrt{d^2 + 16H^2} + \frac{d^2}{8H} \bullet \ln\left(\frac{4H\sqrt{d^2 + 16H^2}}{d}\right) \tag{6.20}$$

Where: SL = Length of high-voltage discharge with arc of travel in inches = cell (B6), calculated value.

d = Distance between discharge terminals in inches = cell (B2), enter value.

H = Distance between base of line between discharge terminals and top of arc of travel in inches = cell (B3), enter value.

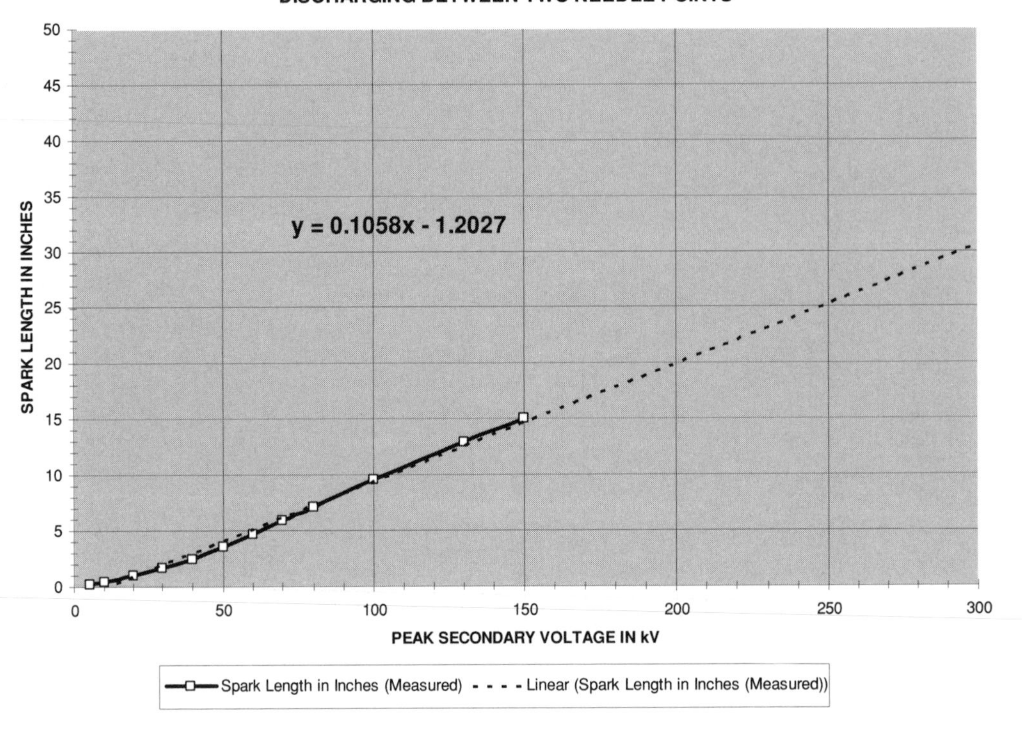

Figure 6-46 Additional detail of measured data in Figure 6-45.

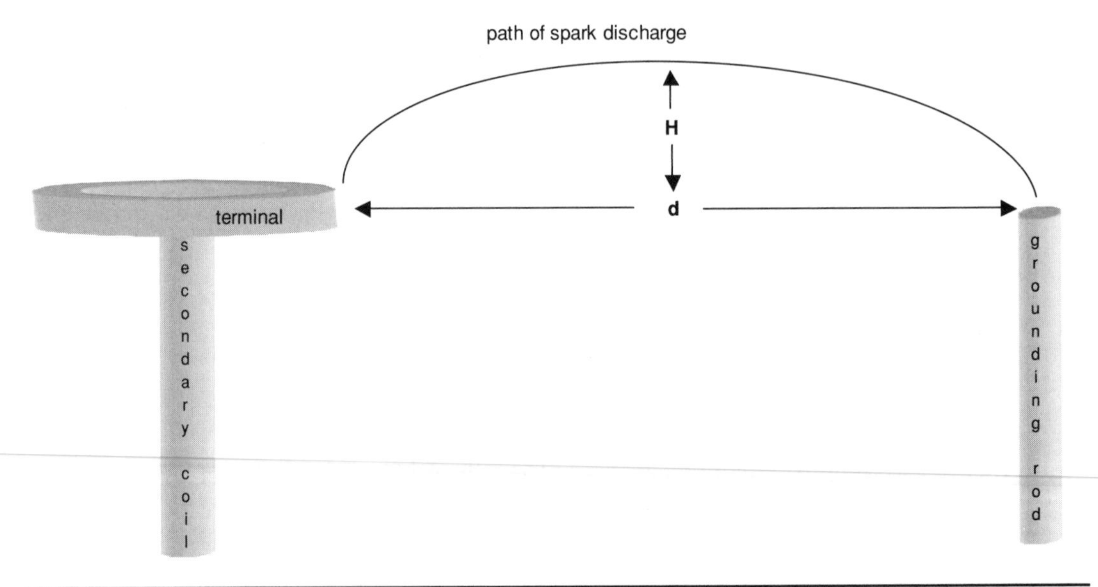

Figure 6-47 Parameters for calculating spark length with an arc of travel between terminal ends.

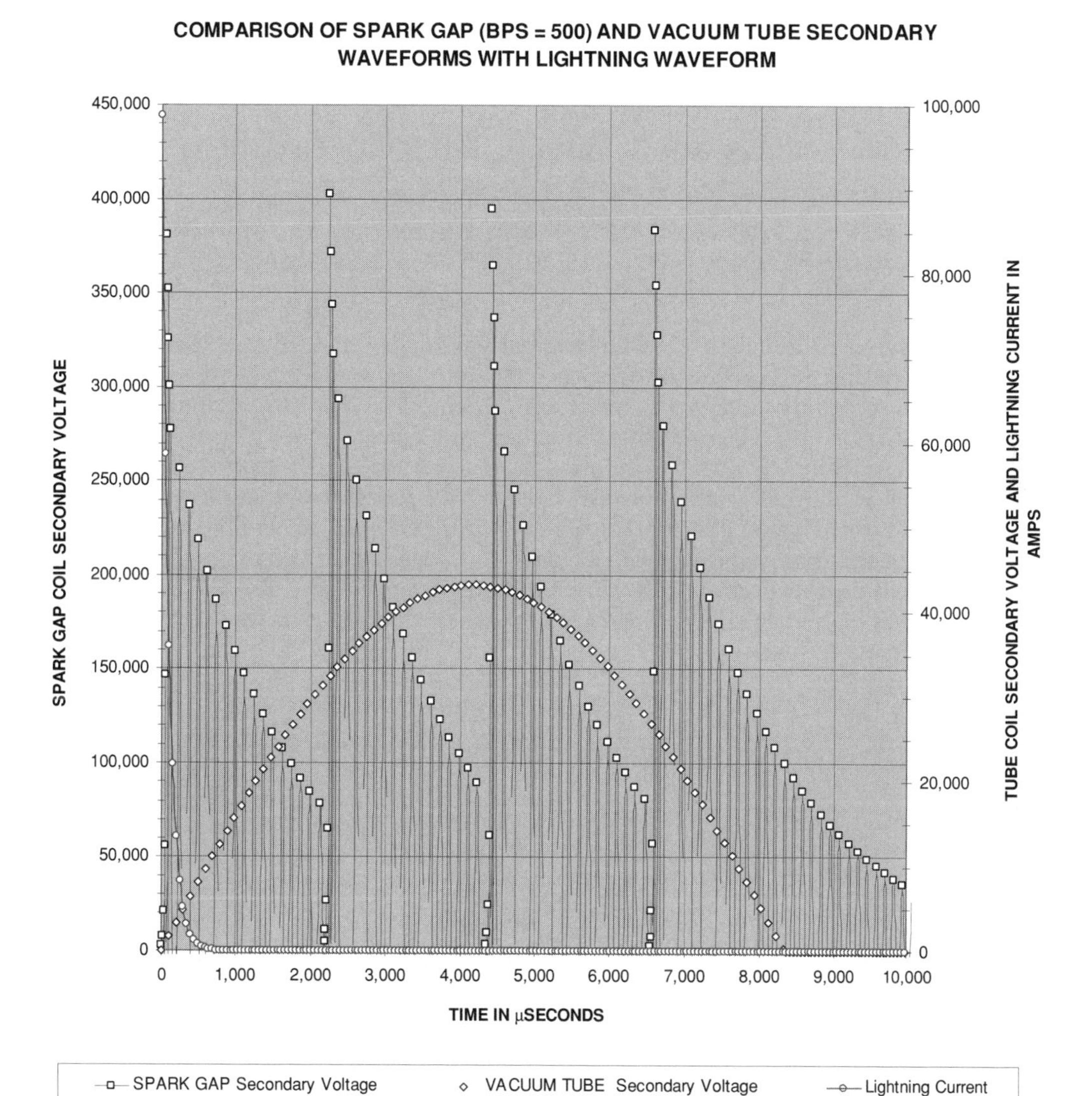

FIGURE 6-48 Comparison of spark gap and tube coil secondary waveforms with lightning waveform.

6.14 Comparison of Spark Produced in the Spark Gap Coil, Tube Coil, and Lightning

If you have seen the spark produced by both a spark gap coil and a tube coil the difference is quite noticeable. The spark gap coil produces a spark comparable with that of lightning.

This is no coincidence. Let's compare this output with lightning characteristics. Reference (11) describes lightning as follows:

- 1 to 10 μs exponential rise time.
- 50 to 1,000 μs exponential fall time.
- Currents of a few kA to 250 kA.
- 30 kV/cm (75 kV/inch) voltage required to break down air between cloud and ground (streamer formation). Only 10 kV/cm (25 kV/inch) is required in the presence of water droplets.

Excel was used to generate a lightning current waveform with an exponential 10 μs rise to 100 kA and an exponential decay of 1,000 μs. This waveform is compared to the spark gap coil secondary voltage waveform from Chapter 2 and the vacuum tube coil secondary voltage waveform in Figure 6-48. The voltage and current waveforms will follow the same characteristic, the choice of waveforms made were for ease of calculation. Only the peak envelope of the tube coil waveform is shown to prevent covering the spark gap waveform. As clearly seen the spark gap waveform possesses similarities to the exponential lightning waveform, therefore its similarity in appearance to lightning is easy to follow. The tube coil waveform is quite different, thus its appearance will be quite different. The tube coil's high-voltage output is closer in visual appearance to the 60-Hz plasma arc in a Jacob's ladder. Considering the secondary peak voltage in a tube coil forms a modulated

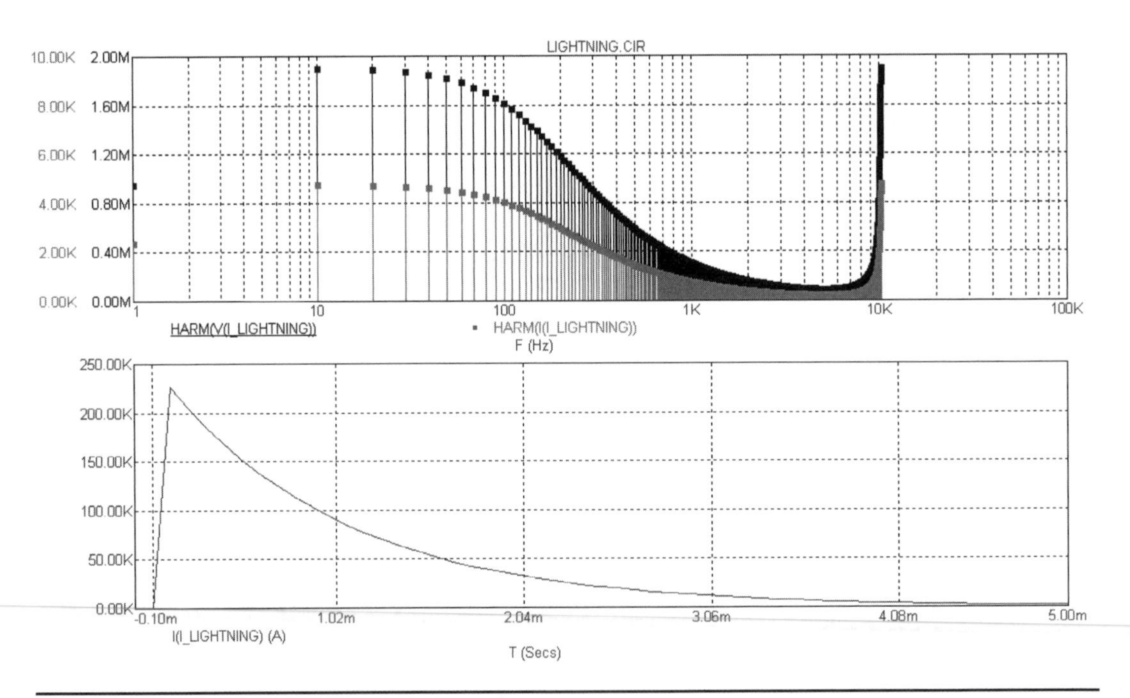

FIGURE 6-49 Harmonic content of lightning waveform.

60-Hz envelope it is not surprising the visual appearance more closely resembles a 60-Hz plasma arc.

The lightning discharge is rich in harmonics, which is why it produces so much Electromagnetic Interference (EMI). The harmonic content in lightning was analyzed using a Spice model. The transient characteristics of the lightning are shown in the lower graph of Figure 6-49 with the harmonic content shown in the upper graph. Much of the energy is contained in frequencies below 10 kHz. A harmonic content analysis of the spark gap Tesla coil Spice model used in Chapter 8 indicates the primary oscillations are also rich in harmonics. It is not unlike the lightning or any other disruptive discharge. In contrast, the high Q and selectivity of the Tesla secondary circuit produces oscillations in a very narrow band of frequencies with much less harmonic content. Harmonics are attenuated by the inverse square law in oscillating or discharge waveforms regardless of their frequency. In small and medium coils the EMI is often negligible when measured only a few meters from the source.

References

1. Plastic materials data sheets found on Industrial Plastic Supply, Inc.'s Web site: www.indplastic.com/polycarb.html.
2. Plastic materials data sheets found on A.L Hyde Co.'s Web site: www.alhyde.com
3. Plastic materials data sheets found on Westlake Plastic's Web site: westlakeplastics.com
4. From Jim Lux's Web site: www.home.earthlink.net/ jimlux/hv/rotgap.html. Reference cited: Naidu & Kamaraju, *High Voltage Engineering.*
5. *Handbook of Chemistry and Physics*, Editor: Charles D. Hodgeman, 30th Ed: 1947. Chemical Rubber Publishing Co, Boca Raton, FL. "Based on results of the American Institute of Electric Engineers."
6. Dr. J. Zenneck, Translated by A.E. Seelig. *Wireless Telegraphy*. McGraw-Hill: 1915.
7. John C. Devins and A. Harry Sharbaugh. *The Fundamental Nature of Electrical Breakdown*. Electro-Technology, Feb. 1961, pp. 103–122.
8. Lawrence Martin, M.D. *Scuba Diving Explained: Questions and Answers on Physiology and Medical Aspects of Scuba Diving*. 1997. Section D. Found on website: http://www.mtsinai.org/pulmonary/books/scuba/sectiond.html.
9. Saturated Water Vapor Pressure in mmHg for various air temperature data from website: http://hyperphysics.phy-astr.gsu.edu/hbase/kinetic/watvap.html.
10. *Reference Data For Radio Engineers*. H.P. Westman, Editor. Federal Telephone and Radio Corporation (International Telephone and Telegraph Corporation), American Book. Fourth Ed: 1956, pp.920–922.
11. E. Kuffel and W.S. Zaengl. *High Voltage Engineering: Fundamentals*. Pergamon Press, Oxford: 1984, pp. 391–392, 463–473.
12. *Basics of Design Engineering: Plastics*. Machine Design, June 1991, Penton Publishing. Pp. 794–902.
13. Found on *The Resident's Electronic Handbook*, SI Conversions website: http://www.medana.unibas.ch/eng/amnesix1/si_1.html.
14. E. Kuffel and W.S. Zaengl. *High Voltage Engineering: Fundamentals*. Pergamon Press, Oxford: 1984, pp. 94–112, 354–361.
15. S.L. Herman and W.N. Alerich. *Industrial Motor Control*. Third Ed. Delmar Publishers, Inc., Albany, NY: 1993, pp. 104–105.
16. Formula found in Mathsoft's Mathcad reference files, section 3.1.13 (parabolic segment) referenced to: *Mathematical Handbook of Formulas and Tables*. Murray Spiegel. Mcgraw-Hill: 1968.

17. Dr. J. Zenneck, Translated by A.E. Seelig. *Wireless Telegraphy*. McGraw-Hill: 1915, p. 392, Table V. Source cited: D. Roschansky. Ann. Phys.: 1911, pp. 36, 281.
18. Nikola Tesla. *Colorado Springs Notes: 1899–1900*. Nikola Tesla Museum (NOLIT), Beograd, Yugoslavia: 1978.
19. *Materials Technology Series: Vacuum Metallurgy*. Edited by Rointan F. Bunshah. Reinhold Publishing Co. NY: 1958. Chapter 6: Arc Phenomena, Basic and Applied, pp. 101–112.
20. James Dillon Cobine. *Gaseous Conductors: Theory and Engineering Applications*. McGraw-Hill, NY: 1941.
21. Ibid, pp. 292–295. Source for Ayrton equation cited: Hertha Ayrton. *The Electric Arc*. The Electrician Series, D. Van Nostrand Company, Inc., NY: 1902, pp. 120, 130.
22. J.J. Thomson. Cambridge Physical Series: *Conduction of Electricity through Gases*. Second Ed. Cambridge University Press: 1906. Chapter XV: Spark Discharge, pp. 430–527.
23. John. D. Ryder. *Electronic Engineering Principles*. Prentice-Hall, Inc. NY: 1947, pp. 273–276.
24. U.S. Department of Commerce, National Bureau of Standards, *Radio Instruments and Measurements, Circular 74*. U.S. Government Printing Office. Edition of March 10, 1924—reprinted Jan.1, 1937, with certain corrections and omissions.

CHAPTER 7

Control, Monitoring, and Interconnections

Evidently the soil lets the water run through easily and being extremely dry as a rule it is very difficult to make a good ground connection. This may prove troublesome.

Nikola Tesla. Colorado Springs Notes: 1899-1900, pp. 37.

Tesla laments on the difficulty of obtaining a low-resistance earth ground in the dry Rocky Mountain climate.

It is based on my observation that by passing through a rarefied gas a discharge of sufficient intensity, preferably one of high frequency, the resistance of the gas may be so diminished that it falls far below that of the best conductors.

Nikola Tesla. Colorado Springs Notes: 1899-1900, pp. 45–46.

Tesla observes the low resistance of a rarefied gas.

A well-designed control scheme is essential to ensure safe and reliable operation of the coil. Your control cabinet should contain the following minimum features:

- Adequate input filtering to eliminate EMI, noise, and high-voltage transients from the line service.
- Voltage and current indications to monitor the line voltage and current under operation.
- Short-circuit protection such as a circuit breaker or fuse to disconnect the line service from the cabinet.
- A variable autotransformer to allow the line voltage to be varied from 0 V to the maximum line input value.

Additional features can be added such as:

- Rotary spark gap speed and/or phase control.
- Power indications to monitor the line power under operation.
- Over-current shutdown to disconnect the line service if the line current exceeds a preselected value.

- A panic switch and contactor to enable immediate disconnect of the control cabinet from the line service in case of a malfunction or operating emergency.
- Bells, whistles, lights, and any other ephemeral fluff that is desired to attract attention, increase awareness, or add to the overall esthetics of the coiling experience.

This chapter discusses the design and operation of control components and their interconnection. Let us begin with voltage, current, and power indications.

7.1 Apparent Power

Apparent power in volt-amperes (VA) is measured with a single volt-amp meter or separate volt and ampere meters. A volt-amp meter measures total voltage and current *without regard* to the phase angle between the voltage (V) and current (I). Also known as reactive power it is the power used by the load to maintain its electromagnetic field, as typical loads are inductive. Reactive power is not usable. In an AC circuit the predominately inductive or capacitive load will cause a lag or lead of the current with respect to its phase relationship with the voltage.

As seen in Figure 7-1a, the impedance (Z) is the hypotenuse formed by the DC resistance (base or R) and the AC reactance (altitude or X). X is the vector sum of the inductive reactance (X_L) and capacitive reactance (X_C). X_L and X_C are always present in AC loads and oppose each other. X_L is usually predominant in AC loads since a transformer or motor is included in the load. Using the Pythagorean theorem the hypotenuse (Z) in a predominately inductive load is found:

$$R^2 + X^2 = Z^2 \quad \text{or} \quad Z^2 = \sqrt{R^2 + (X_L - X_C)^2}$$

$$Z = \frac{\text{Adjacent}}{\cos \angle\theta} \quad \text{or} \quad Z = \frac{\text{Opposite}}{\sin \angle\theta}$$

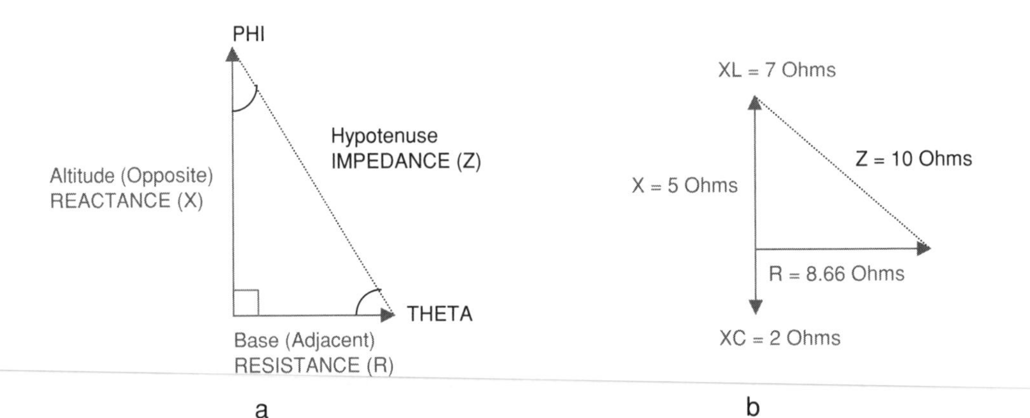

FIGURE 7-1 Relationships of *R*, *X*, and *Z*.

The reactance can be found:

$$X_L = 2\pi fL \quad \text{and} \quad X_C = \frac{1}{2\pi fC}$$

Where: X_L is the AC reactance of the inductive component in ohms.
X_C is the AC reactance of the capacitive component in ohms.
f is the applied AC frequency (60 Hz in U.S.).

For the example in Figure 7-1b: $(8.66\,\Omega)^2 + (5\,\Omega)^2 = Z^2 = 100\,\Omega$, $Z = \sqrt{Z^2} = 100 = 10\,\Omega$.

The cosine of $\angle\theta$ (angle theta) is the power factor. To find $\angle\theta$ use one of the following formulae to obtain the sine (sin), cosine (cos), or tangent (tan) of $\angle\theta$:

$$\sin\angle\theta = \frac{\text{Opposite}}{\text{Hypotenuse}} \quad \text{and} \quad \cos\angle\theta = \frac{\text{Adjacent}}{\text{Hypotenuse}} \quad \text{and} \quad \tan\angle\theta = \frac{\text{Opposite}}{\text{Adjacent}}$$

Next use the calculator's inverse function to find $\angle\theta$, e.g., divide the reactance (opposite) by the resistance (adjacent) to find the tangent of $\angle\theta$, then use the inverse function to obtain $\angle\theta$, and finally use the cosine function to find the cosine of θ. This is the power factor. Note that the resistance (adjacent) divided by the impedance (hypotenuse) will also calculate the cosine of $\angle\theta$.

For the example in Figure 7-1a: $5\,\Omega \div 8.66\,\Omega = 0.5774 =$ tangent of $\angle\theta$, $\angle\theta =$ inverse tangent $= 30°$, PF = cosine of $30° = 0.866$ or 86.6%.

These calculations are automatically performed in all Excel worksheets where needed.

7.2 True Power

True power in watts (W) is measured with a wattmeter. A wattmeter measures instantaneous values of voltage (V) and current (I) *with regard* to the phase angle between V and I. A wattmeter measures an average value of power equivalent to the heating power in a DC resistive circuit. True power is the power usable to the load. With a 60-Hz sine wave applied to the line of the circuit in Figure 7-2 and a resistive load such as seen in a well tuned Tesla coil, the AC voltmeter will measure the RMS equivalent or 0.707. The AC ammeter also measures the RMS equivalent or 0.707 of the peak value. The power in watts is the product of the current and voltage or $0.707 \times 0.707 = 0.5$, which coincidently is equivalent to the average electrical value. Thus two RMS equivalent measurements produce an average power measurement. RMS power is an incorrect term. Power is an average value equivalent to the heating power in a DC resistive circuit also known as I^2R losses. The correct terms are: RMS current and voltage, and average power.

7.3 Power Factor

Power Factor is a ratio of the true and apparent power:

$$\text{Power Factor(PF)} = \frac{\text{True Power}}{\text{Apparent Power}}$$

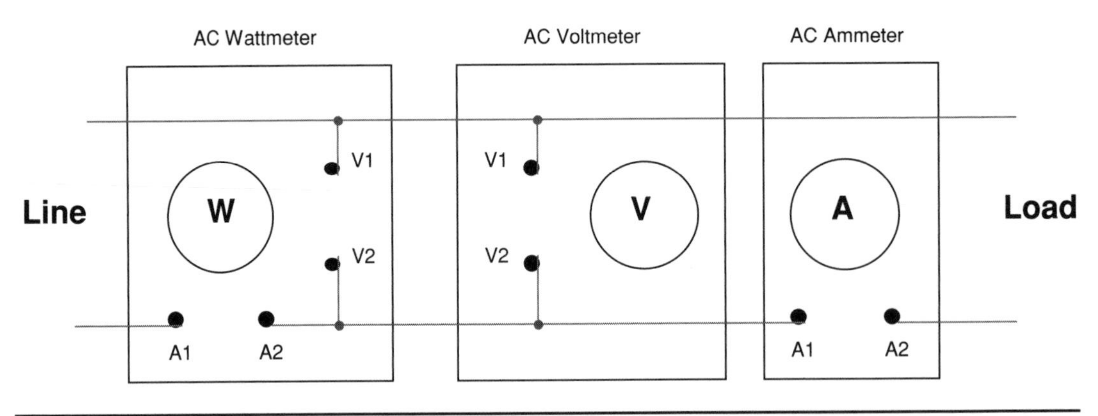

FIGURE 7-2 True and apparent power measurement in an AC circuit.

In DC circuits the PF is 1.0, its highest attainable value. In AC circuits the reactance will cause the generation source, transmission lines, and transformers to become fully loaded once the power factor percentage is reached. For example, instead of delivering 100% power in the example in Section 7.1, the source becomes fully loaded when only 86.6% power is delivered, the remaining 13.4% is used in maintaining the electromagnetic fields. Series RLC circuits that are tuned to resonance have minimum reactance using little reactive power, the power factor being close to 1.0 requires no correction. This can be verified with your coil running if your control cabinet includes a line ammeter, voltmeter, and a wattmeter. The only appreciable reactive power in a well-tuned running coil would be the rotary spark gap motor if one is used.

7.4 Power Indications

An AC wattmeter can be included in your control cabinet. Connected as shown in Figures 2-1 and 7-2, it will display the true power since it monitors instantaneous voltage and line current with respect to any phase difference. It may be difficult to obtain a wattmeter that measures several kW. To extend the range of a wattmeter, use a current transformer to sample the line current. For instance, I find a wattmeter at a Hamfest with a maximum range of 2 kW and want to use it with 240 V line service. It is rated for a line voltage of 250 VAC (an RMS value) and will produce a reading of 1.2 kW with a current of 5 A and voltage of 240 V. I can extend the maximum indication to 20 kW by using a 50/5 current transformer (or any other combination producing a 10:1 turns ratio) to sample the high line current, but still monitor the instantaneous voltage and line current. Multiplying the meter indication by 10 gives a 20-kW range. This produces no safety problems if the meter's rated line voltage is not exceeded. Using this meter and current transformer in a 240 VAC supply, line currents up to 83 A (20 kW/240 V) can be monitored. This meter and current transformer could also be used in a 120-VAC supply line monitoring up to 167 A (20 kW/120 V) of line current.

Using a wattmeter and a voltmeter–ammeter combination the power factor can be monitored. The true power indicated by the wattmeter, divided by the apparent power indicated by the voltmeter–ammeter combination equals the power factor. Power factor meters can even be found at Hamfests, which directly indicate the power factor when connected in the line circuit as directed by the manufacturer.

7.5 Current Indications

An AC ammeter can be included in your control cabinet. Connected as shown in Figures 2-1 and 7-2, it will display the line current. It may be difficult to obtain a direct reading ammeter that will measure over 20 A. Most panel type magnetic vane (or iron vane) ammeters are designed for use with a current transformer or CT that samples the line current and provides an indication of the RMS value of the line current. In this manner high line currents do not have to run through the meter, only currents up to 5 A. Most current transformers have a line voltage rating of 600 V and high isolation resistance. Most provide 0 to 5 A to the indicator from the sampled line current.

The primary (magnetizing winding) of the current transformer is a single turn of wire that is run through the hole in the center, which carries the line current. The secondary is either the two wires protruding from the case or two terminal connections on the case. The numerical ratio printed on these transformers is the primary amps (line side)–to–secondary amps (meter side). For example, a transformer marked 100:5 would indicate a 20:1 turns ratio (20 secondary turns:1 primary turn). 100 A of line current in our example will produce 5 A of sampled current to the indicator and 20 A in the line produces 1 A to the indicator. To maintain this current ratio a single turn of wire is run through the center of the toroidal current transformer as shown in Figure 7-3. The turns ratio of the transformer will then transform the line current to 0–5 A to drive the ammeter. Generally, ammeters designed for use with current transformers will specify a current of 5 A for a full-scale (FS) indication. The meter face will usually be labeled "FS = 5A."These transformers are made from tape wound cores with high permeability, and follow the transformer relationships shown in equation (4.40). These transformers have a typical accuracy of ±2% for 10:1 ratios and better than ±1% for ratios of 50:1 or greater.

Universal current transformers have more than one secondary output for use with different current transformer ratios. These are handy when constructing a new cabinet as once they are installed different meter ranges can be connected without reinstalling another current transformer. Since the source of ammeters and current transformers is generally surplus outlets and Hamfests, an ammeter with a matching ratio current transformer may be difficult

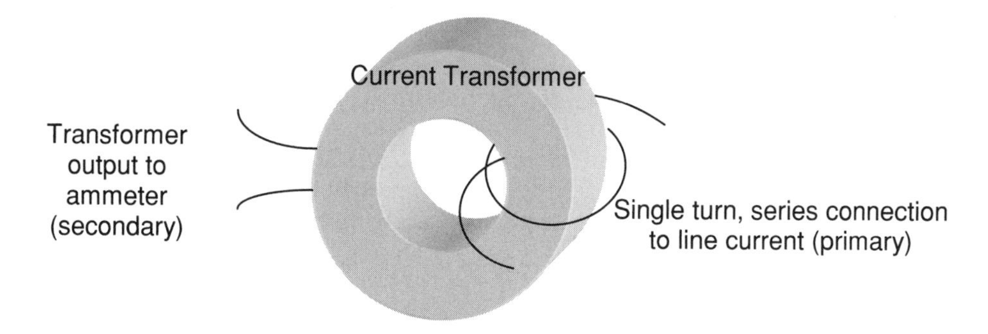

FIGURE 7-3 Current measurement in an AC circuit using a current transformer.

Labeled Current Transformer Ratio	Number of Primary Turns Used (Np)	Secondary-to-Primary Turns Ratio (Ns:Np)	Full Scale Ammeter Value (FS = 5A)
50:5	1	10:1	50A
50:5	2	10:2	25A
60:5	1	12:1	60A
60:5	2	12:2	30A
100:5	1	20:1	100A
100:5	2	20:2	50A
100:5	4	20:4	25A
150:5	1	30:1	150A
150:5	2	30:2	75A
200:5	1	40:1	200A
200:5	2	40:2	100A
200:5	4	40:4	50A

TABLE 7-1 Current transformer ratios and meter indications for selected number of primary turns.

to find. In this case additional turns can be added to the primary winding to obtain an alternate current ratio. For example, at a Hamfest I find a 50 A meter and a 200:5 current transformer, the winding ratio being 40:1. Using a one turn winding in the primary would require 200 A of line current to obtain a 5 A meter current, which would indicate 50 A. However, if four primary turns are used the winding ratio becomes 40:4 and a 50 A line current produces 5 A of meter current and an indication of 50 A. The increased number of primary turns produces a higher magnetizing force (*H*), but for a small number of turns the winding voltages will remain relatively unchanged, presenting no safety concerns. At least one turn is required in the primary therefore a 50:5 current transformer could not drive a 100 A meter by reducing the primary winding to one half of a turn. Table 7-1 lists several current transformer ratios that can be obtained for a selected number of primary turns and the line current required to produce a 5 A meter current in the secondary winding. The table can be consulted to match a current transformer to an unmatched 5A ammeter.

Never leave the secondary of a current transformer open when it is carrying line current. A very high, very dangerous 60-Hz voltage will exist in the secondary winding due to the large turns ratio, and if accidentally touched.... *ALWAYS CONNECT AN AMMETER TO THE CT SECONDARY*. Even more dangerous is disconnecting the secondary with current flowing through the primary. An inductive kickback effect will generate an even higher voltage than previously described.

The ammeters designed to work with current transformers are magnetic vane types, the same as the direct-reading AC ammeter. The current flows into the ammeter through a winding of several turns. The electromagnetic field (EMF) produced by this winding develops a magnetic polarity on a metallic deflector. This deflects a similar piece of metal, which is attached to a meter movement assembly. The EMF increases as the current through the meter increases, producing more magnetic deflection, which moves the vane further through its arc. By opening the covers to these meters, a lot can be learned from a visual inspection. Buy a junk one at your next Hamfest for dissection and spare parts.

DC ammeters cannot be used in AC circuits unless the line current is sampled and rectified. Since DC current cannot be coupled through a transformer (remember the Edison–Westinghouse current wars of the 1890s), a resistive current shunt is used to sample the line current, much the same as a current transformer. The d'Arsonval meter requires a rectified line current as it will accept only DC current, providing an indication of the *peak* value of line current. This is of little use and leads to confusion in estimating the coil's performance where the RMS value is needed. Actually a d'Arsonval meter is an incorrect term as identified in reference (27). The entrepreneur and scientist Edward Weston, whose meters bear his name, was the inventor of the modern electrical meter. When he invented the first truly portable electrical meters in 1888 he held several fundamental patents that quickly grew in number. These DC meters were so superior they dominated the industry for the next three decades using a proprietary meter movement that was erroneously identified as a d'Arsonval movement shortly after their entry into the industry. The term has stubbornly remained. Any DC volt or ammeter uses the fundamental Weston patents in its construction and will not resemble a true d'Arsonval movement as it is extremely inaccurate. We will hereafter credit the true inventor, Weston, by referring to these meter types as Weston movements.

There are also older, electrodynamic meter types that require no rectification but still use the resistive current shunt, and generally used to measure DC currents. The sampled current in these meters drove a special capacitive type meter deflection. These can be used, but give an *average* indication of the line current. Again, the RMS value is simpler in evaluating the performance. Don't add complexity to an already abstract concept unless you understand the differences.

The ammeters mentioned above are generally limited to frequencies under 1 kHz. For measuring the grid and plate currents in a tube coil, an RF ammeter is required. These are thermocouple (hot-wire) meters designed to measure high-frequency currents. They are usually labeled "RF"or "Radio Frequency"on the face.

DC ammeters can be used to measure the grid current in a tube coil. When the meter is placed in the circuit as shown in Figure 7-4, the rectified grid current will drive a DC ammeter because the meter is essentially at ground potential and the grid circuit has rectified the grid oscillating currents at this point. Advantages to using a DC ammeter to measure the grid current in a tube coil is their surplus availability and they are scaled to measure typical grid currents found in tubes.

7.6 Voltage Indications

An AC voltmeter can be included in your control cabinet and connected as shown in Figures 2-1 and 7-2 to provide an indication of the rms value of the line voltage. Most panel type, magnetic vane AC voltmeters have a high internal impedance to limit the meter current using meter movements as described in Section 7.5. I have used Weston DC meters with some success. By rectifying the line voltage with a diode the DC voltmeter can be used in an AC circuit. The Weston meter is generally limited to small currents (50 μA or 50 mA typical). The

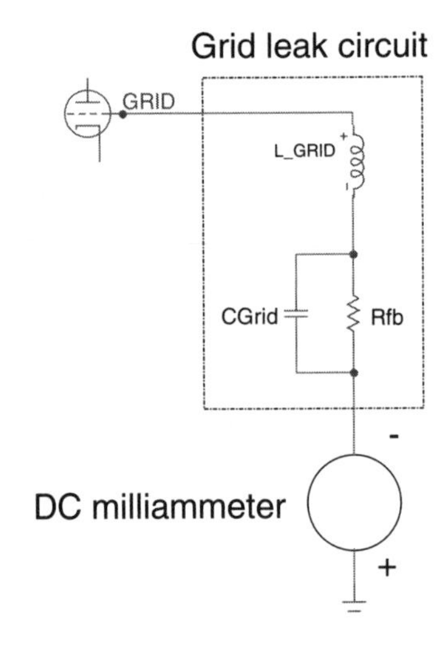

FIGURE 7-4 Using a DC ammeter in a tube coil grid circuit.

meter face should provide an indication of the current needed to provide a full-scale (FS) meter movement. For example, in the lower right corner of the meter is written: FS $= 50\,\mu$A. When the rectified current into the meter is $50\,\mu$A, the meter will indicate its full-scale value. Let's modify a 150-V DC meter for use in a 120-V AC circuit.

- The rectifying diode must have a PIV (Peak Inverse Voltage) rating of at least: 120-Vrms line voltage $\times$ 1.42 $\times$ 2 $=$ 340 V. Our meter requires a maximum current of $50\,\mu$A so a big diode is not required. A 1N4944/6 or 1N5617/9 diode will do, or scrounge around in an old TV set as there are plenty of high PIV diodes in the power supply.
- Ohm's law is used to calculate the resistance value that is used in series with the meter and the source to limit the current to $50\,\mu$A, $R = E/I$. To obtain a 150-Vrms full-scale indication (DC) with an applied 150-V peak waveform (AC), the peak value for an applied voltage of 150 Vrms is found: 150 Vrms $\times$ 1.414 $=$ 212.1V. The applied 212.1 V peak limited to $50\,\mu$A requires a series limiting resistance of: 212.1 V$/50\,\mu$A $= 4.242\,\mathrm{M}\Omega$.
- Construct as shown in Figure 7-5. The applied 120-Vrms line voltage will produce a meter current of: 120 Vrms $\times$ 1.414 $=$ 169.7 V peak/4.242 MΩ $= 40\,\mu$A. Note the $40\,\mu$A$/50\,\mu$A, 120 V/150 V ratios coincide.

These parameters can be calculated using worksheet 7 of the CH_10.xls file (see App. B). RMS and average equivalents are also explained in Chapter 10.

7.7 Resistive Current Limiting

The following discussion assumes the load resistance remains constant in the circuit. In an operating Tesla coil this is not necessarily the case. To limit circuit current there are both resistive and inductive schemes. In resistive current limiting the resistor is connected in series with the line current. This increases the total line resistance, which decreases the line current (when the load resistance remains constant). Since the line current flows through the resistor, it must

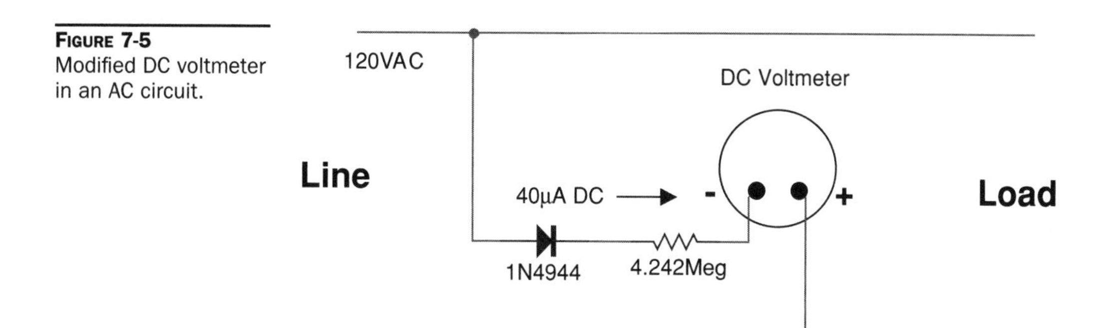

FIGURE 7-5 Modified DC voltmeter in an AC circuit.

be capable of handling this current. It must also have a large power handling capability. The resistive limiter can be thought of as a voltage divider with the load being the other resistance. To be of any use, the resistor should be a variable type or a network of fixed resistors with taps providing a variable range of resistance. The physical size does not necessarily indicate the power handling capability. As seen in the example circuits shown in Figure 7-6, a variable limiting resistance of small value (0 to 2.0 ohms) must be able to dissipate large power levels (up to 1.4 kW in the example).

The only drawback to resistive limiting is the voltage drop resulting from the series resistance and heat produced in the resistor. Refer to the examples in Figure 7-6. When the variable resistor in the medium load is set to 2 Ω and the load is drawing 10 A, the line voltage will drop: 10 A × 2 Ω = 20 V. The 240-V line voltage drops 20 V across the resistor leaving a 220-V supply for the load.

The heat produced in the resistor can become a problem. Heat generated in electrical parts from power dissipation must be removed from the part to keep it below the manufacturer's maximum rated temperature. Heat is removed by three mechanisms: radiation, convection, and conduction. Radiation is the heat lost mostly through infrared radiation from the source of the heat. Convection is heat loss through an interaction with a cooling medium such as airflow or a liquid coolant. Conduction is heat lost through direct contact with a solid material of either low thermal impedance such as aluminum or copper (heat sink) or high thermal impedance such as ceramic or plastic (heat insulator). Radiation and convection are the chief mechanisms in removing heat from these resistors and are affected by the airflow surrounding the resistor. Conduction is not very effective in these devices. In order for the heat to get to a heat sink it must first pass through the ceramic core, which is not a good thermal conductor.

The resistor used in Figure 7-6 is a power rheostat with "2.0 ohms, 600 V, 32 A" marked on the case. The 600 V @ 32 A is a free-air watt rating as defined in reference (26). The 600-V rating is conservative for every inch of resistive wire wound on the ceramic core. The rated current is used to determine the rated power dissipation in these resistors. The resistor has a conservatively rated power dissipation of: $I^2R = (32\text{ A})^2 \times 2.0\,\Omega = 2,048$ W continuous at 350°C. The 2-kW rating must be derated for certain applied environmental conditions.

These effects can be calculated by opening the CH_7.xls file, Resistive Current Limit worksheet (1). The maximum power handling capability of the current limiting resistor is:

$$P = I^2 \bullet R \tag{7.1}$$

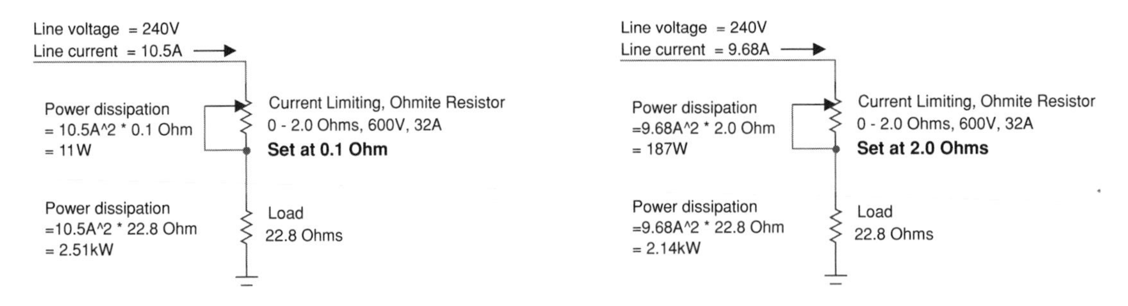

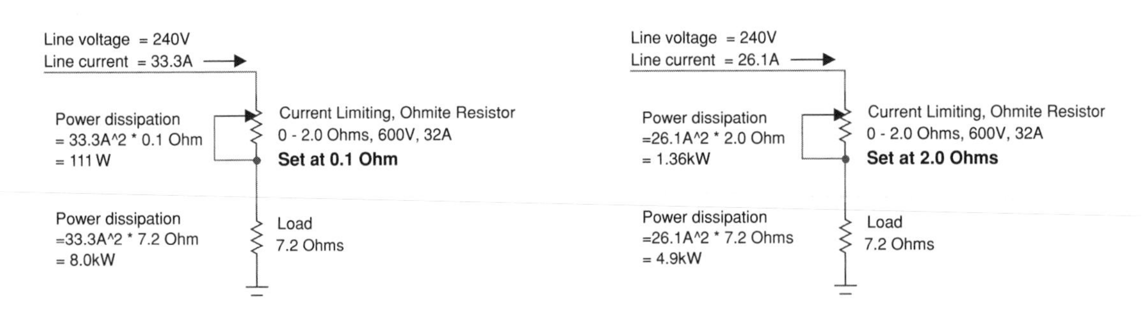

FIGURE 7-6 Resistive current limiting in an AC circuit.

Where: P = Maximum power dissipation ability of resistor or potentiometer in watts = cell (E3), calculated value converted to kW using a 1e-3 multiplier.

I = Maximum rated line current of resistor in amps = cell (B6), enter value. This is an rms value.

E = Maximum resistance value of resistor in ohms = cell (B4), enter value.

The total load impedance with the coil's load and current limiting resistance is:

$$Z_L = R_{\text{LOAD}} + R_{\text{LIMIT}} \tag{7.2}$$

Where: Z_L = Total load impedance with the load and current limiting resistance in ohms = cell (E7), calculated value.

R_{LOAD} = Load impedance in ohms = cell (B5), enter value.

R_{LIMIT} = Series line impedance of current limiting resistor or potentiometer setting in ohms = cell (B4), enter value.

The line current in the circuit with the total load impedance is:

$$I = \frac{Vs}{Z_L} \tag{7.3}$$

Where: I = Line current (rms) in amps = cell (E6), calculated value.
Vs = Line supply voltage (rms) = cell (B3), enter value.
Z_L = Total load impedance of circuit in ohms = calculated value in cell (E7) from equation (7.2).

The line current through the limiting resistance will drop a portion of the supply voltage, leaving less voltage supplied to the load:

$$Vd = I \bullet R_{\text{LIMIT}} \tag{7.4}$$

Where: Vd = Voltage drop across series limiting resistance = cell (E8), calculated value.
I = Line current (rms) in amps = calculated value in cell (E6) from equation (7.3).
R_{LIMIT} = Series line impedance of current limiting resistor or potentiometer setting in ohms = cell (B4), enter value.

The line current through the limiting resistance will also dissipate power in the form of heat:

$$Pd = I^2 \bullet R_{\text{LIMIT}} \tag{7.5}$$

Where: Pd = Power dissipated in series limiting resistance in watts = cell (E4), calculated value converted to kW using a 1e-3 multiplier.
I = Line current (rms) in amps = calculated value in cell (E6) from equation (7.3).
R_{LIMIT} = Series line impedance of current limiting resistor or potentiometer setting in ohms = cell (B4), enter value.

The series limiting resistance has dropped a portion of the supply voltage. The voltage remaining to supply the load is:

$$V_L = Vs - Vd \tag{7.6}$$

Where: V_L = Line voltage remaining to supply the load = cell (E9), calculated value.
Vs = Line supply voltage = cell (B3), enter value.
Vd = Voltage drop across series limiting resistance = calculated value in cell (E8) from equation (7.4).

The power used by the load is:

$$P_L = \frac{(Vs - Vd)^2}{R_{\text{LOAD}}} = I^2 \bullet R_{\text{LOAD}} \tag{7.7}$$

Where: P_L = Power used by the load in watts = cell (E5), calculated value converted to kW using a 1e-3 multiplier.
Vs = Line supply voltage = cell (B3), enter value.

Vd = Voltage drop across series limiting resistance = calculated value in cell (E8) from equation (7.4).

R_{LOAD} = Load impedance in ohms = cell (B5), enter value.

I = Line current (rms) in amps = calculated value in cell (E6) from equation (7.3).

The graph in Figure 7-7 was created using equations (7.1) thru (7.7). Column (H) lists the series limiting resistor values or potentiometer setting in 0.1-Ω increments to a maximum value of

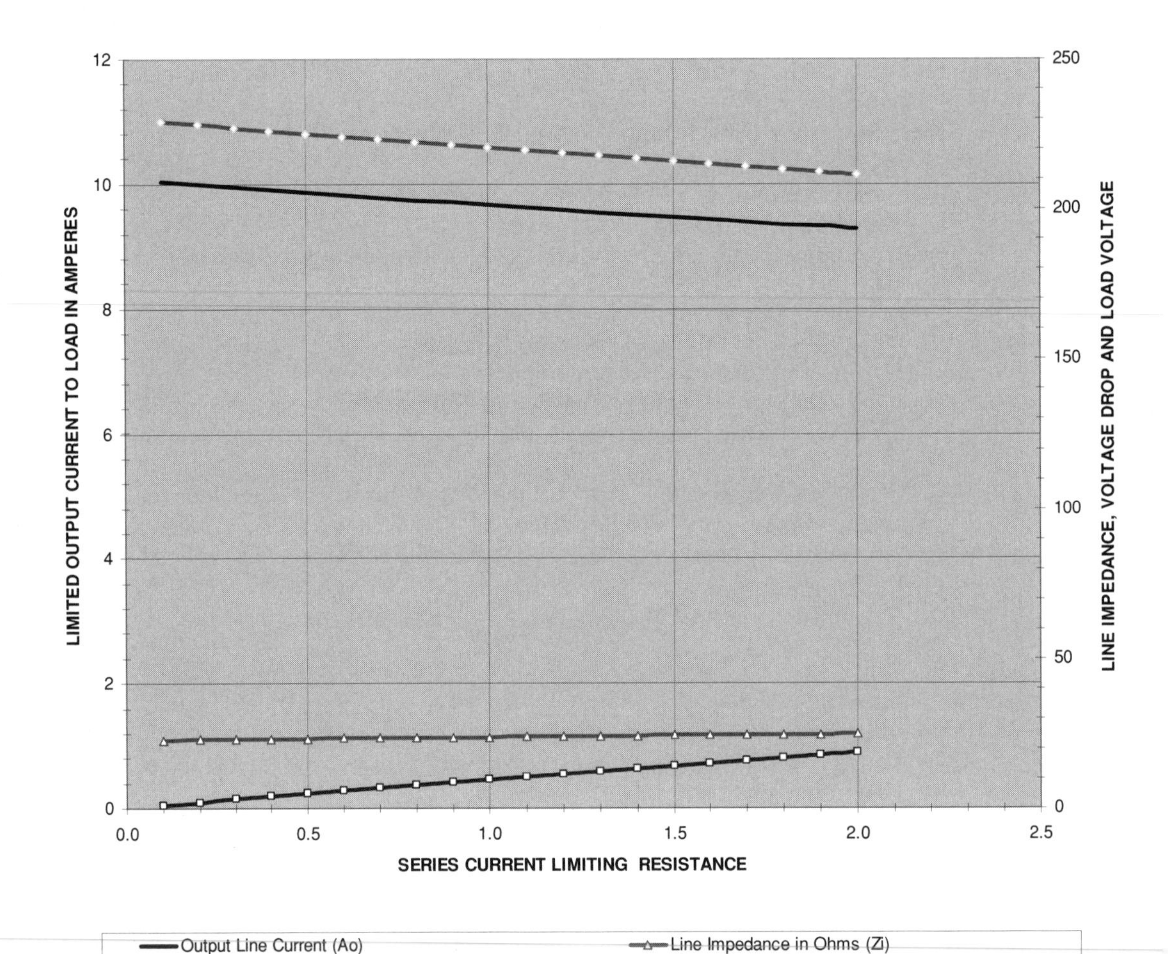

FIGURE 7-7 Current limiting vs. series resistance or potentiometer setting.

24 Ω. For each 0.1 Ω increase, column (K) calculates the total load impedance using equation (7.2), column (I) calculates the line current using equation (7.3), column (L) calculates the line voltage drop using equation (7.4), column (M) calculates the line voltage supplied to the load using equation (7.6), and column (J) calculates the load's power dissipation using equation (7.7). Ensure that the limiting resistor's rated power in cell (E3) can handle the applied power dissipation in cell (E4).

The maximum rated power dissipation capability calculated in equation (7.1) must be derated for a variety of applied operating conditions. The derated power dissipation capability can be estimated using CH_7.xls file, RESISTOR DERATING worksheet (10). The worksheet uses methodology and data contained in reference (26) to derate large power rheostats (potentiometers) and other wirewound resistors on a ceramic base for the applied operating conditions. By entering the required ratings and operating conditions in cells (C4) through (C26) the rated power dissipation in cell (F6), is derated to the power in cell (F28). The power dissipation calculated in equation (7.5) should not exceed the derated power capability calculated in cell (F28). Directions for performing the derating estimation require eight steps listed in rows 31 to 39 of the worksheet. Providing cooling airflow over the resistor, or pulsed (duty cycled) operation, allows the resistor to dissipate more than the rated power. When operated at high ambient temperatures or altitudes, in enclosures, or when stacked in groups the derating values lower the resistor's ability to dissipate power.

Using resistive current limiting on a line feeding a Tesla coil will not effectively limit the line current. The Tesla coil is not a purely resistive load but an impedance (reactive load) and will draw as much current as it needs to operate, leaving the resistor to drop the line voltage to the coil. As the resistance increases so does its voltage drop. At some point the voltage drop will be sufficient enough to prevent the gap from ionizing and the coil will not operate. Up to this point the resistor is not actually limiting current but dissipating power, which must be furnished by the control devices. This is really a waste of power. A similar effect results from using an inductor in series with the line as detailed in Sections 7.8.1 and 7.8.2. The only effective current limiting devices for use in a Tesla coil are either current-limited transformers (e.g., neon sign) or the welding transformer methods detailed in Section 7.8. Adequate circuit breaker or fuse protection should be included in all applications to limit short circuit current to safe values as detailed in Section 7.10. The information in this section and Sections 7.8.1 and 7.8.2 was retained as it may be useful with other types of loads.

7.8 Inductive Current Limiting Using a Welding Transformer

Inductive current limiting can be done in two ways. The first method is to use an arc welder, also known as a stick welder or buzz box for the buzzing noise the core laminations make when heavily loaded. A transformer is used to convert the 240-V line to a lower secondary voltage. A survey of Lincoln Electric's stick welders indicate this secondary voltage ranges from 22 V to 40 V. The transformer also has a gapped core. The gap in a core is typically a separation in the core's magnetic path a few fractions of an inch wide. The gap surface is usually machined and polished to remove surface imperfections in the core laminations. The gap stores energy, the wider the gap becomes the less energy is transferred to the load leaving more energy to be stored (dissipated) in the core's gap. The stick welder uses a shunt that slides in and out of the core gap. This has the same effect as narrowing or widening the gap in the core. Neon

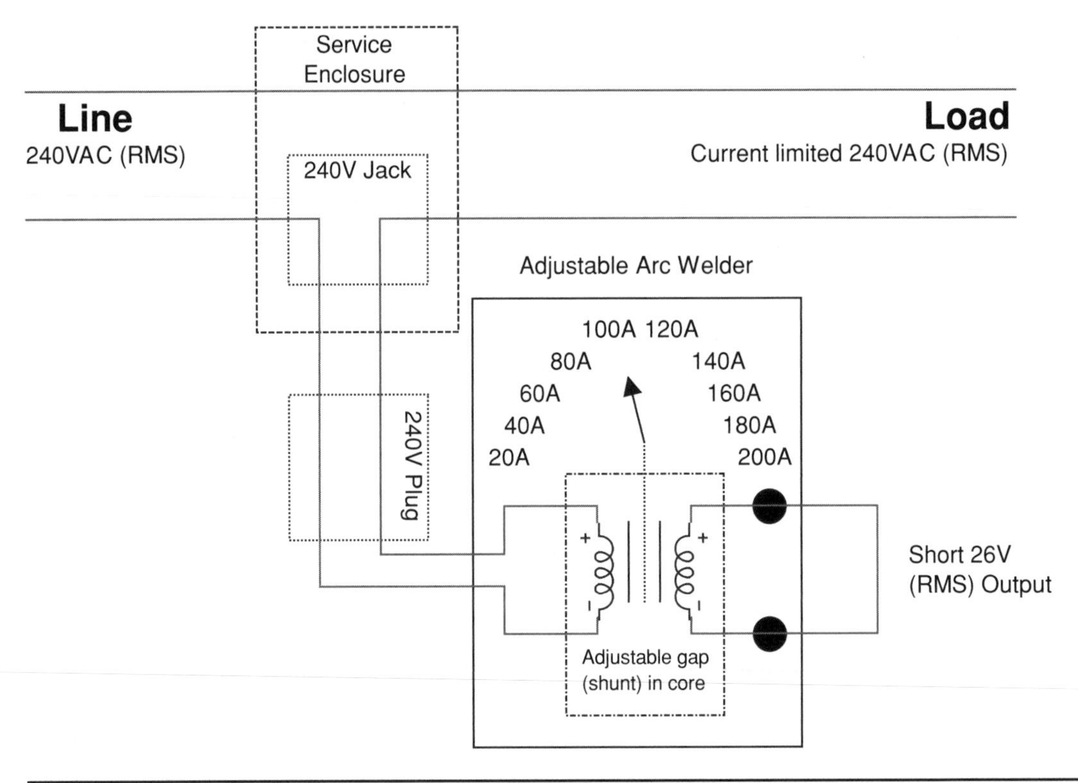

FIGURE 7-8 Using an arc welder for inductive current limiting in an AC circuit.

sign transformers limit output current in much the same manner, as they use a core gap and a non-adjustable shunt. The core is made of silicon steel, iron, or other ferromagnetic material. The shunt is made of material different than the core such as aluminum, to either decrease or increase the magnetic permeability of the core. The permeability is a measure of the core's ability to transfer energy from primary-to-secondary. The energy not transferred is stored or dissipated. Depending on the shunt and core material, the energy stored in the gap is increased or decreased by moving the shunt further into the gap. This limits the current to the secondary winding.

By wiring the arc welder to the line as shown in Figure 7-8, each arc welder front panel amp setting on the horizontal (X) axis in Figure 7-9 will limit the load current to the vertical (Y) axis value intersecting the diagonal line at the same point. A graph of the limited output current to the load for each arc welder amp setting can be constructed using the CH_7.xls file, Inductive Current Limiting worksheet (2).

When the welder is connected to the line voltage in parallel (as a welder) it will operate as a load. Adding a gap to a core will always reduce its efficiency by increasing the leakage inductance. At full rated output (amps setting at 225) the calculated efficiency of model AC-225 is 49%, meaning the secondary will draw 5.63 kW, the welder's transformer will dissipate 5.87 kW, drawing a total power from the source of 11.5 kW. But when the welder is connected to the line in series it offers a small line impedance to limit the line current up to an operating

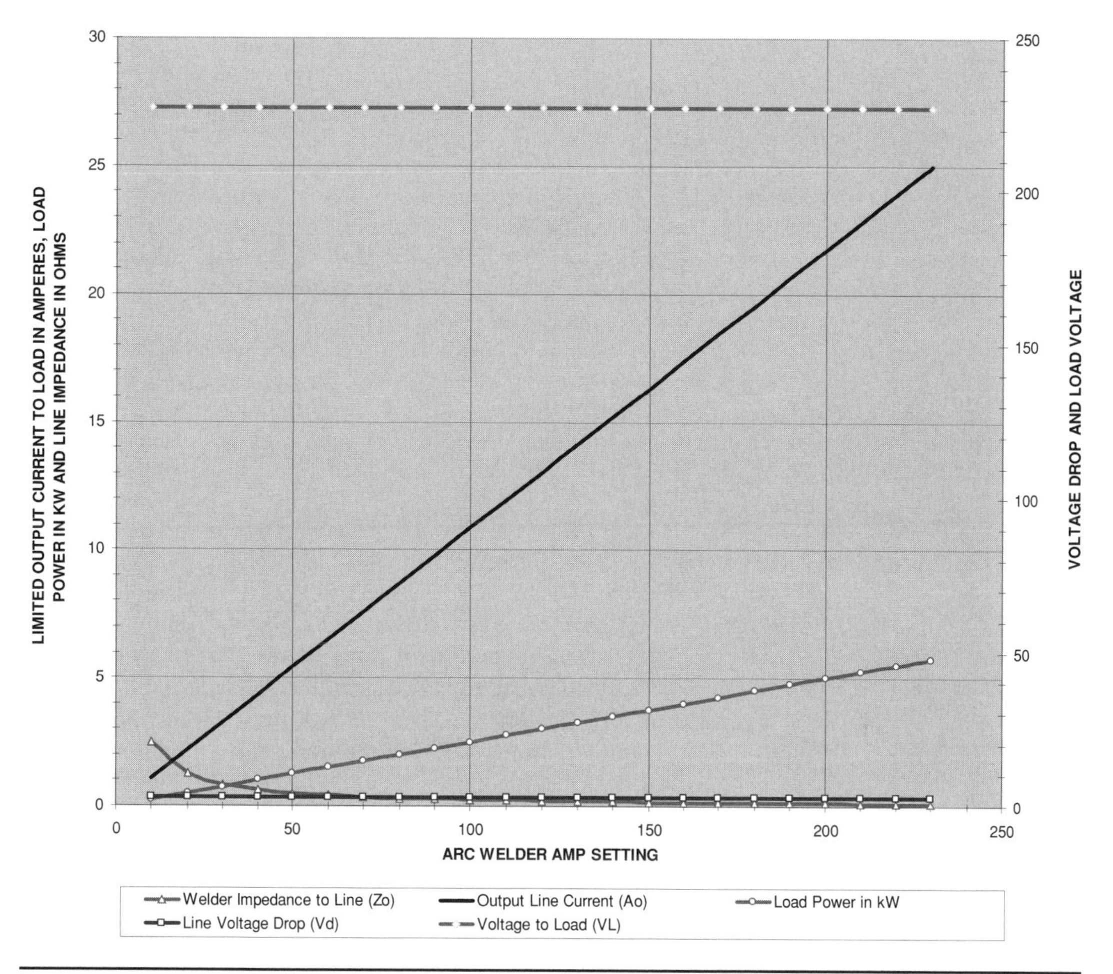

FIGURE 7-9 Current limiting vs. arc welder amps setting when operated within rated power level.

point where the load power (impedance) is equal to the rated output power of the welder as shown in Figure 7-9. Beyond this point the welder's transformer core becomes saturated and any further power drawn from the load will increase the welder's impedance, dropping more voltage. Increasing the load beyond the welder's power rating will produce line effects similar to that shown in Figure 7-10. Note the welder now drops more line voltage leaving less for the load. If higher load power is needed, use a bigger welder or wire more than one in parallel.

These effects can be calculated by opening the CH_7.xls file, Inductive Current Limiting worksheet (2). The maximum power handling capability of the welder is:

$$P = I \bullet E \tag{7.8}$$

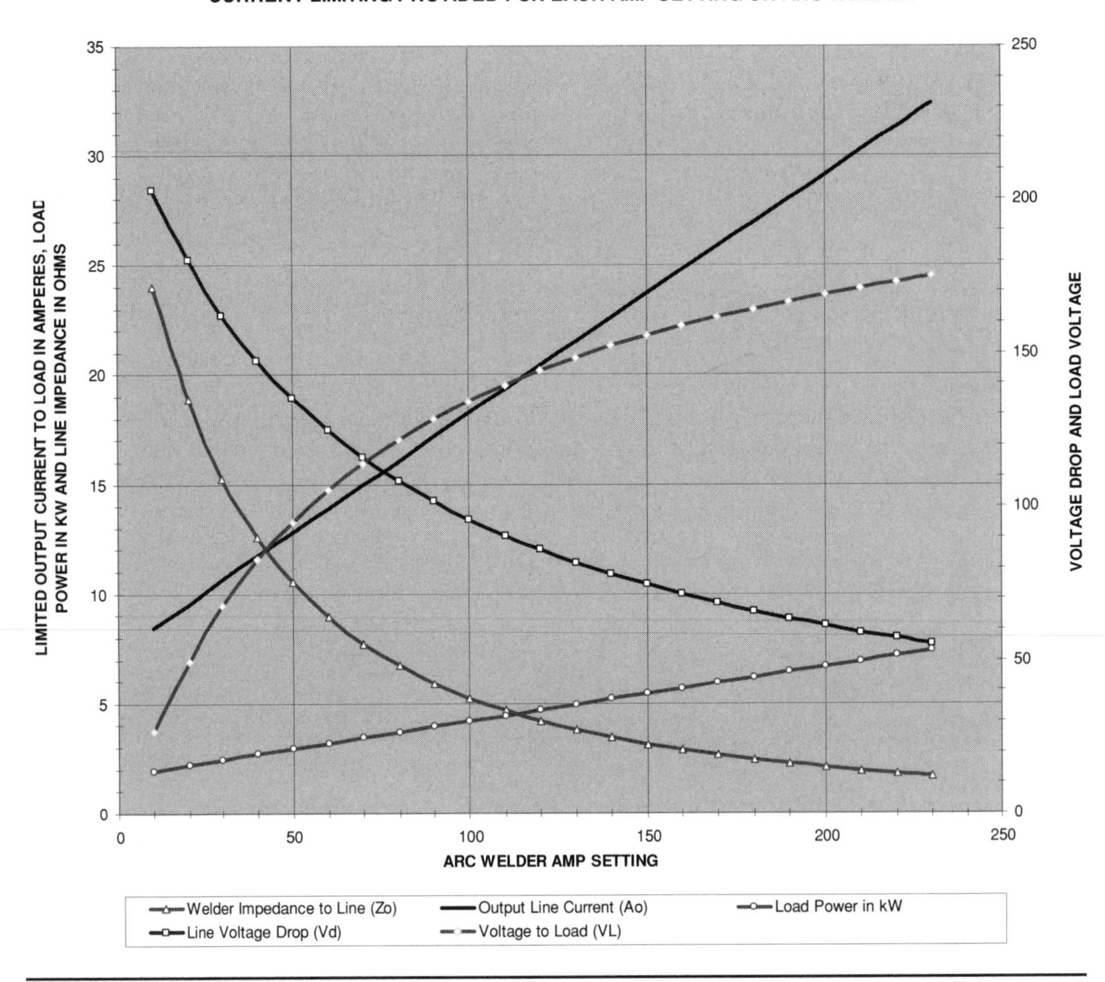

FIGURE 7-10 Current limiting vs. arc welder amps setting when operated beyond rated power level.

Where: P = Maximum power handling ability of welder in watts = cell (E4), calculated value.
I = Maximum output amp setting of welder = cell (B6), enter value.
E = Output voltage of welder in Vrms = cell (B5), enter value.

By entering the line voltage in cell (B3) and the input current in cell (B4) that is required to produce the maximum rated output entered in cell (B6), the power drawn from the line is shown in cell (E3) and the welder's efficiency in cell (E5). The following equations apply only to the welder connected in series with the line for current limiting.

When connected in series the welder presents a variable impedance to the line dependent on the amp setting. When the lowest amp setting is selected the welder is limiting as much current as possible and offers the most impedance to the line. If the load draws enough power the welder's core will saturate, dissipating more energy and presenting a higher impedance to the line. To represent this operating mode conditional statements are used in the Excel worksheet formulae. The total power drawn from the line by the welder and the load is:

$$Po = Io \bullet Vo \tag{7.9}$$

Where: Po = Total limited line power available to load impedance in watts = calculated value in column (J) for output amp setting in same row of column (H).
Io = Output amp setting of welder = value in column (J) for output amp setting in 10 A increments from 10 A to 630 A.
Vo = Output voltage of welder in Vrms = cell (B5), enter value.

NOTE: *If output amp settings fall between the values in column (H) the calculations will be rounded up or down to the nearest 10 A value.*

The welder impedance to the line is:

$$Zo = \frac{Vo^2}{Po} \tag{7.10}$$

Where: Zo = Welder impedance to the line in ohms = calculated value in column (K) for output amp setting in same row of column (H).
Vo = Output voltage of welder in Vrms = cell (B5), enter value.
Po = Total limited line power available to load impedance in watts = calculated value in column (J) from equation (7.9).

The limited line current in the circuit with the total load impedance is:

$$Io = \frac{Po}{V_{LINE}} \tag{7.11}$$

Where: Io = Line current in amps = calculated value in column (I) for output amp setting in same row of column (H).
Po = Total limited line power available to load impedance in watts = calculated value in column (J) from equation (7.9).
V_{LINE} = Line supply voltage = cell (B3), enter value.

The line current through the limiting impedance will drop a portion of the supply voltage, leaving less voltage supplied to the load:

$$Vd = Io \bullet Zo \tag{7.12}$$

Where: Vd = Voltage drop across welder in series with line = calculated value in column (L) for output amp setting in same row of column (H).
Io = Line current in amps = calculated value in column (I) from equation (7.11).
Zo = Welder impedance to the line in ohms = calculated value in column (K) from equation (7.10).

The series limiting impedance has dropped a portion of the supply voltage. The voltage remaining to supply the load is:

$$V_{LOAD} = V_{LINE} - Vd \tag{7.13}$$

Where: V_{LOAD} = Line voltage remaining to supply the load = calculated value in column (M) for output amp setting in same row of column (H).
V_{LINE} = Line supply voltage = cell (B3), enter value.
Vd = Voltage drop across welder in series with line = calculated value in column (L) from equation (7.12).

The total circuit impedance to the line is:

$$Z_{LINE} = Zo + R_{LOAD} \tag{7.14}$$

Where: Z_{LINE} = Impedance of welder and load in ohms = cell (E5), calculated value converted to kΩ using a 1e-3 multiplier.
Zo = Welder impedance to the line in ohms = calculated value in column (K) from equation (7.10).
R_{LOAD} = Load impedance in ohms = cell (B7), enter value.

An IF-THEN conditional statement is used to determine if the load impedance will draw more power than the power handling capability of the series welder determined in equation 7.8. Enter a load value that draws less power than the welder's power handling capability as shown in Figure 7-9. The line voltage drop is very low and the load is limited to the power provided by the welder. Decrease the load value until it draws more power than the welder's power handling capability as shown in Figure 7-10.

Once the welder's core saturates it can no longer limit the line current. The voltage drop across the welder increases, which decreases the voltage available to the load. This is not always a desirable condition. The welder may be damaged and the line current is no longer being limited, which was the purpose of installing the welder. This can be solved by using a welder with a higher power handling capability or more than one welder connected in parallel with each other and in series with the line. To simulate multiple welders in line, multiply the Maximum Output Amp Setting (Am) of one welder by the number of welders connected in parallel and enter this value in cell (B6). Due to the inefficient operation of most welders there will be little accuracy between these calculated values and those experienced in operation. Expect up to a 50% difference between the two.

7.8.1 Inductive Current Limiting Using a Reactor with a Single AC Winding

The second inductive limiting method uses a series inductance with an iron core forming a reactor. The iron core of the reactor will typically follow one of the three basic shapes listed in Figure 7-11.

Unlike the arc welder method (transformer) where it is not desirable to operate the core in saturation, the reactor core (inductor) must be operated in saturation to limit the load current. The reactor can use either a selected number of turns in a single AC winding, an additional DC winding to set the operating point of the core's hysteresis curve (detailed in Section 7.8.2) or an air gap in the core to decrease the core's permeability (detailed in Section 7.8.3), either method

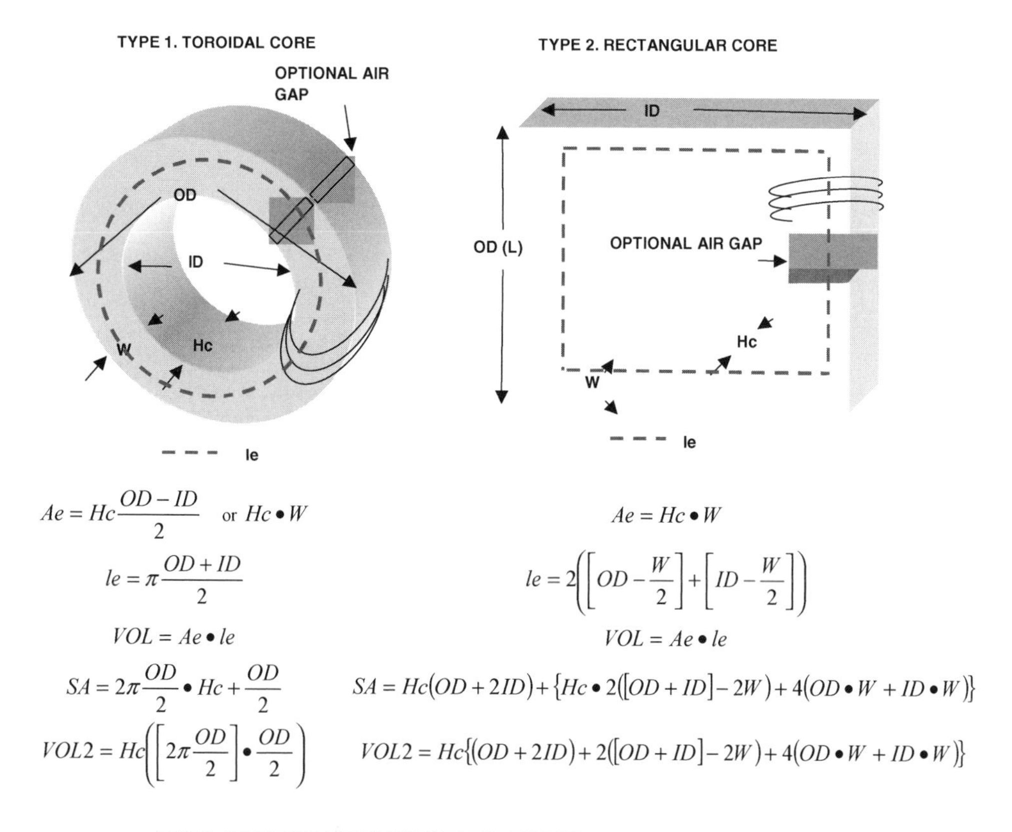

TYPE 3. OPEN ENDED CORE WITH HELICAL WINDING

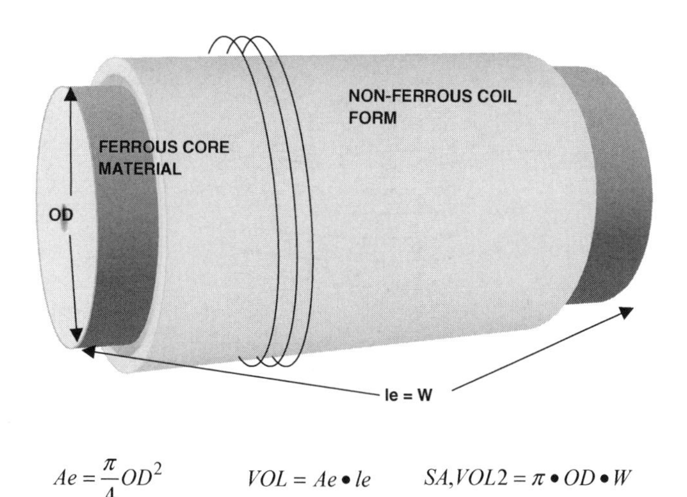

FIGURE 7-11 Core topologies used in a reactor. *(continued)*

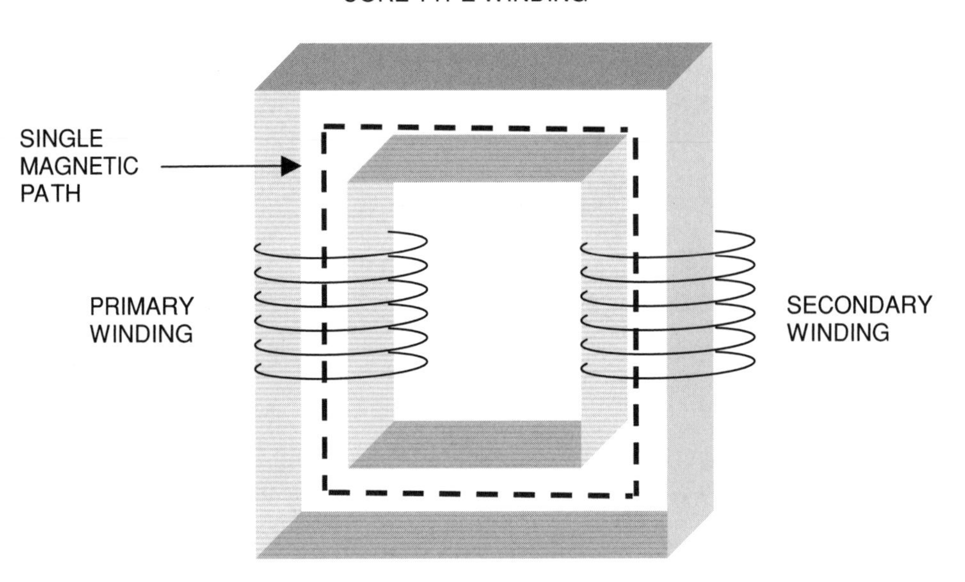

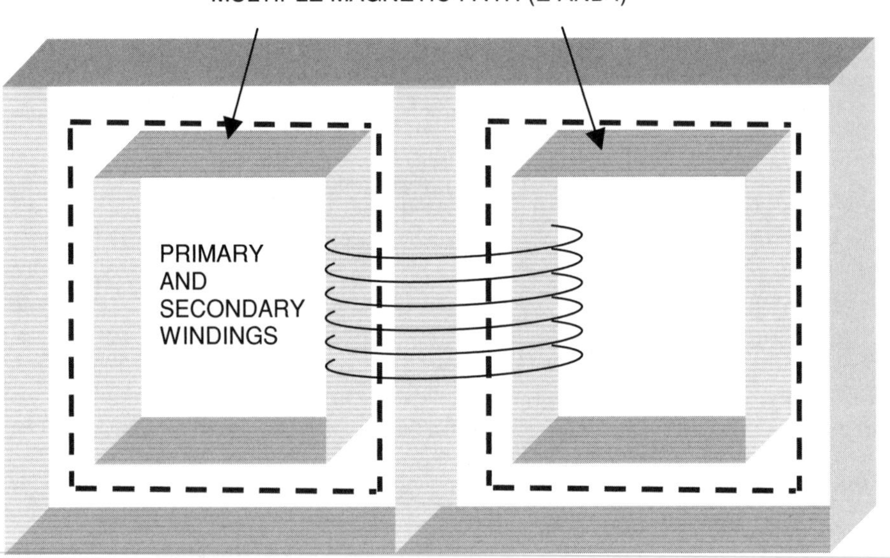

Figure 7-11 Core topologies used in a reactor. (*continued*)

driving the core further toward saturation. As the load draws enough AC line current the reactor core reaches saturation and the load current is limited to the value at this operating point.

To design a reactor using a single AC winding open the CH_7A.xls file, IRON CORE SATURATION worksheet (1). The cells requiring an input and those performing the calculations are shown in the worksheet in Figure 7-18. To design a reactor follow the steps listed below:

1. Enter the LINE characteristics in cells (B4) thru (B6). Dependent on your selected core material enter a 1 in cell (B9) for a TOROID, enter 2 for a RECTANGULAR or 3 for an OPEN ENDED core with helical winding. Refer to the drawings in Figure 7-11. Your core material will most likely originate from surplus materials. Variable autotransformers will provide a toroidal iron core made up of a long thin strip of silicon steel, which is wound like a roll of tape (called tape wound). The thin strip reduces the eddy current losses that would be found in a solid core material. Most core wound power transformers will provide a rectangular iron core and shell wound transformers provide a modified version of the rectangular core in the form of E's and I's (see Figure 7-11). The reactor is not a transformer but a saturable inductor, so use a core wound type not a shell wound type for best results. The least effective method is the open-ended core, which uses bulk iron rods and will suffer from severe eddy current losses if the diameter of the individual rods is not kept very small. Antique electrical appliances containing an electro-mechanical vibrator (interrupter) used this type of saturable core reactor to produce a break rate (on–off switch). The core's electromagnetic field was used to hold a circuit contact closed until the core saturated, which could no longer magnetically hold the contact closed and the circuit opened. With the open circuit the core dropped out of saturation, the contact closed, and the cycle started again. Break rates of over 1000 BPS could be obtained, limited only by the mechanical speed of the contact and hysteresis characteristics of the core material used.
2. Enter the core material's dimensions in cells (B12) thru (B14) and (B15) if it is a type 2 or 3 core. Enter the AC winding characteristics in cells (B16) thru (B18). The AC winding is the series inductance that connects the line to the load.
3. The magnetizing characteristics of the core material must be known. A hysteresis curve, which plots the magnetic flux density (B) vs. magnetizing force (H) on a graph, is generally used to interpret these characteristics. If this curve cannot be obtained, which is generally the case with surplus material, the measured data from reference (24) shown in Figure 7-12 (chart 2 in the worksheet) for 1% silicon steel can be used for our application. This is typical of most low loss silicon steel materials (transformer cores). Table 7-2 lists the electrical characteristics of iron core materials derived from reference (24). The sixth-order polynomial equation (blue trace) shown in Figure 7-12 was derived from the red permeability curve's 2 oersted (Oe) to 24 Oe characteristic and used to calculate the material's magnetic permeability (μ) at the applied magnetizing force in oersteds (H, calculated in step 4) and shown in cell (C10). The resulting magnetic flux density (B, calculated in step 4) is calculated in cell (C57). Unless you have your own interpreted data enter the calculated value in cell (C10) into cell (B10).
4. Enter the estimated rms load current the Tesla coil will draw from the line in cell (B21). Ensure the DC bias current entered in cell (B25) is 0 A. Keep your eye on the magnetic permeability (μ) shown in cell (C10) during the calculations. If the AC load current

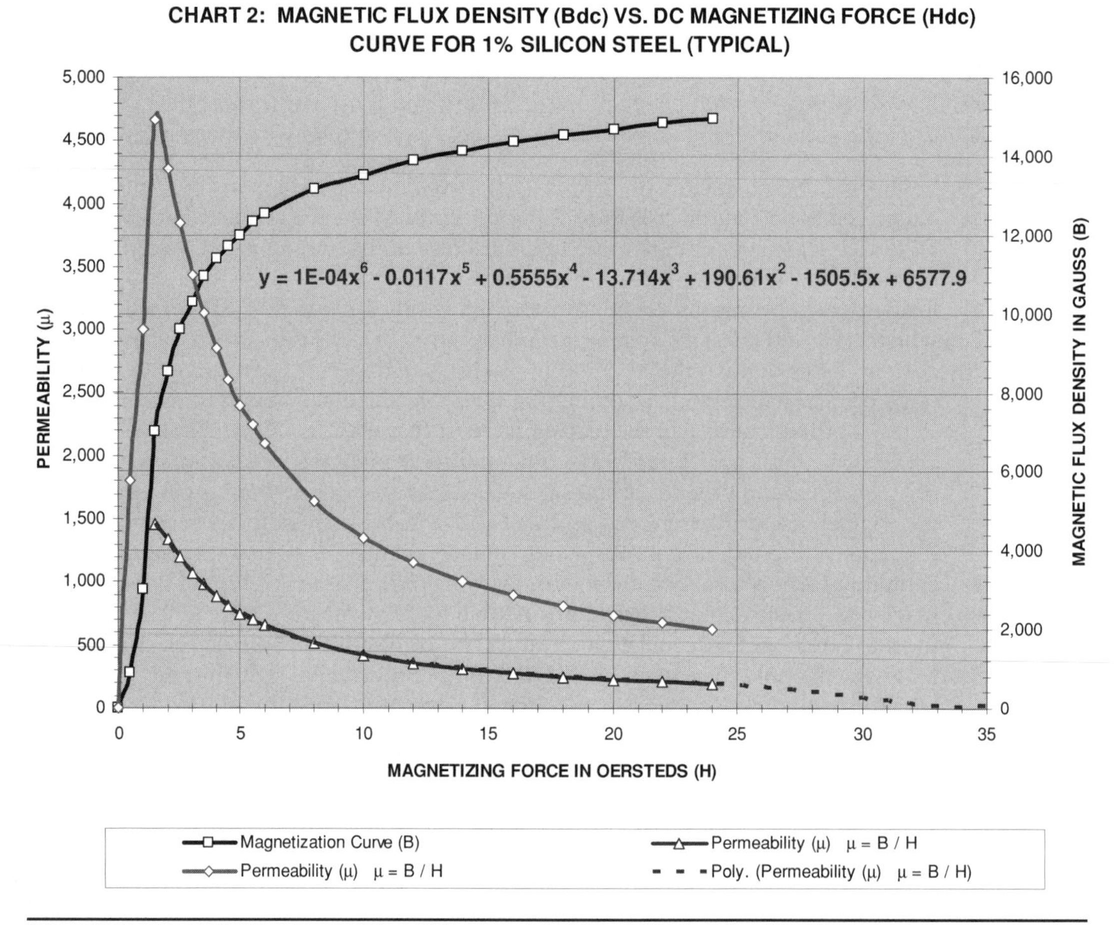

FIGURE 7-12 Magnetic flux density (*B*) vs. DC magnetizing force (*H*) curve for 1% silicon steel.

in cell (B21) is changed you will need to change the value in cell (B10) to the value shown in cell (C10). Table 1 in columns (J) through (T) was constructed to calculate the operating parameters for the applied number of AC windings shown in column (J).

5. The load current entered in cell (B21) uses Ampere's Law to calculate the magnetizing force (H) in column (K) of Table 1. The magnetizing force for the applied parameters entered in cells (B4) to (B22) is shown in cell (B51). Refer to equation 4.11 for Ampere's Law. The sixth-order polynomial equation shown in Figure 7-12 (detailed in step 3) and the magnetizing force in column (K) is used to calculate the permeability (μ) in column (O) of Table 1. The permeability for the applied parameters entered in cells (B4) to (B22) is shown in cell (B53). The inductance of the core and AC winding is calculated in column (L) of Table 1. Refer to equation 4.1 for the inductance formula. The inductance for the applied parameters entered in cells (B4) to (B22) is shown in

Material Property	Permalloy (Western Electric)	Hipernik (Westinghouse)	Si Steel (0.6% Loss)	Si Steel (0.66% Loss)	Si Steel (0.72% Loss)	Si Steel (0.82% Loss)	Si Steel (1.01% Loss)	Si Steel (1.17% Loss)	Si Steel (1.3% Loss)	Pure Iron	Low Carbon Sheet Steel	Cast Steel (Annealed)	Cast Iron (Annealed)
Typical maximum permeability (μ) in kgauss	100	90	8	8	6.5	6	5.5	5.2	5	4.5	2.5	1.5	0.5
Saturation flux density (B) in kgauss	11.	15.6	19.5	19.5	19.8	20.2	20.5	21	21.2	21.6	21.2	21	14
Typical initial permeability (μi)	9,000	6,000	750	700	600	500	400	350	325	275	250	175	125
Maximum core loss aging		0	0	0	0	+2%	+3%	+5%					
Typical coercive force (H) in oersteds (Bmax = 10 kG)	0.04	0.06	0.32	0.36	0.39	0.46	0.56	0.7	0.85	1	2	5	11
Maximum power loss of core material in watts/lb*		0.25	0.6	0.66	0.72	0.82	1.01	1.17	1.3				
Steinmetz hysteresis coefficient	$1e^{-4}$	$1.5e^{-4}$	$4.6e^{-4}$	$1.5e^{-4}$	$5.6e^{-4}$	$6.5e^{-4}$	$8.1e^{-4}$	$8.8e^{-4}$	$1e^{-3}$	$2e^{-3}$	$3e^{-3}$	$5e^{-3}$	$1.2e^{-2}$
Resistivity in μW/cc	22	45	60	58	56	48	41	26	18	10.7	13	15	100
Weight in lbs/sqft using 29 Gauge sheet	0.63	0.6	0.54	0.54	0.54	0.54	0.54	0.55	0.55	0.56	0.56		
Typical application	Telephone Equipment	Relays, Instrument and Audio Transformers	Distribution and Power Transformers		Rotating Machines and Small Transformers		Small Motors, Electromagnets, Starting Transformers			Pole Pieces, Relays	Frames and Fields in DC Machines	Frames and Solid Poles	Frames

*Maximum power loss in watts/lb at 10 kG, 60-Hz, 29 Gauge Si steel (standard used to designate the grade of electrical sheet steel).

TABLE 7-2 Electrical properties of iron core materials.

cell (B52). Now that L, H, and μ are known, Faraday's Law can be used to calculate the magnetic flux density (B) in column (N) of Table 1. Refer to equation (4.12) for Faraday's Law. The magnetic flux density for the applied parameters entered in cells (B4) to (B22) is shown in cell (B56) in kG and cell (B57) in gauss.

6. Enter the estimated power factor of the load in cell (B22). The reactor will present a reactance to the line. The reactor impedance to the line is calculated in column (P) of Table 1 and shown in cell (B54) for the applied parameters entered in cells (B4) to (B22). Refer to equation (3.6) for the inductive reactance formula. The load current entered in cell (B21) times the calculated reactor impedance in column (P) of Table 1 and cell (B54) equals the voltage drop produced across the reactor shown in column (Q) of Table 1. The applied voltage entered in cell (B5) minus the calculated line drop in column (Q) of Table 1 equals the voltage left to the load (primary of step-up transformer) in column (R) of Table 1 and shown in cell (B60). The power factor of the reactor is calculated in cell (B62) using the methodology in Section 7.1.

   The table calculations are plotted on Chart 1 in rows 97 to 150. The figure displays the calculated permeability (μ, blue trace), reactor impedance (X_L, green trace), magnetic flux density (B, black trace), and load voltage (red trace) for the 1 to 140 AC windings shown in column (J) of Table 1 and applied core and AC winding parameters entered in cells (B4) to (B22). A large [+] is shown in the plot in Figure 7-13 to indicate the point of saturation in the core material. A large [X] marks the applied flux density at the load current entered in cell (B21). A large [O] also marks the line voltage available to the load after the resulting reactor voltage drop. The graph is used to estimate the number of AC windings needed to produce the desired reactor characteristics.

Refer to the worksheet calculations shown in Figure 7-18. For the applied parameters entered in cells (B4) to (B22) the flux density is 13.4 kG for 30 AC windings and a load current of 20 A as shown in Figure 7-13. The reactor will drop 25 V of the applied 120-V line leaving 95 V available to the load. In this application the spark gap of an operating Tesla coil was set to ionize at a line voltage of 70 V therefore the 95 V was sufficient to power the coil. The reactor acts as a ballast device which will begin to saturate at the estimated load current of 30.5 A shown in cell (B61). If the load current were to increase to 70 A the resulting load voltage drops to only 6 V as shown in Figure 7-14. The coil would stop firing long before this stage is reached as the line voltage must be >70 V to operate the spark gap. As the load current increases the magnetic flux density also increases until it reaches the saturation point. Once the core saturates it begins to operate in the non-linear region shown in the figures. The permeability increases, which increases the reactor inductance with a corresponding increase in reactor impedance. The reactor drops more of the line voltage until the primary of the step-up transformer (load) can no longer maintain ionization of the spark gap.

The use of this device in a Tesla coil is questionable as it wastes power and does not provide effective current limiting. The only effective current limiting devices for use in a Tesla coil are either current-limited transformers (e.g., neon sign) or the welding transformer methods detailed in Section 7.8. My intent for discussing these devices was to demonstrate their inability to control the line current. These schemes have long been thought by coilers to limit line current, which presents a false sense of security and may present a safety hazard if some form of overcurrent protection is not included in the design. Adequate circuit breaker or fuse protection should be included in all applications to limit short circuit current to safe values as

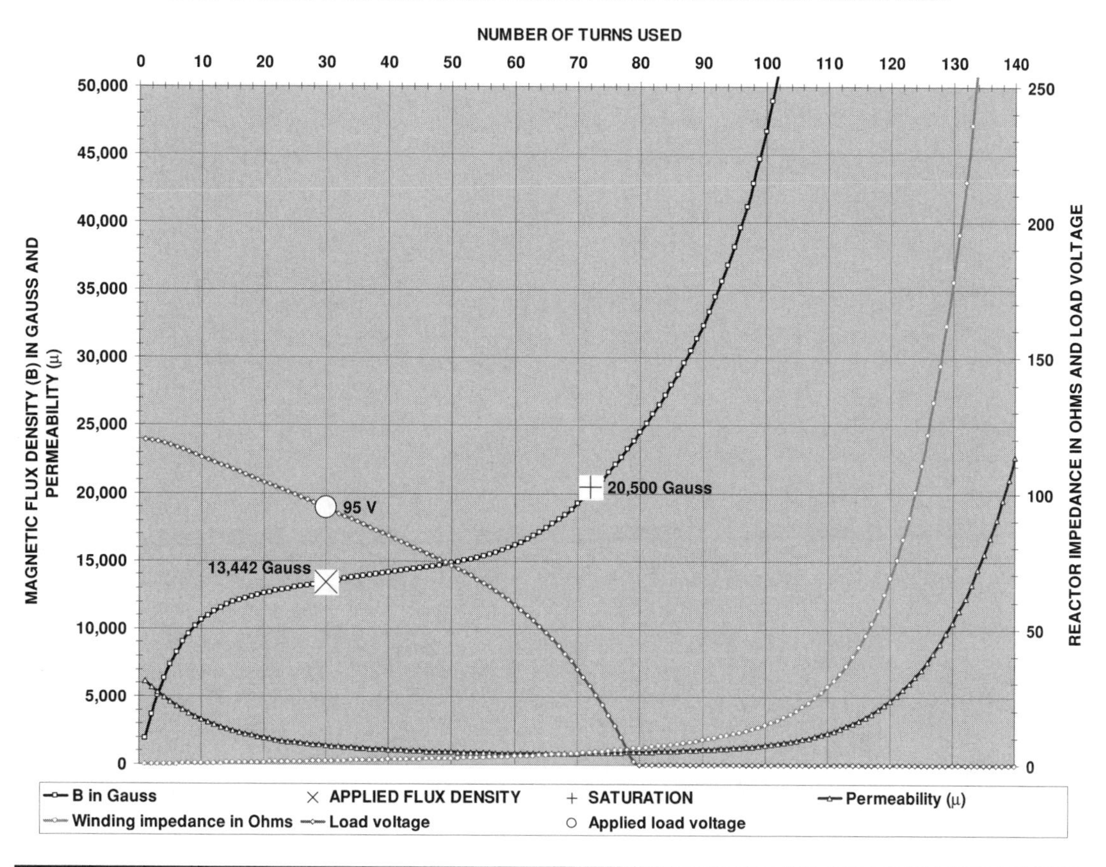

FIGURE 7-13 Operating characteristics for a reactor below saturation using an AC winding only (load current 20 A).

detailed in Section 7.10. The information in Sections 7.7, 7.8.1, and 7.8.2 was retained as it may be useful with other types of loads.

To calculate the thermal rise in the reactor the transformer weight must first be estimated from its dimensions:

$$Wc = \frac{VOL2}{12} \bullet W \tag{7.15}$$

Where: Wc = Reactor core weight in lbs = calculated value in cell (B41).
$VOL2$ = Core volume in inches^2 = calculated value in cell (C41) from formulae shown in Figure 7-11 and core dimensions entered in cells (B12) thru (B15). Converted to sqft by dividing by 12.
W = Weight of silicon sheet steel in lbs/sqft = 0.54 lb/sqft from Table 7-2 for typical 29 gauge silicon steel transformer core material.

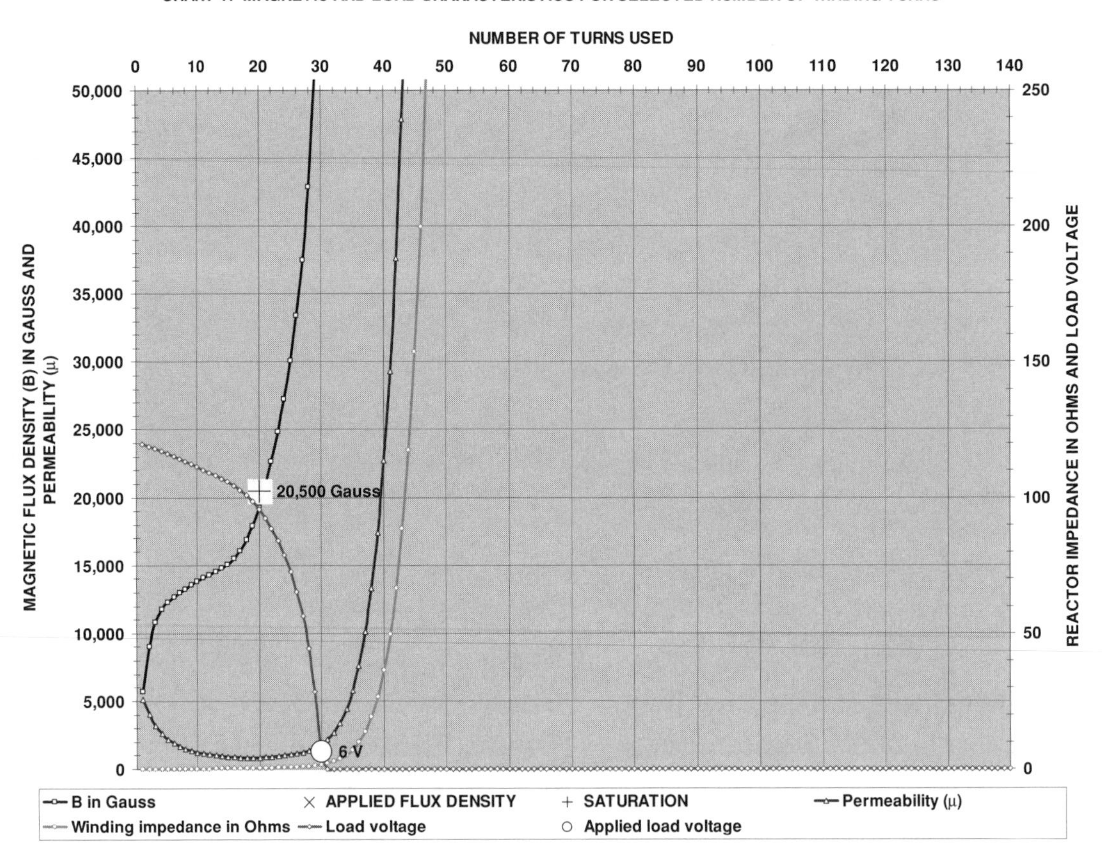

FIGURE 7-14 Operating characteristics for a reactor in saturation using an AC winding only (load current 70 A).

Figure 7-15 (chart 4 in worksheet) was constructed from data found in reference (25) and displays the curve fit formula used in cell (B70) to calculate the power dissipated in the core (PDc) due to hysteresis and eddy current losses for selected line frequencies.

The calculated power dissipation in the windings (I^2R losses) is performed in cells (B43) thru (B54) and totaled in cell (B71). The total power dissipation in the windings is:

$$PDw = (Iac^2 \bullet R_{\mathrm{ACW}}) + (Idc^2 \bullet R_{\mathrm{DCW}}) \tag{7.16}$$

Where: PDw = Power dissipation in the AC and DC windings in watts = calculated value in cell (B71).

Iac = Line/Load current in AC winding in rms amps = cell (B21), enter value.

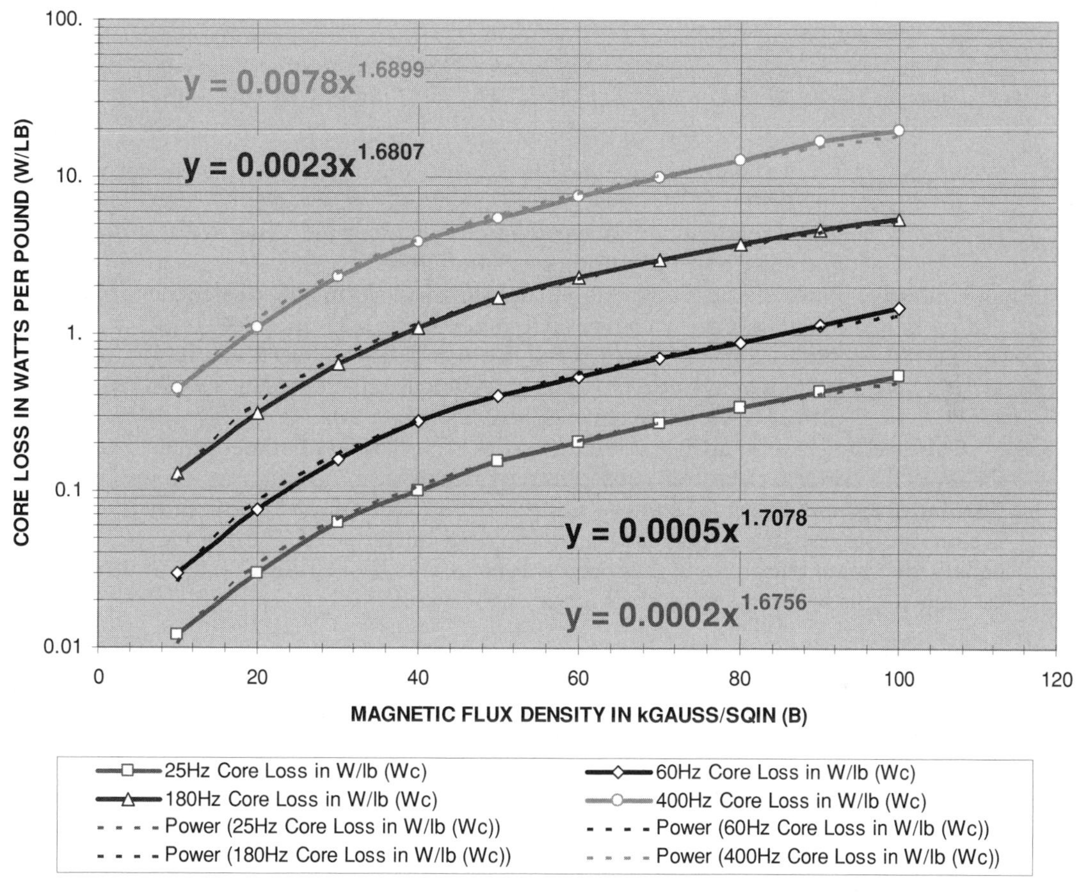

FIGURE 7-15 Loss characteristics for 1% to 4% silicon steel cores.

R_{ACW} = DC resistance of AC winding in ohms = calculated value in cell (B48) from values entered in cells (B16) through (B18) and calculations performed in cells (B43) through (B47). See Section 4.2 for methodology of calculating the DC resistance of magnet wire.

Idc = Current in DC winding in amps = cell (B25), enter value.

R_{DCW} = DC resistance of DC winding in ohms = calculated value in cell (B54) from values entered in cells (B26) through (B28) and calculations performed in cells (B50) through (B53).

The total power dissipation of the reactor is the core dissipation (PDc) plus the winding dissipation (PDw). The positive and negative alternations of an AC electromagnetic field continuously realign the magnetic particles in the core. This magnetic movement produces hysteresis losses and a resulting temperature rise in the core. The AC field also produces circulating eddy currents in the core material producing losses. The thermal rise produced by the core's hysteresis and eddy currents and the winding losses is:

$$\Delta tc = \left(\frac{PDc + PDw}{SA}\right)^{0.833} \tag{7.17}$$

Where: Δtc = Thermal rise from ambient in Reactor core in °C = calculated value in cell (B73).

PDc = Power dissipation in the core in mW = calculated value in cell (B70) in W converted to mW using a 1e3 multiplier.

PDw = Power dissipation in the AC and DC windings in mW = calculated value in cell (B71) in W converted to mW using a 1e3 multiplier.

SA = Surface area of Core in cm^2 = calculated value in cell (C40) from formulae shown in Figure 7-11 and core dimensions entered in cells (B12) thru (B15). Converted to cm in cells (C12) thru (C15).

The mechanism to remove heat from the core is primarily radiation from its outer surface therefore the core's surface area is used in the calculations. The windings should be equally distributed around the core and are generally too few to contribute to the core's surface area as in some transformer design calculations, so they were not included. The thermal rise will not be equally distributed throughout the core resulting in a hot spot within the core. A conservative estimate for the hot spot temperature in the application is:

$$TH = (TA + 1.22) \bullet \Delta tc \tag{7.18}$$

Where: TH = Hot spot temperature in core in °C = calculated value in cell (B74).

TA = Ambient temperature in °C = cell (B6), enter value. Converted to °F in cell (C6) = (°C × 9/5) + 32.

Δtc = Thermal rise from ambient in reactor core in °C = calculated value in cell (B73).

The hot spot temperature calculated in cell (B74) is the highest temperature found in the application and should be lower than the manufacturer's limit for all materials used in the reactor. If the limits are unknown keep this temperature within 55°C above the ambient temperature for short operating times or 25°C above the ambient temperature for continuous operation to be safe. The reactor thermal rise must not be taken for granted. If the core temperature (hot spot) rises above the Curie point the core will lose its magnetic properties and the reactor will cease to limit the current. The Curie point for ferrite cores is generally between 90°C and 130°C. If the winding insulation breaks down the inductance decreases or disappears altogether and the reactor again ceases to function. If the core is using a gap, the gap will dissipate much of this power in the surrounding air with a corresponding temperature rise. Using data found in reference (25) the chart shown in Figure 7-16 (chart 5 in worksheet) was constructed and can also be used to estimate the power dissipation in an iron core and its windings for an uncased transformer with a 40°C temperature rise from ambient. If the transformer is still cased in a tar-based substance or an epoxy resin (potted) this will not apply. The curve fit formula is used to calculate the total power dissipation in cell (B72) for a core operating at 40°C above ambient.

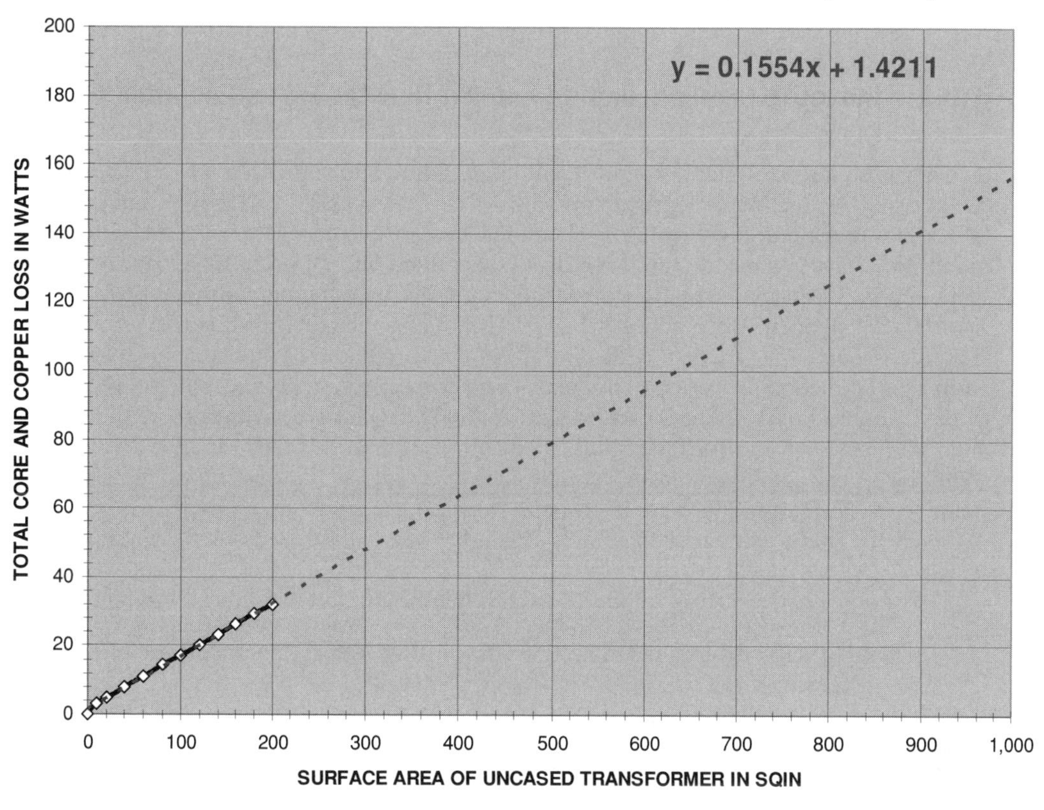

FIGURE 7-16 Typical core and copper losses in an uncased transformer with 40°C thermal rise from ambient.

Steps 1 through 6 above are repeated in worksheet rows 87 to 110 for quick reference. Once the design passes the thermal rise evaluation, it can be constructed using the winding and core parameters entered in the worksheet. The windings that came on the core may be used or distribute new windings equally around the core; do not bunch them together in one spot.

My apologies to Nikola Tesla for not expressing all magnetic flux density (B) values in their correct form, the tesla. One tesla is equivalent to 10,000 gauss, therefore any cells calculating a value of B have their equivalent value in teslas in an adjacent cell. The tesla was not adopted as the SI unit for B until 1956, which was commemorated in Yugoslavia with a stamp issue. Magnetic engineering data have been slow to adopt, if at all. Magnetic device calculations and data are expressed in a variety of standard measurement systems including cgs (centimeter-gram-seconds) units and SI (Systeme International) units. The data do not readily convert to other standards and introduces a plethora of confusion unless one system is adopted throughout the calculations.

7.8.2 Inductive Current Limiting Using a Reactor with an Additional DC Winding

I constructed a reactor using an additional DC winding and connected it to a load. As the DC winding shares the same magnetic core, the AC load current will inductively couple onto the DC winding making it difficult to isolate the DC supply current from the AC current in the winding. I have reviewed many electrical texts which outline using a DC winding without detailing how to isolate the AC current. This method may be only a theoretical possibility, which is not obtainable in practice. When a DC bias is applied to a ferrite core the bias produces a residual level of magnetic flux density (B) in the core known as remanence. Even if an AC current is applied to the core the DC current and remanence of the core material will keep B at some value, and the hysteresis of the AC current will drive B in one direction with the positive alternation, then in an equal magnitude in the opposite direction with the negative alternation. With the operating point of the hysteresis curve set to some value above 0 using a DC bias and remanence, the saturation point of the core material can be achieved with a known value of AC current. This is analogous to what is attempted in using the additional DC winding in the reactor.

To design a reactor using an additional DC winding follow the steps listed below:

1. Repeat steps 1 through 4 in Section 7.8.1. In step 4 the DC bias current will not be 0 A when using a DC winding.
2. Enter the desired number of DC turns used in cell (B28), the wire AWG into cell (B26), and the number of layers into cell (B27). To set the operating bias of the core, enter the value of DC winding current in cell (B25).
3. The load current can be limited by using either an additional DC winding to push the core into saturation or an air gap in the core to store the additional energy beyond the saturation point. If the extra DC winding is used enter a 2 in cell (B31) and the DC winding characteristics in cells (B25) thru (B28). Remember to check cells (B10) and (C10) to ensure they match. A large [+] is shown in the plot in Figure 7-13 to indicate the point of saturation in the core material. When the AC magnetizing (load) current reaches this point during its positive excursion it cannot increase beyond the saturation point. The curve shows magnetization beyond the saturation point in the calculations but it does not exist in the application. This is the region where the current is limited. During the negative excursion there is no current limiting. The range between the full current at the negative alternation and the limited current in the positive alternation is averaged in cell (B64), which is the maximum current the load can draw from the line for the selected characteristics.
4. Note in Figure 7-13 (chart 1 in worksheet) the large [X] on the plot that indicates the operating point (DC bias) of the core. The DC bias is constant unless the DC current is changed and determines the center of the AC hysteresis curve. Note that both curves in Figures 7-13 and 7-14 are centered about the DC bias point, which is 0 for the case

shown. When the AC load current is on a positive alternation it will produce a magnetizing force (H) and magnetic flux density (B) in the positive quadrant of the BH curve. When the AC load current is on a negative alternation the B and H will fall into the negative quadrant.

Once the design passes the thermal rise evaluation it can be constructed using the winding and core parameters entered in the worksheet. The windings that came on the core may be used or distribute new windings equally around the core, do not bunch them together in one spot. It will not matter much whether you wind the AC winding around the core first. Cover the AC winding with insulating material and then wind the DC winding over that, or vice versa.

7.8.3 Inductive Current Limiting Using a Reactor with an Air Gap

To design a reactor using an air gap in the core follow the steps listed below:

1. Repeat steps 1 through 4 in Section 7.8.1.
2. Enter a 1 in cell (B31) and the gap distance in inches into cell (B32). Enter a 0 in cell (B25).

From reference (24) is found the formula for calculating the magnetic flux density (B) for an iron core with air gap:

$$B = \frac{3.2\sqrt{2} \bullet NI}{G} \tag{7.19}$$

Where: B = Magnetic flux density in the core in gauss = calculated value in cell (B78), converted to kG using a 1e-3 multiplier. Shown as a black trace in Figure 7-17 (chart 3 in worksheet) from calculated values in cells (E197) thru (E226).
N = Number of turns used in the AC winding = cell (B18), enter value.
I = AC load current in rms amps = cell (B21), enter value.
G = Air gap distance in inches = cell (B32), enter value.

The inductance is:

$$L = \frac{3.2N^2 Ae}{G \bullet 10^8} \tag{7.20}$$

Where: L = Reactor inductance in henries = calculated value in cell (B78), converted to mH using a 1e3 multiplier. Also the calculated values in cells (F197) thru (F226) for use in calculating μ in Figure 7-17 (chart 3 in worksheet).
N = Number of turns used in the AC winding = cell (B18), enter value.
Ae = Core cross-sectional area in inches^2 = calculated value in cell (B37) from formulae shown in Figure 7-11 and core dimensions entered in cells (B12) thru (B15).
G = Air gap distance in inches = cell (B32), enter value.

Using Ampere's Law the magnetizing force (H) in oersteds is:

$$H = \frac{0.4\pi NI}{\text{le}} \tag{7.21}$$

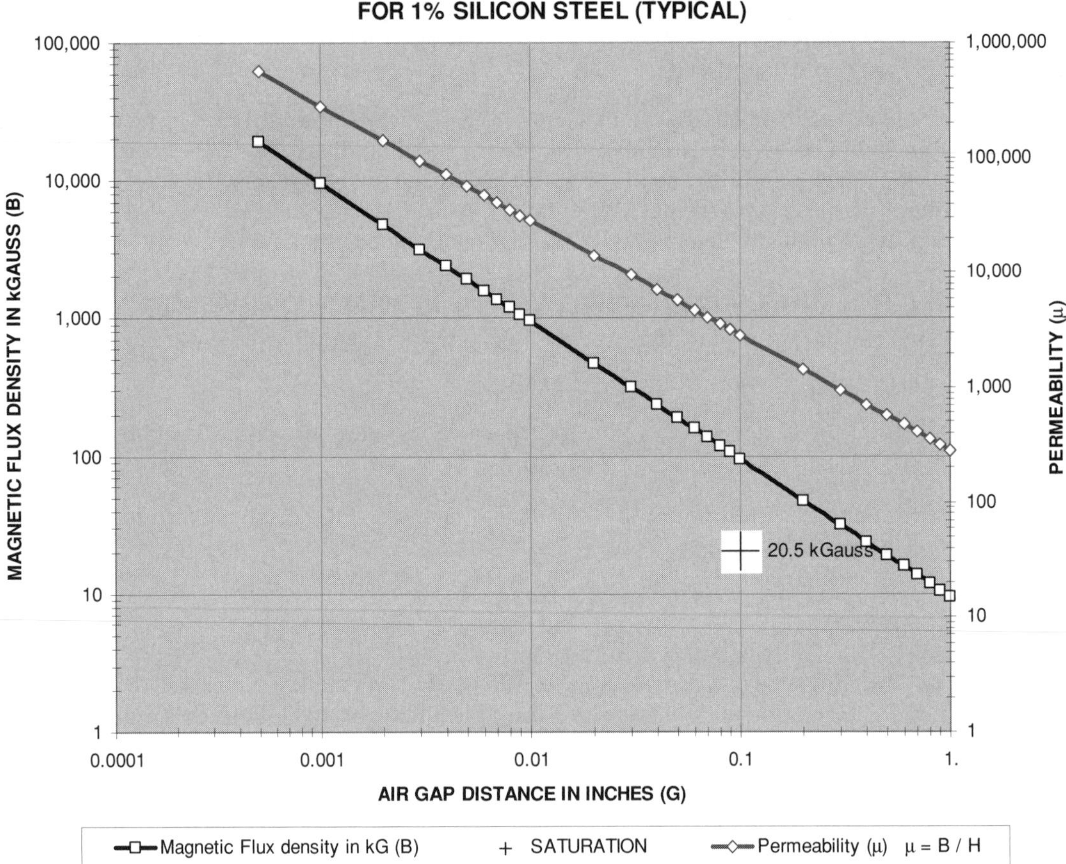

FIGURE 7-17 Magnetic flux density (B) and permeability (μ) in a reactor using an air gap for the core and winding characteristics listed in Figure 7-18.

Where: H = Magnetizing force (H) in oersteds = calculated value in cell (G196) for use in calculating μ in Figure 7-17 (chart 3 in worksheet).

N = Number of turns used in the AC winding = cell (B18), enter value.

I = AC load current in rms amps = cell (B21), enter value.

le = Core magnetic path length in cm = calculated value in cell (C38) from formulae shown in Figure 7-11 and core dimensions entered in cells (B12) thru (B15). Cell (B38) displays calculated le in inches.

And the core's permeability can be found:

$$\mu = \frac{B}{H} \tag{7.22}$$

	ENTER HERE			ROW
LINE CHARACTERISTICS				
Enter LINE frequency in Hz (**f**)	60 Hz	0.0167 sec		4
Enter LINE voltage (**VL**)	120 V	169.68V		5
Ambient temperature (**TA**)	25.0°C	77.0 °F		6
Ambient temperature (**TA**)	70 V	Calculate saturation characterisitcs at this voltage		7
				8
				9
CORE AND AC WINDING CHARACTERISTICS				10
Enter core type: (1) TOROID, (2) RECTANGULAR, (3) OPEN ENDED	2	see diagrams in rows 159 to 183.		11
Enter magnetic permeability of core material (μ)	1,404 ←	1800	green cell (C10) calculates μ at applied H for 1% si steel from Chart 2 curve fit formula.	12
Enter saturation flux density of core material in Gauss (**Bsat**)	20,500	20.5 kGauss	**2.050 TESLA**	13
Enter outside diameter in inches of core types 1, 2, and 3 (**OD**)	11.0 inches	27.9 cm		14
Enter inside diameter of core in inches of core types 1, 2, and 3 (**ID**)	7.0 inches	17.8 cm		15
Enter height of core in inches of core types 1, 2, and 3 (**Hc**)	2.6250 inches	6.7 cm		16
Enter width of core type 2 or length of core type 3 in inches (**W**)	2.5 inches	6.4 cm		17
Enter desired wire gauge used in AC winding (**AWGac**)	10			18
Number of strands (layers) of wire (**NSac**)	1			19
Enter desired number of turns used in AC winding (**Nac**)	30			20
				21
LOAD CHARACTERISTICS				22
Enter AC rms load current in Amps (**Iac**)	**70.0 A**			23
Enter power factor of loading effect (**PFL**)	100%			24
				25
DC WINDING CHARACTERISTICS (set current value to 0 if dc winding is not used)				26
Enter DC winding current in Amps (**Idc**)	**0.0 A**			27
Enter desired wire gauge used in DC winding (**AWGdc**)	10			28
Number of strands (layers) of wire (**NSdc**)	1			29
Enter desired number of turns used in DC winding (**Ndc**)	48			30
				31
AIR GAP CHARACTERISTICS (if using an air gap in the core)				32
Enter: (1) if a gap is used in the core or (2) if no gap is used	2			33
Enter gap distance if used in inches (**G**)	0.1000 inches	0.2540 cm		34
				35
CALCULATED VALUES FOR REACTOR CORE				36
CORE DIMENSIONS				37
Core cross-sectional area in square inches (**Ae**)	6.6 in^2	16.7 cm^2	Ae = see diagrams in rows 195 to 224.	38
Core magnetic path length in inches (**le**)	31.0 inches	78.7 cm	le = see diagrams in rows 195 to 224.	39
Core volume in cubic inches (**VOL**)	203.4 in^3	516.7 cm^3	VOL = see diagrams in rows 195 to 224.	40
Core surface area in square inches (**SA**)	342.8 in^2	870.6 cm^2	SA = see diagrams in rows 195 to 224.	41
Core weight in lb (**Wc**)	28.6 lbs	VOL2 = 635.3	Wc = (VOL2/12)* 0.54lbs/sqft, where VOL2 is core volume in sqin.	42
				43
CALCULATED VALUES FOR REACTOR USING AC WINDING ONLY				44
AC WINDING DIMENSIONS				45
AC winding wire diameter in inches (**d**)	0.1019		nAWG = (0.0050) x (1.1229322)^ (36-n)	46
AC winding wire area in cirmils (**da**)	10383		da = (d*1000)^2	47
AC winding wire resistance in **Ω / ft**	0.00099		Ω/ft = 10.3 / da	48
AC or DC winding wire length per turn in inches (**LW/T**)	10.3		LW/T = type 1: (OD-ID)/2)*Hc*2, type 2: (Hc+W)*2, type 3: π*d	49
AC winding wire length total (**Ltac**) in feet	25.6 ft		Ltac = (LW/T * Nac) / 12	50
AC winding DC Resistance in Ohms (**Racw**)	0.03 Ω		Racw = (Ω/ft * Lt) / NSac] * 1 + (TA -20 * 0.00393)	51

FIGURE 7-18 Iron core reactor worksheet.

A	B	C	D	E	F	G	COLUMN
							52
OPERATING POINT							
Applied magnetizing force (**H**)	33.51Oe		Lookup from Table 1 column K.				53
Applied inductance (**L**)	0.004310 H		Lookup from Table 1 column L.				54
Applied permeability (µ)	1,800		Lookup from Table 1 column O. Uses 6th order polynomial equation in Chart 2.				55
Winding impedance at applied flux density (**XL**)	1.62 Ω		Lookup from Table 1 column P.				56
Winding inductance in mH at applied flux density (**L**)	4.310 mH		Lookup from Table 1 column M.				57
Applied flux density in kG (**B**)	60.331 kG		Lookup from Table 1 column N.				58
Applied flux density in Gauss (**B**)	60,331 Gauss		Lookup from Table 1 column N.				59
Saturation flux density (**Bsat**)	20,500 Gauss		Entered in cell (B11)				60
Winding turn at saturation (**Nsat**)	**20**		Lookup from Table 1 column T.				61
Load Voltage at applied flux density (**VLoad**)	**6 V**		Lookup from Table 1 column R.				62
Find AC rms load current limit for application (**ILmax**)	**48.0 A**		Lookup from Table 3 column AQ.				63
Find Power Factor of load and reactor (**PF**)	73%		PF = cos[arctan{XL / ((Vp / IL)*PFL)}]				64
							65
							66
CALCULATED VALUES FOR REACTOR USING DC AND AC WINDING							67
DC WINDING DIMENSIONS							68
DC winding wire diameter in inches (**d**)	0.1019		nAWG = (0.0050) x (1.1229322)^ (36-n)				69
DC winding wire area in cirmils (**da**)	10383		da = (d*1000)^2				70
DC winding wire resistance in **Ω / ft**	0.00099		Ω/ft = 10.3 / da				71
DC winding wire length total (**Ltdc**) in feet	41.0 ft		Ltdc = (LW/T * Ndc) / 12				72
DC winding DC Resistance in Ohms (**Rdcw**)	0.04 Ω		Rdcw = (Ω/ft * Lt) / NSdc] * 1 + (TA -20 * 0.00393)				73
OPERATING POINT							74
Find DC Winding inductance in mHenries (**Ldc**)	8.605 mH	0.000000 H	Ldc in Henries = (0.4*π*µ*Ndc^2*Ae) / (le*1e^8)				75
Find AC Winding inductance in mHenries (**Lac**)	3.361 mH	0.003361 H	Lac in Henries = (0.4*π*µ*Nac^2*Ae) / (le*1e^8)				76
Total DC magnetic field strength (**Hdc**) in Oersteds	0.00Oe		Hdc = (0.4*π*Ndc*Idc) / le(in cm)				77
Total AC magnetic field strength (**Hac**) in Oersteds	33.51Oe		Hac = (0.4*π*Nac*Iac) / le(in cm)				78
Find DC flux density in kGauss (**Bdc**)	0.000 kG	**0.000 TESLA**	Bdc = (Ldc*Idc*10^8) / (Ndc*Ae(in cm))				79
Find AC flux density in kGauss (**Bac**)	47.054 kG	**4.705 TESLA**	Bac = (Lac*Iac*10^8) / (Nac*Ae(in cm))				80
Find Total flux density in kGauss (**B**)	47.054 kG	**4.705 TESLA**	B = Bac + Bdc				81
Find surplus flux density beyond saturation in kGauss (**Bsat'**)	26.554 kG	14.27Oe	Bsat' = B - Bsat				82
Find AC rms load current limit for application (**ILmax**)	30.5 A		ILmax = [(Bsat-Bdc)/(Lac*1e8)]*Nac*Ae Transposed from Faraday's Law in Mathcad				83
Find reactance of reactor (**XL**)	1.27 Ω		XL = 2*π*f*Lac				84
Find line voltage available to load (**Vp**)	31 V		Vp = VL - (Iac * XL)				85
Find Power Factor of load and reactor (**PF**)	80%		PF = cos[arctan{XL / ((Vp / IL)*PFL)}]				86
							87
POWER DISSIPATION AND OPERATING TEMPERATURE FOR ALL APPLICATIONS							88
Find power dissipated in reactor core (**PDc**)	10.3 W		PDc = (W/lb @ B and f) x Wc . Curve fit formula from Chart 4 (row 305 to 340).				89
Find power dissipated in AC and DC windings in W (**PDw**)	127.0 W		PDw = (Iac^2 * Racw) + (Idc^2 * Rdcw)				90
Find total power dissipated in core and windings for a 40°C rise from ambient (**PD5**)	54.7 W		Uses curve fit formula in chart 5 (row 342 to 377).				91
Find core temperature rise from ambient (**Δtc**)	**68 °C**		Δtc = ([Pd (in mW) / SA (in cm^2)]^0.833				92
Find maximum device temperature (Hot Spot Temp) (**TH**)	108 °C		TH = TA + 1.22 * Δtc				93
							94
CALCULATED OPERATING POINT FOR REACTOR USING AIR GAP (dimensions are calculated above)							95
Find AC Winding inductance in mHenries (**Lac**)	1.890 mH	0.001890 H	Lac in Henries = (3.2*Nac^2*Ae) / (G*1e^8),	Ae and G in inches			96
Find AC flux density in kGauss (**Bac**)	95.035 kG	**9.504 TESLA**	Bac = [3.2*sqrt(2)*Nac*Iac] / G,	G in inches			97

FIGURE 7-18 Iron core reactor worksheet. (*continued*)

Where: μ = Core permeability = calculated value in cells (H197) thru (H226) shown as red trace in Figure 7-17 (chart 3 in worksheet).

B = Magnetic flux density in the core in gauss = calculated value in cell (B78) from equation 7.19, converted to kG using a 1e-3 multiplier. Shown as a black trace in Figure 7-17 (chart 3 in worksheet) from calculated values in cells (E197) thru (E226).

H = Magnetizing force (H) in oersteds = calculated value in cell (G196) from equation 7.21, for use in calculating μ in Figure 7-17 (chart 3 in worksheet).

Figure 7-17 (chart 3 in the worksheet) displays the calculated magnetic flux density (B) and permeability (μ) for a range of gap distances (G) from 0.0005″ up to 1.0″. The load, core, and AC winding characteristics are also used in the chart calculations as explained above. The core material saturation is also shown by a large [+] in the graph so the optimum gap distance can be interpreted from the B vs. G line (black trace), approximately 0.13″ in this case. All other calculations are detailed in Section 7.8.1.

7.9 Variable (Voltage) Transformers

Variable autotransformers commonly referred to as Variacs, Powerstats, and other trade names function as a variable output voltage, non-isolated power transformer or autotransformer. Steel laminate strips are wound to look like a roll of tape producing a toroid-like shape, which the primary winding is wound around. The secondary output uses the common (neutral) connection with the primary and a variable tap with some portion of the primary inductance as shown in Figure 7-19. In an autotransformer the primary and secondary use the same common connection. The autotransformer's primary is not isolated from the secondary as in a conventional power transformer where separate windings are used. The core of an autotransformer generally saturates at a current that is much higher than the rated output current, meaning it will safely handle more output current than the manufacturer's ratings for short periods of time; however, the voltage ratings should not be exceeded. Higher output currents yield higher output power.

The transformer can operate at 100% duty cycle at the rated power, meaning you can run it all day without interruption, and it can safely dissipate the heat produced in the core. When the rated output current is exceeded it is not guaranteed that the heat produced in the core will be safely dissipated. If too much current is run through the transformer the core can saturate and the device will malfunction. A generally accepted rule among coilers is the autotransformers

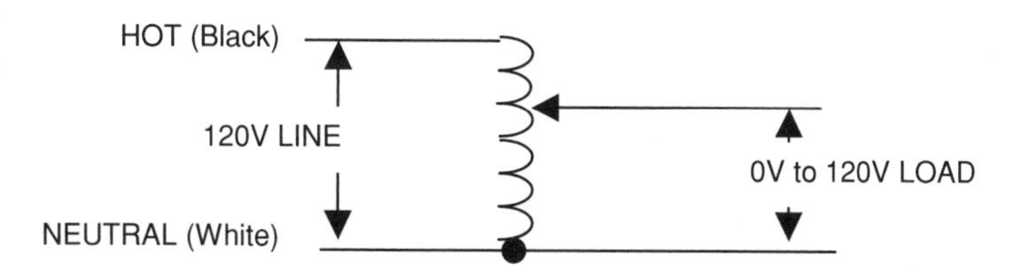

FIGURE 7-19 Circuit connections for variable autotransformers.

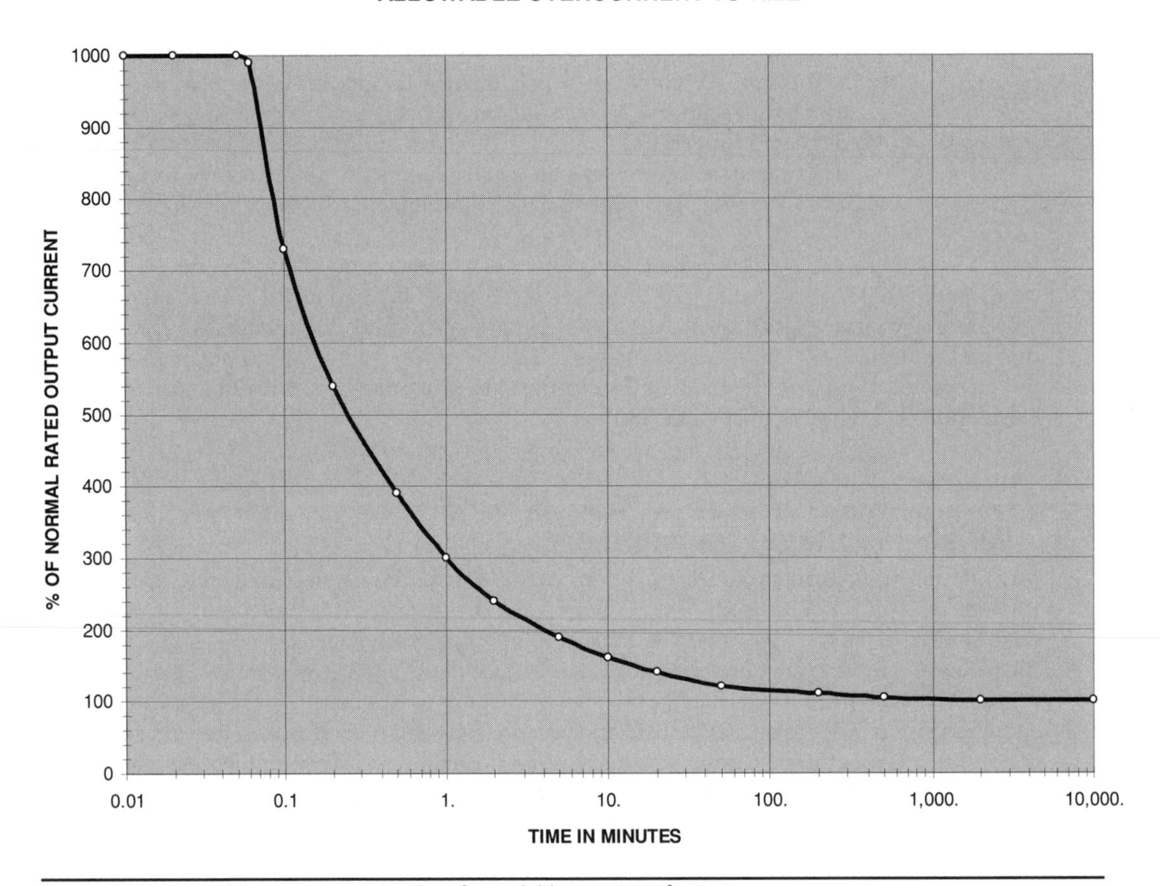

FIGURE 7-20 Allowable overcurrent vs. time for variable autotransformers.

can handle 2 to 2.5 times the rated output current for short coil runs. Reference (19), which is typical for any variable autotransformer, provides the allowable overcurrent for short time periods as displayed in Figure 7-20.

Note that the rated output current can be safely exceeded up to 300% if the run time is limited to 1 minute and up to 200% for 4 minutes. The manufacturer's ratings are very conservative. I have run these transformers at even higher power levels for longer periods of time without damage. The values shown in the figure are good for ambient temperatures up to 50°C. If your lab is in the Mojave Desert with an enclosed equipment cabinet and higher ambient temperatures are expected, derate the allowable run time period by 2.5%/°C. A new chart can be constructed for higher ambient temperatures and the applied temperature derating by entering the higher ambient temperature in cell (B3) of CH_7.xls file, AUTOTRANSFORMER worksheet (6).

7.10 Circuit Protection

Small Tesla coils that typically use NSTs will not require circuit protection as the NST is current-limited and will draw less line current than the 15-A rating that is typical for building electrical outlets. If a short circuit occurs in the operating coil the electrical outlet is assumed to have some form of circuit protection (e.g., a fuse or circuit breaker in the service box). Circuit protection can, however, be added for additional safety but it is generally not required. If you are constructing a medium or large spark gap coil your controls *must* include some form of circuit protection. The use of adequate circuit protection eliminates the need for current limiting schemes when using non–current-limited step-up transformers. Adequate circuit protection also isolates a short circuit from the line in case of catastrophic part failure during a run, something the current limiting schemes will not do. You have several choices from which some or all (recommended) can be selected:

- Motor start controls. I like these because they incorporate a motor stop feature. This is usually a large red push button switch that can be easily "hit" in an emergency. This "panic button" is often labeled "emergency stop," and when wired as shown in Figure 2-1, will disconnect the line voltage entering the control cabinet when the panic button is hit. The motor start control is reliably designed so you can count on everything shutting down the instant you hit the panic button. The exception to this is the rotary gap will still be spinning from its inertia and may continue to supply the tank with stored energy in the tank capacitor. Other features often include interconnects requiring a key to connect power, and "on" and "off" buttons. I replace the integral thermal fuses with shorting bars, relying on other components for breaking the circuit in overcurrent conditions. These can usually be found at Hamfests.
- Relays and contactors. These devices should be used to make and break circuits that carry large currents. Circuit breakers are generally not designed to perform this function with the repetition of an on–off switch. Relays and contactors that use 120-VAC coils are the easiest to adapt in your control scheme. The contact material in these devices is specially designed to handle the rated current, which is a DC or rms value. When a circuit opens there is an immediate rise in voltage, often high enough to generate an arc. Relays and contactors are designed to handle these transients, often to 100,000 actuations. On a microscopic scale there is a very destructive and complex mechanism involved in contact operation. Depending on the materials used there is usually an arc established each time the contact cycles, with associated melting of some of the contact surface. They are designed to handle it but the more severe it is the quicker it wears out. Be careful when cleaning the contact surfaces as very thin coatings and complex mechanical wiping actions are inherent in the designed and are easily damaged, resulting in accelerated wear out and failure. If you can find surplus mercury relays consider using them. Except for the glass case they are virtually indestructible, the contact surfaces never need cleaning and the liquid mercury provides a fresh metallic contact surface with each actuation. The mercury poses no hazard unless the glass envelopes which contain the mercury are accidentally broken.
- Safety gap. A safety gap *must* be incorporated to protect your step-up transformer. Because of the load on the step-up transformer, the primary coil's di/dt and the tank

capacitor's dv/dt are constantly changing and high-voltage transients in excess of 100 kV can exist on the lines from the transformer to the tank circuit. The only way to protect your transformer is to use a safety gap that is adjusted to ionize (fire) at some voltage above your transformer's step-up voltage. See Chapter 6 for details on construction and adjustment.

- Fuses. A one-shot deal and expensive to replace unless you stock up at the next Hamfest. They offer a guaranteed method of disconnecting the line voltage (service) from your control cabinet and the load. If you can obtain manufacturer's data sheets or application information on the fuses you intend to use, get them. Reference (21) provides several informative application notes.
- There are a myriad of considerations when designing with fuses. The current rating is the DC or symmetrical AC rms current (equal + and − amplitudes centered on 0 A) the fuse can handle under continuous load conditions. When the overcurrent required to melt the fuse element is reached (blow), it does so and an arc is established. Current will continue until the arc is quenched. Some fuses contain materials to increase quenching ability. The voltage rating of the fuse is an indication of the quenching ability, therefore a fuse with a lower voltage rating than the supply voltage it is protecting *should not* be used; however, a fuse with a higher voltage rating than the supply voltage it is protecting can be used. If the proper fast blow fuse is used all of this will happen within a few thousandths of a second when a short circuit condition appears. Slower blowing fuses incorporate a time delay to accept higher currents before they blow, e.g., as with inrush currents drawn by the starting of motors. Dual element fuses incorporate features of both fast- and slow-blow fuses. Section 7.13 details the fusing current required to melt different conductor materials, some of these are commonly used in fuses and were used exclusively as fuses in early electrical equipment until the cartridge fuses became commercially available. If a fast-blow fuse is placed in series in the line end of a circuit breaker, the fuse will protect the circuit breaker if a serious short circuit condition develops, and sacrifice itself to save the circuit breaker.
- Circuit breakers. Better than fuses because they can be reset after an overcurrent condition requiring no additional expense to reset the circuit. There are four types of circuit breakers:

  1. Thermal. The most commonly found and most reliable type for use with Tesla coils. The current rating is the DC or symmetrical AC rms current the breaker can handle under continuous load conditions. An overcurrent condition in a bimetallic strip will heat one side quicker than the other and the strip will bend away from its contact, opening the circuit. The time required to heat the bimetallic strip is dependent on I^2R heating and provides an integral time delay to allow short duration inrush currents and prevent nuisance tripping, e.g., as with inrush currents drawn by the starting of motors. Generally, thermal circuit breakers will sustain twice their rated current for about 50 seconds before the bimetallic strip is heated. With 6 times the rated current about 1 second elapses before the breaker trips and 10 times the rated current is required to reduce the time delay to fractions of a second. Remember that circuit breakers are not current limiters but safety devices designed to disconnect the line from the load when a short circuit exists. I have run up to 75 A rms of current

through a known good 20 A thermal circuit breaker for about one-half minute before it tripped in an operating Tesla coil. If your coil design draws more current than your control and indicator system is designed to handle, redesign with parts rated to handle the intended load. Do not count on a circuit breaker to break the circuit unless a short-circuit condition exists. You may need to run some tests to determine what will work.

When the bimetallic strip opens the contact an arc may form. Current will continue until the arc is quenched. The voltage rating of the circuit breaker is an indication of the quenching ability; therefore a breaker with a lower voltage rating than the supply voltage it is protecting *should not* be used; however, a breaker with a higher voltage rating than the supply voltage it is protecting can be used.
Derating factors, which change the value of current that will trip the breaker, include:

- The gauge of conductor used to connect the breaker to the line and load acts as a heat sink with thermal breakers and the manufacturer's datasheet should be followed to ensure the correct conductor gauge is used. The bimetallic strips are designed to be matched to a specific gauge conductor. Reducing the size of the conductor in the application allows the bimetallic strip to heat quicker (from the lower thermal conductance of the wire) and the breaker will trip at a lower current than normal. Increasing the size of the conductor slows the heating process in the bimetallic strip and the breaker will trip at a higher current than normal (unsafe condition). This feature can be used to customize the breaker's performance but you need to know what you are doing.
- As the ambient temperature surrounding the breaker increases the amount of current required to trip the breaker will decrease. At low temperatures this current increases.
- Using a breaker with a different line frequency than its rating will affect the current required to trip the breaker. Generally this will be 60 Hz and most of the surplus breakers you will find are rated for 60-Hz line frequencies.
- Thermal breakers are designed in air density conditions that are close to 0 feet in altitude (sea level). As the altitude of the application increases the air density decreases. This lowers the density of the air surrounding the bimetallic strip and its thermal radiation ability. As the altitude increases the amount of current required to trip the breaker will decrease.
- The type of load and duty cycle will affect the amount of current required to trip the breaker.

Reference (20) was used to develop the CH_7.xls file, CIRCUIT BREAKER worksheet (8) to calculate the circuit breaker current rating needed for the derating applications above. It includes explanations for each step in the calculations. An example application is shown in Figure 7-21.

2. Magnetic. An overcurrent condition energizes a steel–copper solenoid which separates the contacts, opening the circuit. There is an additional trip current inaccuracy of +10% when used in supply frequencies below 60 Hz and −10% when used in supply frequencies above 60 Hz due to increased or decreased reactance and heating

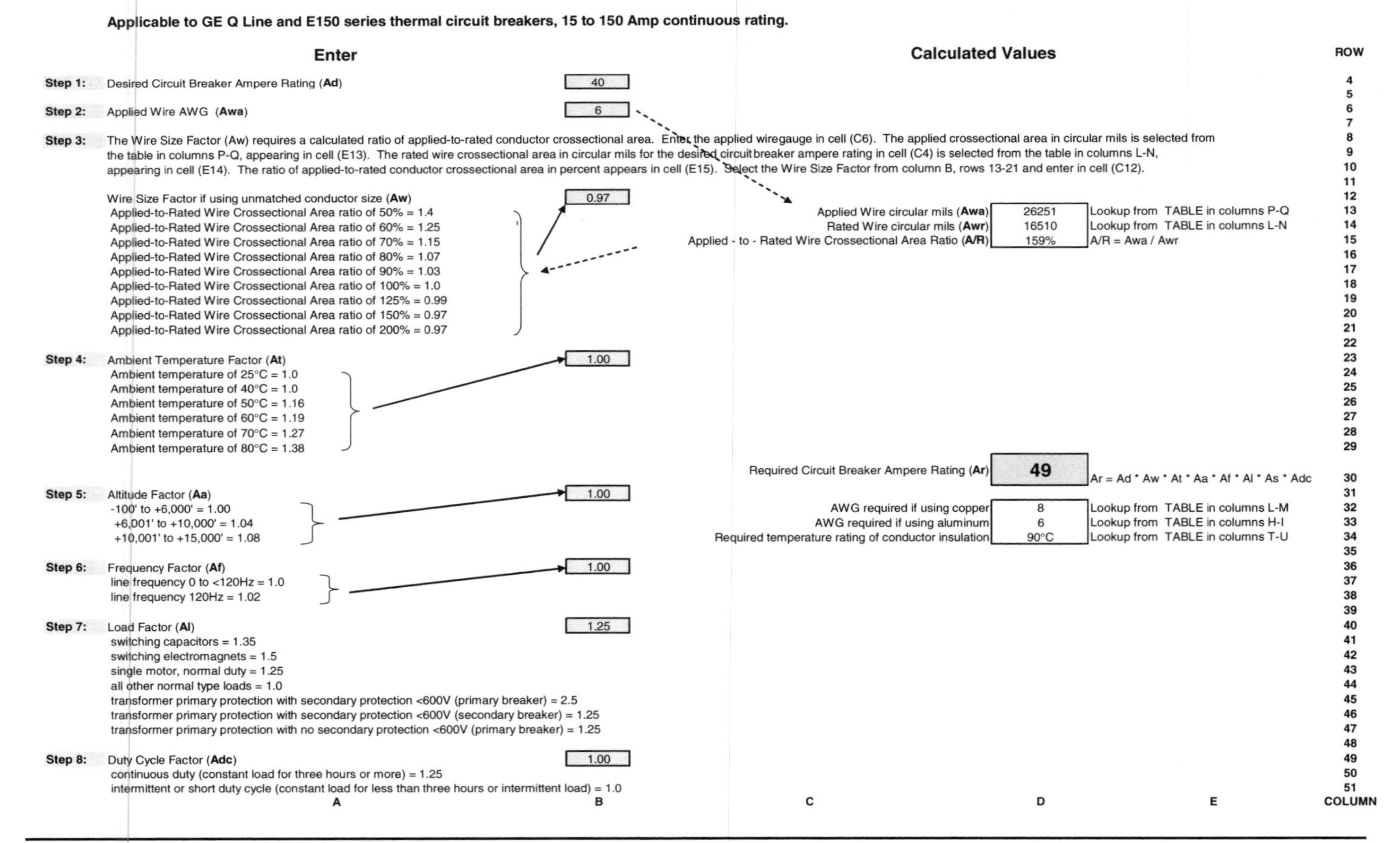

FIGURE 7-21 Thermal circuit breaker calculations.

within the solenoid. Many magnetic circuit breaker types have an adjustable trip current.

3. Thermal–Magnetic. Combines elements of both thermal and magnetic breaker construction.
4. Solid-State. A control section uses solid-state electronics to sense an overcurrent condition, which opens contacts in a load section by means of a relay or contactor. These being solid state in nature should not be used in Tesla coil construction.

7.11 Interconnections

Interconnections are the conductors between electrical parts such as seen in Figure 2-1. The interconnections must support the current being carried from one part to all others on that connection and offer as little resistance as possible. They must also be separated by enough distance to prevent corona and flashover or covered with high-voltage insulation.

Table 7-3 is derived from reference (14) for typical interconnecting materials used in electrical part manufacturing. A material's ability to conduct electricity (conductance) is the opposite (reciprocal) of its electrical resistivity, shown in the third column in the table.

- Silver is the best electrical and thermal conductor, has good resistance to corrosion, and continues to conduct well even when pitted from arcing on contact surfaces. It is the material of choice for relay contact surfaces. It is expensive, but not prohibitively so when used as a thin plating over a base material. You can find a lot of older high-quality parts at Hamfests, which use silver plating on the conducting surfaces. It is easy to find, as silver will turn black as it oxidizes. Described in reference (22) silver is also prone to another oxide: sulfidation, accumulating at the rate of 70 micrograms per square centimeter per day under typical indoor conditions. To clean silver conductor use a high-quality silver polish. The polish will remove only trace amounts of the silver as well as the oxidation and a good polish will keep the silver removal to a minimum. Remember it is probably not solid silver, just a few thousandths of an inch of plating.
- Copper and aluminum are the most common and least expensive conductive material. If you built a Tesla coil of solid silver conductor and one of copper I doubt you would notice any difference in the performance; however, when considering the cost, copper makes a good compromise. Unlike silver, copper and aluminum corrode more easily and are prone to surface pitting. This will increase your corona losses (remember you may not see the corona losses but they can still be there). With the exception of the spark gap, your material oxidation will be retarded or accelerated by the air quality and humidity in your lab. In most environments copper corrodes more slowly than aluminum. The copper oxidation will first appear as a darker brown color increasing to a greenish color powder, whereas the aluminum oxidation will appear as a white powder.

Material Type	Periodic Symbol	Electical Resistivity in $\mu\Omega$/cm	Magnetic Permeability $\times 10^{-6}$ henry/meters	Relative Resistance to Copper (k1 factor in equation 8.18)	Temperature Coefficient of Resistivity (%/°C or PPM)	Thermal Conductivity @ 25°C BTU/ft^2 h in °C	Linear Coefficient of Thermal Expansion per °F $\times 10^{-6}$	Melting Point	Modulus of Elasticity (Tension) lb/in^2 $\times 10^6$
Silver	Ag	1.59		0.95	0.0038	2,900	10.9	961°C/1,761°F	11
Copper	Cu	1.673	1.256	1.00	0.00393	2,730	9.2	1,083°C/1,981°F	16
Gold	Au	2.19		1.416	0.0034	2,060	7.9	1,063°C/1,945°F	12
Aluminum	Al	2.655		1.64	0.0039	1,540	13.3	660°C/1,220°F	10
Tungsten	W	5.5		3.25	0.0045	900	2.4	3,410°C/6,150°F	50
Zinc	Zn	5.92		3.4	0.0037	784	9.4-22	787°F	12
Nickel	Ni	6.84	7.54e-4	5.05	0.0047	639	7.4	1,452°C/2,651°F	30
Iron	Fe	9.71	6.28e-3	5.6	0.0052	523	6.5	2,802°F	28.5
Platinum	Pt	9.83		6.16	0.0030	494	4.9	1,769°C/3,224°F	21
Tin	Sn	11.5		6.7	0.0042	464	13	449°F	6
Lead	Pb	20.65		12.78	0.0039	241	16.3	621°F	2.6
Steel	–	17–19	5.03e-2	7.6	–	27	8.1-8.4	1,510°C	

TABLE 7-3 Properties of common interconnecting materials.

- Your chances of encountering any parts suitable for a Tesla coil containing gold are remote. Copper is a better conductor but cannot match gold's superior resistance to oxidation.
- Many of the other materials listed will be found in vacuum tube construction. Notice the high resistance of tin and lead, which are the two components in electrical solder. They are not very good conductors. Your solder connections should incorporate a crimp to increase the conductivity of the connection. Solder is merely a filler, adding rigidity and strength to electrical connections. As long as your copper conductor is touching the copper in the connection being soldered through a crimp, there will be good conduction even if it is soldered. If the solder somehow melts during a coil run the crimp will also hold the connection. You do not want any interconnections coming loose and shorting during a coil run.

Do not use ferrous metals (containing iron) for interconnections anywhere near the primary circuit. The intense high-frequency electromagnetic (EM) fields being produced in a primary tank circuit will induce strong eddy currents in any ferrous metal in proximity to it. The eddy currents cause heating of the metal. Running a medium sized coil for about 2 minutes through a steel spring clamp (similar to those on a set of battery jumper cables) produced enough heat to melt plastic insulation materials and melt solder. When the steel clamp was replaced with copper materials the same run time produced no perceptible heating. Another connection problem surfaced from using a steel fastener. It got hot enough during the run to melt the PVC end cap used in the coil form. When replaced with a brass fastener the connection ran cool to the touch (felt after the power was shut down of course). A few soldered connections used in the secondary remained cool, as did all of the connections in the control cabinet.

For optimum power transfer and safety, run your coil for increasingly longer periods and check the interconnections between runs when there is no power to the coil. Don't forget to discharge the tank capacitor before touching the primary. Section 5-12 details how to safely discharge the high-voltage tank capacitor.

Because of skin effect in conductors, I like to use 1/4, 3/8, or 1/2 inch copper tubing for interconnections. These have been the choice of many veteran coilers for some time. It is very malleable and more easily bent and shaped than solid copper conductor of the same diameter. The ends can be flattened in a vise and a hole drilled through the center of the flattened portion to provide a nice solderless lug type terminal for connection to the circuit. I like to use lug type terminals since making any connection in the circuit requires no additional soldering. Most of the surplus parts you obtain will have large bolts or studs (e.g., 10–32, 1/4–20, 3/8–16) for circuit connections. This makes connections a simple matter of bending some tubing, making a terminal lug at the ends and tightening some nuts.

Occasionally check all interconnections for tightness and corrosion. Plated connecting hardware will help but all interconnections *will* oxidize causing degradation in coil performance. Using different fastener metals in connections will form dissimilar metal junctions causing galvanic currents to flow, just as in a battery. Generally this situation cannot be avoided but it does increase corrosion and loosens connections after a number of thermal cycles (cold-to-hot-to-cold…). Loose connections during a high-power run are unsafe and can damage the control or tank components. The rotary or stationary spark gap surfaces will also erode/corrode sooner if you are not using tungsten rod. For sharp performance, check and clean everything often.

If a conductor's surface is not copper or aluminum it is likely plated with a thin layer of a more precious metal. To clean all but silver and gold platings the most effective method I have found is a trick from old Marine Corps radio operators. Remove oxidation with a big pink eraser like the one you used in grade school. Light rubbing removes the oxidation and a negligible amount of the plating material. More pressure will remove even the most stubborn corrosion from copper, brass, and nickel. A shiny, smooth surface will result with a minimum of corona loss. Most of the parts you obtain at a Hamfest or surplus outlet will need a rigorous cleaning, which includes complete disassembly. I built a vacuum tube coil out of surplus parts and it did not perform to expectations. *With just a detailed cleaning of the interconnections and parts the efficiency and output increased by 50%.*

7.11.1 Distributed Effects and Skin Effect in Straight Conductors

Skin effect calculations for straight conductors differ slightly from those presented in Chapter 4 for helically wound inductors using magnet wire. The difference is attributed to the reduced or negligible proximity effect in straight conductors, which are separated from other AC current carrying conductors. Open the CH_7.xls file, INTERCONNECT SKIN EFFECT worksheet (3) and follow along through the calculations that follow. The worksheet calculations are shown in Figure 7-22. From reference (7) is found the following methodology:

The general formula for calculating skin depth for straight conductors is:

$$\delta(\text{meters}) = \sqrt{\frac{\lambda}{\pi \sigma \mu c}} \tag{7.23}$$

Where: δ = Depth of current penetration from outer skin in meters (mks rationalized units).
λ = Free-space wavelength in meters.
$1/\sigma$ = 1.724×10^{-8} p/pc ohm-meter for copper = cell (B7), enter value.
μ = Magnetic permeability of conductor $4\pi \times 10^{-7}$ μr henry/meter = cell (B8), enter value.
c = Velocity of light in vacuum = 2.998×10^{8} meters/second.

The reference provides more useful conversions of equation (7.23):

$$\delta(\text{inches}) = (1.50 \times 10^{-4} \bullet \sqrt{\lambda})k1 = \frac{2.60}{\sqrt{f}}k1 \tag{7.24}$$

$$\delta(\text{cm}) = (3.82 \times 10^{-4} \bullet \sqrt{\lambda})k1 = \frac{6.61}{\sqrt{f}}k1 \tag{7.25}$$

$$\delta(\text{mils}) = \frac{2.60}{\sqrt{\text{f(MHz)}}}k1 \tag{7.26}$$

Enter			Calculated Values	STANDARD	METRIC	ROW

Rac Loss calculation for SOLID conductors. δ < (d / 8)

Enter	Value	
Ambient Temperature in °C (T_A)	25°C	77.0 °F
Frequency in kHz (**f**)	100.000KHz	100000Hz
Resistivity of conductor material in Ω-meters (**p**)	1.7240E-08	
Magnetic permeability of conductor material in Henry/meters (μ)	1.2566E-06	
Conductor diameter in inches (**d**)	0.2500	0.6350 cm ← From E5
For wire 0000 thru 60 AWG, enter AWG value in cell (B12) below, then enter inches displayed in cell (E5) into cell (B9).		
Wire guage (**ga**)	2	→ To E5
Number of strands (N_S)	1	
Total conductor length in feet (**L**)	5.00	
Spacing between conductor and ground return in feet (**s**)	3.00	
Effective kV Applied between conductor and ground return (**e**)	25.00 KV	
Conductor irregularity Factor (**i**)	0.87	
Visual Factor (**v**)	1.00	
Barometric Pressure of Surrounding Air in inches (**BP**)	29.90	

Calculated Values	STANDARD	METRIC	ROW
			4
Wire diameter (**d**) w/o insulation	0.2576 in	0.6544 cm	5
Wire area in circular mils (**A**)	62500		6
DC resistance (Ω/ft)	0.0002		7
DC resistance of entire conductor (**Rdc**)	0.0008		8
			9
Wavelength of frequency (λ)	9840.00 ft	2998.000 meters	10
Depth of penetration or one skin depth (δ)	0.0082 in	0.0209 cm	11
Resistance per square in Ohms (**Rsq**)	0.00033	0.00013	12
Resistance per foot of cylindrical conductor in Ohms (**R**)	0.00126	0.004133 /meter	13
AC resistance of entire conductor (**Rac**)	0.0063		14
			15
Rac / Rdc ratio (**Fr**)	7.497		16
D sqrt(f) (**Df**)	79		17
Correction coefficient (**kA**)	1.05		18
			19
Rac with Correction Coefficient applied (**RAC**)	**0.0066**		20
Total Inductance of conductor (**Lt**)	**1.80 μH**		21
Total Capacitance of conductor (**Ct**)	**0.000015μF**		22
For conductor diameters >0.2 inches:			23
Air Density Factor (**A**)	1.000		24
a	79.229		25
Apparent Strength of Air in kV/inch effective with applied sine wave (**gv**)	82.54 KV		26
Visual Critical Voltage in kV effective (**ev**)	**32.79 KV**		27
Disruptive Critical Voltage in kV (**eo**)	**18.59 KV**		28
Applied Corona Loss in kW (**P**)	**0.31KW**		29
			30

Rac Loss calculation for TUBULAR conductors. (OD - ID) < (OD / 8)

Enter	Value	
Ambient Temperature in °C (T_A)	25°C	77.0 °F
Frequency in kHz (**f**)	100.000KHz	100000Hz
Resistivity of conductor material in Ω-meters (**p**)	1.7240E-08	
Magnetic permeability of conductor material in Henry/meters (μ)	1.2566E-06	
Conductor outside diameter in inches (**OD**)	0.2500	0.6350 cm
Conductor inside diameter in inches (**ID**)	0.2100	0.5334 cm
Total conductor length in feet (**L**)	5.00	
Spacing between conductor and ground return in feet (**s**)	3.00	
Effective kV Applied between conductor and ground return (**e**)	25.00 KV	
Conductor irregularity Factor (**i**)	0.87	
Visual Factor (**v**)	1.00	
Barometric Pressure of Surrounding Air in inches (**BP**)	29.90	

Calculated Values	STANDARD	METRIC	ROW
			31
Thickness of conductor (**T**)	0.0400 in	0.1016 cm	32
Area in circular mils (**A**)	62500		33
DC resistance (Ω/ft)	0.0002		34
DC resistance of entire conductor (**Rdc**)	0.0008		35
			36
Wavelength of frequency (λ)	9840.00 ft	2998.000 meters	37
Depth of penetration or one skin depth (δ)	0.0082 in	0.0209 cm	38
Resistance per square in Ohms (**Rsq**)	0.00033	0.00013	39
Resistance per foot of cylindrical conductor in Ohms (**R**)	0.00126	0.004133 /meter	40
AC resistance of entire conductor (**Rac**)	0.0063		41
			42
Rac / Rdc ratio (**Fr**)	7.497		43
T sqrt(f) (**Tf**)	13		44
Correction coefficient (**kA**)	1.00		45
			46
Rac with Correction Coefficient applied (**RAC**)	**0.0063**		47
Total Inductance of conductor (**Lt**)	**1.80 μH**		48
Total Capacitance of conductor (**Ct**)	**0.000015μF**		49
For conductor diameters >0.2 inches:			50
Air Density Factor (**A**)	1.000		51
a	79.229		52
Apparent Strength of Air in kV/inch effective with applied sine wave (**gv**)	82.54 KV		53
Visual Critical Voltage in kV effective (**ev**)	**32.79 KV**		54
Disruptive Critical Voltage in kV (**eo**)	**18.59 KV**		55
Applied Corona Loss in kW (**P**)	**0.31KW**		56
			57

Rac Loss calculation for RECTANGULAR conductors. δ < (Height / 8)

Enter	Value	
Ambient Temperature in °C (T_A)	25°C	77.0 °F
Frequency in kHz (**f**)	100.000KHz	100000Hz
Resistivity of conductor material in Ω-meters (**p**)	1.7240E-08	
Magnetic permeability of conductor material in Henry/meters (μ)	1.2566E-06	
Conductor Height in inches (**H**)	0.0500	0.1270 cm
Conductor Width in inches (**W**)	0.3450	0.8763 cm
Total conductor length in feet (**L**)	5.00	
Same formulae are used. Equivlent diameter becomes (H+W)*2/π		
Spacing between conductor and ground return in feet (**s**)	3.00	
Effective kV Applied between conductor and ground return (**e**)	25.00 KV	
Conductor irregularity Factor (**i**)	0.87	
Visual Factor (**v**)	1.00	
Barometric Pressure of Surrounding Air in inches (**BP**)	29.90	

Calculated Values	STANDARD	METRIC	ROW
			58
Equivalent Conductor diameter (**d**)	0.2515 in	0.6387 cm	59
Wire area in circular mils (**A**)	63235		60
DC resistance (Ω/ft)	0.0002		61
DC resistance of entire conductor (**Rdc**)	0.0008		62
			63
Wavelength of frequency (λ)	9840.00 ft	2998.000 meters	64
Depth of penetration or one skin depth (δ)	0.0082 in	0.0209 cm	65
Resistance per square in Ohms (**Rsq**)	0.00033	0.00013	66
Resistance per foot of rectangular conductor in Ohms (**R**)	0.00125	0.004109 /meter	67
Rac / Rdc ratio (**Fr**)	7.541		68
			69
AC resistance of entire conductor (**Rac**)	**0.0063**		70
Total Inductance of conductor (**Lt**)	**1.80 μH**		71
Total Capacitance of conductor (**Ct**)	**0.000015μF**		72
For conductor diameters >0.2 inches:			73
Air Density Factor (**A**)	1.000		74
a	79.460		75
Apparent Strength of Air in kV/inch effective with applied sine wave (**gv**)	82.45 KV		76
Visual Critical Voltage in kV effective (**ev**)	**32.89 KV**		77
Disruptive Critical Voltage in kV (**eo**)	**18.67 KV**		78
Applied Corona Loss in kW (**P**)	**0.30KW**		79

Columns: A B C D E F COLUMN

FIGURE 7-22 Skin effect calculations worksheet for straight conductors.

Where: f = Applied frequency = cell (B6), enter value in kHz, converted to Hz in cell (C6).
λ = Free-space wavelength in meters.
$k1$ = 1.0 or unity for copper from equation (7.29).

The resistance in ohms per square is found:

$$Rsq = \frac{1}{\delta\sigma} \tag{7.27}$$

Where: Rsq = Resistance in ohms per square (in any units) = calculated value in cell (E12).
δ = Depth of current penetration from outer skin in inches = calculated value in cell (E11) from equation (7.24). Depth of current penetration from outer skin in cm = calculated value in cell (F11) from equation (7.25).
$1/\sigma$ = 1.724×10^{-8} p/pc ohm-meter for copper = cell (B7), enter value.

A more usable form of equation (7.27) was developed in the reference:

$$Rsq = \frac{4.52 \times 10^{-3}}{\sqrt{\lambda}} k2 = 2.61 \times 10^{-7} \bullet \sqrt{f} \bullet k2 \tag{7.28}$$

k1 and k2 are multipliers used to compare copper with other conductor materials. The k1 and k2 value for copper is 1.0 or unity. This is similar to comparing the magnetic permeability of air with other magnetic materials in Section 4.9.

$$k1 = \left(\frac{1}{\mu_r} \bullet \frac{\rho}{\rho_c}\right)^{\frac{1}{2}} \tag{7.29}$$

$$k2 = \left(\mu_r \bullet \frac{\rho}{\rho_c}\right)^{\frac{1}{2}} \tag{7.30}$$

Where: μ_r = Relative permeability of conductor to copper (Cu). Cu and other non-magnetic materials are 1.0 = cell (B8), enter value.
ρ = Resistivity of conductor at any temperature.
ρc = Resistivity of copper at 20°C = 1.724 μohm/cm = cell (B7), enter value.

NOTE: *The expression $n^{(\frac{1}{2})}$ or $n^{0.5}$ is the same as $\sqrt{n}$.*

The reference also provides a formula using the *Rsq* calculated in equation (7.28) to determine the resistance per foot in a *solid* cylindrical conductor:

$$\Omega/ft = \frac{12}{\pi d} Rsq = \frac{(0.996 \times 10^{-6} \bullet \sqrt{f})}{d} \tag{7.31}$$

Where: Ω/ft = AC resistance per foot of conductor at applied frequency = cell (E13), calculated value.

f = Applied frequency = cell (B6), enter value in kHz, converted to Hz in cell (C6).

d = Diameter of cylindrical conductor in inches = cell (B9), enter value. As an aid to design the NEMA wire table can be referenced, by entering a wire gauge in cell (B12). The wire diameter appearing in cell (E5) from the lookup table can be used to determine the conductor diameter entered in cell (B9).

The AC resistance for the applied length of the conductor is:

$$Rac = \Omega/\text{ft} \bullet L \tag{7.32}$$

Where: Rac = AC resistance of selected length of conductor at applied frequency = cell (E14), calculated value.

Ω/ft = AC resistance per foot of conductor at applied frequency = calculated value in cell (E13) from equation (7.31).

L = Length of conductor in feet = cell (B14), enter value.

A correction factor is then applied for a total AC resistance in a solid cylindrical conductor:

$$RAC = Rac \bullet k \tag{7.33}$$

Where: RAC = Total AC resistance of solid cylindrical conductor at applied frequency = cell (E20), calculated value.

Rac = AC resistance of selected length of conductor at applied frequency = cell (E14), calculated value.

k = Correction coefficient from table below = cell (E18), automatic lookup from calculated $d\sqrt{f}$ value in cell (E17).

$d\sqrt{f}$ value	k
<3.0	Rac ≈ Rdc
5	2.0
10	1.3
13	1.2
26	1.1
48	1.05
98	1.02
160	1.01
220	1.005
370	1.0
>370	1.0

$d\sqrt{f}$ is abbreviated from:

$$d\sqrt{f} = d\sqrt{f} \bullet \sqrt{\mu r \frac{\rho c}{\rho}}$$

Where: d = Diameter of cylindrical conductor in inches = cell (B9), enter value.

f = Applied frequency in Hz = cell (B6), enter value in kHz, converted to Hz in cell (C6).

$\sqrt{\mu r \frac{\rho c}{\rho}}$ = 1.0 or unity for copper and other non-magnetic materials.

NOTE: *The reference states the correction factor produces accurate Rac values for solid conductors spaced apart from each other by a distance of at least 10 times the diameter of the conductor. If the separation is only 4 times the diameter of the conductor the Rac will be increased another 3% from the calculated value.*

A similar correction factor is applied for a total AC resistance in a *tubular* conductor:

$$RAC = Rac \bullet k \tag{7.34}$$

Where: RAC = Total AC resistance of tubular conductor at applied frequency = cell (E47), calculated value.
Rac = AC resistance of selected length of conductor at applied frequency = cell (E41), calculated value.
k = Correction coefficient from table below = cell (E45), automatic lookup from calculated $T\sqrt{f}$ value in cell (E44).

T$\sqrt{f}$ value	k	Rac/Rdc
<1.3	2.6	1.0
1.31	2.0	1.0
1.77	1.5	1.02
2.08	1.3	1.04
2.29	1.2	1.06
2.6	1.1	1.1
2.85	1.05	1.15
3.15	1.01	1.23
3.5	1.00	1.35
>3.5	1.00	0.384

$T\sqrt{f}$ is abbreviated from:

$$T\sqrt{f} = T\sqrt{f} \bullet \sqrt{\mu r \frac{\rho c}{\rho}}$$

Where: T = Thickness of tubular conductor wall in inches = cell (E32), calculated value from Outside Diameter entered in cell (B36) – Inside Diameter entered in cell (B37).
f = Applied frequency in Hz = cell (B33), enter value in kHz, converted to Hz in cell (C33).
$\sqrt{\mu r \frac{\rho c}{\rho}}$ = 1.0 or unity for copper and other nonmagnetic materials.

NOTE: *The reference states the correction factor produces accurate Rac values for tubular conductors when spaced apart from each other by a distance of at least 10 times the diameter of the conductor. If the separation is only 4 times the diameter of the conductor the Rac will be increased another 3% from the calculated value. The conductor thickness must also be less than 1/8 of the outside diameter of the tubular conductor if the current is flowing on the outside diameter ($T < d/8$) or less than 1/8 of the inside diameter of the tubular conductor if the current is flowing on the inside diameter.*

The reference also provides a formula using the *Rsq* calculated in equation (7.20) to calculate the resistance per foot in a *rectangular* conductor:

$$\Omega/ft = \frac{12}{\pi d} Rsq = \frac{(0.996 \times 10^{-6} \bullet \sqrt{f})}{d} \tag{7.35}$$

Where: Ω/ft = AC resistance per foot of conductor at applied frequency = cell (E67), calculated value.
f = Applied frequency in Hz = cell (B60), enter value in kHz, converted to Hz in cell (C60).
d = Equivalent diameter of rectangular conductor in inches = cell (E59), calculated value using:

$$d = (H + W)\frac{2}{\pi}$$

Where: H = Rectangular conductor height dimension in inches = cell (B63), enter value.
W = Rectangular conductor width dimension in inches = cell (B64), enter value.

The AC resistance for the applied length of the conductor is:

$$Rac = \Omega/ft \bullet L \tag{7.36}$$

Where: Rac = AC resistance of selected length of conductor at applied frequency = cell (E70), calculated value.
Ω/ft = AC resistance per foot of conductor at applied frequency = calculated value in cell (E67) from equation (7.35).
L = Length of conductor in feet = cell (B65), enter value.

Contrary to the correction factor applied to the solid cylindrical conductor, the rectangular conductor requires no correction factor.

Distributed effects can also be calculated for straight conductors. Reference (9) provides formulae for calculating the inductance of one wire of a two-wire, single-phase transmission system. This was used assuming the single wire must also have an equal length wire for the ground return. This generally applies to all facets of coil construction, except the wire used in the primary and secondary windings. The formula was simplified for DC current, but is accurate for commercial AC power frequencies (60 Hz).

To calculate the inductance of a straight *solid* conductor:

$$L = \left(0.74114 \bullet \log_{10}\left(\frac{s}{r}\right) + 0.0805\right)10^{-3} \text{ Henry/mile} \tag{7.37}$$

Where: L = Inductance of straight wire in two wire system in mH per mile.
s = Spacing between centers of both conductors in two wire system in feet = cell (B16), enter value. If an actual ground (plane) is used for the return instead of a similar conductor, s equals twice the distance between the conductor to the ground (plane) return.
r = Radius of conductor in feet = cell (B9), enter diameter in inches. Converted in worksheet to radius by (d/2) and inches to feet by (d/12).

NOTE: *The calculated inductance in cell (E21) uses the result of equation (7.37) in mH per mile and converts it as follows:*

$$Lt = \left(\frac{L}{5280} \bullet \text{Length}\right) \bullet 1000$$

Where: Lt = Total inductance of applied length of straight wire in μH = cell (E21), calculated value.
L = Inductance of straight wire in two wire system in mH per mile from equation (7.37).
Length = Total length of conductor in feet = cell (B14), enter value.

To calculate the capacitance of a straight wire:

$$C = \frac{0.03883}{\log_{10}(\frac{s}{r})} \times 10^{-6} \text{ Farad/mile} \tag{7.38}$$

Where: C = Capacitance of straight wire in two wire system in μF per mile = cell (E22), calculated value.
s = Spacing between centers of both conductors in two wire system in feet = cell (B16), enter value. If an actual ground (plane) is used for the return instead of a similar conductor, s equals twice the distance between the conductor to the ground (plane) return.
r = Radius of conductor in feet = cell (B9), enter diameter in inches. Converted in worksheet to radius by (d/2) and inches to feet by (d/12).

NOTE: *The calculated capacitance in cell (E22) uses the result of equation (7.38) in μF per mile and converts it as follows:*

$$Ct = \frac{C}{5280} \bullet \text{Length}$$

Where: Ct = Total Capacitance of applied length of straight wire in μF = cell (E22), calculated value.
C = Capacitance of straight wire in two wire system in μF per mile from equation (7.38).
Length = Total length of conductor in feet = cell (B14), enter value.

Equations (7.37) and (7.38) are used for calculating the inductance (cells E48 and E71) and capacitance (cells E49 and E72) of tubular and rectangular conductors respectively. Tables were included in the reference for calculated and measured inductance and capacitance (distributed effects) for various lengths of conductors. There was a good degree of correlation therefore the formulae should predict distributed inductance and capacitance with acceptable accuracy.

For comparison to a contemporary methodology (circa 1995) page 11-31 of reference (17) gives the following formula for distributed inductance in a solid copper conductor:

$$L(\mu H) = 0.0002Lg\left[\log_{10}\left(\frac{2Lg}{r}\right) - 0.75\right]$$

Where: L = Inductance of solid copper wire in μH = calculated value in cell (E141).
Lg = Length of conductor in mm = enter length in feet in cell (B138), cell (C138) converts to cm using × 2.54 multiplier and converted to mm using × 10 multiplier.
r = Radius of conductor in mm = enter AWG in cell (B136), diameter appears in cell (E135) from wire table, converts to cm in cell (F135) using × 2.54 multiplier and converted to mm using ×10 multiplier.

This methodology does not account for spacing from the return conductor, so exact correlation will not be obtained. The newer method calculates only about one half the actual inductance when using large diameter conductors such as used in Tesla coil interconnections. Page 11-19 of reference (17) gives the following formula for distributed (stray) capacitance in a circuit:

$$C(pF) = \frac{0.00885k\,A}{d}$$

Where: C = Distributed (stray) capacitance in pF = calculated value in cell (E142).
k = Dielectric constant relative to air = cell (B141), air = 1.0.
A = Plate area of conductor in mm^2 = [calculated diameter in cm in cell (F135) converted to mm using ×10 multiplier × calculated length in cm in cell (C138) converted to mm using ×10 multiplier] ^{2}.
d = Plate separation in mm = cell (B139), enter separation in feet, converted to cm in cell (C139) using × 2.54 multiplier and converted to mm using ×10 multiplier.

Again this methodology yields only one half the calculated distributed capacitance for larger diameter conductors. As the conductor diameter and spacing are reduced and the length increased, the two methods produce better correlation. At a 22AWG diameter with 0.1 feet of spacing and 40 feet of length the two methods produce about the same results. The older methodology is the best I have found and should be used to account for the larger diameter conductors used in Tesla coil construction.

Pages 43 through 46 of reference (9) provide formulae for calculating the effects of corona. Refer to Figure 7-22 for *solid* conductors. The same conductor with ground return is applicable. When the voltage between these two conductors reaches a critical value the air surrounding them breaks down and becomes conductive. Luminescent corona is produced along with power loss. To evaluate the effects of corona the voltage threshold (lowest voltage) where corona effects may start and known as the disruptive critical voltage (e_o) is calculated:

$$e_o = 2.303 g_o i A r \bullet \log_{10}\left(\frac{s}{r}\right) \tag{7.39}$$

Where: e_o = Disruptive critical voltage in kV = cell (E28), calculated value.
g_o = Disruptive gradient of air in kV = 53.5 kV per inch effective. This is an rms equivalent for an applied sine wave = 53.5 kV × 1.414 = 75.6 kV peak. This approximates the 75 kV/inch rule.
i = Conductor surface irregularity factor = cell (B20), enter value. For highly polished wires i is 1.0. For roughened or weathered wires it is in the range of 0.93 to 0.98. For stranded cables i is in the range of 0.83 to 0.87. Use the lowest i value for conservative results.
s = Spacing between centers of both conductors in two wire system in feet = cell (B16), enter value. If an actual ground (plane) is used for the return instead of a similar conductor, s equals twice the distance between the conductor to the ground (plane) return.

r = Radius of conductor in feet = cell (B9), enter diameter in inches. Converted to radius in worksheet by (d/2) and inches to feet by (d/12). A = Air density factor = cell (E24), calculated value from the equation below:

$$A = \frac{17.92BP}{459 + T_A}$$

Where: BP = Barometric pressure in inches = cell (B23), enter value.
T_A = Ambient temperature of air surrounding conductor = cell (B5), enter value in °C. Temperature in cell (C5) is cell (B5) converted to °F for use with this equation. The resistance calculations in equations (7.23) thru (7.36) use °C.

NOTE: *Air density factor (A) = 1.0 for an ambient temperature of 77°F (25°C) and a barometric pressure of 29.90 inches, which are standard laboratory conditions.*

The voltage threshold (lowest voltage) where visual corona effects start is the visual critical voltage (e_v) and calculated:

$$e_v = 2.303 g_v v r \bullet \log_{10}\left(\frac{s}{r}\right) \qquad (7.40)$$

Where: e_v = Visual critical voltage in kV = cell (E27), calculated value.
v = Visual factor = cell (B21), enter value. For wires v is in the range of 0.93 to 1.0. For cables with decided corona along the length of the conductor, v is 0.82. For cables with local corona along the length of the conductor, v is 0.72. Use the lowest v value for conservative results.
s = Spacing between centers of both conductors in two wire system in feet = cell (B16), enter value. If an actual ground (plane) is used for the return instead of a similar conductor, s equals twice the distance between the conductor to the ground (plane) return.
r = Radius of conductor in feet = cell (B9), enter diameter in inches. Converted to radius in worksheet by (d/2) and inches to feet by (d/12).
g_v = Apparent strength of air (surface gradient at rupture) in kV per inch effective = cell (E26), calculated value from the equation below:

$$g_v = g_o A\left(1 + \frac{0.189}{\sqrt{Ar}}\right)$$

Where: g_o = Disruptive gradient of air in kV = 53.5 kV per inch effective. This is an rms equivalent for an applied sine wave = 53.5 kV × 1.414 = 75.6 kV peak. This approximates the 75 kV/inch rule.

r = Radius of conductor in feet = cell (B9), enter diameter in inches. Converted to radius in worksheet by (d/2) and inches to feet by (d/12).

A = Air density factor = cell (E24), calculated value from the equation below:

$$A = \frac{17.92BP}{459 + T_A}$$

Where: BP = Barometric pressure in inches = cell (B23), enter value.

T_A = Ambient temperature of air surrounding conductor = cell (B5), enter value in °C. Temperature in cell (C5) is cell (B5) converted to °F for use with this equation. The resistance calculations in equations (7.23) thru (7.36) use °C.

NOTE: *Air density factor (A) = 1.0 for an ambient temperature of 77°F (25°C) and a barometric pressure of 29.90 inches, which are standard laboratory conditions.*

Theoretical corona losses will not occur if the voltage applied to the conductor and its ground return (neutral) is below the calculated visual critical voltage (e_v). Surface imperfections such as dirt, oxidation, and manufacturing variations will always lower the threshold (e_v) where this occurs. For applied voltages above e_v the resulting power losses from corona are calculated:

$$P = a(f + 25)(e - e_o)^2 \bullet 1 \times 10^{-5}\ \text{kW/mile} \tag{7.41}$$

Where: P = Power lost in conductor from corona in kW per mile.

f = Applied frequency = cell (B6), enter value in kHz. Converted to Hz in cell (C6).

e = Effective voltage (rms) applied to both conductors in two-wire system in kV = cell (B18), enter value in kV.

e_o = Disruptive critical voltage in kV = calculated value in cell (E28) from equation (7.39).

a = Calculated value from the equation below:

$$a = \frac{388}{A} \bullet \sqrt{\frac{r}{s}}$$

Where: r = Radius of conductor in feet = cell (B9), enter diameter in inches. Converted to radius in worksheet by (d/2) and inches to feet by (d/12).

s = Spacing between centers of both conductors in two wire system in feet = cell (B16), enter value. If an actual ground (plane) is used for the return instead of a similar conductor, s equals twice the distance between the conductor to the ground (plane) return.
A = Air density factor = cell (E24), calculated value from the equation below:

$$A = \frac{17.92BP}{459 + T_A}$$

Where: BP = Barometric pressure in inches = cell (B23), enter value.
T_A = Ambient temperature of air surrounding conductor = cell (B5), enter value in °C. Temperature in cell (C5) is cell (B5) converted to °F for use with this equation. The resistance calculations in equations (7.23) thru (7.36) use °C.

NOTE: *The calculated power dissipation in cell (E29) uses the result of equation (7.33) in kW per mile and converts it as follows:*

$$Pt = \frac{P(10^{-5}\,\text{kW/mile})}{5280} \bullet \text{Length}$$

Where: Pt = Power dissipation of applied length of straight wire in kW = cell (E29), calculated value.
P = Power lost in conductor from corona in kW per mile from equation (7.41).
Length = Total length of conductor in feet = cell (B14), enter value.

The corona calculations are most accurate with conductors of radius greater than 0.1 inch. Generally the conductors in the primary and line circuits will be of larger radius to transfer the power levels used in Tesla coils. Tables were included in the reference for calculated and measured power losses in various lengths of conductor. There was a good degree of correlation; therefore, the formulae should predict design performance with accuracy.

If the applied voltage produces corona and cannot be reduced, cover the conductor with an insulating sleeving such as clear plastic tubing. This will provide some additional dielectric strength and increase the apparent strength of the air surrounding the conductor. A better approach is to change the conductor's dimensions until a safer margin is calculated. Notice that increasing the spacing between the conductor and its ground return, the conductor diameter, or the irregularity factor (reduce the conductor surface imperfections) will increase the critical voltage threshold, thereby reducing or eliminating the power lost through corona. To minimize corona loss follow these design rules:

- Keep the conductors clean and polished.
- Avoid bends of sharp radius. The bend radius should be several times greater than the conductor diameter. Avoid sharp edges and points.
- Humidity, smoke, fog, increasing altitude (lower barometric pressure), increasing temperature and frequency will lower critical voltages and increase corona losses. Some of these may not be avoidable but other parameters can be varied to compensate.

So what is the best conductor shape to use? If doesn't matter as long as the conductor's dimensions exceed the depth of penetration at the resonant frequency. If tubular conductors are used, the conductor's wall thickness must be greater than the depth of penetration, or the resistance per foot will increase. If a rectangular conductor is used the calculated equivalent diameter (equation 7.35) must also be greater than the depth of penetration, or the resistance per foot will increase.

The skin effect in the tank circuit conductors (primary winding, tank capacitor, and spark gap) will be more pronounced than those in the control circuit and supply line. The oscillating frequency will be in the RF range of 50 kHz–500 kHz. Figure 7-23 shows the skin effect in a 0.25, 0.5, and 1.0 inch diameter copper (or other non-magnetic) conductor. Note that at 60 Hz the depth of penetration is 0.336 inch. At 100 kHz the depth of penetration is only 0.0082 inch. This means the current flows on the outer 0.0082 inch of the 0.25, 0.5, and 1.0 inch conductor. If a solid conductor is being used to make these connections, the inner 0.2418, 0.4918, and 0.9918 inch is unused material. If a tubular conductor such as copper tubing is used, there will be less wasted material. One might think 25 AWG magnet wire could be used to complete the connections since the current is only being carried in a 0.0165 inch diameter ($\delta \times 2$). But the resistance per foot is also affected. The resistance of one foot of 25 AWG magnet wire at 100 kHz is 17.6 mΩ, but the resistance of one foot of 1/2 inch copper tube at 100 kHz is only 0.63 mΩ. The increased resistance will dissipate more power and generate more heat, perhaps enough to melt a solder connection, insulation, or the conductor. The increased resistance will also drop the circuit voltage perhaps to inoperable levels.

High-voltage power transmission lines experience the same effect. At 60 Hz the depth of penetration is only 0.336 inch yielding a theoretical optimum conductor diameter of 0.672 inch. However, larger diameter cables also provide less resistance per foot. Weight and mechanical stresses are also a factor, but the end result is less than 50% of the power generated at the hydroelectric plant ever reaches the load. The rest is dissipated in the line's resistance and using massive conductor diameters provide a little compensation for the skin effect. Don't take these interconnections for granted as a lot of power can be dissipated in them from the increased resistance, leaving smaller sparks in the secondary.

7.11.2 Using RF Coaxial Cable

RF coaxial cable also works quite well for interconnections if it is large enough to carry the AC currents. Some terms used by manufacturers to rate RF coax cable performance are defined in the paragraphs that follow.

The velocity of propagation indicates how fast an electromagnetic wave will travel through a cable relative to propagation in free space (vacuum), which is 100%. An air dielectric is the fastest dielectric medium, but offers less dielectric strength than modern fillers. Typically, as

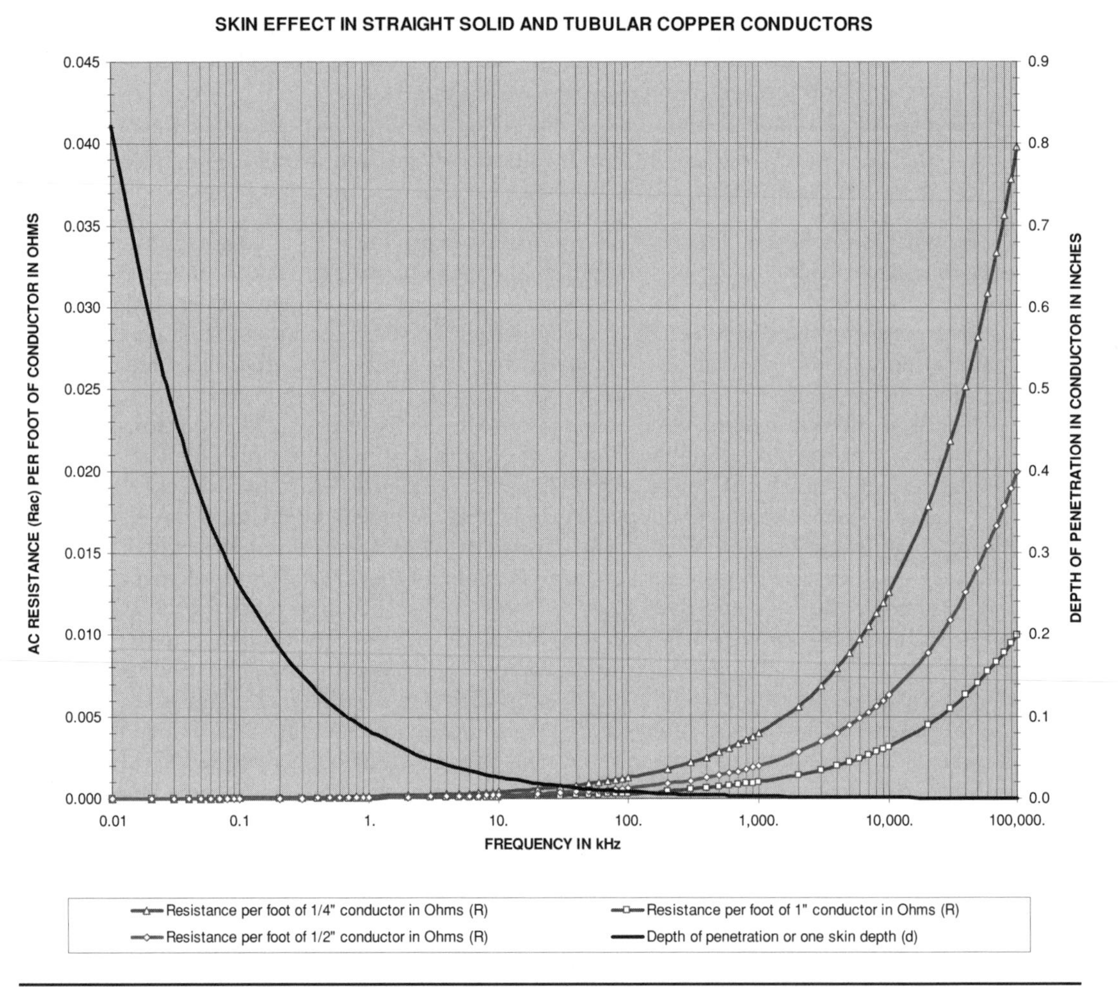

FIGURE 7-23 Typical skin effect in solid conductor.

the density of the dielectric material increases so does its dielectric strength, at the cost of propagation velocity. The velocity of propagation in RF coaxial cable is determined by the type of dielectric material used for the insulation:

$$v = \frac{c}{\sqrt{e}} \tag{7.42}$$

Where: v = Velocity of propagation of signal through RF coaxial cable.
c = Speed of light $= 3 \times 10^8$ m/sec $= 1.18 \times 10^{10}$ inches/sec.
e = Dielectric constant of insulation material from Table 7-4.

Dielectric Material	Dielectric Constant	Velocity of Propagation in % of Speed of Light
Air (atmospheric)	1.0	100
TFE (Teflon)	2.1	69
Polyethylene	2.3	66
Polyurethane	6.5	39
Polypropylene	2.25	67
Rubber	4.0	50
Rubber (silicone)	3.1	57

TABLE 7-4 Properties of insulating materials used in RF coaxial cables.

The characteristic impedance of RF coaxial cable is determined by the type of dielectric material used in the insulation and the geometry of the conductors:

$$Zo = \frac{138}{\sqrt{e}} \bullet \mathrm{Log}_{10}\left(\frac{D}{d}\right) \tag{7.43}$$

Where: Zo = Characteristic impedance of RF coaxial cable at specified frequency in ohms.
e = Dielectric constant of insulation material from Table 7-4.
D = Inside diameter of outer conductor. See Figure 7-24.
d = Outside diameter of inner conductor. See Figure 7-24.

Each type of Radio Grade (RG) cable has specific dimensions used in its construction. When the cable's inductance and capacitance per unit length are specified by the manufacturer a more efficient formula can be used to calculate the characteristic impedance:

$$Zo = \sqrt{\frac{L}{C}} \tag{7.44}$$

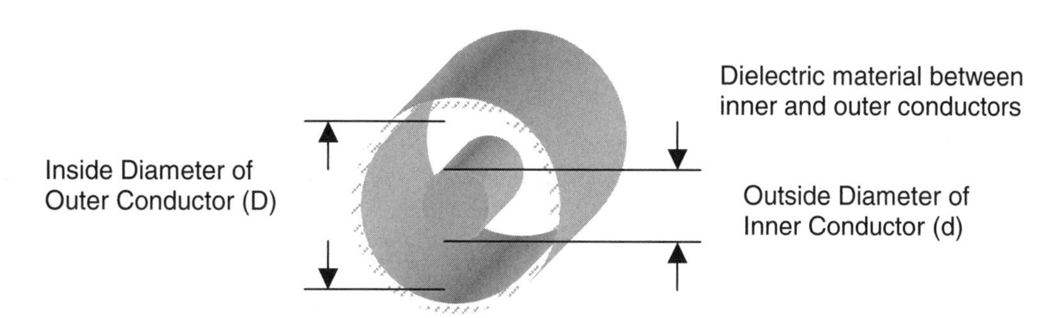

FIGURE 7-24 RF coaxial cable dimensions.

Where: Zo = Characteristic impedance of RF coaxial cable at specified frequency in ohms.
L = Inductance per unit length of cable from Table 7-5 or from formula below:

$$L = 0.140 \bullet \log_{10}\left(\frac{D}{sd}\right)$$

Where: L = Inductance per foot of cable in μH.
s = strand factor from number of strands used in center conductor found in the table below:

Number of Strands	s
1	1.000
7	0.939
19	0.970
37	0.980
61	0.985
91	0.988

D = Inside diameter of outer conductor. See Figure 7-24.
d = Outside diameter of inner conductor. See Figure 7-24.
C = Capacitance per unit length of cable from Table 7-5 or from formula below:

$$C = \frac{7.36e}{\log_{10}(\frac{D}{sd})}$$

Where: C = Capacitance per foot of cable in pF.
e = Dielectric constant of insulation material from Table 7-5.
D = Inside diameter of outer conductor. See Figure 7-24.
d = Outside diameter of inner conductor. See Figure 7-24.
s = strand factor from number of strands used in center conductor found in the table above.

Table 7-5 lists manufacturer's specifications for several types of Radio Grade RF coaxial cable. Cables may be labeled with the RG-x/U or MIL-C-17/x designations, which can be found in the table. Commercial manufacturers that do not conform to the military specifications necessary for RG or MIL-C-17 may still possess the electrical properties in the table as long as the geometries and dielectric are the same. The maximum specified operating voltage gives an indication of how well the cable is insulated. Only the larger diameter conductors suitable for coil building are included in the table.

Generally the coaxial cable will be used only as a one-wire conductor (center conductor) with high-strength insulation (dielectric). The braided outer conductor (shield) can be stripped off or left on but unconnected in the circuit. It is not critical to shield the interconnections from the RF oscillations in Tesla coils, as it will make a negligible difference in performance. Used in this manner the characteristic impedance no longer applies. The cable's impedance when only one conductor is used is equal to the AC resistance of the center conductor, namely the DC resistance plus the calculated skin effect as detailed in Section 7.11.1. The inductance and capacitance per foot will be equivalent to the calculated values in Section 7.11.1. Remember the objective is to fabricate a single high-voltage conductor, the coaxial shield is not needed as a return.

RG Number	Number of Strands in Center Conductor	Typical Outside Diameter of Center Conductor in inches (d)	Typical Inside Diameter of Outer Conductor in inches (D)	Typical Inductance in μH/foot (Calculated)	Typical Capacitance in pF/foot (Specified by Manufacturer)	Typical Impedance in ohms (Specified by Manufacturer)	Maximum Rated Operating Voltage RMS (Specified by Manufacturer)
RG-4/U (1)							
RG-58C/U (1)*	1	0.0320	0.116	0.0783	30.8	50	1,900
RG-5A/U (1)							
RG-212/U (1)*	1	0.0508	0.181	0.0773	30.8	50	3,000
RG-8(A)/U (1)							
RG-213/U (1)*							
RG-10(A)/U (1)							
RG-215/U (1)*	7	0.0855	0.285	0.077	29.5	52	4,000
							5,000 (A)
RG-9(A)/U (1)							
RG-214/U (1)*	7	0.0855	0.280	0.076	30.8	51	4,000
RG-11(A)/U (1)							
RG-12(A)/U (1)	7	0.0477	0.285	0.1125	20.6	75	4,000
							5,000 (A)
RG-13A/U (1)							
RG-216/U (1)*	7	0.0477	0.280	0.1114	20.8	52	4,000
RG-14/U (1)							
RG-217/U (1)*	1	0.1020	0.370	0.0783	29.5	52	5,500
RG-14A/U (1)	1	0.1020	0.370	0.0783	29.5	76	7,000
RG-15/U (1)	1	0.0571	0.370	0.1136	20.0	52	5,000
RG-16/U (1)	1	0.1250	0.460	0.0792	29.5	52	6,000
RG-17(A)/U (1)							
RG-218/U (1)*							
RG-18(A)/U (1)							
RG-219/U (1)*	1	0.1880	0.680	0.0782	29.5	52	11,000
RG-19(A)/U (1)							
RG-220/U (1)*							
RG-20(A)/U (1)							
RG-221/U (1)*	1	0.2500	0.910	0.0786	29.5	52	14,000
RG-25A/U (2)							
RG-26A/U (2)	19	0.0585	0.288	0.0988	50.0	48	10,000
RG-27A/U (2)							
RG-28A/U (2)	19	0.0925	0.455	0.0987	50.0	48	15,000
RG-29/U (1)							
RG-58/U (1)*	1	0.0320	0.116	0.0783	28.5	53.5	1,900
RG-55A/U (1)							

TABLE 7-5 Electrical properties of RF coaxial cables.

RG Number	Number of Strands in Center Conductor	Typical Outside Diameter of Center Conductor in inches (d)	Typical Inside Diameter of Outer Conductor in inches (D)	Typical Inductance in μH/foot (Calculated)	Typical Capacitance in pF/foot (Specified by Manufacturer)	Typical Impedance in ohms (Specified by Manufacturer)	Maximum Rated Operating Voltage RMS (Specified by Manufacturer)
RG-223/U (1)*	1	0.0350	0.116	0.0729	30.8	50	1,900
RG-59(A)/U (1)	1	0.0253	0.146	0.1066	21.0	73	2,300
RG-64/U (2)	19	0.0585	0.308	0.1029	60.0	48	10,000
RG-77(A)/U (2)							
RG-78(A)/U (2)							
RG-88(A)/U (2)	19	0.0585	0.288	0.0988	50.0	48	8,000 (peak)
							8,000 (peak)
							10,000
RG-115(A)/U (3)							
RG-225/U (3)*	7	0.0840	0.255	0.0713	29.4	50	5,000
RG-116/U (3)							
RG-227/U (3)*	7	0.0960	0.280	0.0689	29.4	50	5,000
RG-117(A)/U (3)							
RG-211A/U (3)*							
RG-118(A)/U (3)							
RG-228A/U (3)*	1	0.1880	0.620	0.0726	29.4	50	7,000
RG-119/U (3)							
RG-120/U (3)	1	0.1020	0.332	0.0718	29.4	50	6,000
RG-144/U (1)	7	0.0537	0.285	0.1053	19.5	75	5,000
RG-147/U (1)	1	0.2500	0.910	0.0786	29.5	52	14,000
RG-149/U (1)							
RG-150/U (1)	7	0.0480	0.285	0.1121	20.6	75	5,000
RG-156/U (1)	7	0.0855	0.285	0.0770	32.0	50	10,000
RG-157/U (1)	19	0.1005	0.455	0.0937	38.0	50	15,000
RG-158/U (1)	37	0.1988	0.455	0.0516	78.0	25	15,000
RG-164/U (1)	1	0.1045	0.680	0.1139	20.6	75	10,000
RG-165/U (1)							
RG-166/U (1)	7	0.0960	0.285	0.0700	29.4	50	5,000
RG-177/U (1)	1	0.1950	0.680	0.0759	30.8	50	11,000
RG-190/U (2)	19	0.0585	0.380	0.1156	50.0	50	15,000
RG-191/U (2)	1	0.4850	1.065	0.0478	85.0	25	15,000
RG-192/U (2)	1	1.0550	2.200 #	0.0447	175.0	12.5	15,000 (peak)
RG-193/U (2) Silicon							
RG-194/U (2) Silicon	1	1.0550	2.100 #	0.0419	159.0	12.5	30,000 (peak)

Source: Reference (13). (1) denotes solid polyethylene dielectric, (2) rubber, (3) PTFE.
* denotes comparable performance. # denotes outside diameter of entire cable.

TABLE 7-5 Electrical properties of RF coaxial cables. *(continued)*

If both the center conductor and shield are used as a transmission line the cable's resistance will become the typical impedance listed in column (7) of Table 7-5. This will be independent of length, in other words it will have the same impedance whether it is 1′ long or 1,000′ long.

I have used 5-foot sections of MIL-C-17/164 (RG-164/U) coaxial cable as an interconnection between the rotary spark gap and tank capacitor and the primary winding at 16.8 kV, drawing about 5 kW from the line. Although the coax cable was placed directly on the ground there was no loss of power or voltage arc over, illustrating how conservative the manufacturer's specifications are. This cable is designed to carry RF currents with minimum losses so you will be hard pressed to find a better substitute when making interconnections in the primary tank circuit in small and medium coils. It works especially well in tube coil construction. Large coils may exceed the current carrying capacity of the RF cable you find at Hamfests. I can't remember a Hamfest that didn't have at least one rolled-up bundle of coax cable for a few dollars.

7.12 Ground

The purpose of an electrical ground system is to maintain a return for current flow (conventional theory) and a zero voltage point in the circuit. Any electrical connection that is grounded will have zero volts potential difference with any other connection that has zero ohms of impedance to ground. As shown in Figure 7-25 the neutral bus bar in the circuit breaker or fuse panel is at ground potential, the same as the center tap of the high-voltage distribution transformer on the power grid. In this manner two 120-V circuits are formed as well as a 240-V circuit, all from a three-wire input.

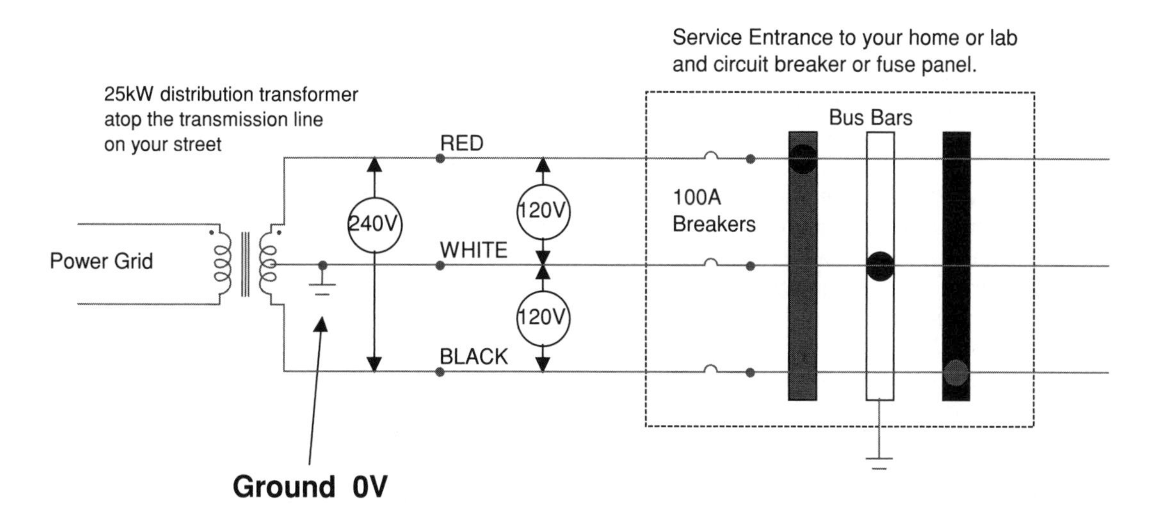

FIGURE 7-25 Line connections and ground in a 240-V service entrance.

7.12.1 Equipment Ground and Shielding

The control cabinet shown in Figure 2-1 should be made of metal, providing a shield for the controls and indicators. The chassis of the variable autotransformers and EMI filters are typically provided a connection to ground. Ground in this case is the metal control cabinet, which also should be grounded. Any *grounded* conducting material that completely surrounds a part will provide effective shielding against electrostatic (*E*) fields, known as a Faraday shield. Every ungrounded conductor in your lab forms a capacitor with every other ungrounded conductor using the air as a dielectric, which can store a charge.

I watched a convincing demonstration by Richard Hull of the Tesla Coil Builders of Richmond (TCBOR). Richard connected a picoammeter to a spherical capacitor of approximately 10″ diameter. The picoammeter would register any E field influence on the spherical capacitor. An amber rod was charged by friction and was inducing a deflection in the picoammeter when waved toward the capacitor, even 20 feet away! If you try this yourself and don't have a picoammeter, a gold leaf electroscope may be sensitive enough to pick up the charge in the capacitor.

The radiated electrostatic (*E*) and electromagnetic (*H*) fields are attenuated by the square of the distance from their source and will have little effect on electrical circuits outside your lab area, but don't neglect the effects of even a few volts of field strength on radio receivers, which are affected by μV fields. How well your coil is tuned will affect the level of radiated emissions interference.

But what of the H field? Faraday shields have little effect on strong electromagnetic fields and a large Tesla coil produces very strong H fields. The same skin effect, which lowers the Q in our coil windings at high frequencies, plays a large part in shielding from radiated H field effects. The high-frequency currents cannot penetrate very far into a shielding material where it is absorbed and unable to induce currents in your controls.

Reference (15) provides data on the effects different materials have in shielding from electromagnetic fields. Figure 7-26 shows the depth of penetration (δ) of the H field for a range of frequencies in three materials. The symbol δ should not be confused with the decrement symbol. In electronics many of the Greek symbols are used to represent several different terms and some familiarity with the subject is required to keep the intended representation separate. The field strength at δ is attenuated to 37% of its free air value.

If our control cabinet is made of 1/16″ steel (0.0625″ or 62.5 mils) a 60-Hz H field is attenuated by 63% before it gets halfway through it. The remainder will be attenuated another 63% as it travels through the other half of the 0.0625″ thickness for a total attenuation of about 86%. See Section 10.2 for assistance in exponential calculations. This may not seem like enough shielding, but remember we are not radiating 60-Hz fields of any appreciable strength. The strong H fields being radiated from the primary winding to the secondary winding at the higher frequency oscillations of the Tesla coil, say 100 kHz, will not penetrate more than halfway into our 0.0625″ thick control cabinet before they are fully attenuated. Expensive materials like Mumetal are not recommended. While these are effective at 60 Hz, they are less effective than steel, aluminum, or copper at 100 kHz. Figure 7-27 shows how Mumetal's attenuation of magnetic fields begins to fall off at 10 kHz, but does not entirely account for the effect.

Our 0.0625″ thick control cabinet will effectively shield our controls and indicators from the strong H fields of the primary winding even in close proximity to it. This same shielding method can be used to reduce radiated emissions from escaping your lab if you are operating

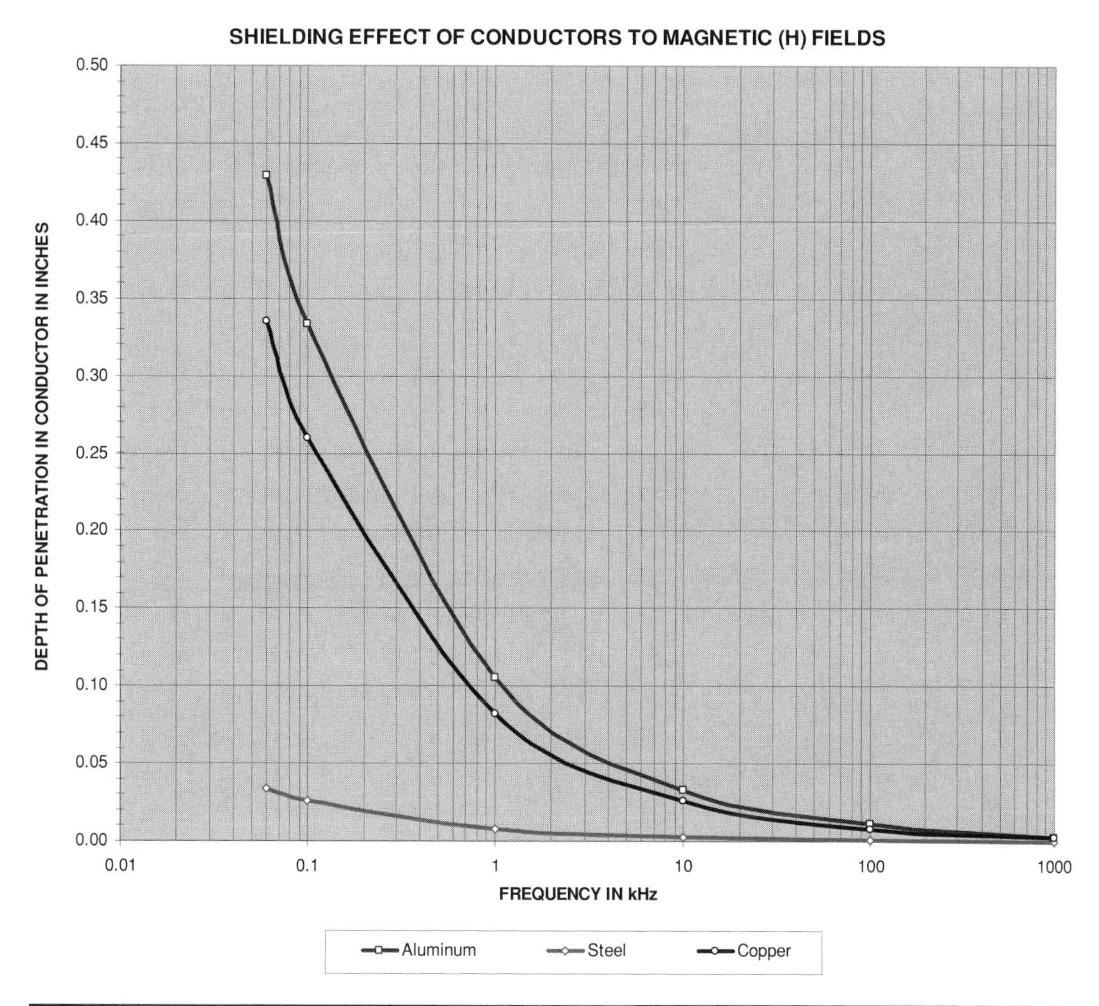

FIGURE 7-26 Shielding effect of conductors to magnetic fields.

a large Tesla coil and your neighbors are in close proximity of your lab. It should be obvious at this point not to build a control cabinet using wood if you are driving a large coil.

7.12.2 Tesla Coil (Secondary) Ground

For a medium or large Tesla coil a dedicated earth ground should be used in the secondary circuit. This means constructing a separate earth ground just for the coil, not just using existing telephone/electrical service grounds or water pipes. Follow the 240-V service entrance out of your house, back to the distribution transformer on its transmission line pole. A wire is usually running out of the transformer to a grounding rod at the bottom of the pole. This

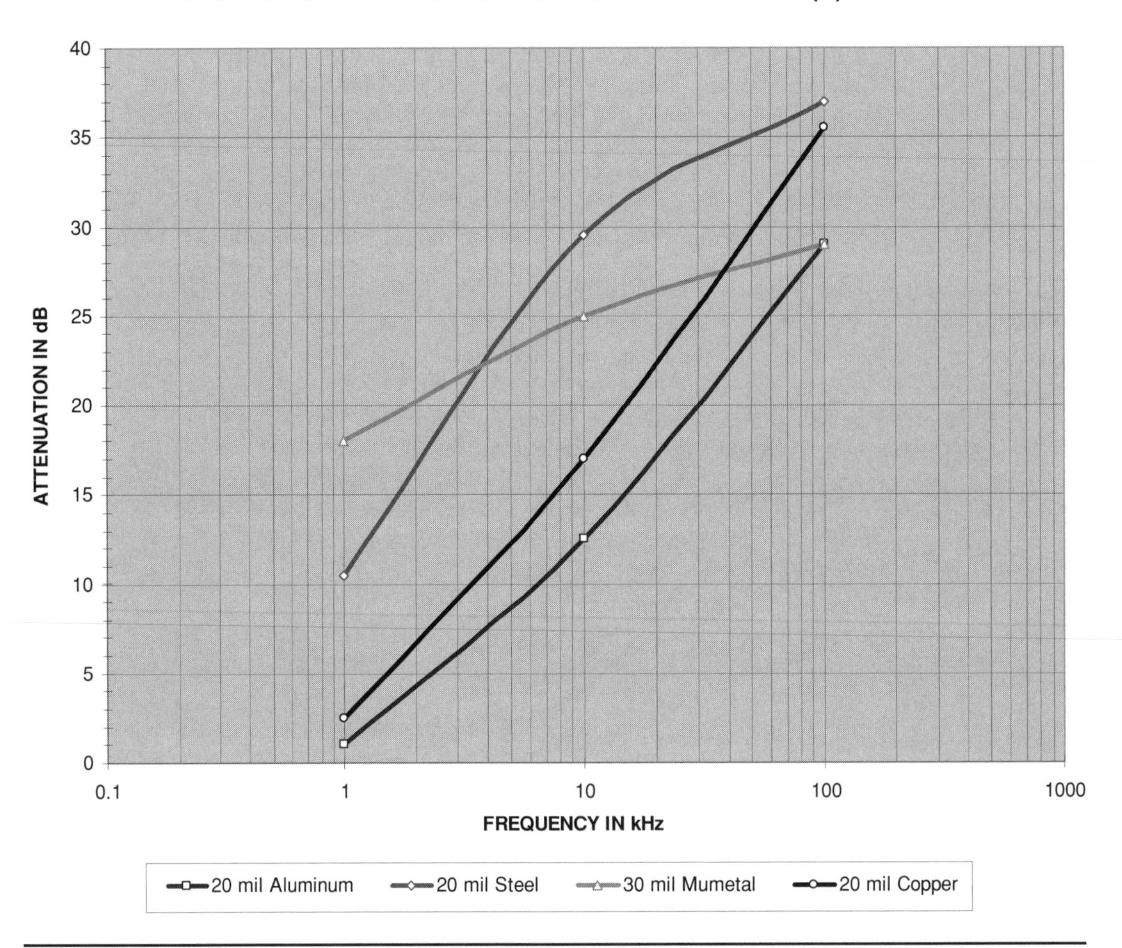

FIGURE 7-27 Attenuation of magnetic fields in conductors.

keeps the center tap of the transformer at 0 V. The dedicated Tesla ground will use this to keep the bottom of the secondary coil at 0 V with respect to the source of power (power grid).

Ground resistance data for various grounding schemes and conditions were found in reference (10) and presented in Tables 7-6 and 7-7, and Figures 7-29 thru 7-31. Notice in Table 7-6 that mixing the ground soil with fills reduces the resistance. Table 7-7 shows a marked increase in ground resistance during winter months as well as the effectiveness of using ground plates instead of ground rods (pipes).

The most economical approach is to use standard eight-foot copper-clad grounding rods or a galvanized or iron pipe. I prefer the 1/2″ copper clad grounding rods which provide approximately 35 Ω as seen in Figures 7-29 and 7-30. The ground rods should be driven into the

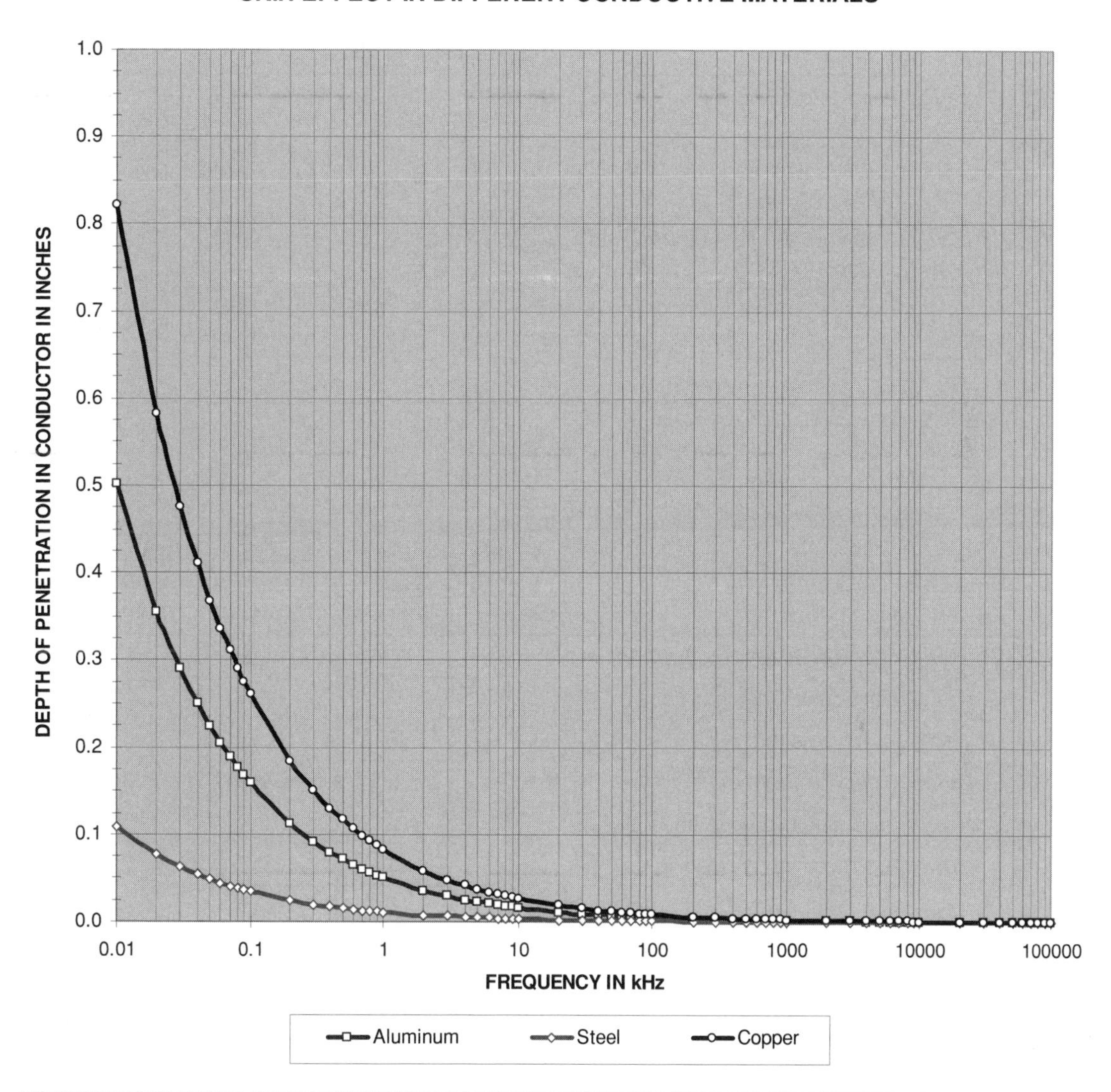

FIGURE 7-28 Skin effect of different conductor materials.

earth a short distance from where your coil is set up or a convenient distance if low-resistance, large-diameter cabling is used to connect the bottom of the coil to the grounding rod. I placed one of these just outside a garage window where a 00 gauge cable runs from the rod to the bottom of the secondary coil. The ground is more conductive if the earth is wet so I placed the rod just under where the water runs off the roof of the lab. Ninety percent of the resistance-to-ground is within a 6 foot radius of the rod.

Ground (Soil) Type	Number of Samples	Minimum Resistance in ohms	Average Resistance in ohms	Maximum Resistance in ohms
Fills and ground containing refuse (ashes, cinders, brine)	24	3.5	14	41
Clay, shale, adobe, gumbo, loam, and sandy loam (no gravel or stones)	205	2.0	24	98
Clay, adobe, gumbo, and loam mixed with sand, gravel, and stones	237	6.0	93	800
Sand, gravel, or stones with little or no clay or loam	72	35	554	2,700

TABLE 7-6 Ground resistance for type of soil.

Ground Type	Depth of Buried Ground in feet	Number of Samples	Average Resistance in ohms on August 31	Average Resistance in ohms on December 1
Paragon ground cones (2') in coke	5	2	39.5	62.2
Paragon ground cones (1') in coke	5	2	46.4	72.1
Maxum ground boxes in clay	5	3	33.2	57.4
Large L.S. Brach hydroground in clay	5	1	39.8	61.2
Standard L.S. Brach hydroground in clay	5	3	42.2	57.1
Medium L.S. Brach hydroground in clay	5	1	37.5	52.7
Disk type (Lord Mfg. Co.) hydroground in clay	5	2	76.1	124
Federal Sign System cartridge ground plate (22") in clay	5	2	42.6	71.9
Federal Sign System cartridge ground plate (10") in clay	5	2	47.3	80.6
2′ × 4′ plates on edge in clay	6	3	20.3	29.9
2′ × 4′ plates flat in clay	5	2	25.4	35.6
3/4″ galvanized-iron pipes	10	8	35.6	47.1

TABLE 7-7 Ground resistance for various types and conditions.

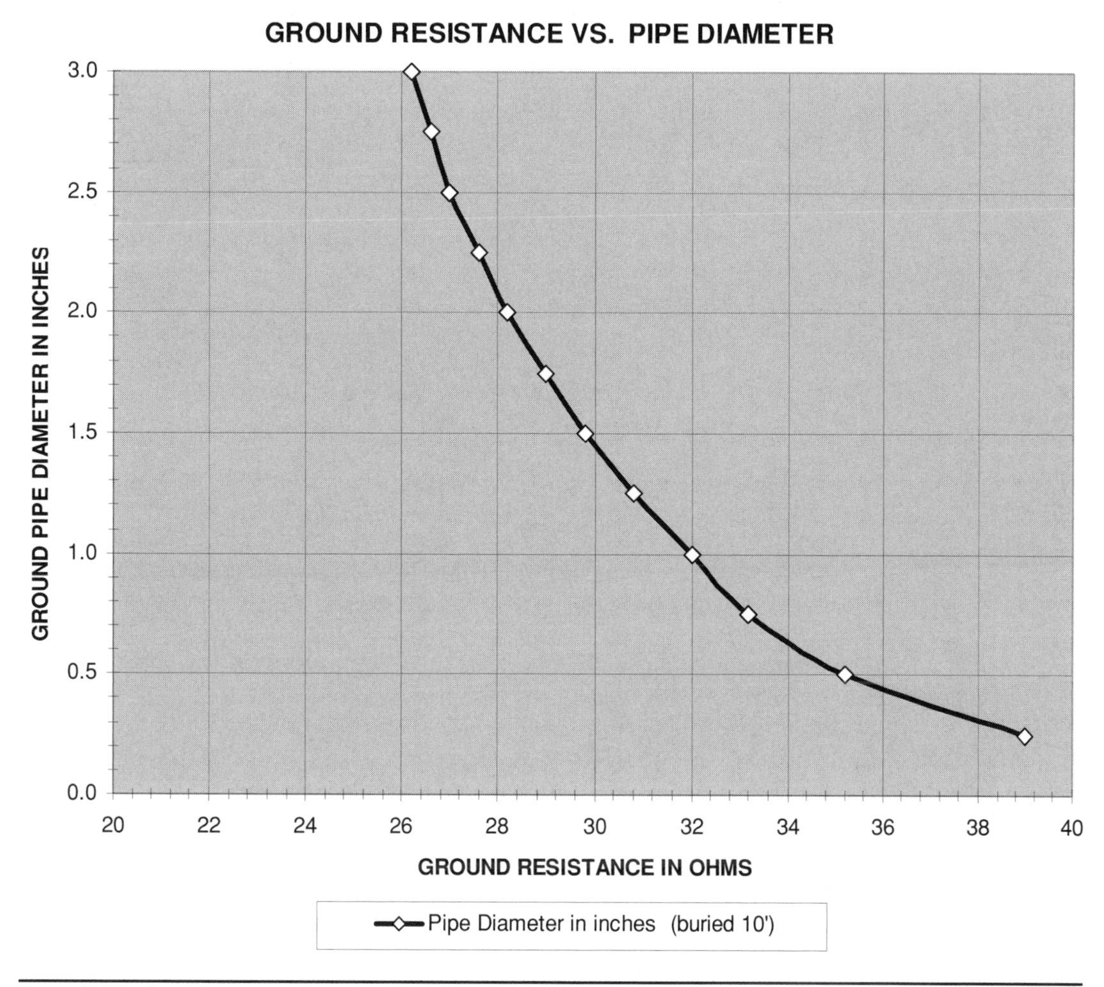

FIGURE 7-29 Resistance vs. diameter of pipe buried 10 feet in ground.

To decrease the ground resistance, add more rods in a semicircular or circular pattern with more than one foot of separation. As shown in Figure 7-31, the greater the separation between rods, the more effective the ground. Connect all rods together using heavy cable or similar conductor. The previously mentioned 35 Ω can be decreased to 3.5 Ω by adding 9 additional rods in a semicircular pattern from the house foundation with about 2 feet separation between rods. Adding even more rods can become rather excessive and you will be hard pressed to achieve a ground resistance of less than 1.0 Ω under even the best conditions. Tesla probably achieved a ground resistance <1.0 Ω in his elaborate Wardenclyffe experiment.

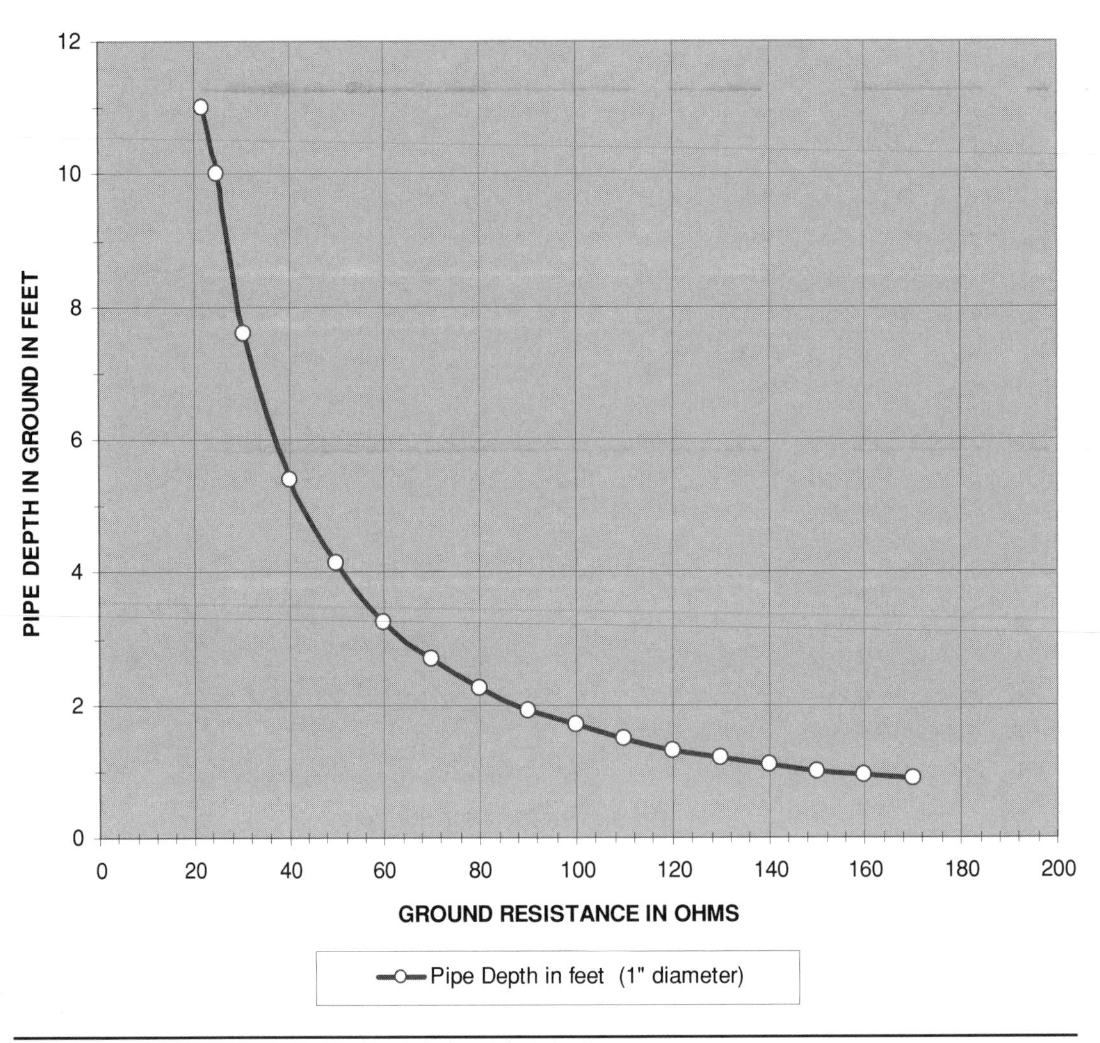

FIGURE 7-30 Resistance vs. depth of buried pipe 1 inch in diameter.

A copper sheet equivalent of three square feet in area can also be used if buried several feet in the earth. When attaching the cable to the copper sheet or rod, ensure all contacting surfaces are clean and tight. As the connection from the cable to the sheet is buried use some combination of oxidation inhibitor and sealant to keep the interconnection conductive. Salts or coke added to the soil will increase the ground conductivity but may also ruin your lawn

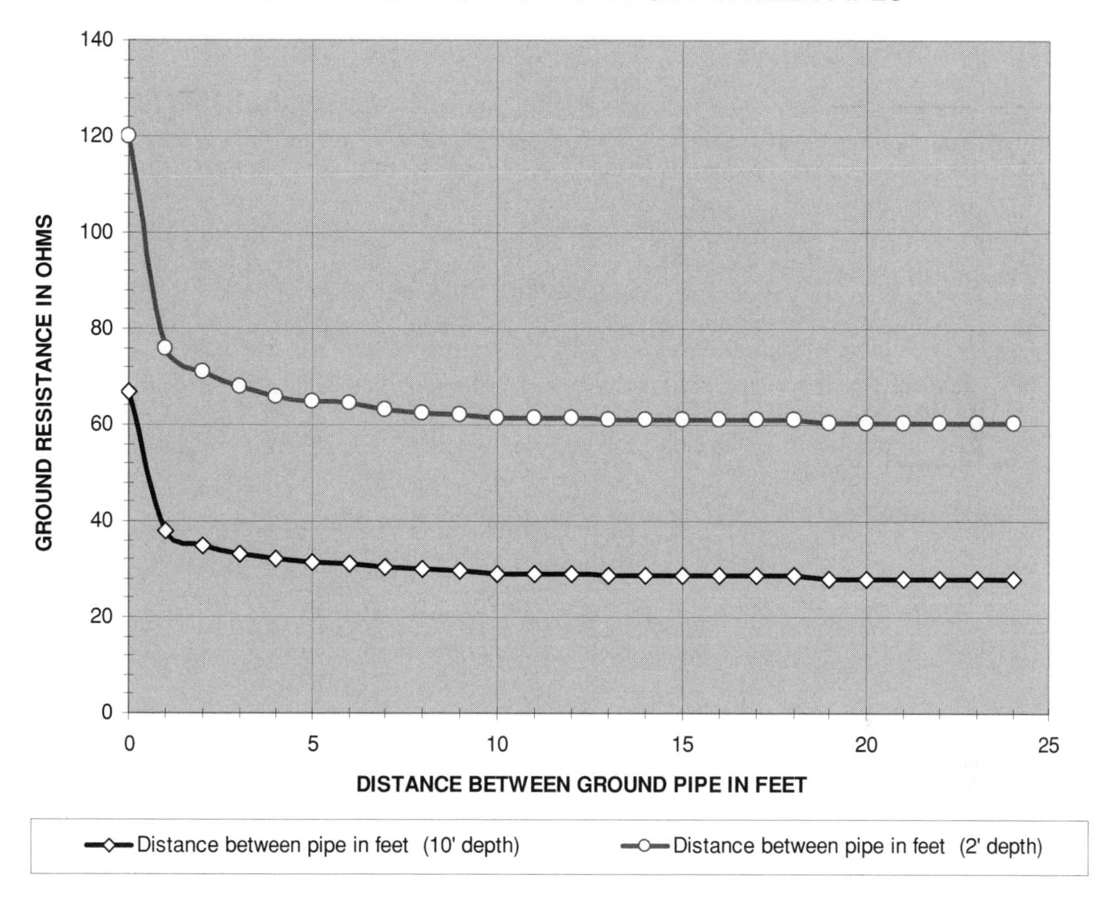

FIGURE 7-31 Resistance vs. distance between two buried pipes.

and could be prohibited by local law. I don't use them; however, Tesla found them necessary in his Colorado Springs experiments.

7.13 Fusing Current of Wire

The current required to melt a wire of given diameter is known as the fusing current. The fusing current data can be used to construct your own fuses or detect an unsafe condition in your conductors. Figures 7-32 thru 7-34 show the calculated fusing currents for selected wire diameters and materials.

The CH_7.xls file, FUSING CURRENT OF WIRE worksheet (5) uses the data provided in reference (16) for calculating the fusing current for selected wire material using W.H. Preece's

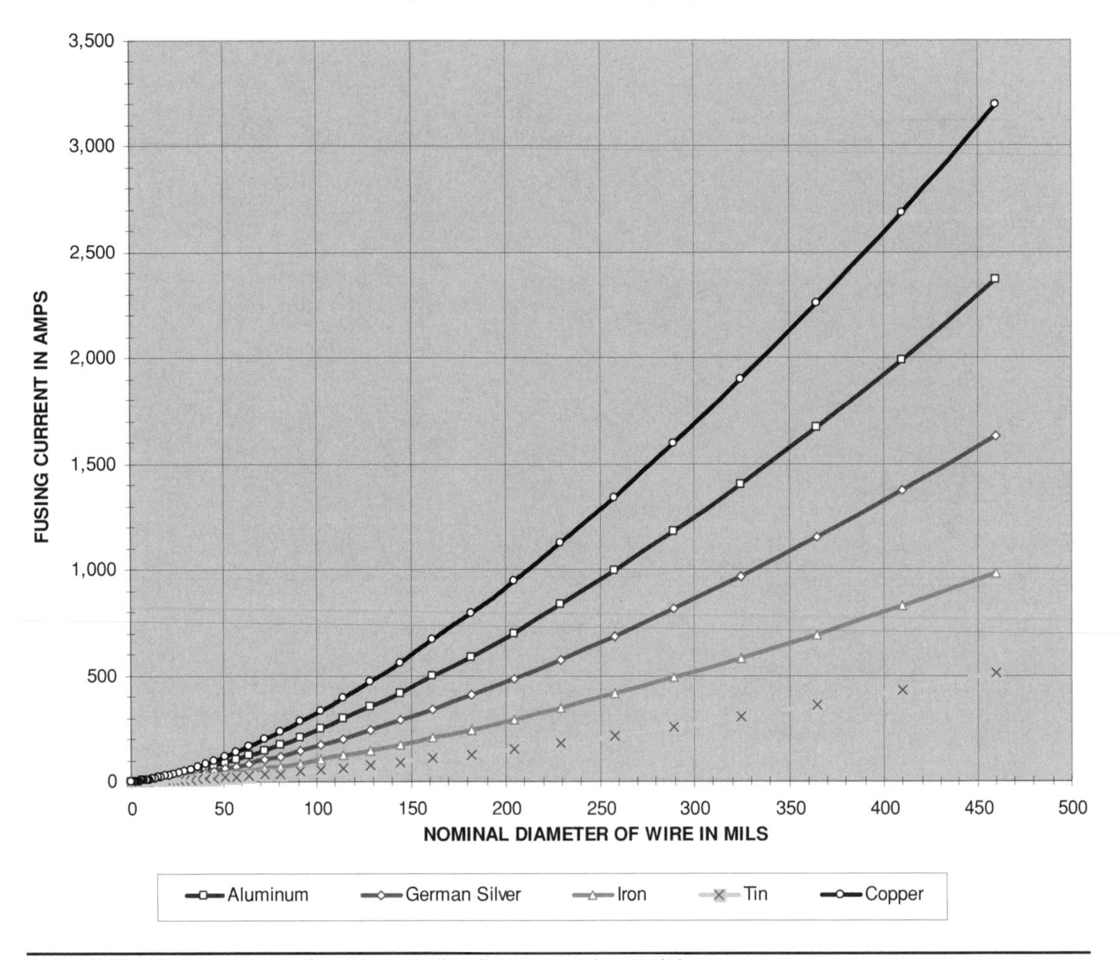

FIGURE 7-32 Fusing currents for selected wire diameter and material.

formula:

$$A = K \bullet d^{\frac{3}{2}} \tag{7.45}$$

Where: A = Fusing current of wire in amps = calculated value in cell (E5).
d = Outside diameter of wire = calculated value in cell (E4) for AWG value entered in cell (B4). See Section 4.7.
K = Constant for wire type entered in cell (B5) from table below:

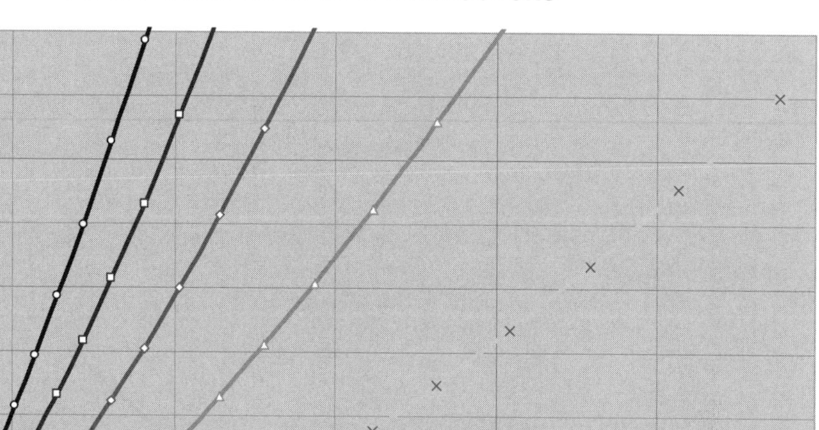

FIGURE 7-33 Fusing currents for 7 to 50 AWG wire of selected material.

Wire Material	K
Copper	10,244
Aluminum	7,585
German Silver	5,230
Iron	3,148
Tin	1,642

This can be taken one step further using M. Onderdonk's equation to calculate the time required to melt a conductor of specified area and current. The CH_7.xls file, FUSING CURRENT (Onderdonk) worksheet (12) uses Onderdonk's formula from reference (28) to calculate the fusing current and time for selected wire AWG:

$$If = A\sqrt{\frac{\text{Log}_{10}[1 + \frac{Tm-Ta}{234+Ta}]}{33t}} \tag{7.46}$$

Where: If = Fusing current of wire in amps for selected time duration and AWG = calculated value in cell (E5).

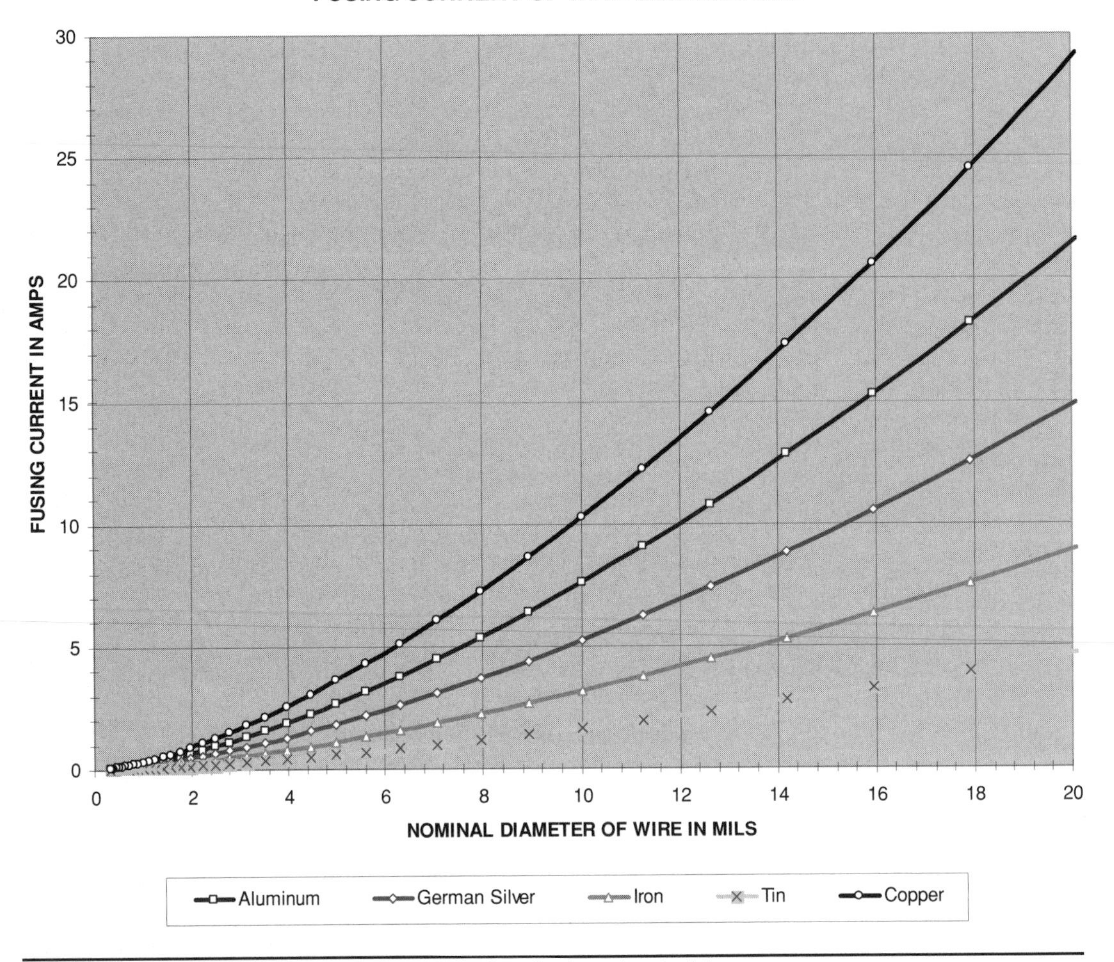

FIGURE 7-34 Fusing currents for 24 to 50 AWG wire of selected material.

A = Wire area in cirmils = calculated value in cell (E3) for AWG value entered in cell (B4). See Section 4.7.

Ta = Ambient temperature in °C = enter value in cell (B4).

Tm = Melting temperature of conductor (1083°C for Cu) = enter value in cell (B5).

t = time in seconds = enter value in cell (B6).

Shown in Figure 7-35 is the calculated fusing time vs. current for selected AWG using the Onderdonk formula. A 24 AWG copper wire requires 3 seconds to fuse at a current of 28 A as shown by the [X] marker in the figure. A fusing time of about 3 seconds is required to correlate Preece's formula with Onderdonk's solution.

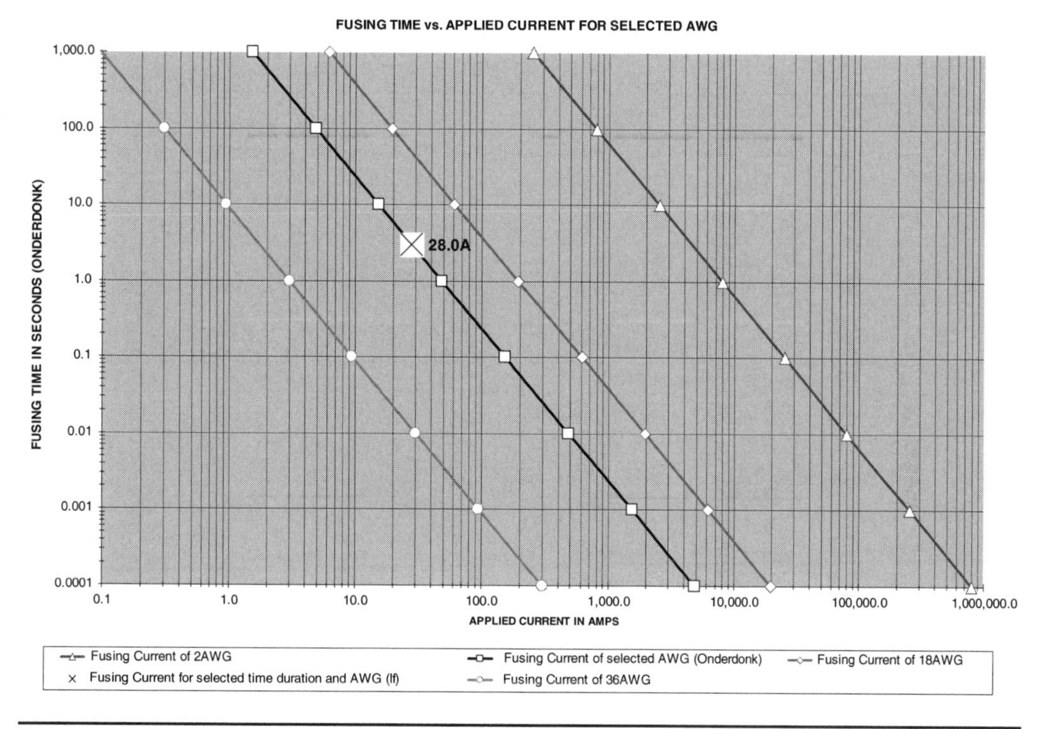

FIGURE 7-35 Fusing time vs. current for selected AWG (copper).

7.14 Safe Current Carrying Capacity of Wire

Table 7-8 lists the ampacity for wire AWG 0000 through 30 using four current standards. Note as the wire is increased in size by 3 AWG the area in circular mils is doubled, which also doubles the ampacity.

The current that a conductor can safely carry is less than the fusing current. Once the conductor is carrying the fusing current the insulation has already melted away and the load appears as a short circuit to the line. The maximum safe RMS (DC) current a wire can carry, known as ampacity, is dependent on the circular mil per amp standard. Some common industry standards are 1,000, 750, 500 and 250 cirmils/amp. The ampacity of wire using these standards for AWG 0000 through 30 are listed in Table 7-8. The National Electrical Code uses a standard comparable to 750 cirmils/amp or less for copper conductor. If you want to bulletproof your control interconnections use the 1,000 cirmils/amp standard but follow manufacturer recommendations for interconnections to protective devices (e.g., fuses and circuit breakers).

Using 2 AWG copper interconnections throughout the control cabinet and a 1,000 cirmils/amp standard would provide a safe current carrying capacity of 66.4 A and the system could provide a controllable output of 66.4 A × 120 V = 8 kW for a 120 V line or 66.4 A × 240

AWG	Area in Circular Mils	Nominal Diameter of Bare Wire (inches)	Ampacity for 1,000 Nominalcirmil/amp	Ampacity for 750 Cirmil/amp	Ampacity for 500 Cirmil/amp	Ampacity for 250 Cirmil/amp
0000	211,600	0.4600	211.6	282.1	423.2	846.4
000	167,806	0.4096	167.8	223.7	335.6	671.2
00	133,077	0.3648	133.1	177.4	266.2	532.3
0	105,535	0.3249	105.5	140.7	211.1	422.1
1	83,693	0.2893	83.7	111.6	167.4	334.8
2	66,371	0.2576	66.4	88.5	132.7	265.5
3	52,635	0.2294	52.6	70.2	105.3	210.5
4	41,741	0.2043	41.7	55.7	83.5	167.0
5	33,102	0.1819	33.1	44.1	66.2	132.4
6	26,251	0.1620	26.3	35.0	52.5	105.0
7	20,818	0.1443	20.8	27.8	41.6	83.3
8	16,510	0.1285	16.5	22.0	33.0	66.0
9	13,093	0.1144	13.1	17.5	26.2	52.4
10	10,383	0.1019	10.4	13.8	20.8	41.5
11	8,234	0.0907	8.2	11.0	16.5	32.9
12	6,530	0.0808	6.5	8.7	13.1	26.1
13	5,178	0.0720	5.2	6.9	10.4	20.7
14	4,107	0.0641	4.1	5.5	8.2	16.4
15	3,257	0.0571	3.3	4.3	6.5	13.0
16	2,583	0.0508	2.6	3.4	5.2	10.3
17	2,048	0.0453	2.0	2.7	4.1	8.2
18	1,624	0.0403	1.6	2.2	3.2	6.5
19	1,288	0.0359	1.3	1.7	2.6	5.2
20	1,022	0.0320	1.0	1.4	2.0	4.1
21	810	0.0285	0.81	1.08	1.62	3.24
22	642	0.0253	0.64	0.86	1.28	2.57
23	509	0.0226	0.51	0.68	1.02	2.04
24	404	0.0201	0.40	0.54	0.81	1.62
25	320	0.0179	0.32	0.43	0.64	1.28
26	254	0.0159	0.25	0.34	0.51	1.02
27	202	0.0142	0.20	0.27	0.40	0.81
28	160	0.0126	0.16	0.21	0.32	0.64
29	127	0.0113	0.13	0.17	0.25	0.51
30	101	0.0100	0.101	0.134	0.201	0.402

TABLE 7-8 Wire ampacity for selected AWG.

V = 16 kW for a 240-V line, provided our control components can handle these levels. The 2 AWG wire would not fuse until about 1,320 A is drawn through the load. This unsafe condition is avoided by incorporating circuit protection in the design as outlined in Section 7.10. For another comparison the same size wire is used in automobiles as the battery (+) and (−) interconnections and to the starter. On a very cold morning the starting system may have loads of

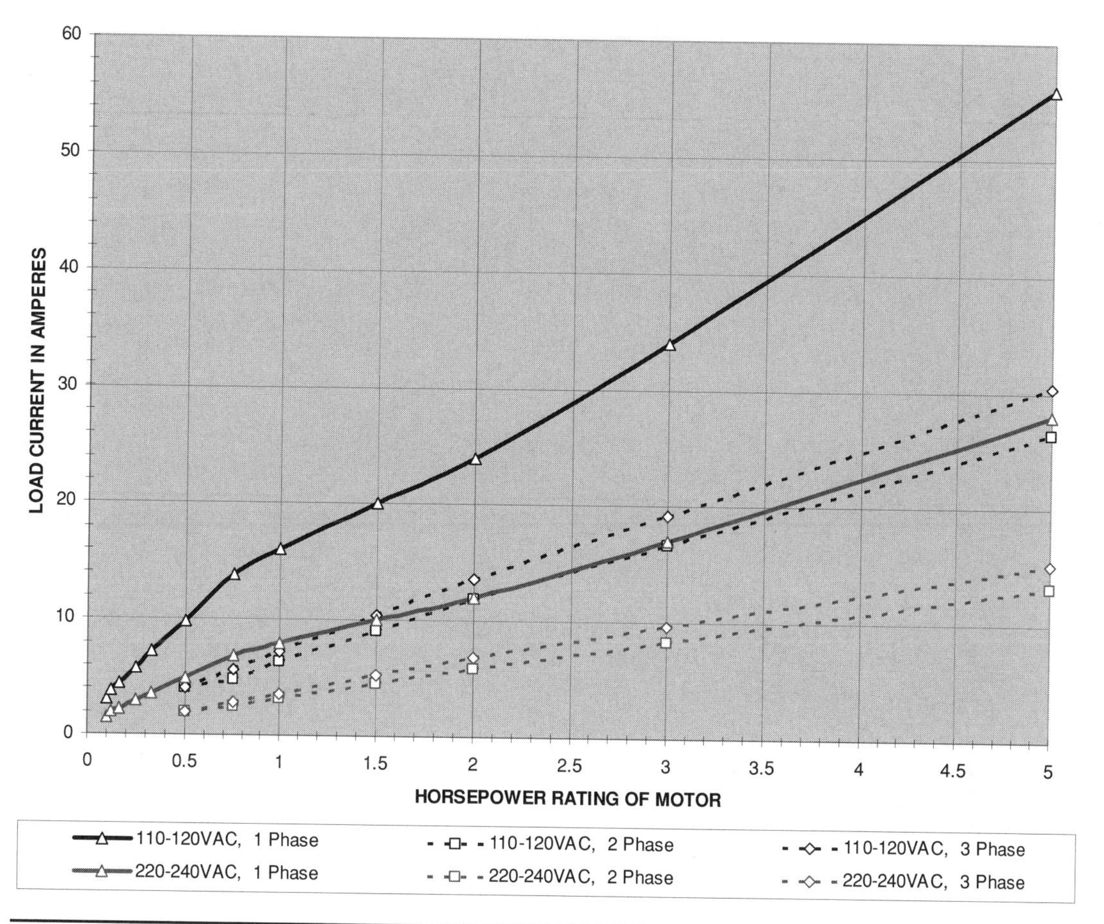

FIGURE 7-36 Typical full load current for AC motors.

several hundred amperes placed on it. If you feel the cable after a hard start the heating effects are only slightly perceptible.

7.15 Equipment Weight

The step-up and autotransformers increase in weight as the power handling ability increases. To estimate the weight involved at the power levels in your control design use the approximation below:

$$lb = P - P\eta \tag{7.47}$$

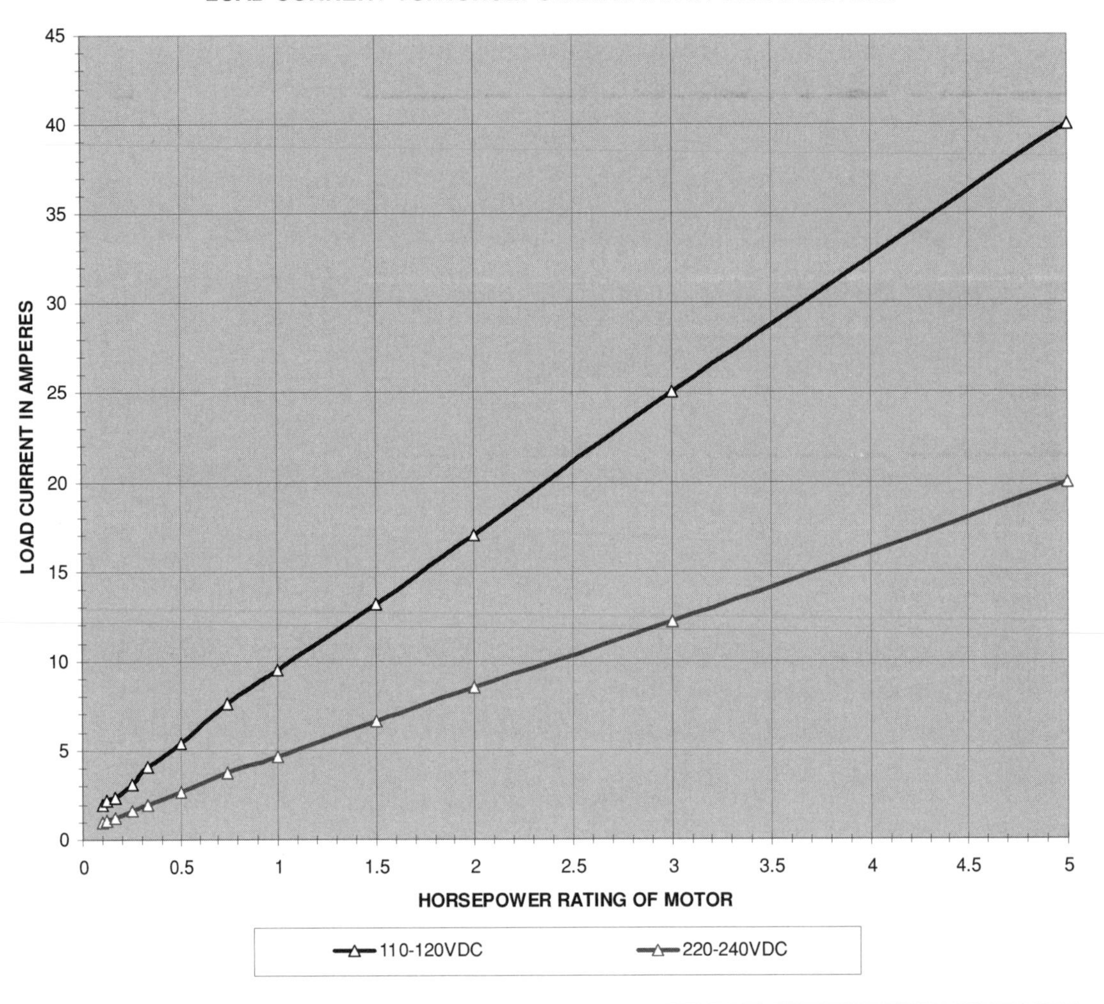

FIGURE 7-37 Typical full load current for DC motors.

Where: lb = Weight of transformer in pounds.

P = Power level of transformer in watts.

η = Typical efficiency coefficient of transformer. Page 280 of Reference (16) estimates silicone steel laminated transformer at 60 Hz yields a generic loss of 1 W/lb of core material. Estimating additional loss for copper weight the generic efficiency coefficient is listed in the table below:

Type Transformer	η
Power (isolation or step-up)	0.97
Power (autotransformer)	0.99

For example, I want a 3-kW high-voltage step-up transformer and a 240 V/25 A (6 kW) autotransformer in my design. The HV transformer would weight 3,000 W – (3,000 W × 0.97) = 90 lb. The autotransformer would weight 6,000 W – (6,000 W× 0.99) = 60 lb. This is a fair estimate when compared to equipment I have weighed. If a control cabinet is built to handle 6 kW with both 120 and 240-V variable voltage transformers the cabinet will gain 120 lbs just for the autotransformers. Adding current limiting will add another 60 to 90 lbs. The wire, indicators, protection circuits, filters, etc., will add more weight. As you can see it is easy to end up with a 200-lb cabinet, which is why a steel cabinet with castered wheels is almost essential for power levels above a few kilowatts. The tank circuit for a medium or large coil can be built on the same wheeled steel frame incorporating the step-up transformer, a high-voltage capacitor (perhaps oil filled), and a 1 HP rotary spark gap. If you are building a small coil with power levels under 1 kW the weight is probably not a concern.

7.16 Motor Current

If your coil contains a rotary spark gap there is an AC or DC motor driving it. The motor's load current must be included in your control and monitoring design. The motor you have selected for use in your rotary spark gap should have a data plate indicating the full load current. If the data plate is missing, the load current can be estimated for an AC motor using Figure 7-36, and for a DC motor using Figure 7-37. The figures were constructed using reference (23). Note the motor current increases fairly linearly as the horsepower is increased by a relative amount. The motor current also increases linearly as the line voltage is decreased by a proportional amount. During motor start-up these currents will typically increase 6 times the full load indicated in the figures until the nominal motor speed is reached and the start-up torque returns to a steady state value. Using a fractional horsepower (less than 1 HP) motor may result in the motor taking several seconds to start up in which time the line breakers or fuses may activate from the high start-up current. To avoid this design concern, use enough horsepower to drive the rotor to a start-up time of one second or less. A one horsepower motor is usually sufficient to drive most rotary gap designs.

References

1. Stafford H.E. *Troubles of Electrical Equipment: Their Symptoms, Causes, and Remedy*. McGraw-Hill Book Co., 1940. Chapter 13: Power Factor, pp. 263–280.
2. Terman F.E. *Radio Engineer's Handbook*. McGraw-Hill Book Co., 1940. Chapter 13: Power Factor pp. 263–280.
3. Magnetics, Inc. *Tape Wound Cores Design Manual*, TWC-400, pages 24–25.
4. Clement, P.R. and Johnson, W.C. *Electrical Engineering Science*. McGraw-Hill Book Co., 1960. Chapter 6: The Measurement of Electrical Quantities, pp. 198–226.
5. *Voltage Measuring Devices and Their Use*. 29FR-1. National Radio Institute. 1931.
6. Arc Welder information found on Lincoln Electric's website: www.lincolnelectric.com.
7. Skin Effect formulae from *Reference Data for Radio Engineers*. H.P. Westman, Editor. Federal Telephone and Radio Corporation (International Telephone and Telegraph Corporation), American Book. Fourth Ed: 1956, pp. 128–132.

8. ESD Association Standard *ANSI/ESD S20.20-1999* Electrostatic Discharge Association, Rome, NY: 1999.
9. Lewis, W.W. *Transmission Line Engineering*. McGraw-Hill: 1928, pp. 5–8.
10. Ibid, pp. 341–355.
11. RF coaxial cable information found on Madison Cable Corporation's website: www.madisoncable.com.
12. RF coaxial cable information found on General Cable's website: www.generalcable.com.
13. Amp Incorporated Product Catalog, Appendix A: Typical Coaxial Cable Specifications.
14. Baumeister, et. al. *Mark's Standard Handbook for Mechanical Engineers*. 8th ed. McGraw-Hill: 1987.
15. Alan Rich. *Shielding and Guarding: How To Exclude Interference Type Noise*. Best of Analog Dialog 17-1, 1983, Analog Devices, Inc., pp. 124–129.
16. *Reference Data For Radio Engineer's*. H.P. Westman, Editor. Federal Telephone and Radio Corporation (International Telephone and Telegraph Corporation), American Book. Fourth Ed: 1956. Page 55.
17. "Linear Design Seminar."Analog Devices, Inc.: 1995, pp. 11–17, 11–31.
18. Nadon, J.M., Gelmine, B.J. and McLaughlin, E.D. *Industrial Electricity*, Fifth Ed. Delmar Publishers, Inc: 1994.
19. Derived from Variac W10 autotransformer datasheet 3060-0140-F.
20. Circuit Breaker information found in GET-2779 Thermal Magnetic Application Guide on General Electric Industrial System's website: www.geindustrial.com/industrialsystems/circuitbreakers/notes/GET-2779J.pdf.
21. Fuse information found in online technical library at Cooper Bussmann's website: www.bussmann.com/library/.
22. Tyco Electronics Corporation, P&B division application note IH/12-00: "Relay Contact Life."
23. *Engineer's Relay Handbook*, Fifth Ed. National Association of Relay Manufacturers: 1996, pp. 17–15, 17–16. Data from UL STD 508.
24. H. Pender, W.A. Del Mar. *Electrical Engineer's Handbook*, Third Ed. John Wiley & Sons, Inc. 1936. Magnetic Materials, pp. 2–38 to 2–49.
25. R. W. Landee, D. C. Davis and A. P. Albrecht. *Electronic Engineer's Handbook*. McGraw-Hill: 1957, pp. 14–15 to 14–34.
26. Ohmite Manufacturing Company application note and datasheet for power resistors and rheostats found in product catalog.
27. David O. Woodbury. *A Measure for Greatness: A Short Biography of Edward Weston*. McGraw-Hill: 1949.
28. *Fusing Current: When Traces Melt Without a Trace*. By Douglas Brooks. Printed Circuit Design, a Miller Freeman publication, December, 1998. Cites source as *Standard Handbook for Electrical Engineers*. 12th Ed. McGraw-Hill, p. 4–74.

CHAPTER 8

Using Computer Simulation to Verify Coil Design

One of the few practical uses I have for computers is simulating design changes before spending the time and money to build them. Spice is the term used to describe the computer program used to perform analog electronic circuit simulations. A Spice model is a schematic representation of the analog circuit to be analyzed by the Spice program. If you are experienced with Spice the models will present no challenge and the circuits will aid in understanding the operation of Tesla coil circuits. If you have no Spice experience this chapter will expose you to it with easy-to-run circuit models. The calculations in the preceding chapters are actually more accurate for design and performance evaluation; however, an intuitive understanding of circuit operation is developed using the models and they generally approximate the calculations. The work is already done so you might as well try it.

8.1 Using Spice-Based Circuit Simulation Programs

Spice models were created that run in Spectrum Software's Micro-Cap 9 program. You can download an evaluation (student) version of this program for free that will run the Spice models shown in this chapter. To get the free program and run the Spice models in this chapter, do the following:

1. Access the Spectrum Soft website @ http://www.spectrum-soft.com/.
2. Select the Demo (demonstration) option. Fill out the demo request form registration. Select: Submit. Spectrum Software will send a download link to the email address provided in the registration form.
3. Open the download link "Download the Working Demo/Student Version - 7.3M" from the email received from Spectrum Software. Your computer should prompt you to save as a file. Save the download as demo.zip in a folder of your choice (creating a new folder to contain the unzipped files is recommended). Remember the folder name.
4. After downloading access the folder containing the demo.zip file. Unzip (PKUNZIP) this file using WinZip or other utility program. Run SETUP.EXE or the Setup Application file from the unzipped file if the unzip utility does not automatically install MC9

(if a new folder was created in the previous step to contain the unzipped files this will be easier to find). MC9 is compatible with any modern Windows operating system including Windows 95, 98 and Windows NT 4.0 (2000 and XP) compatible. The demo is limited as follows: circuit size limited to 50 components (all layers), anywhere from 0 to 300% slower than the professional version, limited component library, no Model program, and some of the advanced features are not available. The circuits shown in this chapter will run in the MC9 demo version.

5. Download the following PDF files from the Support→Manuals option: MC9 Users Guide (4.6M) and MC9 Reference Guide (7.3M). This will give a detailed description of how Micro-Cap works, how to construct circuits and how to run DC, AC, and transient analysis. The help function in the program is also a good source of information.
6. Open the Micro-Cap 9 (MC9) program.
7. In the MC9 File→Open option, open the TCOIL SG (MC9).CIR file from the appropriate drive e.g., C or D. You should open the file from the MC9 program. Your operating system may not recognize the .CIR file or may open it in another program, which will not run. You may also be able to place the mouse on the file and click the right button and select the open with→MC9 option, which will open the file in Micro-Cap 9.
8. After the file is opened in MC9 it should look similar to the circuit in Figure 8-1 with one small exception. The primary inductance is modeled differently and the appearance is slightly altered to that in the figure (this is detailed in the "Set the primary inductance" paragraph of section 8.2). This difference is the result of the new demo upgrade from MC8 to MC9 by Spectrum Software. The TCOIL SG (MC9).CIR circuit's primary inductance was modified to run in MC9. The circuit shown in Figure 8-1 is from the TCOIL SG.CIR file and was constructed to run in MC8 and earlier versions which is retained for users who have the earlier software. If the file appears similar to the figure select the Analysis→Transient option, the Transient Analysis Limits window appears. Select RUN. The program will now run the analysis and display the waveforms shown in Figures 8-2, 8-12 and 8-13.
9. If the TCOIL SG (MC9).CIR file doesn't run, familiarize yourself with the MC9 program using the downloaded PDF files from step 5. Build the circuit model shown in Figure 8-1 and run a transient analysis with the time (X-axis) and voltage (Y-axis) limits shown in Figure 8-2. The circuit shown in Figure 8-1 can also be built in other Spice programs such as P-Spice.

NOTE: *Spectrum-Software frequently upgrades their Micro-Cap product to a new version and makes changes to their website. The above steps were current when this design guide went to print. The above steps may need slight modification as the upgrades and website changes occur.*

8.2 Using the Circuit Simulation Program to Design a Spark Gap Tesla Coil

Once the TCOIL SG (MC9).CIR file is opened and a transient analysis produces the waveforms shown in Figure 8-2, change the circuit values as shown in Figures 8-3 thru 8-13 and observe

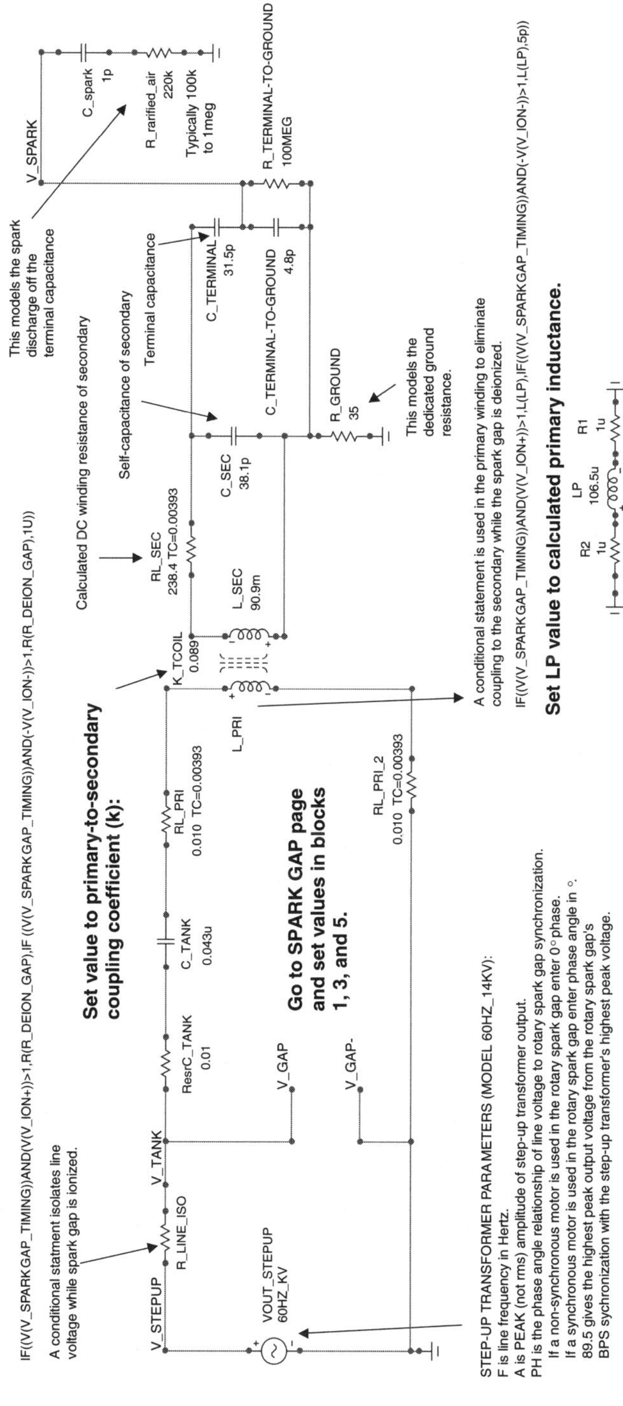

Set R_GROUND value to calculated ground resistance.
Set L_SEC value to calculated inductance.
Set C_SEC value to calculated self-capacitance.
Set RL_SEC value to calculated secondary resistance (with skin and proximity effects).
Set C_TERMINAL and C_TERMINAL-TO-GROUND values to calculated capacitance.
Set C_TANK value to selected capacitance.
Set RL_PRI and RL_PRI_2 values to half the calculated primary winding resistance (split to improve convergence).

FIGURE 8-1 Micro-Cap simulation model for a spark gap Tesla coil.

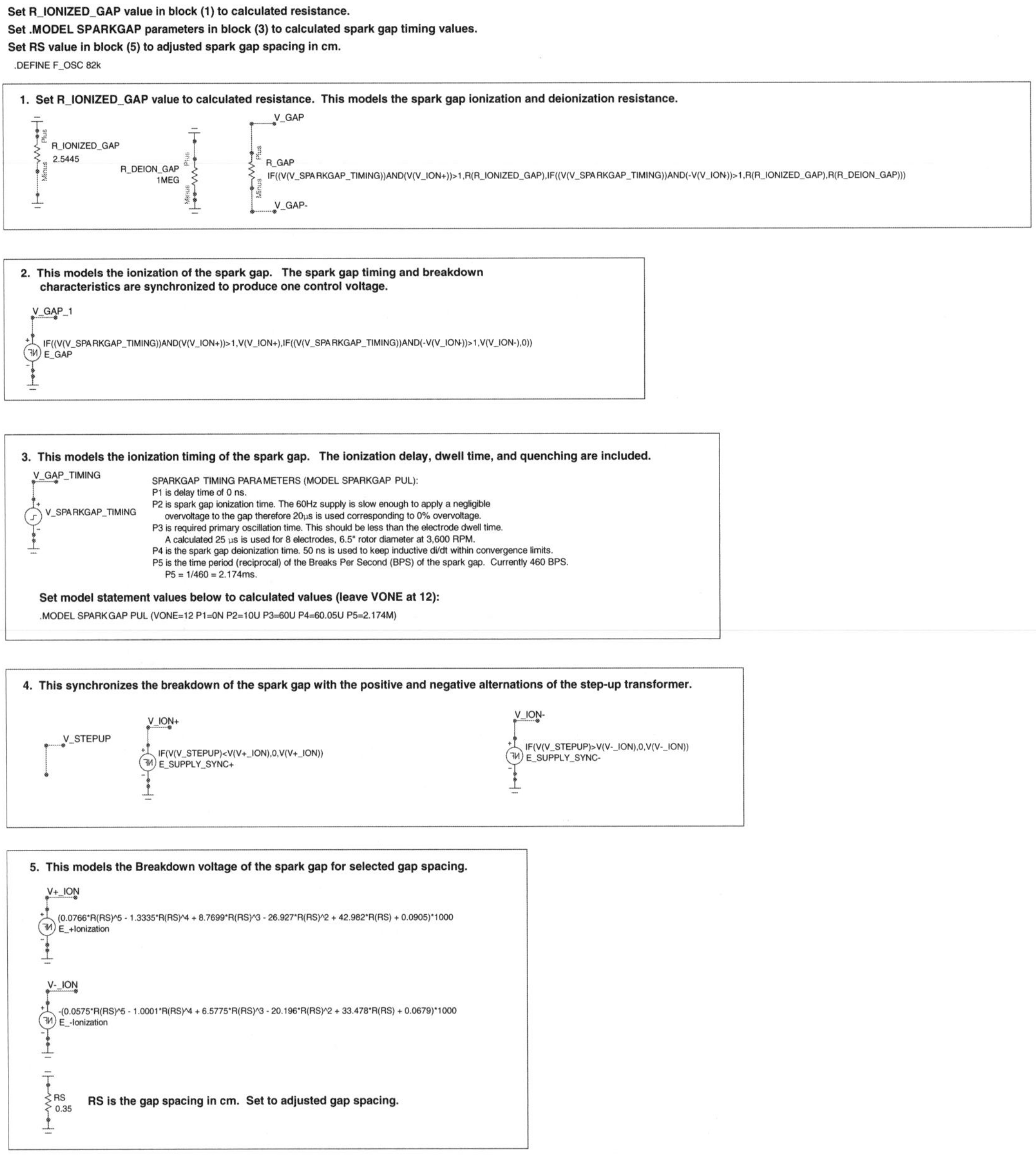

Set R_IONIZED_GAP value in block (1) to calculated resistance.
Set .MODEL SPARKGAP parameters in block (3) to calculated spark gap timing values.
Set RS value in block (5) to adjusted spark gap spacing in cm.

.DEFINE F_OSC 82k

1. Set R_IONIZED_GAP value to calculated resistance. This models the spark gap ionization and deionization resistance.

V_GAP
Plus
R_IONIZED_GAP
2.5445
Minus
R_DEION_GAP
1MEG
Plus
Minus
R_GAP
IF((V(V_SPARKGAP_TIMING))AND(V(V_ION+))>1,R(R_IONIZED_GAP),IF((V(V_SPARKGAP_TIMING))AND(-V(V_ION-))>1,R(R_IONIZED_GAP),R(R_DEION_GAP)))
V_GAP-

2. This models the ionization of the spark gap. The spark gap timing and breakdown characteristics are synchronized to produce one control voltage.

V_GAP_1
IF((V(V_SPARKGAP_TIMING))AND(V(V_ION+))>1,V(V_ION+),IF((V(V_SPARKGAP_TIMING))AND(-V(V_ION-))>1,V(V_ION-),0))
E_GAP

3. This models the ionization timing of the spark gap. The ionization delay, dwell time, and quenching are included.

V_GAP_TIMING
V_SPARKGAP_TIMING

SPARKGAP TIMING PARAMETERS (MODEL SPARKGAP PUL):
P1 is delay time of 0 ns.
P2 is spark gap ionization time. The 60Hz supply is slow enough to apply a negligible overvoltage to the gap therefore 20μs is used corresponding to 0% overvoltage.
P3 is required primary oscillation time. This should be less than the electrode dwell time. A calculated 25 μs is used for 8 electrodes, 6.5" rotor diameter at 3,600 RPM.
P4 is the spark gap deionization time. 50 ns is used to keep inductive di/dt within convergence limits.
P5 is the time period (reciprocal) of the Breaks Per Second (BPS) of the spark gap. Currently 460 BPS. P5 = 1/460 = 2.174ms.

Set model statement values below to calculated values (leave VONE at 12):

.MODEL SPARKGAP PUL (VONE=12 P1=0N P2=10U P3=60U P4=60.05U P5=2.174M)

4. This synchronizes the breakdown of the spark gap with the positive and negative alternations of the step-up transformer.

V_STEPUP

V_ION+
IF(V(V_STEPUP)<V(V+_ION),0,V(V+_ION))
E_SUPPLY_SYNC+

V_ION-
IF(V(V_STEPUP)>V(V-_ION),0,V(V-_ION))
E_SUPPLY_SYNC-

5. This models the Breakdown voltage of the spark gap for selected gap spacing.

V+_ION
(0.0766*R(RS)^5 - 1.3335*R(RS)^4 + 8.7699*R(RS)^3 - 26.927*R(RS)^2 + 42.982*R(RS) + 0.0905)*1000
E_+Ionization

V-_ION
-(0.0575*R(RS)^5 - 1.0001*R(RS)^4 + 6.5775*R(RS)^3 - 20.196*R(RS)^2 + 33.478*R(RS) + 0.0679)*1000
E_-Ionization

RS
0.35
RS is the gap spacing in cm. Set to adjusted gap spacing.

FIGURE 8-1 Micro-Cap simulation model for a spark gap Tesla coil. *(continued)*

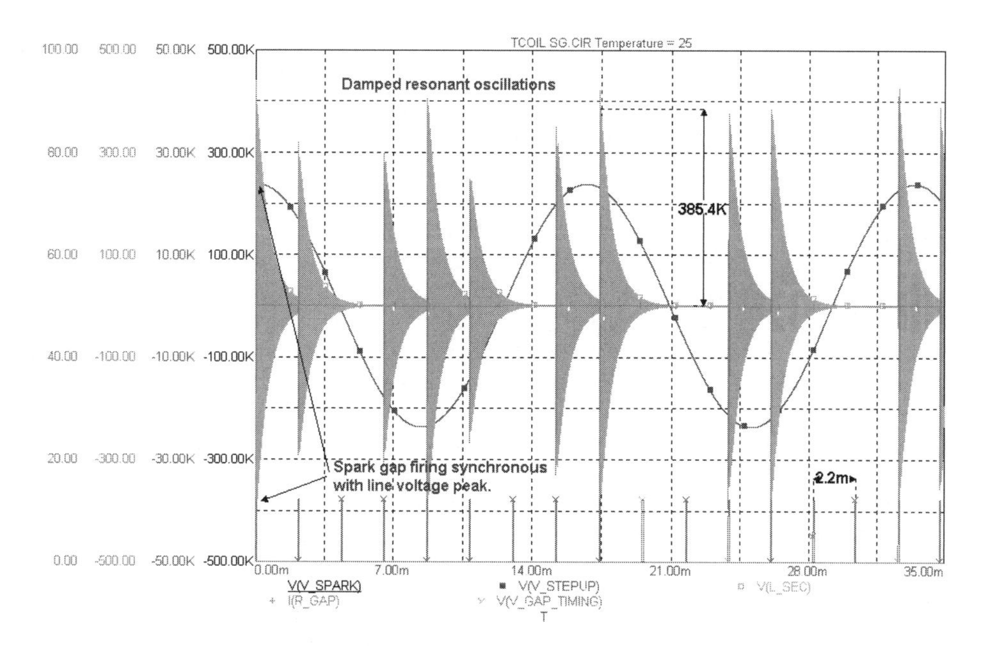

FIGURE 8-2 Simulation model waveforms for a spark gap Tesla coil ($k = 0.089$, terminal capacitance = 31.5 pF, BPS = 460, $\theta = 89.5°$).

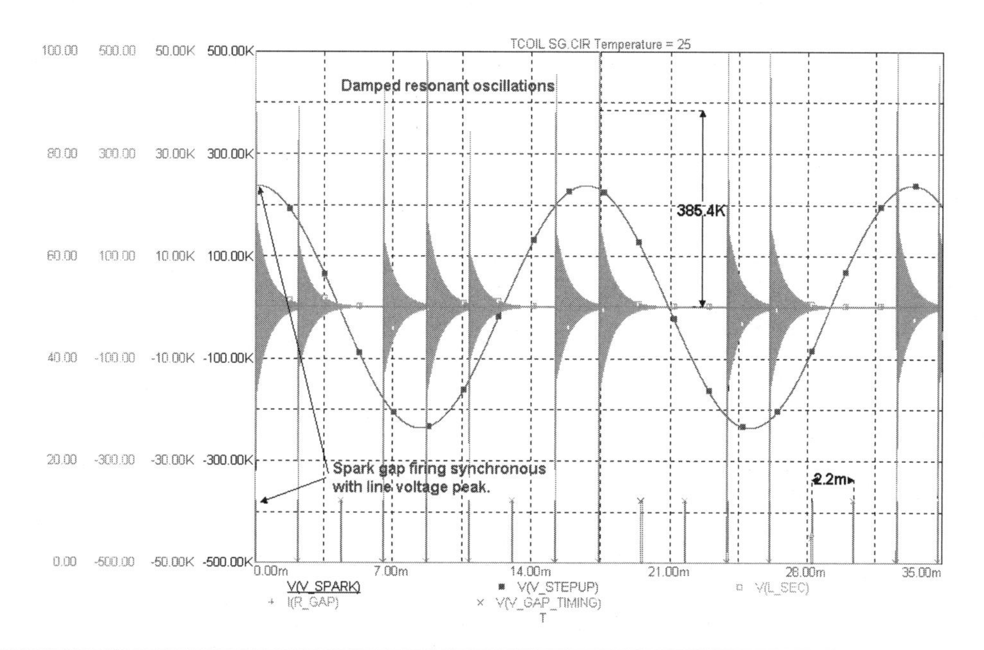

FIGURE 8-3 Simulation model waveforms for a spark gap Tesla coil ($k = 0.2$, terminal capacitance = 31.5 pF, BPS = 460, $\theta = 89.5°$).

the results as shown in the figures. This is how the circuit simulation can be used to model the coil before it is built. The results may not be dynamic enough to exactly replicate a Tesla coil's performance, but it is close enough for design verification. Circuit values in the primary and secondary can be changed and the results quickly analyzed. The notes in the file and Figure 8-1 provide information on how the model operates and deriving circuit values not covered in Chapter 2. The analysis results shown in Figures 8-2 thru 8-13 were generated using the older TCOIL SG.CIR file for MC8 and earlier versions. There will be slight differences between the MC8 results shown in the figures and those generated using the newer MC9 version TCOIL SG (MC9).CIR file. When the file is run without changes four graphs are generated in the transient analysis. To better observe these parameters and generate the graph shown in Figure 8-2, disable graphs 2, 3 and 4 by removing the 2s, 3s and 4s from the transient analysis limits box before clicking the run option. If these changes are saved using a different file name you can easily reproduce Figures 8-2 thru 8-13 and retain the edits each time you open the appropriate file.

Shown in Figure 8-1 are the Spice circuit details. The spark gap coil model is contained on two circuit pages in the file. The line, primary, and secondary circuitry are modeled on the TESLA COIL SCHEMATIC circuit page and the spark gap operating characteristics are modeled on the SPARK GAP circuit page. By double clicking on the part the circuit values are changed to match the calculated values from chapter 2. The circuit values are defined as follows:

- Set the line voltage characteristics. The high-voltage output of the step-up transformer is the source for the primary circuit. The output is modeled using a sine-wave source (VOUT_STEPUP) with a model statement (60 HZ_KV). Change the model statement parameters to match the circuit line frequency in Hz (F), peak voltage amplitude (A), phase angle (PH), and source resistance (RS). In the example $F = 60$, $A = 23{,}755$, PH = 89.5, RS = 1u. The line frequency is typically 60 Hz U.S. The peak voltage output of the step-up transformer is the rms voltage $\times$ 1.414. The phase angle of 89.5 corresponds to a rotary spark gap, which is synchronous to the line voltage. The source resistance is set at 1 $\mu\Omega$ (1u) to prevent loading of the source.
- It was necessary to isolate the line (source) from the primary circuit when the spark gap is ionized. The resistor (R_LINE_ISO) uses a conditional statement to isolate the source from the primary during ionization. While the primary is deionized the tank capacitance follows the line voltage peak and upon ionization the line is isolated allowing the tank capacitance to discharge its energy through the RLC series circuit of the primary. Do not make any changes to (R_LINE_ISO).
- Set the primary-to-secondary coupling coefficient (k). The primary-to-secondary coupling is modeled using a coupling (K_TCOIL) device. Change the device value to the calculated coupling coefficient (0.089 in example).
- Set the primary inductance. The primary inductance is modeled one of two ways depending on whether you are using MC9 or MC8 and earlier versions. For MC8 and earlier versions the primary inductance is modeled using inductors (L_PRI) and (LP). It was necessary to develop a conditional statement for the primary-to-secondary coupling as without it the primary and secondary continued to be coupled after the spark gap deionized. The inductor (L_PRI) uses a conditional statement to change the ionized inductance value of (LP) to the deionized value of 5 pH. This produced values that were close to operating coil characteristics. Do not make any changes to (L_PRI).

Change the model (LP) value to the calculated primary inductance value. The u in the text is μH. 106.5 μH is entered as 106.5u in the part value statement. The DC resistance of the winding is modeled using the resistors (RL_PRI1) and (RL_PRI2). Change the resistance values to one half the calculated primary winding resistance value (0.010 in example). The resistance is split in half so that a small resistance is attached to each end of the primary winding. This prevents convergence problems during transient analysis. Resistors (R1) and (R2) were added to aid convergence and should not be changed. The TC = 0.00393 is the temperature coefficient of copper conductor and is constant for all applications. For MC9 the primary inductance is modeled using the single inductor (L_PRI). Inductor (LP) and its conditional statement are no longer needed for convergence. Change the model (L_PRI) value to the calculated primary inductance value. The DC resistance is modeled the same as in the MC8 model.

- Set the primary tank capacitance. The primary capacitance is modeled using a capacitor (C_TANK). Change the model to the primary tank capacitance value. The u in the text is μF. 0.043 μF is entered as 0.043u in the part value statement. The Equivalent Series Resistance (ESR) of the capacitor is modeled using the resistor (ResrC_TANK). Change the resistance value to the capacitor ESR value if known. See Chapter 5 for an explanation of equivalent series resistance (ESR).
- Set the secondary inductance. The secondary inductance is modeled using an inductor (L_SEC). Change the model to the calculated secondary inductance value. The m in the text is mH. 90.9 mH is entered as 90.9 m in the part value statement. The total resistance of the winding is modeled using the resistor (RL_SEC). Change the resistance value to the calculated total resistance (DC resistance with skin and proximity effect) of the secondary winding (238.4 in example). The TC = 0.00393 is the temperature coefficient of copper conductor and is constant for all applications.
- Set the secondary winding self-capacitance. The self-capacitance of the secondary winding is modeled using a capacitor (C_SEC). Change the model to the calculated value. The p in the text is pF. 38.1 pF is entered as 38.1p in the part value statement.
- Set the terminal capacitance. The terminal capacitance is modeled using a capacitor (C_TERMINAL). Change the model to the calculated value. The p in the text is pF. 31.5 pF is entered as 31.5p in the part value statement.
- Set the terminal-to-ground capacitance. The terminal-to-ground capacitance is modeled using a capacitor (C_TERMINAL-TO-GROUND). Change the model to the calculated value. The p in the text is pF. 4.8 pF is entered as 4.8p in the part value statement. Before the transient analysis runs, a DC operating point is calculated requiring a DC path for all circuit nodes to ground. The resistor (R_TERMINAL-TO-GROUND) provides this path for the (C_TERMINAL) and (C_TERMINAL-TO-GROUND) node. Do not make any changes to (R_TERMINAL-TO-GROUND).
- Set the dedicated ground resistance. The ground resistance is modeled using a resistor (R_GROUND). Change the model to the calculated value (35 in example). The ground resistance is seldom lower than 1 Ω even in elaborate ground systems and is typically much higher. See Chapter 7 for determining the ground resistance.
- The discharge of the high-voltage terminal is modeled using capacitor (C_spark) and resistor (R_rarified_air). The 1 pF and 220 kΩ shown are typical.

- The spark gap operating characteristics are modeled on the SPARK GAP circuit page and connected to the TESLA COIL SCHEMATIC page using nodes (V_GAP) and (V_GAP-).
- Set the resistance of the spark gap while ionized. The resistance of the spark gap while ionized is modeled using resistor (R_IONIZED_GAP). Change the model to the calculated value (2.5445 in example). Resistor (R_DEION_GAP) models the deionized resistance of the gap and (R_GAP) uses a conditional statement to change between the ionized and deionized resistances corresponding to the spark gap timing characteristics. Do not make any changes to (R_GAP) or (R_DEION_GAP).
- The spark gap timing and breakdown voltage characteristics are synchronized in the model using the non-linear function source (E_GAP). Do not make any changes to (E_GAP).
- Set the spark gap timing characteristics. The spark gap timing characteristics are modeled using pulse generator (V_SPARKGAP_TIMING). Change the model statement parameters to match the delay time in seconds (P1), ionization time in seconds (P2), primary oscillation time in seconds (P3), quench time in seconds (P4) and ionization repetition time in seconds (P5). In the example P1 = 0, P2 = 10u, P3 = 60u, P4 = 60.05u, P5 = 2.174m. The delay time (P1) is set to 0 since no delay is desired. The ionization time (P2) is the calculated time period required for the gap to ionize and is typically less than 25 μs. The primary oscillation time (P3) is the calculated time period required for the primary oscillations to dampen from peak to the minimum ionization current. The quench time (P4) is the time required for the gap to deionize after the primary oscillations have dropped below the minimum ionization current. The quench time is not defined but is a very small period. In the example 50 ns is used to allow the pulse generator to drop from 12 V to 0 V and prevent convergence errors. The ionization repetition time (P5) is the reciprocal of the breaks per second (BPS) and is the time period of the break rate or 1/460 BPS = 2.174 ms in the example. A pulse is formed from the statement values: during time (P1) the pulse is 0 V, during time (P2) the pulse rises to 12 V, during time (P3) the pulse is at 12 V, during time (P4) the pulse is falling back to 0 V and stays at 0 V until the pulse is repeated at time (P5). Leave (VONE) in the statement at 12 V. A short primary oscillating time corresponding to a high primary decrement was used in the model and although similar does not match calculated values. Very long primary oscillating times (low decrement) may not model well.
- The line voltage and spark gap breakdown voltage characteristics are synchronized in the model using the non-linear function source (V_ION+) and (V_ION−). Do not make any changes to (V_ION+) and (V_ION−).
- Set the spark gap breakdown voltage characteristics. As the line voltage is alternating positive and negative the spark gap will break down (ionize) at a specified voltage for the selected gap spacing. The curve fit formulae developed in Section 6.11 are used in the model. Non-linear function source (E_+Ionization) determines the spark gap breakdown voltage during the positive line alternation and (E_−Ionization) during the negative line alternation. Do not make any changes to (E_+Ionization) and (E_−Ionization). The breakdown voltage changes with the selected gap spacing in resistor (RS). Change the value of (RS) to the spark gap spacing in cm (0.35 cm used in example, entered as 0.35).

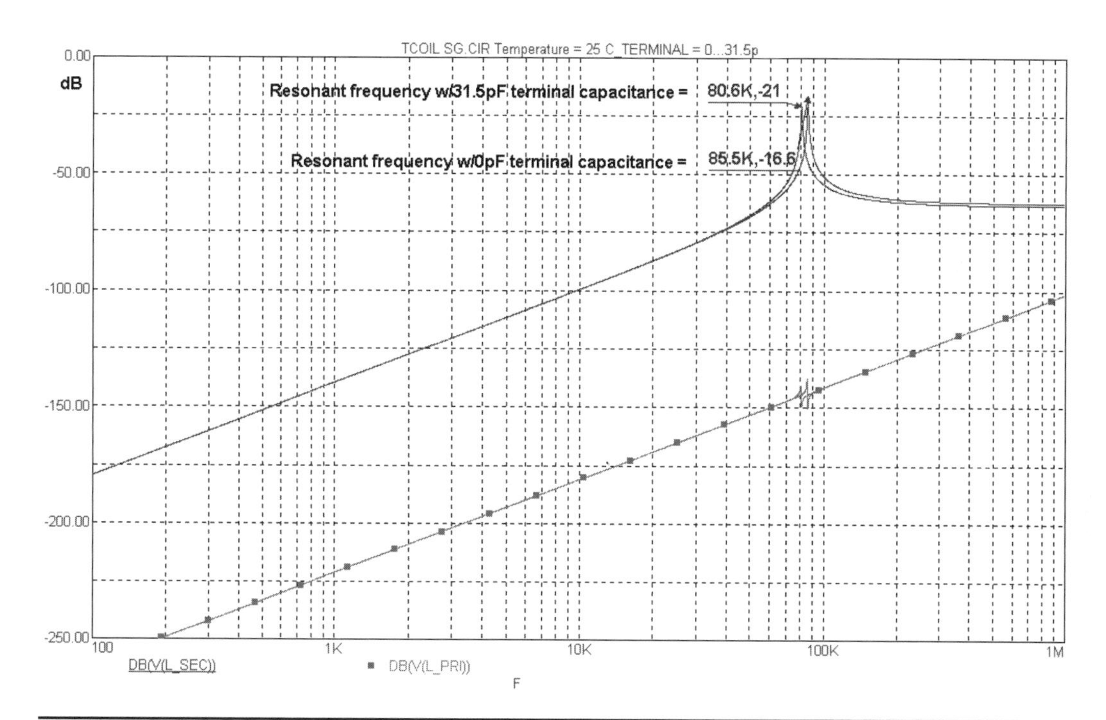

FIGURE 8-4 Simulation model resonant frequency for a spark gap Tesla coil ($k = 0.089$, terminal capacitance = 31.5 pF, BPS = 460, $\theta = 89.5°$).

The transient and AC analysis results are shown in Figures 8-2 through 8-13 for the modeled circuit in Figure 8-1. Figure 8-2 shows the output voltage of the step-up transformer in the blue trace, timing of the spark gap ionization (BPS) in the green trace, oscillating primary current in the red trace, voltage at the top of the secondary winding in the black trace, and voltage at the terminal capacitance in the gray trace. The operating characteristics of the spark gap coil are easily seen in the following modeling runs:

- In Figure 8-3 the primary-to-secondary coupling coefficient is increased from 0.089 to 0.2 in a transient analysis. This is at the theoretical limit for efficient coil operation and the decrease in secondary performance is clearly seen. Theoretically a coupling of $k = 0.2$ would be too tight for this application.
- In Figure 8-4 the terminal capacitance is changed (stepped) from 0 pF to 31.5 pF in an AC analysis. The resonant frequency changes from 85.5 kHz @ 0 pF to 80.6 kHz @ 31.5 pF, illustrating the decreasing resonant frequency of the secondary with increasing terminal capacitance.
- In Figure 8-5 the terminal capacitance is increased from 31.5 pF to 60 pF in a transient analysis. There is a slight change in secondary voltage from the results in Figure 8-2.
- In Figure 8-6 the break rate is doubled from 460 BPS to 920 BPS in a transient analysis. There is a slight change in the secondary voltage but twice as much line power is

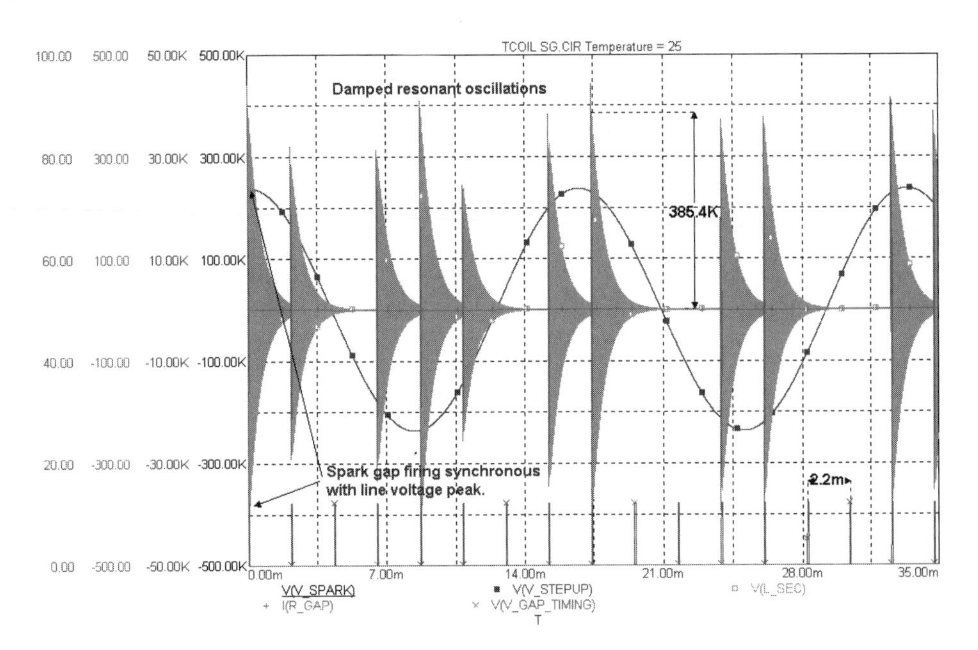

FIGURE 8-5 Simulation model waveforms for a spark gap Tesla coil ($k = 0.089$, terminal capacitance = 60 pF, BPS = 460, $\theta = 89.5°$).

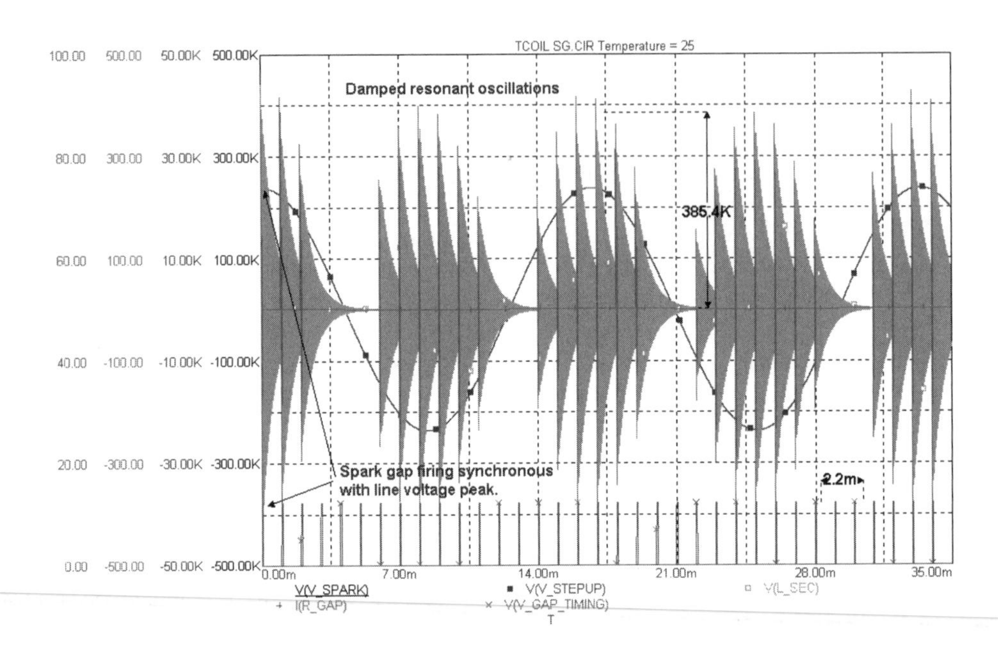

FIGURE 8-6 Simulation model resonant waveform for a spark gap Tesla coil ($k = 0.089$, terminal capacitance = 31.5 pF, BPS = 920, $\theta = 89.5°$).

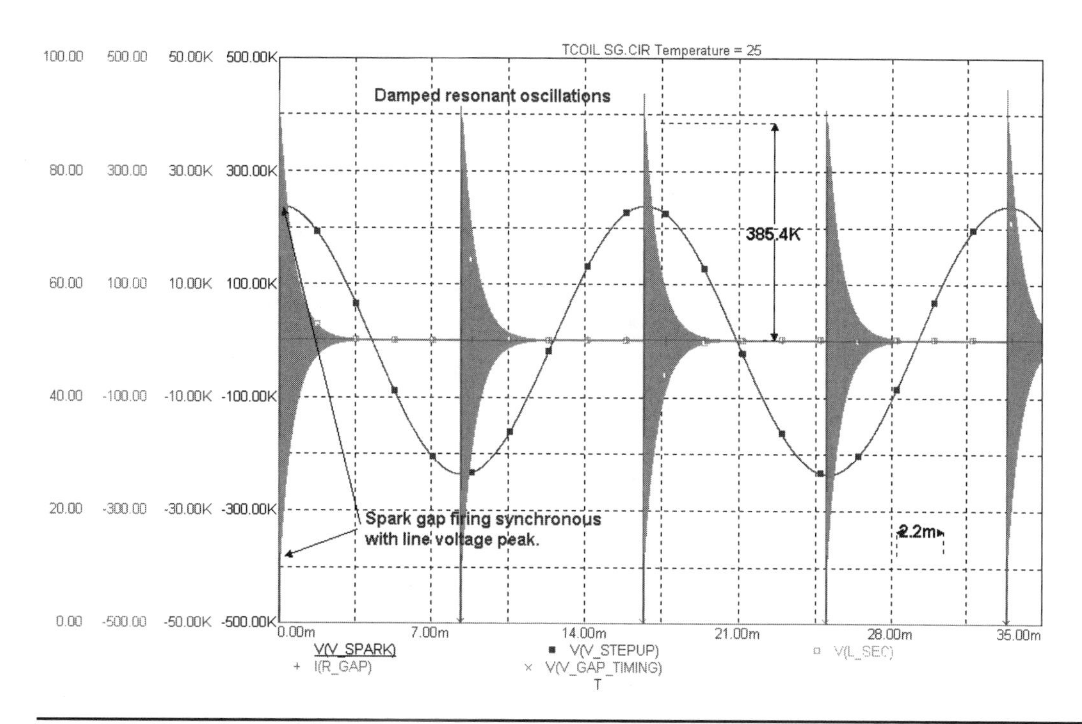

FIGURE 8-7 Simulation model waveforms for a spark gap Tesla coil ($k = 0.089$, terminal capacitance = 31.5 pF, BPS = 120, $\theta = 89.5°$).

required. Note that Figure 8-6 has more area colored than Figure 8-5 indicating more power in the secondary which results in a brighter spark.

- In Figure 8-7 the break rate is decreased from 460 BPS to 120 BPS in a transient analysis. There is a slight change in the secondary voltage. The synchronous fixed gap rate of 120 BPS can produce the same output voltage as a synchronous rotary gap rate of 460 BPS with much less power drawn from the line; however, the spark may loose some intensity.
- In Figure 8-8 the phase angle (θ) of the line voltage is changed from 89.5° to 50° in a transient analysis. There is a slight change in the secondary voltage. A non-synchronous θ of 50° with a high break rate of 460 BPS will produce about the same secondary voltage as a synchronous θ of 89.5°.
- In Figure 8-9 the phase angle (θ) of the line voltage is changed from 89.5° to 50° and the break rate is decreased from 460 BPS to 120 BPS in a transient analysis. There is no output in the secondary, as the spark gap never ionizes. Low break rates in rotary spark gaps require synchronous or near synchronous operation to produce a secondary output.
- In Figure 8-10 the secondary impedance at resonance is determined in an AC analysis. This is the output impedance seen at the source (line) at the resonant frequency of the secondary. The 46.3 kΩ presents a high impedance to the source. The primary impedance is very low and is not shown in the figure.

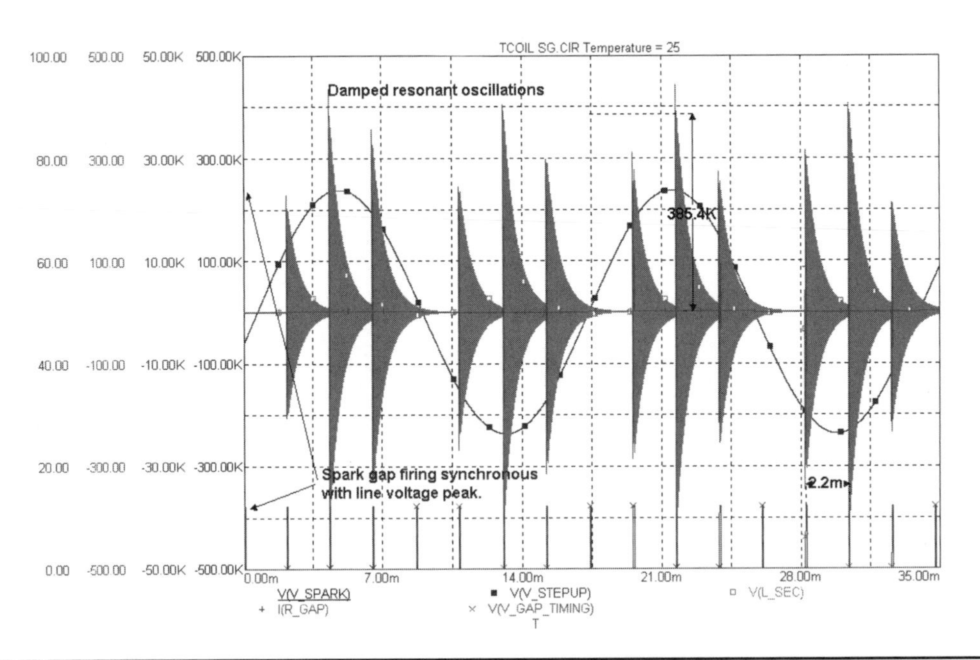

FIGURE 8-8 Simulation model waveforms for a spark gap Tesla coil ($k = 0.089$, terminal capacitance = 31.5 pF, BPS = 460, $\theta = 50°$).

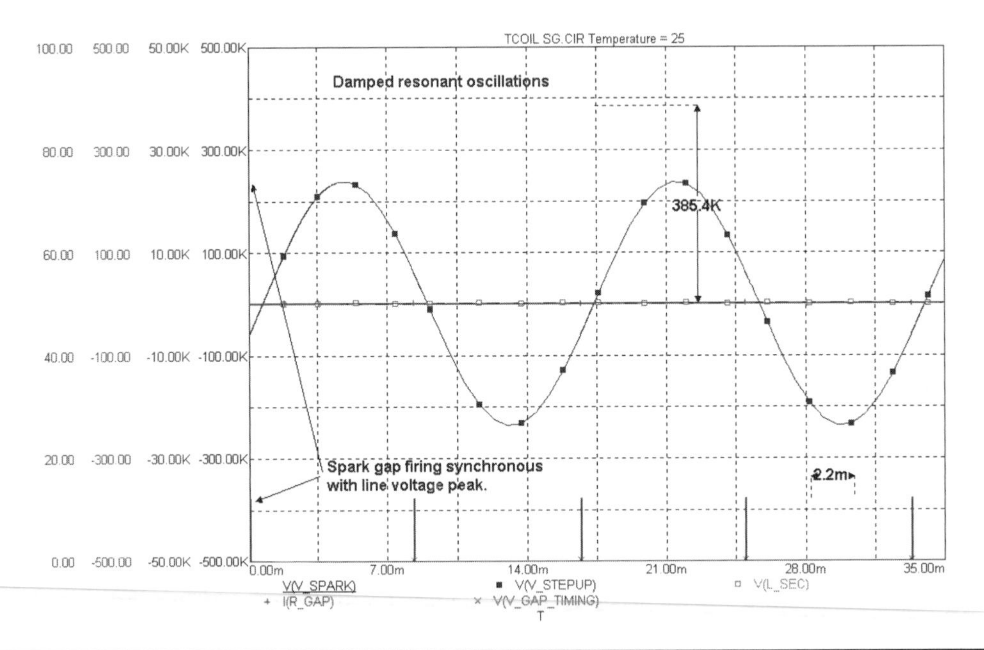

FIGURE 8-9 Simulation model waveforms for a spark gap Tesla coil ($k = 0.089$, terminal capacitance = 31.5 pF, BPS = 120, $\theta = 50°$).

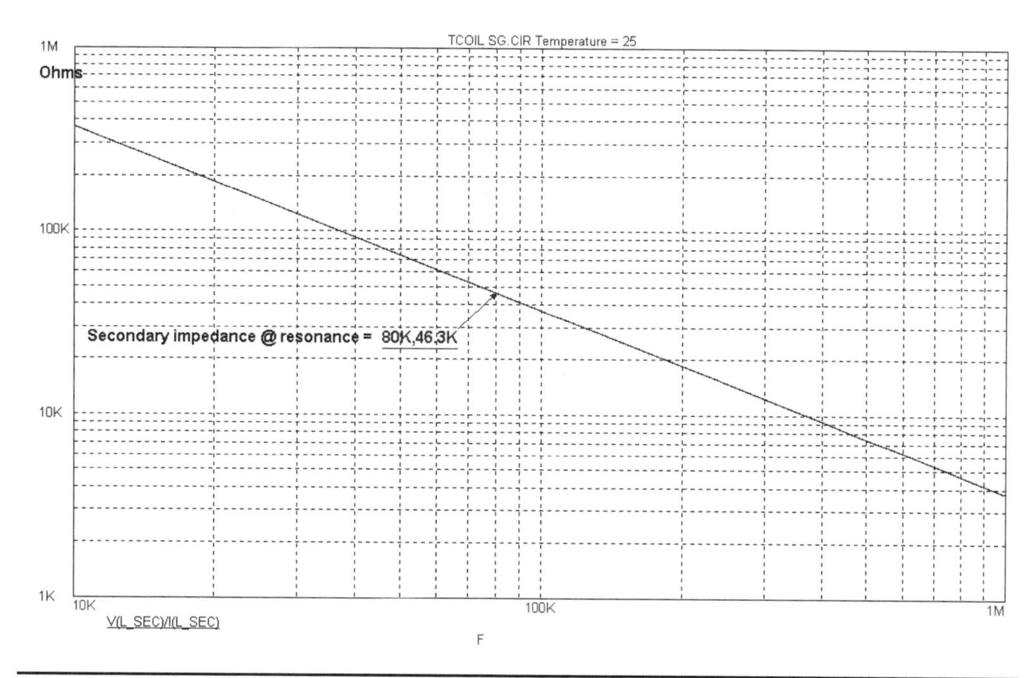

FIGURE 8-10 Simulation model secondary impedance-to-source at resonance for a spark gap Tesla coil.

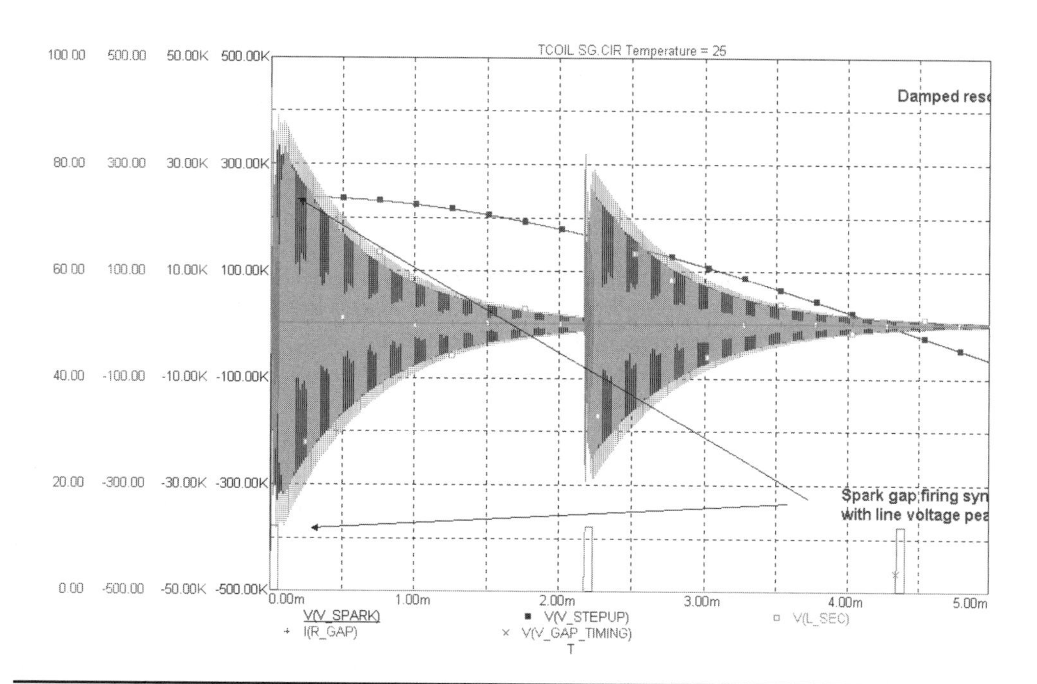

FIGURE 8-11 Simulation model resonant waveform for 0.5 msec/div.

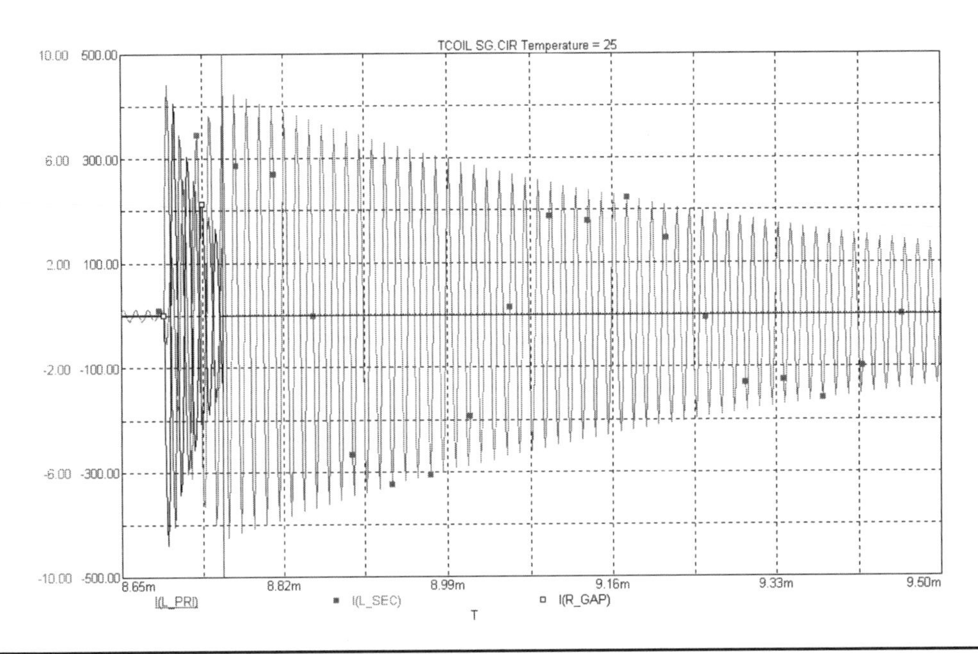

FIGURE 8-12 Simulation model resonant waveform details.

- Waveform details of the nominal circuit values in Figure 8-1 are shown in Figure 8-11. Further detail is shown in Figure 8-12 for the primary current in the black and red traces and secondary current in the gray trace. Note how the primary oscillations dampen between ionization and deionization. The secondary voltage rises from zero at spark gap ionization to the peak value at deionization and then dampens slower than the primary oscillations due to its lower decrement.
- In Figure 8-13 the harmonic content of the primary and secondary oscillations are determined in an AC analysis. Note the wideband response of the primary current in the red trace, narrower band response of the primary voltage in the blue trace and narrow band response at the resonant frequency of 76.3 kHz of the secondary voltage in the black trace. A substantial high-frequency fifth harmonic at 388.9 kHz also appeared in the analysis.

This illustrates some of the utility of Spice based programs in designing and evaluating spark gap coils. Once an accurate model is developed the Spice simulation can evaluate circuit operation faster than manual calculations with a visual display that is more intuitive than calculations. Very complex analysis such as the harmonic analysis or Fast Fourier Transform (FFT) shown in Figure 8-13 is accurately performed and could be done manually only with great difficulty, not to mention the errors induced in all lengthy manual calculations. Practice is required to become familiar with the program operation and circuit details. See Chapter 2 for a more detailed description on optimizing the design.

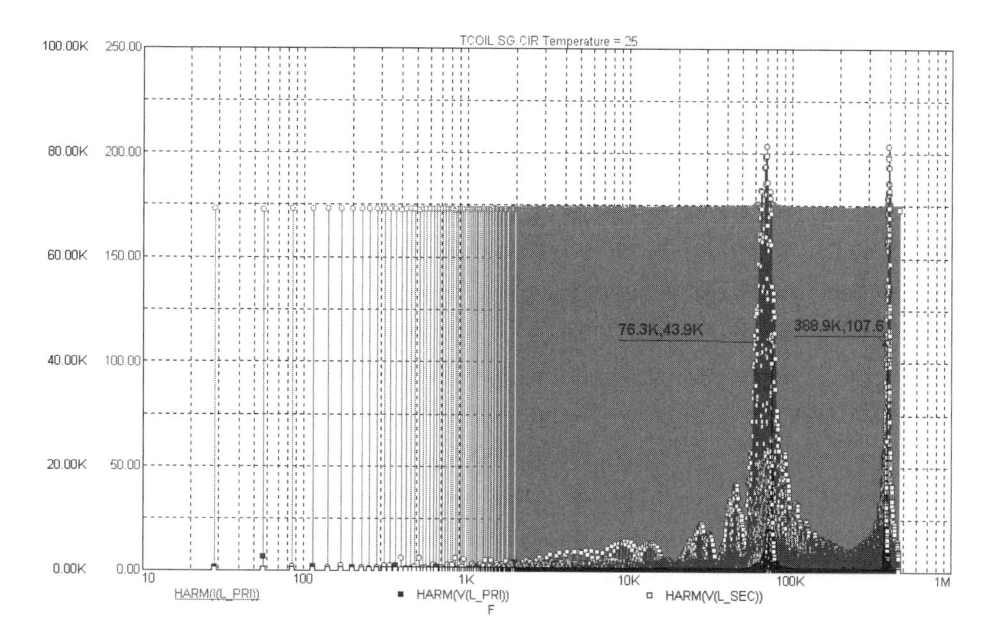

FIGURE 8-13 Simulation model harmonic analysis.

8.3 Circuit Simulation Results and Performance Characteristics of EMI Filters

This section will not address EMI filter design. With the availability of surplus commercial filters there is no need. Instead we will analyze how well a commercial filter works in filtering high-voltage, high-frequency transients generated in the Tesla coil out of the incoming 60 Hz line. A good portion of the available surplus filters are made for switch mode power supplies (SMPS) which take incoming AC or DC current, convert it to a high-frequency switched current, increase or decrease the voltage using a transformer, and rectify it back to a DC output. Any current that is pulsed, switched on and off, or changes over time is essentially an AC signal. The SMPS EMI filters keep these high-frequency switching transients out of the line input as well as input noise out of the filter's output (load). The switching frequencies of SMPS are generally between 50 kHz to 1 MHz, the same range of interest for a Tesla coil. It makes sense to use a surplus EMI filter that has been designed to keep these high-frequency transients out of the filter's input (line) while handling high current loads. Visit: http://www.schaffner.com/filters/index.html or http://www.cor.com/ for good EMI filter websites.

Input (EMI) Filters come in various topologies. There are single-stage, two-stage, and three-stage filters that may be acquired at Hamfests. There are also RF filters that typically filter a single line instead of all the lines in a single-phase or three-phase AC input. The Schaffner filters were chosen for analysis since their datasheets provide schematic and part values that can be used to create Spice models. A Schaffner single-stage filter was constructed in the INPUT_RIPPLE.CIR file as shown in Figure 8-14.

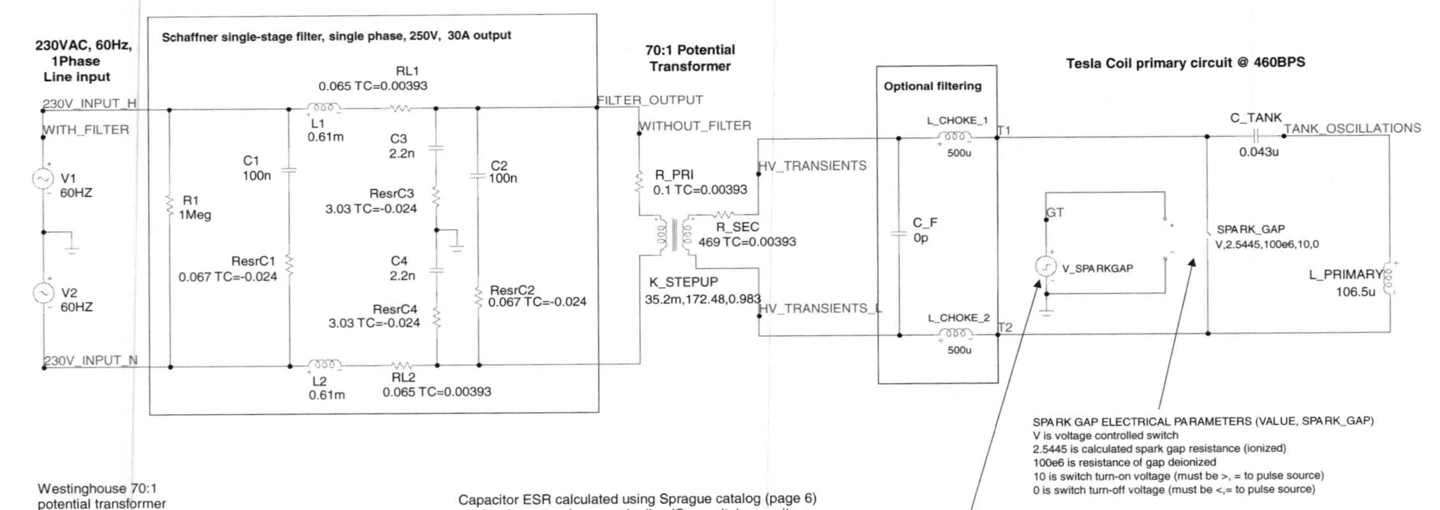

FIGURE 8-14 Micro-Cap simulation model for a single-stage EMI filter.

I have found no reason for not using three-phase filters, the three inputs and three outputs paralleled for increased current handling capability. Schaffner's recommendations allow an overcurrent level of only 1.5 times the rated current level for 1 minute per hour. This means you should not exceed the manufacturer's rated current-handling specification for each line. Instead, use more filters paralleled together to increase your filter's current handling ability.

Some coilers connect these filters backwards in the circuit under the presumption that it will more efficiently isolate the high-voltage transients. When the line input and load output of the filter are reversed (connected backwards) an AC and transient analysis of the Spice model in Figure 8-14 shows no change in system performance. This presumption is incorrect and I suggest using the filter as the manufacturer intended.

When reactive parts are added to the line they provide either a gain or attenuation of the transients generated in the oscillating primary tank circuit. The largest transients occur at the moment of ionization and deionization of the spark gap. The effects of using commercial single-stage filters and optional HV filter capacitors and inductors on the step-up transformer output are modeled in this section.

The two optional air core chokes and HV filter capacitor shown in Figure 8-14 provide the line almost no isolation to the high-voltage transients. The transient analysis displayed in the lower graph of Figure 8-15 shows the damped primary oscillations and ionization transients in the blue trace and the high-voltage transients on the step-up transformer output in the red trace (the red and green traces overlap in the analysis and only the green trace is shown in

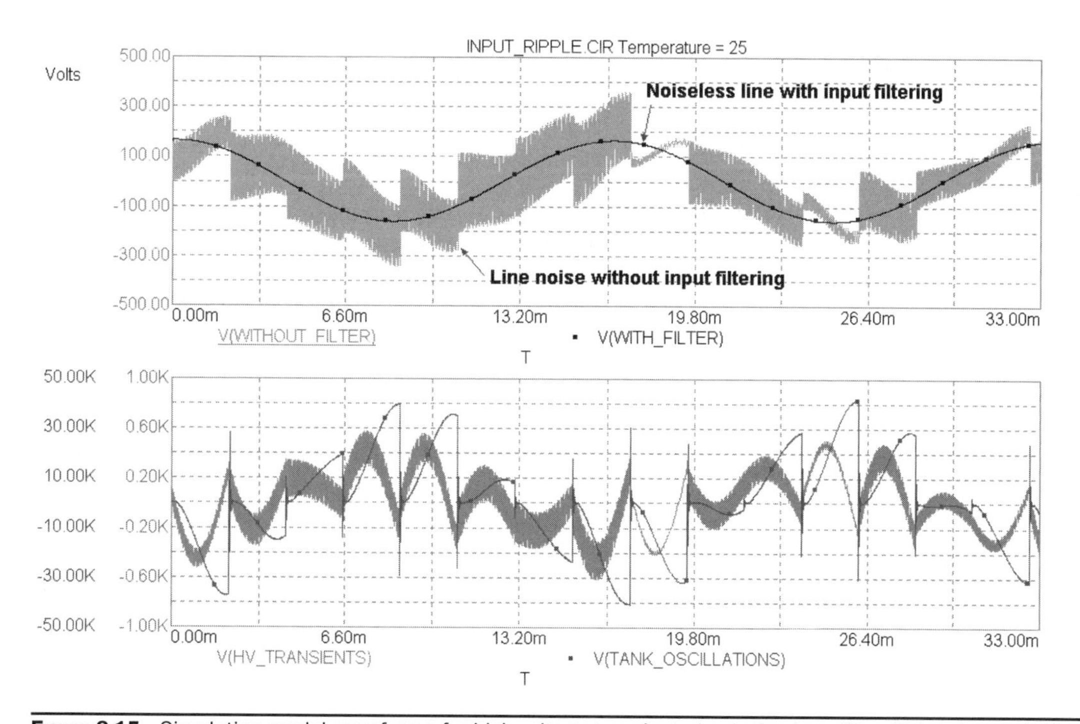

FIGURE 8-15 Simulation model waveforms for high-voltage transients in single-stage filter (RF chokes = 500 μH, HV filter capacitor = 0 pF).

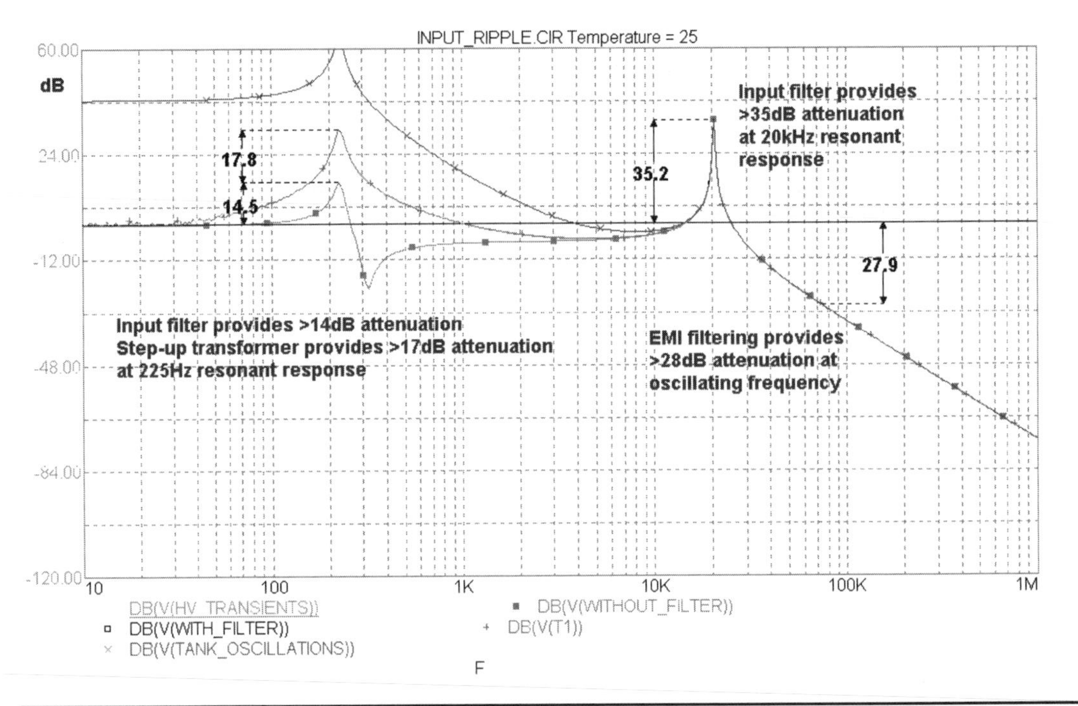

FIGURE 8-16 Simulation model AC response for high-voltage transients in single-stage filter (RF chokes = 500 μH, HV filter capacitor = 0 pF).

the figure). The upper graph displays the line noise when no input filter is used in the gray trace and the almost noiseless line with the single-stage input filtering in the black trace. The benefits of using an input filter are already apparent. The AC analysis in Figure 8-16 provides additional detail of where the noise is isolated in the circuit. The AC source in the model is the 230-V (rms), 60-Hz generator on the left end of the circuit shown in Figure 8-14. All gain or attenuation in the AC analysis is in respect to this source. With the line filter the black trace in Figure 8-16 indicates a 0 dB gain at all frequencies from 10 Hz to 1 MHz. This means there is adequate isolation of the transients to the line input and no appreciable transient noise. The difference in dB between the black trace and the gray trace indicates the isolation provided by the input filter. The difference in dB between the gray trace and the red trace indicates the isolation provided by the step-up transformer. The difference in dB between the red trace and the green trace indicates the isolation provided by the optional air-core inductor and filter capacitor on the step-up transformer output. As no difference can be seen between the red and green traces (overlap each other with only the green shown in the figure) there is negligible isolation provided by the optional inductor-capacitor filtering on the step-up transformer output. The predominant attenuation is provided by the step-up transformer and input filter. There is at least 28 dB of attenuation at the primary oscillating frequency of 74 kHz provided by the input filter where all traces overlap each other. A resonant response at 20 kHz would see a noise gain of 35 dB if the input filter were not used. Another resonant response of the circuit

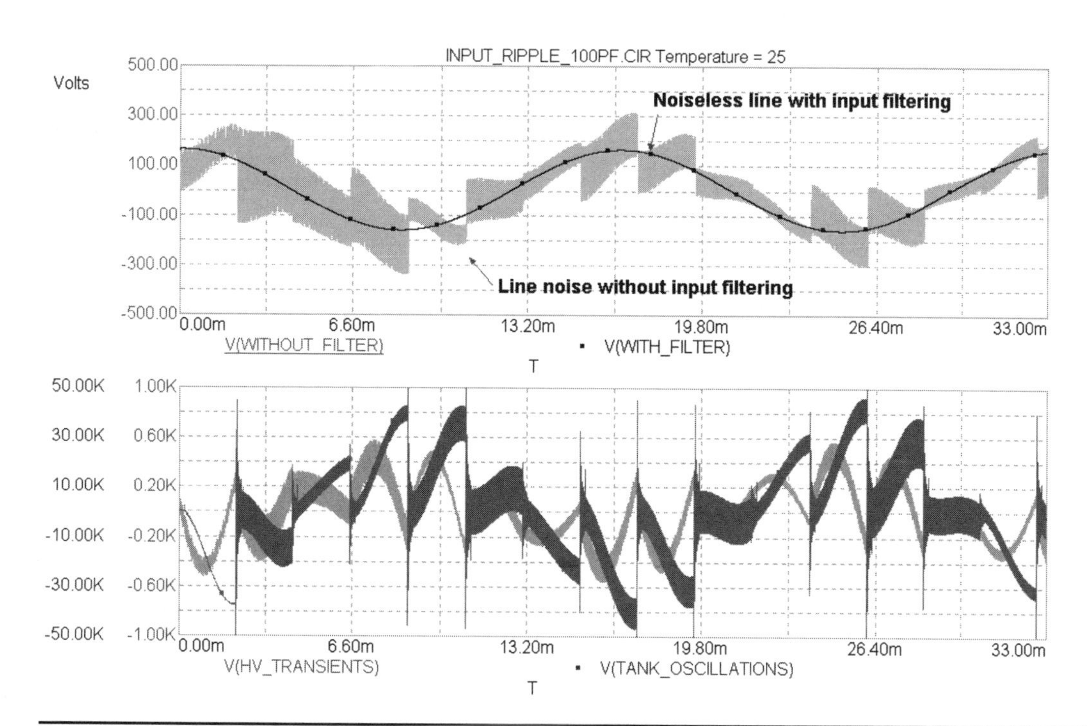

FIGURE 8-17 Simulation model waveforms for high-voltage transients in single-stage filter (RF chokes = 500 μH, HV filter capacitor = 100 pF).

at 225 Hz is attenuated 17.8 dB by the step-up transformer and another 14.5 dB by the input filter.

Adding the HV filter capacitor on the step-up transformer output actually increases the high-voltage transients in the primary tank circuit as it forms an additional resonant circuit with the transformer winding and air chokes. This effect is shown in the transient analysis in Figure 8-17. The HV filter capacitor was stepped from 0 pF to 5,000 pF with little change in the isolation characteristics shown in the AC analysis in Figure 8-18. The only observable change is the noise above the 74 kHz resonant frequency where adequate attenuation was seen without the capacitor. Adding this capacitor to the circuit is therefore not recommended as it may do more harm than good. A safety gap should be used to protect the step-up transformer and tank capacitor instead of this HV filter capacitor.

The HV filter capacitor and air chokes were effectively removed from the circuit by assigning values of 0. The transient analysis results shown in Figure 8-19 and AC results in Figure 8-20 indicate the almost unnoticeable effect the chokes have in isolating the transients from the line. Compare the results with Figures 8-15 and 8-16. This verifies that adding RF chokes or filter capacitors across the step-up transformer output have little effect on protecting the transformer from high-voltage transients or isolating them from the AC supply line. A safety gap should be used instead and input filtering is a must. A single-stage filter is quite effective in isolating the high-voltage transients in the primary of the Tesla coil from the line service into your lab.

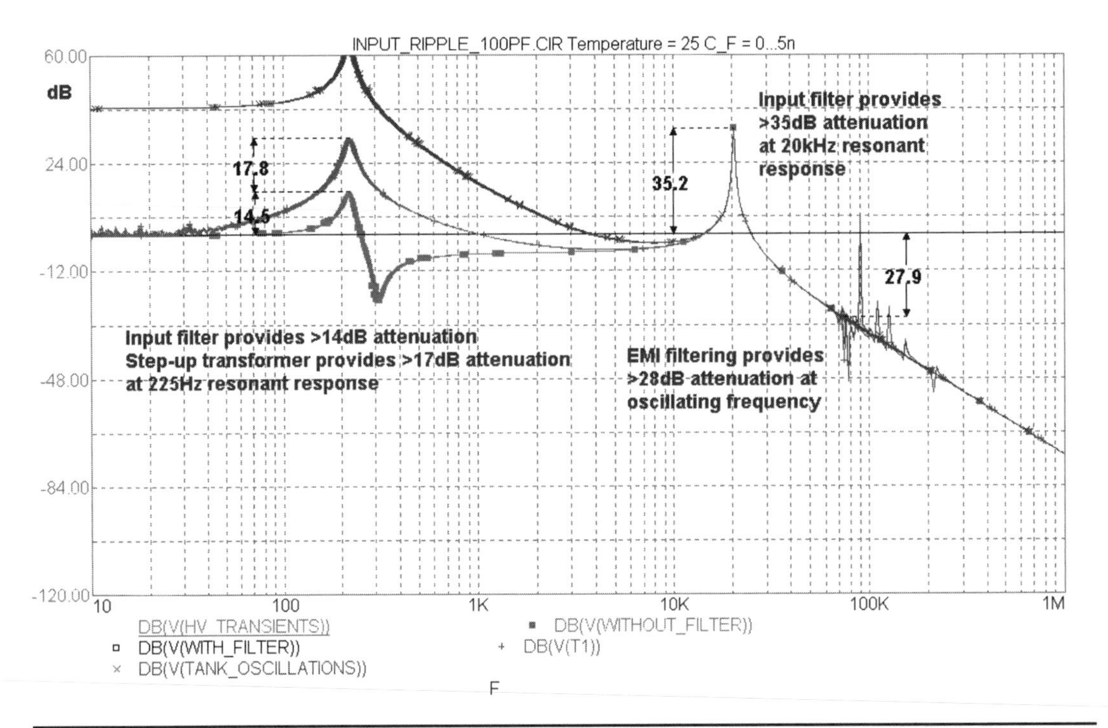

FIGURE 8-18 Simulation model AC response for high-voltage transients in single-stage filter (RF chokes = 500 μH, HV filter capacitor = 0-5,000 pF).

When the three-stage EMI filter shown in Figure 8-21 from the INPUT_RIPPLE_3STAGE.CIR file is used, the line is effectively isolated from the high-voltage transients as shown in the noise-free line input (black trace) in Figure 8-22. The circuit response is different from the single-stage filter. The optional air choke and capacitor filter on the step-up transformer output were effectively removed from the circuit by assigning values of 0 to each. The AC analysis in Figure 8-23 indicates the three-stage input filter provides greater than 73 dB of isolation at the primary oscillating frequency, 45 dB greater than the 28 dB isolation from the single-stage filter. The manufacturer's intent of isolating high frequency switching transients from the line is evident. The resonant response at 31.3 kHz is attenuated 10 dB and at 4.3 kHz, 54 dB. The resonant response at 280 kHz is attenuated 27 dB by the step-up transformer and an additional 9 dB by the input filter.

Note the large voltage transient generated as the spark gap ionizes shown in the blue trace of the transient analyses in Figure 8-22. This was intentionally enlarged in the simulations to ensure conservative results.

Let's evaluate the performance of the single-stage input filter. Open the CH_7.xls file, INPUT FILTER Worksheet (4), shown in Figure 8-25. (See App. B.) For a 200-kV noise transient that is missed by the safety gap, the amplitude in dB is:

$$dB = 20 \bullet \log(V) \tag{8.1}$$

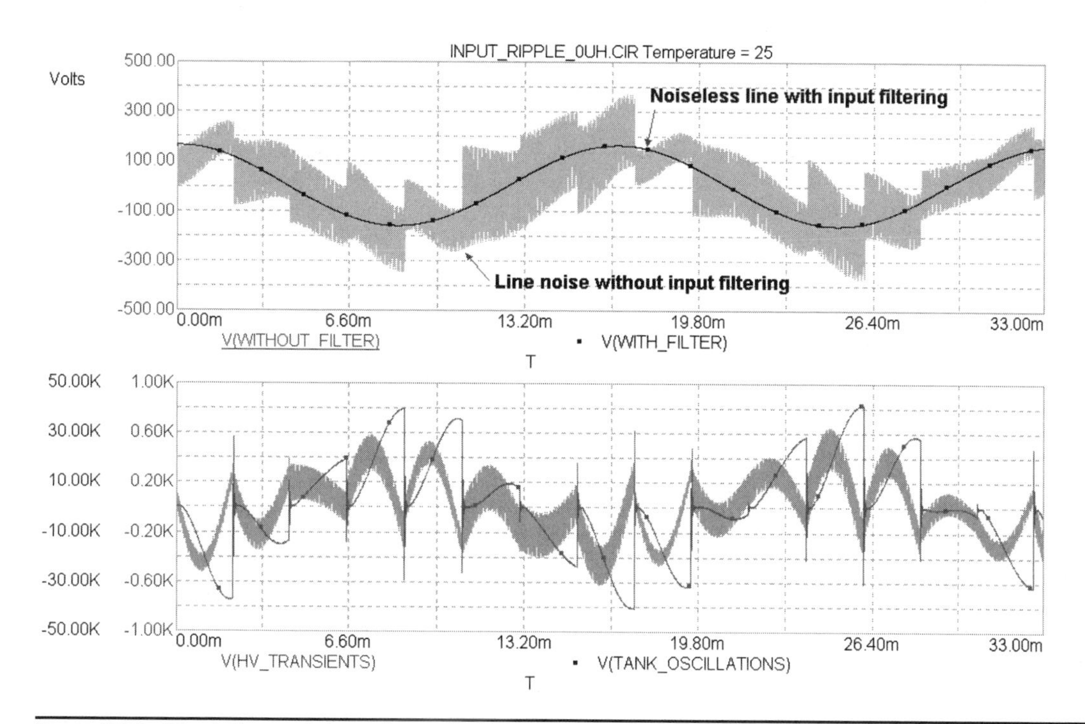

FIGURE 8-19 Simulation model waveforms for high-voltage transients in single-stage filter (RF chokes = 0 μH, HV filter capacitor = 0 pF).

Where: dB = Voltage transient converted to decibel value = cell (E3), calculated value. A decibel is one tenth of a bel. A bel denotes an increase of 10 times above a baseline value. 0 dB = 1 mW or 0.63 V peak.

V = Voltage transient in volts = cell (B3), enter value in kV, converted to volts in cell (C3).

This means our 200-kV noise transient equals 106 dB. The amplitude of the transient at the line end with the single-stage input filter attenuation of 28 dB at 74 kHz is:

$$\text{Line} = \text{Noise} - \text{filter} \tag{8.2}$$

Where: Line = Amplitude of noise transient at the line end in dB = cell (E4), calculated value.

Noise = Amplitude of noise transient in dB = calculated value in cell (E3) from equation (8.1).

Filter = Attenuation provided by input filter and step-up transformer in dB = cell (B4), enter value.

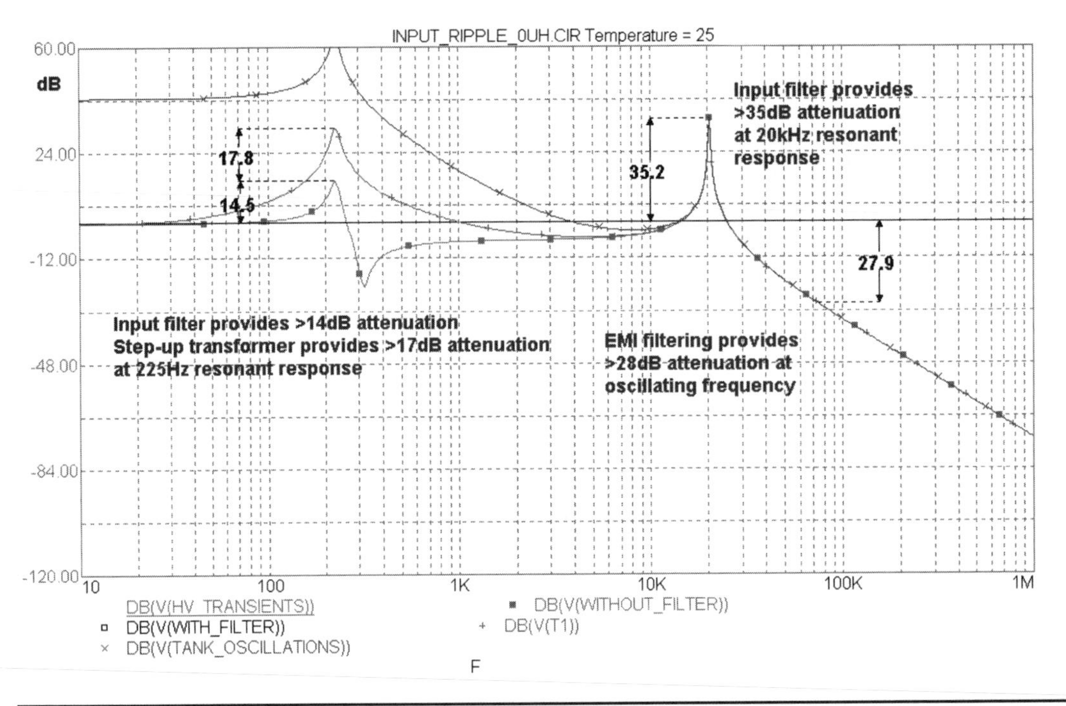

FIGURE 8-20 Simulation model AC response for high-voltage transients in single-stage filter (RF chokes = 0 μH, HV filter capacitor = 0 pF).

The 106-dB noise transient is attenuated 28 dB leaving 78 dB at the line. This may not sound like much but let's convert it back to a voltage:

$$V = \log^{-1}\left(\frac{\text{dB}}{20}\right) \tag{8.3}$$

Where: V = Amplitude of noise transient at the line end in volts = cell (E5), calculated value.

dB = Amplitude of noise transient at the line end in dB = calculated value in cell (E4) from equation (8.2).

NOTE: *$\log^{-1}$ is the antilog function. Excel will not perform this function. However, this formula performs the antilog function in Excel:* $\sqrt{10^{\left(\frac{dB}{10}\right)}}$

The 200-kV (106-dB) transient will only be attenuated to 7,962 V at the line end. This would be unacceptable to the power company and the electrical stresses on your control cabinet so I would suggest using an input filter with more output-to-input attenuation (isolation) or adjusting your safety gap so it will fire at a low enough level to make up the difference. Also note if these transients are feeding back into the line service they are also in your house wiring and

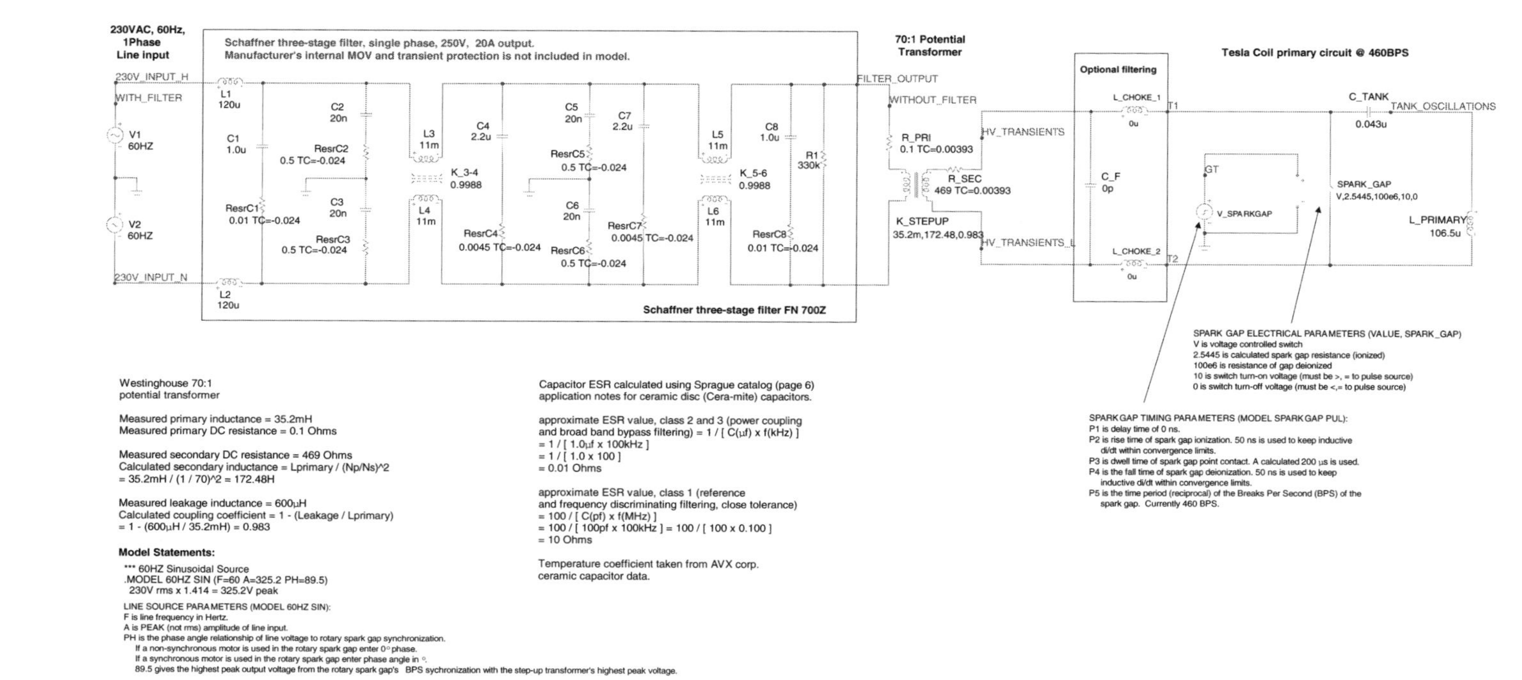

FIGURE 8-21 Micro-Cap simulation model for a three-stage EMI filter.

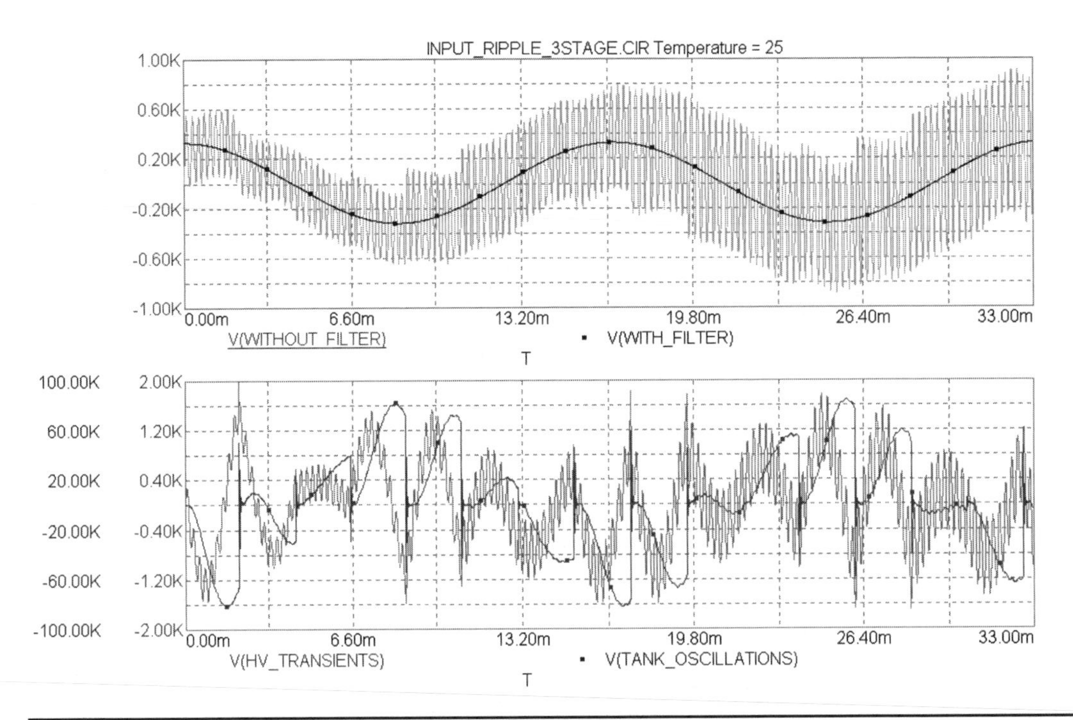

FIGURE 8-22 Simulation model waveforms for high-voltage transients in three-stage filter (RF chokes = 0 μH, HV filter capacitor = 0 pF).

at the input to all of your home appliances. Using Spice to simulate adding another single-stage filter in series with the single-stage filter modeled in Figure 8-14 appears to provide additional isolation at 74 kHz as shown in Figure 8-24. The 62 dB isolation from using two series filters attenuates the 200-kV transient to 159 V, which is still too high, but much safer. The 73-dB attenuation of the three-stage filter would decrease the transient to 45 V and two of these in series would provide adequate filtering for even a 200-kV transient. This should illustrate the need for a safety gap in the tank circuit and effectiveness of using an input filter to keep high-voltage transients and noise off the line service to your lab.

Using adequate input filtering in your power cabinet will keep conducted EMI off the power grid. If your small or medium coil is properly tuned to produce maximum spark output it will radiate negligible emission levels outside a 3-meter area and somewhat more for large coils. To test for radiated emissions, set an AM/FM radio receiver >5 meters away from the coil while it is operating to detect emissions above a negligible level. Without expensive equipment this cannot be accurately measured and the simple AM/FM receiver becomes a cost-effective substitute. Once your coil is properly tuned you can run it at high power and should not pick up any static in your radio receiver. Remember we are not operating a radio transmitter when the energy is converted to a spark.

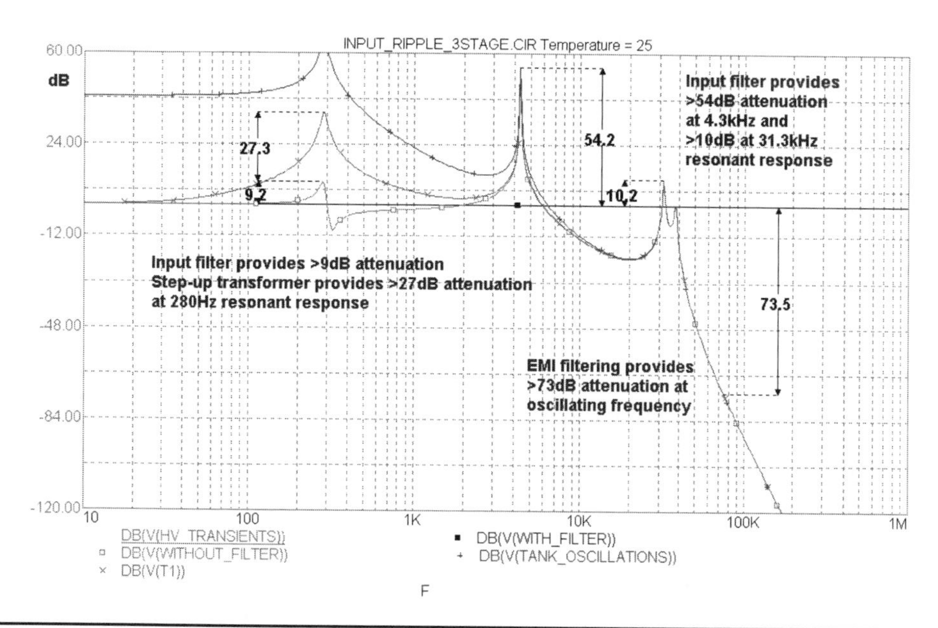

FIGURE 8-23 Simulation model AC response for high-voltage transients in single-stage filter (RF chokes = 0 μH, HV filter capacitor = 0 pF).

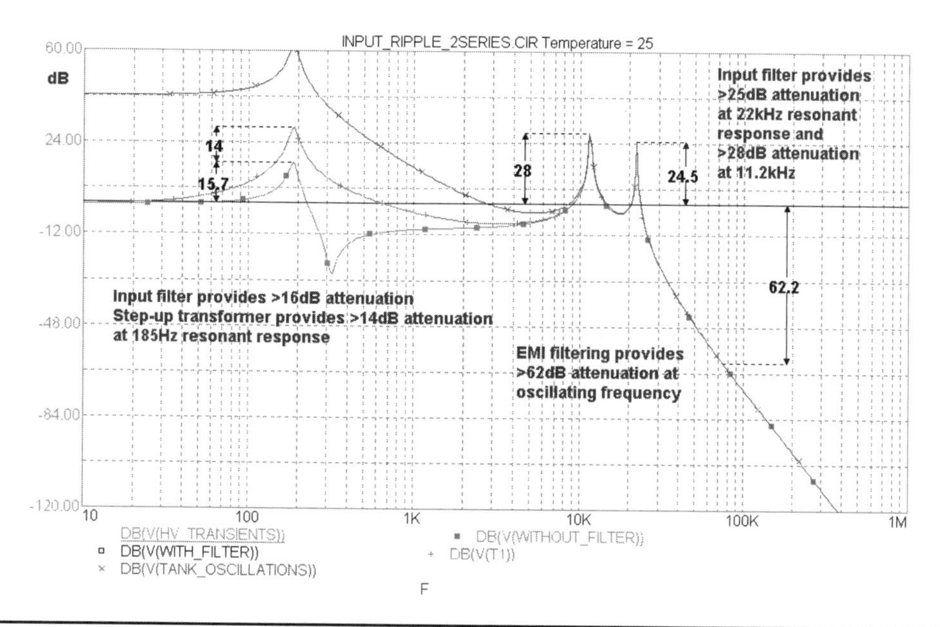

FIGURE 8-24 Simulation model AC response for two EMI filters in series.

ENTER			CALCULATED VALUES			ROW
Transient Voltage or Noise Amplitude in kVolts (**V**)	200.00 KV	200000V	Transient Voltage or Noise Amplitude in dB (**VdB**)	106.0dB	VdB = 20*log(V)	3
Input Filter Attenuation (**A**)	28.0dB		Transient Amplitude at line end in dB (**LdB**)	78.0dB	LdB = VdB - A	4
			Transient Amplitude at line end in Volts (**LV**)	7,962 V	LV = log^-1*(LdB/20) or SQRT(10^(LdB/10))	5
A	B	C	D	E		COLUMN

FIGURE 8-25 Voltage transients and EMI filter response worksheet.

8.4 Using the Circuit Simulation Program to Design a Phase Shift Network for Synchronizing a Rotary Spark Gap to the Peak Line Voltage

The highest secondary voltage output of a Tesla coil can be obtained when the rotary spark gap fires in coincidence with the line voltage peak feeding the tank circuit. This may not be worth the effort with a high break rate but is essential when using low break rates, e.g. >120 BPS to 300 BPS. If a stationary spark gap is used the break rate is 120 BPS; however, medium and large coils do not function well when using only a stationary gap. Figure 8-26 shows a phase shift network used to provide a variable 0° to 90° phase shift of the 60-Hz line voltage feeding the rotary spark gap motor in a Tesla coil to synchronize it with the 60-Hz line voltage feeding the primary tank circuit. Included in the figure are recommended construction and operation notes. The circuit is modified from that originally appearing in reference (2).

The Spice model shown in Figure 8-27 from the PHASE_SYNC.CIR file was used to evaluate the transient performance of the phase shift network. The squirrel cage motor used in the rotary spark gap has two stator windings. A start winding is used with a start capacitor in the motor to provide a phase shift. As the line is connected to the motor start winding the phase is shifted from that on the run winding. This phase shift provides starting torque for the motor. As the motor runs up to about 75% of its nominal RPM a centrifugal switch opens and the start winding is removed from the circuit. A series of switches were added to the model to simulate the start-up, motor run, and shutdown and braking effects of operation when the circuit is constructed as shown in Figure 8-26. A small EMI filter was also added to the circuit to provide additional line filtering and does not appreciably affect the circuit performance. The model is adapted to the application by following the steps below:

Step 1. Measure the total inductance (connection 1-to-3) of the autotransformer used for the phase shift network. As the autotransformer is rotated from its zero stop the inductance provided to the phase shift network is increased. Enter any value from 0 H to the measured maximum inductance value in the L_AUTOTRANSFORMER part value in the schematic.
Enter the value of the phase shift capacitance in the C_LINE part value.

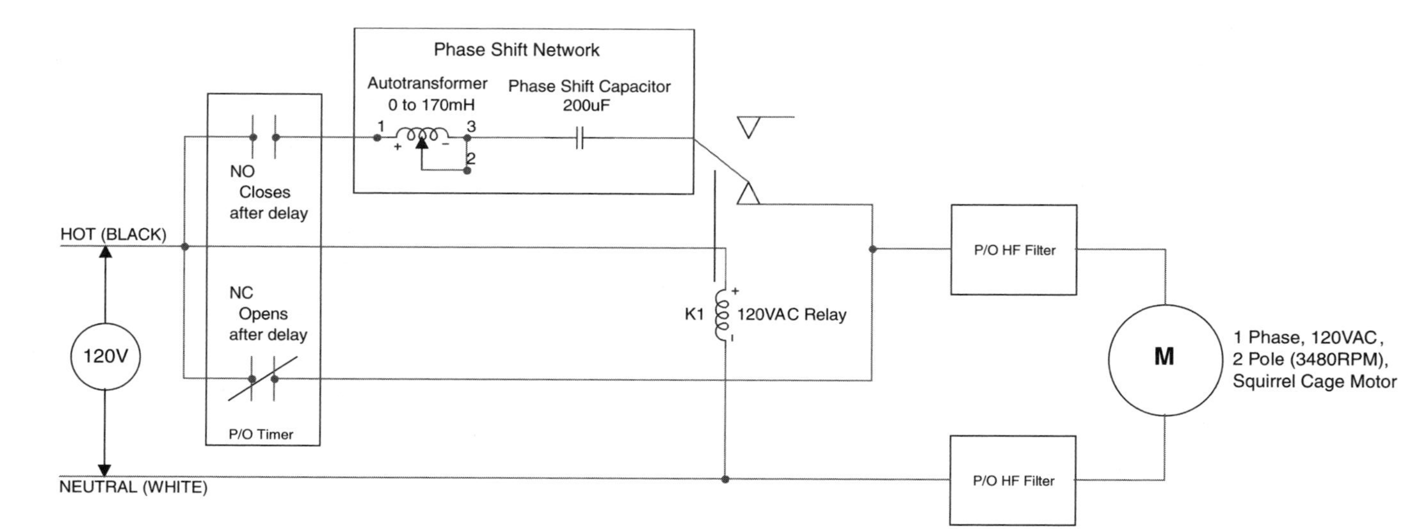

Use an adjustable timer (agastat) to provide delay connections shown. This will remove the reactance and voltage drop of the phase shift network during the motor start cycle (a few seconds) allowing the motor to start efficiently. By using an adjustable timer the motor and load can be fully compensated for on start-up. Start-up delays for different motor-load combinations can also be adjusted. If a timer cannot be obtained a manual switch or relay combination can provide the same function using the electrical connections shown.

When the line voltage is removed from the induction motor's stator the rotor continues to turn from inertia and becomes a generator while it spins down. To eliminate possible high voltage kickback effects the phase shift network is removed from the stator by relay K1 as soon as the line voltage is removed.

High frequency filter can be used to remove coil oscillator frequency and noise without significantly affecting phase shift circuit performance.

Autotransformer will give the same range of phase shift (min to max). The phase shift capacitor (C_LINE) will shift this range from leading to lagging as it's value is changed. Use a motor start type of capacitor, which is unpolarized.

FIGURE 8-26 Phase shift circuit diagram.

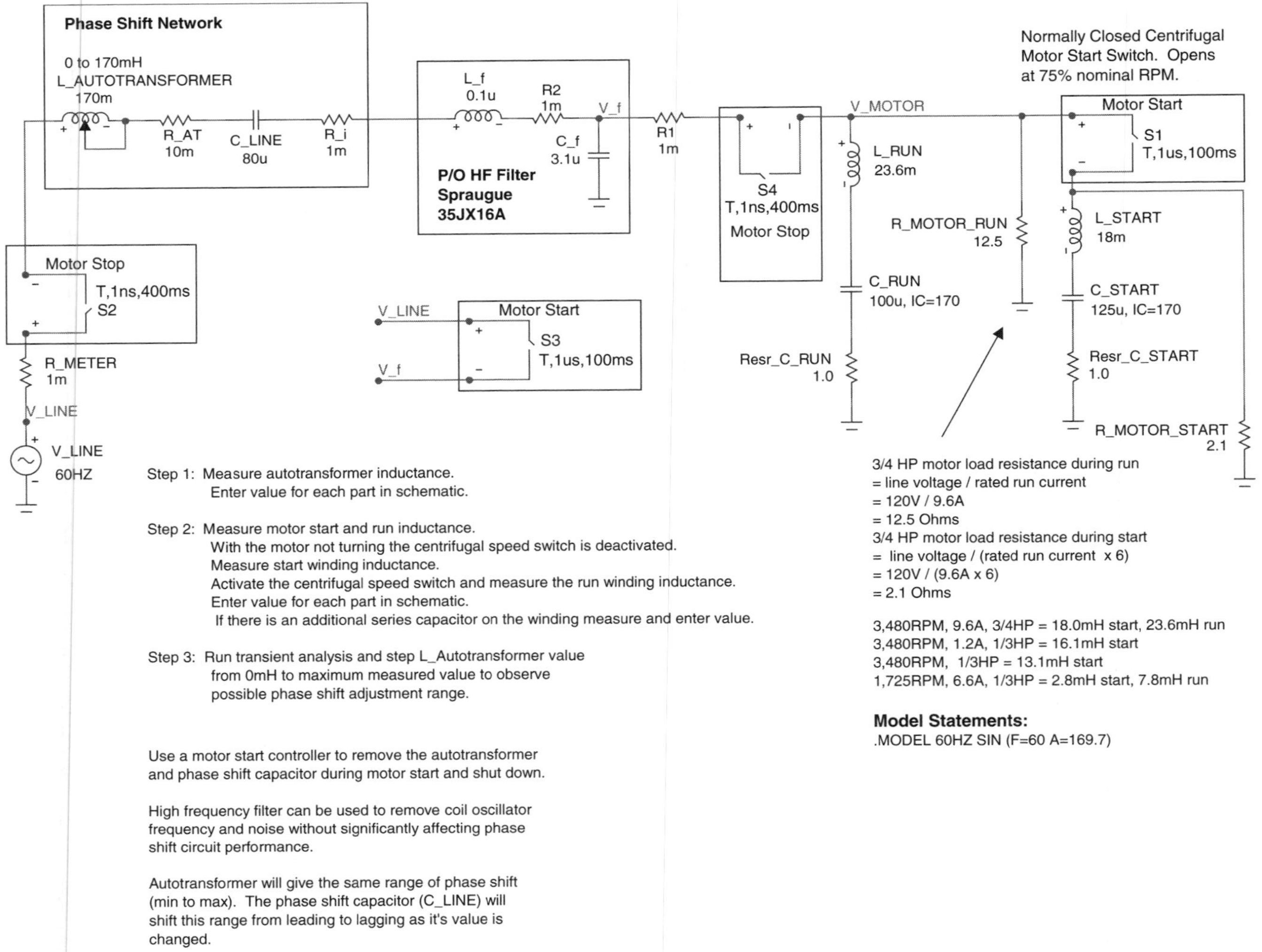

FIGURE 8-27 Phase shift transient model.

The resistances were added to ensure convergence and can be left as they are or enter the measured DC winding resistance of the autotransformer into the R_AT part value.

The optional HF filter can be modified, left as is or removed from the circuit.

The capacitor ESR values in the circuit are not critical to the evaluation and can be changed to a known or calculated value, left as is or removed from the circuit.

The load resistance of the operating run winding is calculated using the line voltage divided by the rated run current of the motor. If the rated current is unknown see Section 7.16. Enter the calculated value in the R_MOTOR_RUN part value. The load current during start is up to 6 times greater than the run current therefore the resistance during start-up is typically 1/6 the run value. Enter the calculated value in the R_MOTOR_START part value.

Step 2. Measure the motor start and run winding inductances. With the motor not turning the centrifugal speed switch is deactivated. Measure the start winding inductance. Enter this value in the L_START part value.

Access the motor start capacitor if included on the motor and measure the capacitance or record the labeled value. Enter this value in the C_START part value.

Access the centrifugal speed switch in the motor activating it by hand and measure the run winding inductance. Enter this value in the L_RUN part value.

If there is an additional series capacitor on the run winding measure the capacitance or record the labeled value. Enter this value in the C_RUN part value.

Step 3. Run transient analysis and step the L_AUTOTRANSFORMER value from 0 mH to the maximum measured value to evaluate the phase shift adjustment range.

The time periods of the transient switches have been selected for an optimum display. The start-up time period may be much longer. To extend the start-up time change the 100 ms value in part S1 and S3 to the desired time period. If a longer run time is desired before the shutdown and braking effects are modeled change the 400 ms value in part S2 and S4 to the desired time period.

A transient analysis was run on the modeled circuit shown in Figure 8-27. The start-up, motor run, shutdown, and braking effects are shown in Figure 8-28 with the autotransformer set to its minimum value of 0 mH. In Figure 8-29 the inductance is changed to its maximum value of 170 mH and the motor voltage phase is shifted from its original position. Note the phase shift that occurs between the voltage driving the motor (black trace in upper graph) and the line voltage (red trace in upper graph), which is also supplying the step-up transformer. As the phase driving the motor is shifted, an arbitrary point on the rotor of the rotary spark gap is also shifted in relation to its position in the 360° arc of rotation. This effectively shifts the position of the rotating electrodes until their alignment with the stationary electrodes coincides with the line voltage peak applied to the step-up transformer. The gap ionization is now synchronous to the line peaks.

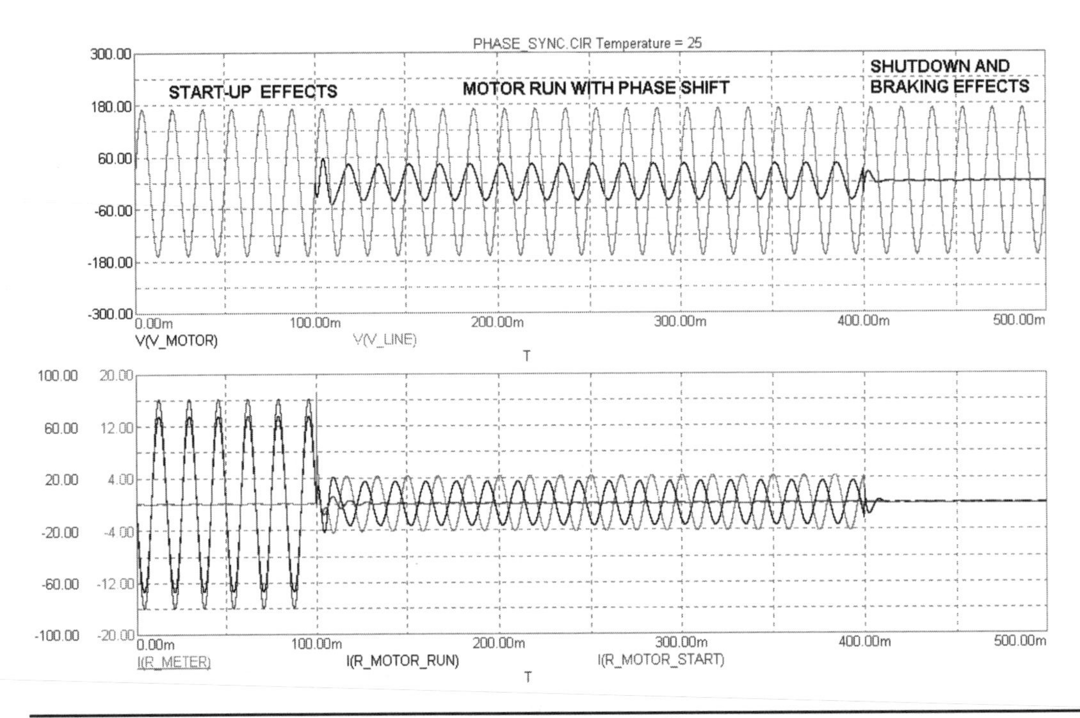

FIGURE 8-28 Results of transient analysis (autotransformer = 0 mH).

As seen in Figures 8-28 and 8-29 the motor start current is about 6 times higher than the run current due to the value of load resistance set in the model. When the value of C_LINE in the phase shift network is set to 40 μF there is much less phase shift between the minimum and maximum autotransformer settings. When the capacitance is increased to 120 μF little additional phase shift is provided. As the capacitance value is increased from 40 μF to 120 μF the voltage drop in the phase shift network increases and at some point the motor will not have enough voltage to run. The capacitance value will have to be selected to provide the desired phase shift with a low enough voltage drop to run the motor. In this example a capacitance between 40 μF and 80 μF would provide the desired operating characteristics and will shift the phase at least 90°.

The AC model shown in Figure 8-30 provides additional insight into the performance of the phase shift circuit. The shift in phase angle of the voltage supplying the run winding in the motor and the line is shown in the AC analysis results in Figure 8-31. The autotransformer inductance is stepped from 0 mH to 170 mH. At 30 mH a phase shift of 43° is provided at 60 Hz up to a phase shift of –108° at 170 mH.

Again, the phase shift circuit is only useful in rotary spark gaps to achieve synchronous operation with the line voltage. If the phase shift circuit does not produce desired results with the motor of your selection don't be dismayed. A higher break rate (>400 BPS) will produce results comparable to a synchronous design as detailed in Chapter 2. The higher break rate

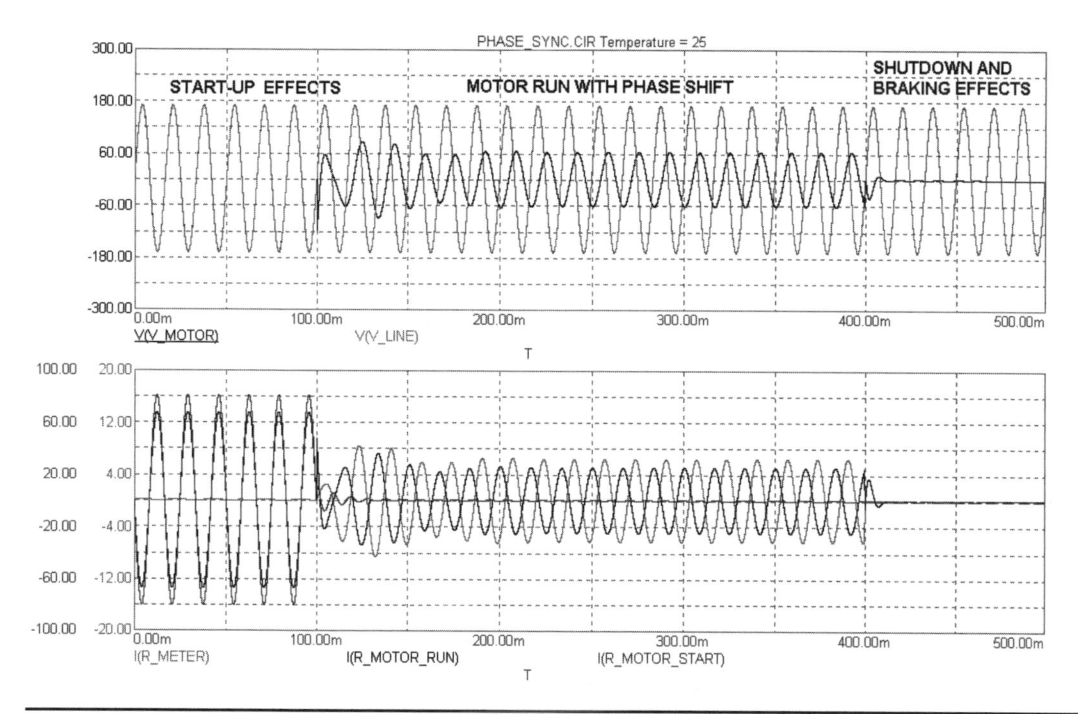

FIGURE 8-29 Results of transient analysis (autotransformer = 170 mH).

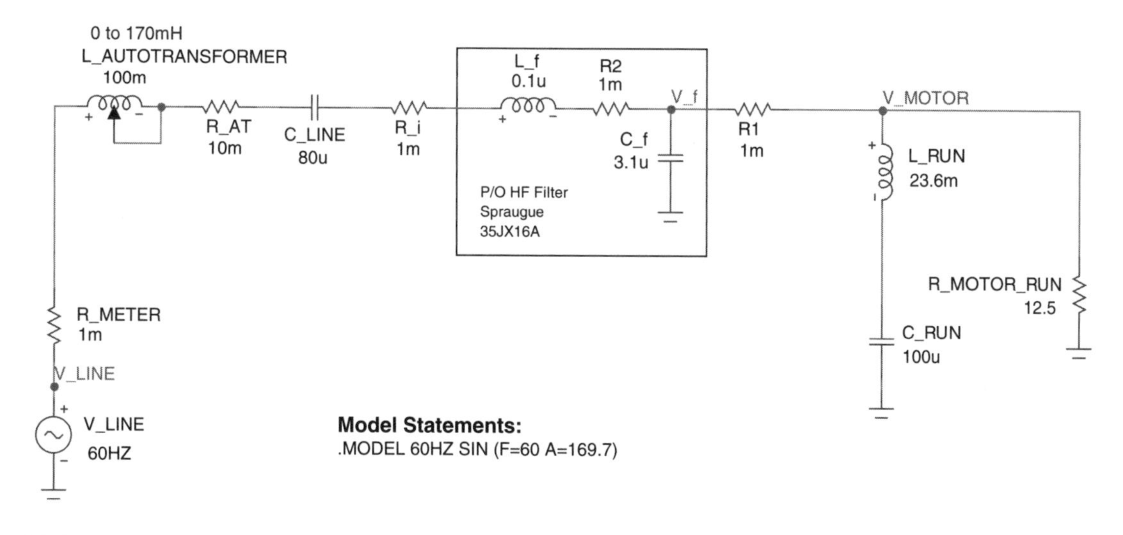

FIGURE 8-30 Phase shift AC model.

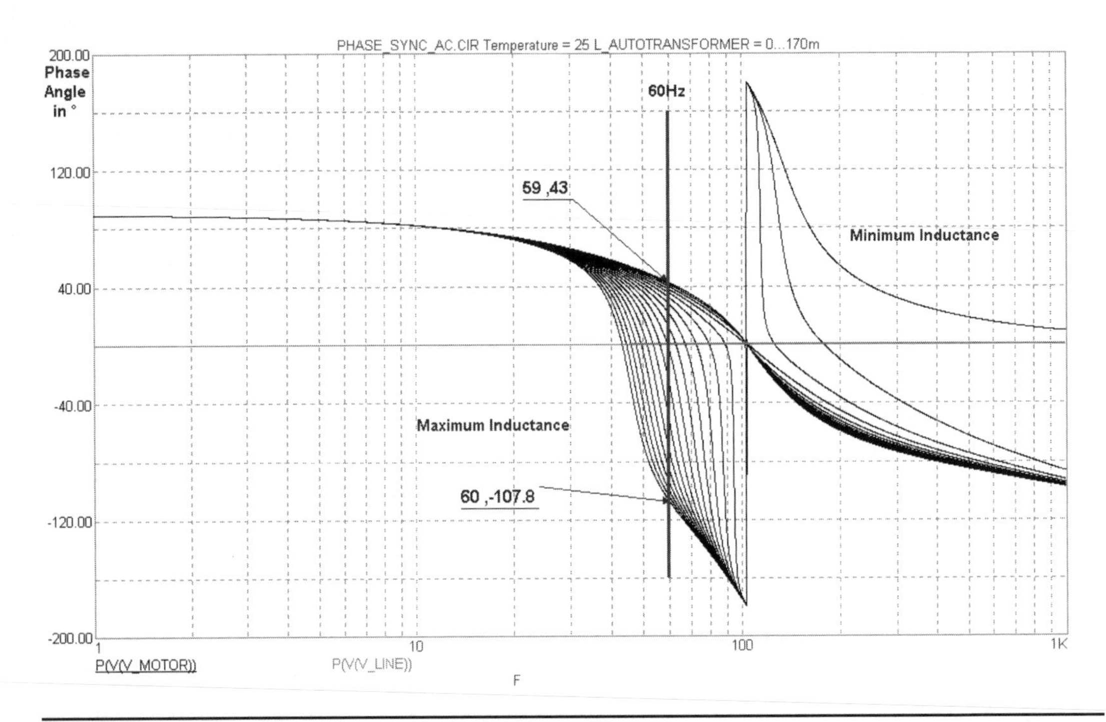

FIGURE 8-31 Results of AC analysis.

will draw more power from the line but the high-voltage spark will be about as long as you can make it by varying the design's BPS.

References

1. Spectrum Soft website @ http://www.spectrum-soft.com/.
2. John Freau. *Easy Remote Sync Gap Phase Adjustment*. TCBA News, Vol. 20, #2: 2001, pp. 13, 14.

CHAPTER 9

Coil Construction

I have produced electrical discharges the actual path of which, from end to end, was probably more than one hundred feet long; but it would not be difficult to reach lengths one hundred times as great. I have produced electrical movements occurring at the rate of approximately one hundred thousand horse-power, but rates of one, five, or ten million horse-power are easily practicable.
Nikola Tesla. The Problem of Increasing Human Energy. Century Magazine: 1900.

Tesla comments on the virtually limitless possibilities of voltage generation in a Tesla coil.

Constructing the Tesla coil presents many challenges. Theory is one thing, transforming your design calculations into a working coil requires manual skills, attention to detail, patience, and trial and error. Here is where veteran coilers and especially machinists stand out in the crowd. The construction techniques presented in this chapter are only examples and recommendations from my own coiling experience. If you are not confident this is within your ability, subcontract the work out to a qualified machinist. As you construct more Tesla coils your level of ability will increase and you will eventually be able to construct your own experimental prototypes.

The primary and secondary windings, spark gap, dedicated ground, control and monitoring, and interconnections will all be constructed using stock materials. Dedicated ground, control and monitoring, and interconnections are detailed in Chapter 7. Figure 2-1 shows the interconnections used to build a medium coil, which are typical for any spark gap coil. Note the output winding of the step-up transformer is shown in parallel (shunt) with the rotary spark gap and in series with the tank capacitor. This is often shown reversed in Tesla coil schematics. If these components are reversed from the arrangement shown in the figure the capacitor and step-up transformer may have additional electrical stresses placed on them. Unless you are experimenting and want to observe the effects, connect these as shown in Figure 2-1.

9.1 Construction Techniques for the Resonant Transformer

The following construction procedures are based on the design calculations performed in the CH_2.xls file, AWG vs. VS worksheet (1). These calculations are detailed in Chapter 2. The design calculations are summarized in CH_2.xls file, CONSTRUCTION SPECS worksheet (3) and shown in Figure 9-6 for the example coil designed in Figure 2–7 (see App. B).

9.1.1 Constructing the Secondary

The secondary winding is constructed as follows:

1. Find a suitable coil form of the desired diameter. This can be made of any insulating material. I generally use PVC pipe as it can be procured from most hardware stores in 2.375″, 3.5″, 4.5″, and 6.5″ diameters. A 10-ft section of 4.5″ PVC costs less than $10, which makes it a cheap source for winding form stock. Standard hardware stock is about 1/4″ thick but I have seen PVC 1/8″ thick in a slightly smaller diameter of 4″, which is very good. The pipe is rigid enough to avoid distortion when the windings are put on and light enough to add little weight to the final assembly. The 6.5″ pipe gets heavy when used in long windings of 3 feet or more. A variety of PVC end caps makes mounting the finished winding and terminal capacitance as easy as putting a cork in a bottle. Another advantage to using PVC is the ease of changing the coupling coefficient. I made a spark gap coil using a helically wound primary. The coupling turned out to be too tight with lots of secondary breakdown and arcing from secondary-to-primary. A 4″ high piece of PVC was cut and using a PVC coupling the secondary was raised a total of six inches from the base of the primary. This decreased the coupling enough to enable smooth operation of the coil. It only took about ten minutes and two pieces of plastic to effect. The PVC is also strong enough to support a lot of weight. This can be advantageous when experimenting with large terminal capacitances.

   Cardboard tubing also makes a suitable form but should not have any ink printing on the outside layer. I tried using the large diameter (8″, 10″, and 12″) cardboard tubes used for concrete molds. Some of the forms exhibited voltage breakdown at the vicinity of the ink printing while other forms exhibited no breakdown. Plastic materials such as plexiglass can also be used. They make nice coil forms but are expensive. Occasionally a bakelite form can be found at the Hamfest. Veteran coiler Tom Vales has noted problems using dark plastics, especially those colored black. The pigments used are often conductive and the form will enhance corona losses and breakdown between the windings. Do not paint the form for the same reason.

   Another material used in old coil construction (before plastics) was wood. Thin strips of wood were used to make an octagonal cage type structure. The windings were not round, but octagonal. Magnet wire really wasn't available yet so an interwinding spacing was used. The open space construction offered excellent form rigidity and insulation when coated with varnish or beeswax. This type construction is very labor intensive but offers an antique look to your coil. You may develop some surprising spark lengths using these older techniques.

   Some coilers go through great lengths insulating the form before the winding is applied. If a cardboard material is used it is usually necessary to apply a coat or two of varnish or polyurethane and let it dry 24 hours before proceeding. I do no preparation work on PVC material other than a light sanding with 320 grit abrasive for a better adhesive surface. Other coilers prefer to apply a coat of varnish or polyurethane and dry until the surface is tacky to facilitate winding the coil. Try different techniques to find which is best for you. Elaborate baking and varnishing techniques are probably not worth the effort and will not significantly impact the performance of the coil.

Cut the form to the desired length. Using the calculations to determine the total winding height add 2 to 3 inches to each end. The required height for the selected resonant frequency is found in cell (F15) in the CH_2.xls file, AWG vs. VS worksheet (1) or in the CH_2.xls file, CONSTRUCTION SPECS worksheet (3) summary. When using PVC pipe I usually add two inches to each end for insertion into end caps. I add another two inches to each end to work with wire connections. If you are unsure what to use take the winding height and add four inches to both ends of the winding and cut the form to this length. It is better to have it too long than too short once the winding is completed!

2. Using the calculations to determine the required wire length, ensure your magnet wire is long enough to finish the winding. If you have been collecting wire or have access to a complete roll of magnet wire this should not present a problem. My first coil had a splice in the middle of the winding as the roll of wire used was not long enough to complete the winding. Corona and spark discharges emanated from this splice no matter how much insulation was applied. When I wound my first coils I did not have a source for magnet wire and was unwinding scrap motors to obtain enough magnet wire. This does not work well as the wire insulation has a tendency to disappear or break during the unwinding and subsequent winding operations.

3. Mark the center of the form and measure up one half the winding height and place another mark. From the center measure down one half the winding height and place another mark. Start the first turn of the winding around the bottom mark using a piece of tape to temporarily fasten the end. Leave at least a foot of the beginning of the wire extending out the bottom of the winding for interconnections. Using an even, moderate tension wind about 20 turns and place another piece of tape to temporarily fasten the other end. I use low-tack masking tape for this and always keep a couple of pieces within reach. Once you let go of the winding the tension is released and the turns will unravel. This means you get to start over. Using a hot glue gun permanently fasten the first few turns of the winding to the form and let cool. Super glue or an epoxy can be used but the hot glue will set much faster allowing you to continue winding. Grab the wire from the roll feeding into the winding with one hand and remove the tape keeping tension on the winding with the other. Continue winding until you get tired. When you need to put the coil down replace the tape to hold the winding in place. It is heartbreaking to get close to the end of a large winding project and have it unravel. Once the winding is completed use the hot glue gun to permanently fasten the top windings to the form. Before cutting the wire off the spool double check the winding height and leave at least three feet of unwound wire extend out the top of the winding for interconnections. Do not apply excessive winding tension, as it will distort the wire and its electrical characteristics. This is a technique learned through experience so don't be disappointed if your first coil doesn't wind properly. Try it again and slow down, paying attention to detail. If you have selected to use an interwinding space try using a nylon or poly type line (e.g., fishing line) of the desired diameter as a spacer. Place the line between the first two magnet wire windings and continue winding both materials together. Plastic line will provide a consistent interwinding space unlike a cotton line material. Winding a coil with an interwinding space presents more of a challenge than one without a space.

I wound my first several coils by hand using this method including a 3,000 turn, 4.5″ diameter winding using 29 AWG magnet wire which took a whole weekend. I finally got around to making an improvised. The jig was made from four 3/8″ × 16 (threads per inch) threaded rods, two aluminum plates, and a DC motor all mounted to a wooden base. The PVC form is laid horizontally and attached to a PVC cap fastened to the DC motor, which is mounted in the center of an aluminum plate. The other end of the form is attached to another PVC cap fastened to a spindle, which is mounted to the center of another aluminum plate. The four threaded rods secure the two aluminum plates using nuts, allowing the width of the assembly to be adjusted for form heights of 12″ to 60″. A coupling nut can be used to join two threaded rods for even longer lengths. Two end caps for the 2.375″, 3.5″, 4.5″, and 6.5″ diameters can be dedicated for mounting the PVC form to the motor and spindle ends. I also cut end caps out of 1/4″ pressboard using a router and tape the coil form to the pressboard caps after they are mounted to the motor and spindle. Taping the coil form to custom cut end caps enables the use of any form diameter and the tape will hold long enough to wind the coil (duct tape works best, many thanks are extended to Red Green). A speed control was constructed for the DC motor using a variable autotransformer, bridge rectifier, and filter capacitor. A toggle switch and relay were also used to switch power supply polarity to the motor and change the direction of rotation. A small motor was selected so the motor torque could be overcome with moderate hand pressure and slow or stop the motor's rotation. This adds extra control when winding. Simply adjust the DC voltage to the motor to obtain a controllable winding speed and grip the form harder with your free hand to slow the rotation down, or loosen the grip to speed it up. This provides a wide variety of coiling control. If the motor is too small it will overheat using this friction method of winding. A lathe can be used but offers too much torque to control the speed with your hand. AC motors offer the same lack of controllability. It now takes about 1 to 2 hours to wind almost any coil instead of a weekend. Construction details for the winding jig are shown in Figure 9-1.

9.1.2 Constructing the Primary

The primary winding is constructed as follows:

1. If a helically wound primary coil was chosen the outside diameter, interwinding distance and number of turns have been selected. A form must be built to place the selected winding geometry on. It must also take into account the selected coupling characteristics (geometrical relationships with the secondary) outlined in Section (4.9). These relationships have been selected during the design calculations and must be considered when building the primary form.
2. Unlike the close-wound magnet wire winding in the secondary an open-air construction should be used in the primary. The conductors used in the winding are typically bare to facilitate a tap used to select any of the windings for tuning. The open-air construction and interwinding distance are necessary to provide insulation between the turns of bare wire. Wood is probably the cheapest material, is easy to form, and provides adequate electrical resistance. The simplest primary form will take the shape of a cube. With the rectangular cube no complex angles need to be calculated. Any number

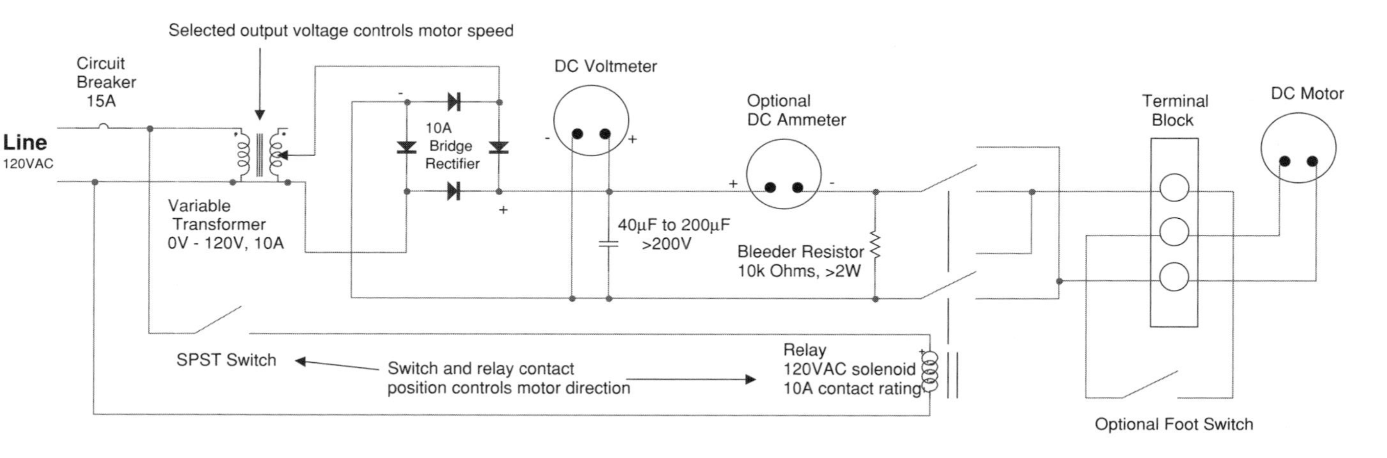

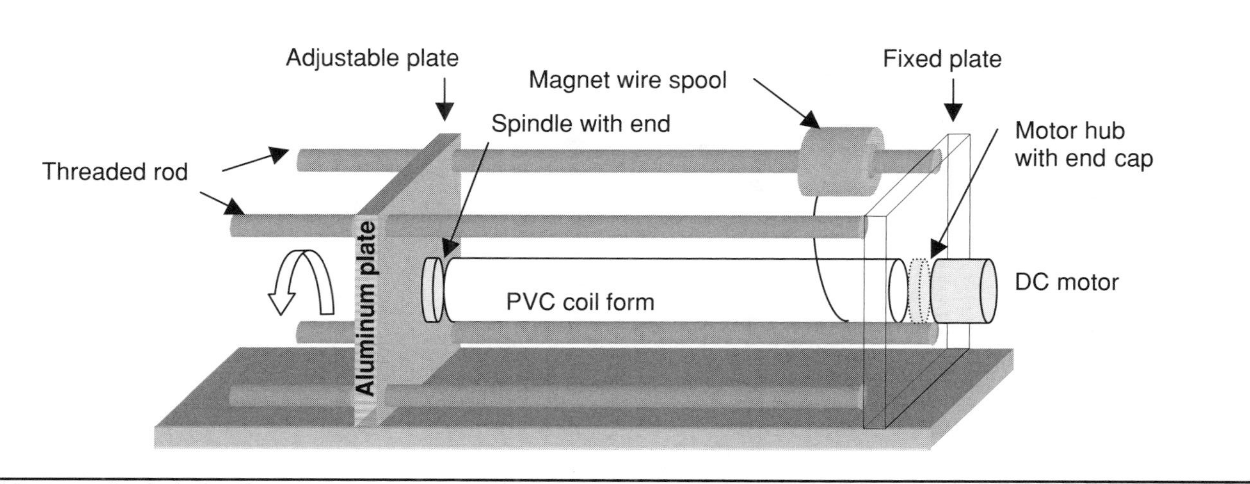

FIGURE 9-1 Improvised winding jig.

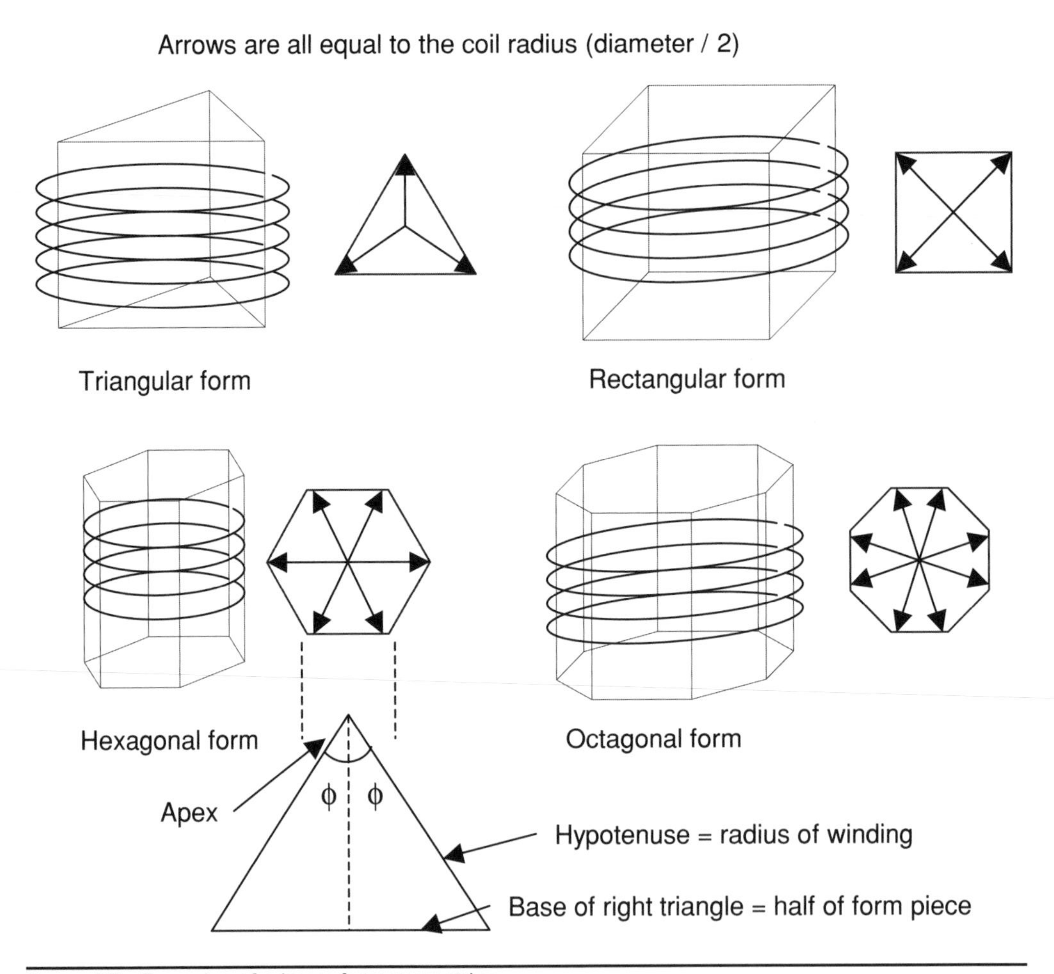

FIGURE 9-2 Examples of primary form geometries.

of vertical supports can be used, some examples are shown in Figure 9-2. The form will require a horizontal base of triangular, square, hexagonal, or octagonal shape. The respective number of vertical supports required is three, four, six, or eight. A duplicate of the horizontal base is used on top of the form to affix the vertical supports. These simple geometries offer the best structural and dielectric strength.

Choose a cube form and determine the dimensions needed for the selected diameter and winding height. Cut the base materials and complete the upper and lower horizontal bases. Table 9-1 provides details for calculating the dimensions of each piece in the horizontal base. Each piece forms the base in an isosceles triangle (two equal sides with sum of interior angles of 180°) shown in Figure 9-2 where the hypotenuse dimensions are equal to the radius of the primary winding (outside diameter/2). When each isosceles triangle is divided in half, each half becomes a right triangle with the base

Form Geometry	Length of each horizontal base piece	Joint angle (apex/2) of each horizontal base piece (ϕ)
Triangular	$\frac{OD}{2} \sin\phi \bullet 2 = 0.866$ radius$\bullet 2$	Apex $= 360° \div 3 = 120°$ $\phi = 120° \div 2 = 60°$
Rectangular	$\frac{OD}{2} \sin\phi \bullet 2 = 0.7076$ radius$\bullet 2$	Apex $= 360° \div 4 = 90°$ $\phi = 90° \div 2 = 45°$
Hexagonal	$\frac{OD}{2} \sin\phi \bullet 2 = 0.5006$ radius$\bullet 2$	Apex $= 360° \div 6 = 60°$ $\phi = 60° \div 2 = 30°$
Octagonal	$\frac{OD}{2} \sin\phi \bullet 2 = 0.3836$ radius$\bullet 2$	Apex $= 360° \div 8 = 45°$ $\phi = 45° \div 2 = 22.5°$

TABLE 9-1 Primary form geometry dimensions.

being half the base of the isosceles triangle and the hypotenuse being the winding radius. We now have two known values in a right triangle, the hypotenuse (radius) and angle phi (ϕ). Pythagorean's theorem is used to solve for the base in a right triangle and twice the base would be the base of the isosceles triangle. The column in Table 9-1 labeled joint angle is one half the apex of the isosceles triangle.

For instance a rectangular form is chosen for an 18″ diameter winding. The length of each horizontal piece is: [OD / 2] × sin(ϕ) × 2 = [18″ / 2] × sin(45°) = [9″] × 0.707 × 2 = 12.7″. The calculated length is actually from the center of the joint at one end of the piece to the center of the joint at the other end of the piece. A paper or cardboard template can be made to use as a guide and double check all your measurements before the wood is cut. Use a good adhesive and dowel pins to join the ends together. I recommend a construction adhesive in the caulking tubes (e.g., Liquid Nails) or Gorilla Glue. It is messy so use disposable latex gloves when working with it. Screws or other metal fasteners can be used to hold the form in place as the adhesive sets; however, remove the fasteners once dry.

CH_4.xls file, PRIMARY DIMENSIONS worksheet (10) performs the calculations used to determine the length of the horizontal supports in a triangular form (F12), rectangular form (F18), hexagonal form (F24), and an octagonal form (F30) for the diameter entered in cell (B3). The minimum height of the vertical supports is calculated in cell (F4) for the number of windings entered in cell (B5) and interwinding distance entered in cell (B4).

3. Determine the height of the vertical supports by the number of windings needed multiplied by the interwinding distance. The height of the primary winding for the number of turns needed to tune to resonance was calculated in cell (F59) of CH_2.xls file, AWG vs. VS worksheet (1) during the design phase of the project; however, a few more turns are typically needed to tune the actual coil so always add a few more holes in each support for additional windings. This will increase the vertical support height. About two more inches of height should be added to each end for attaching the support to the horizontal upper and lower bases. After this final height is determined

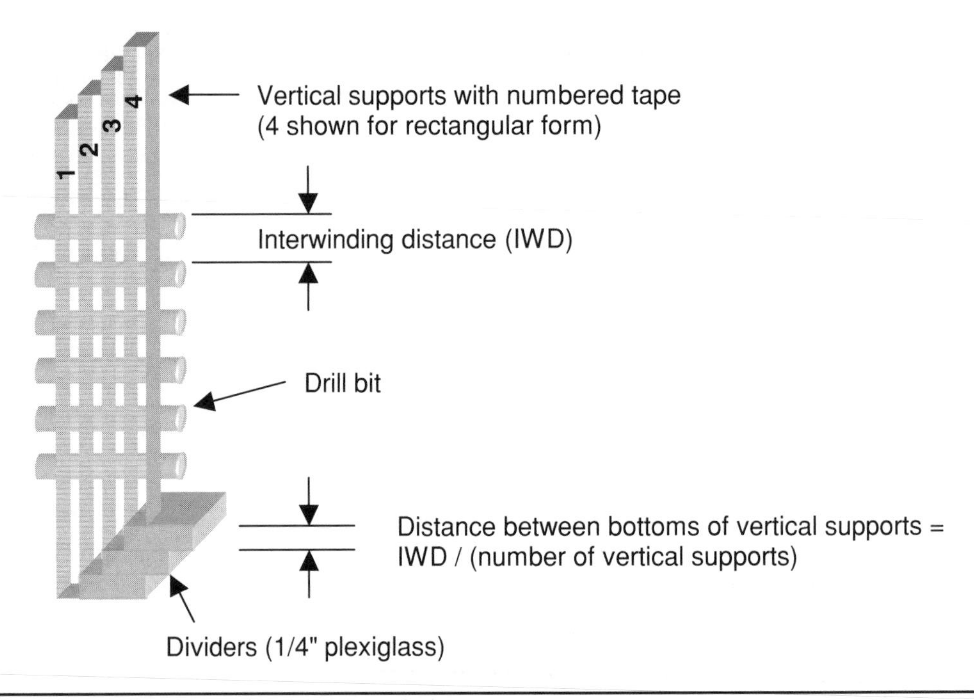

Figure 9-3 Offsetting the holes in the vertical supports.

cut the vertical support material. The vertical supports should have the holes for the winding drilled before they are mounted to the upper and lower horizontal bases. Using a marker and tape, number the support pieces as shown in Figure 9-3 and do not remove until indicated. The holes must be offset from the bottom of each support as shown to enable a continuous helical winding.

For instance a 1″ interwinding distance (IWD) is selected using four vertical supports (#S) in a rectangular cube form. The bottom of each vertical support should be separated by a distance of: IWD/#S = 1″/4 = 0.25″. Using 3 strips of 0.25″ plexiglass or similar material, the strips and vertical supports are stacked as shown in Figure 9-3 and the vertical supports clamped together. Mark the holes and drill in a press to ensure they are straight. This can be done with a hand drill if you can drill a straight hole. The diameter of the hole should be larger than the diameter of the tubing being used for the primary. Try a test hole and ensure the tubing slides easily through the hole. I drill these holes at least 1/8″ larger than the tube diameter. Soft woods such as pine are not recommended for these support pieces. Drill bits tend to chew apart the exit hole in softer wood. Better results are obtained with a hardwood such as maple and the holes made with a router instead of a drill bit. This makes a nice round hole where it is supposed to be and the wood surrounding the hole is intact. If a router is used the holes will come out intact in the pine also. The method used will depend on your woodworking skills.

4. Using dowel pins and adhesive, join the vertical supports to the horizontal upper and lower bases in the order shown in Figure 9-4. Drill out the holes for the dowel pins

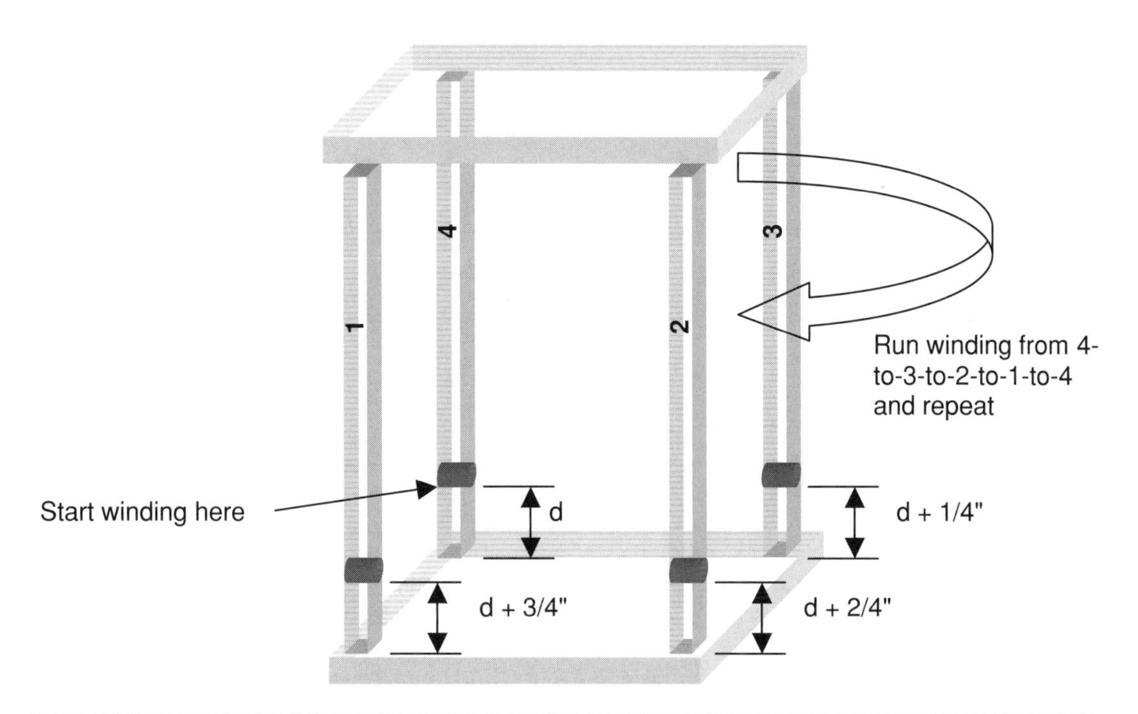

FIGURE 9-4 Vertical support mounting order to horizontal bases.

one at a time as the assembly progresses. Support the entire assembly with clamps if necessary for 24 hours. The adhesive must be completely dry before the windings are applied. As a rule I let everything dry for 24 hours. Once dry, start the tubing in the bottom hole of support number 4 (the numbered tape should still be on the support). Run the tubing through the bottom hole in support number 3, the bottom hole of support number 2, the bottom hole of support number 1 and the next hole up (second to bottom) in support number 4. This completes the first turn. Repeat for the number of turns used. If the correct order is followed and the holes are at least 1/8″ oversized the tubing will slide fairly easily through the holes. As more windings are applied it will take some practice to get the tubing through the holes. The entire winding will take some time so if it gets difficult break up the winding sessions to allow a decompression period. The primary can also be wound starting at the other end of the form, winding in the reverse direction from that shown in the figure.

5. If an Archimedes spiral primary coil was chosen the inside and outside diameter, inter-winding distance, and number of turns have been selected or calculated in the work-sheet. The Archimedes spiral form is built in a similar manner as the helically wound form. The inside diameter and angle of inclination are the only additional concerns. I use a single piece of particle board with a Formica surface as found in shelving, cut in a triangle to form the vertical supports as shown in Figure 9-5. The shelving material is strong and easy to drill and cut; however, it is denser and increases the total weight. The minimum height of the vertical supports (H) is determined by the

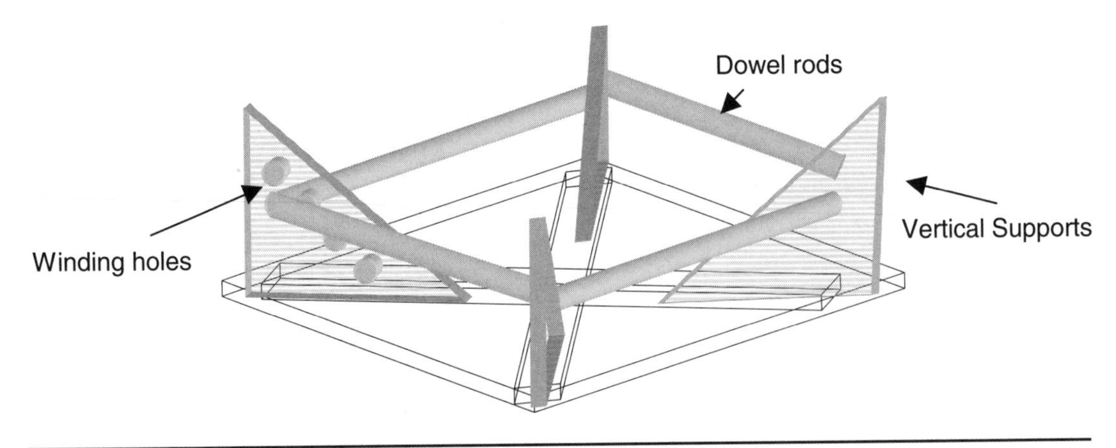

FIGURE 9-5 Vertical and horizontal supports used in an Archimedes spiral.

angle of inclination (θ), interwinding distance (IWD), and the number of windings (N) needed, and is calculated in cell (F59) of CH_2.xls file, AWG vs. VS worksheet (1) using equation (2.21). Add holes for a few extra turns and two inches to the top and bottom of the support for fastening to the horizontal bases. The lower horizontal base includes diagonal support pieces through the center. I use dowel rods as the upper horizontal support pieces. These are easily cut and attached to the vertical supports using adhesive. I have found the tubing is easier to wind if started on the inside diameter and working out than when started from the outside working in, but is a matter of preference. It will take practice to develop a personal technique to wind the tubing on the form.

Shown in Figure 9-6 is a summary of construction details from CH_2.xls file, CONSTRUCTION SPECS worksheet (3) for the design calculations performed in CH_2.xls file, AWG vs. VS worksheet (1), which are shown in Figure 2-7. The summary is useful during the construction phase as it highlights the dimensions needed to build the primary and secondary forms. The number of turns used in each winding is also shown. Other parameters that are often sought during construction or during test runs were also included. The summary can be printed for consultation during construction and test. All parameters shown in the summary were calculated in the design worksheet with two exceptions. Enter the secondary form height in cell (E31), which is addressed in Section 9.1.1 step 1. Also enter the wire gauge used in the primary in cell (B3) to assist in determining the primary height needed in the primary form.

9.2 Construction Techniques for the Rotary Spark Gap

If a single component of the Tesla coil could be named the most important it would be the spark gap. As mentioned in Chapter 7 the stationary spark gap operates at 120 BPS regardless of the gap setting. The stationary gap can be made to handle large primary power levels. On the other hand the rotary gap is designed to operate at break rates above 120

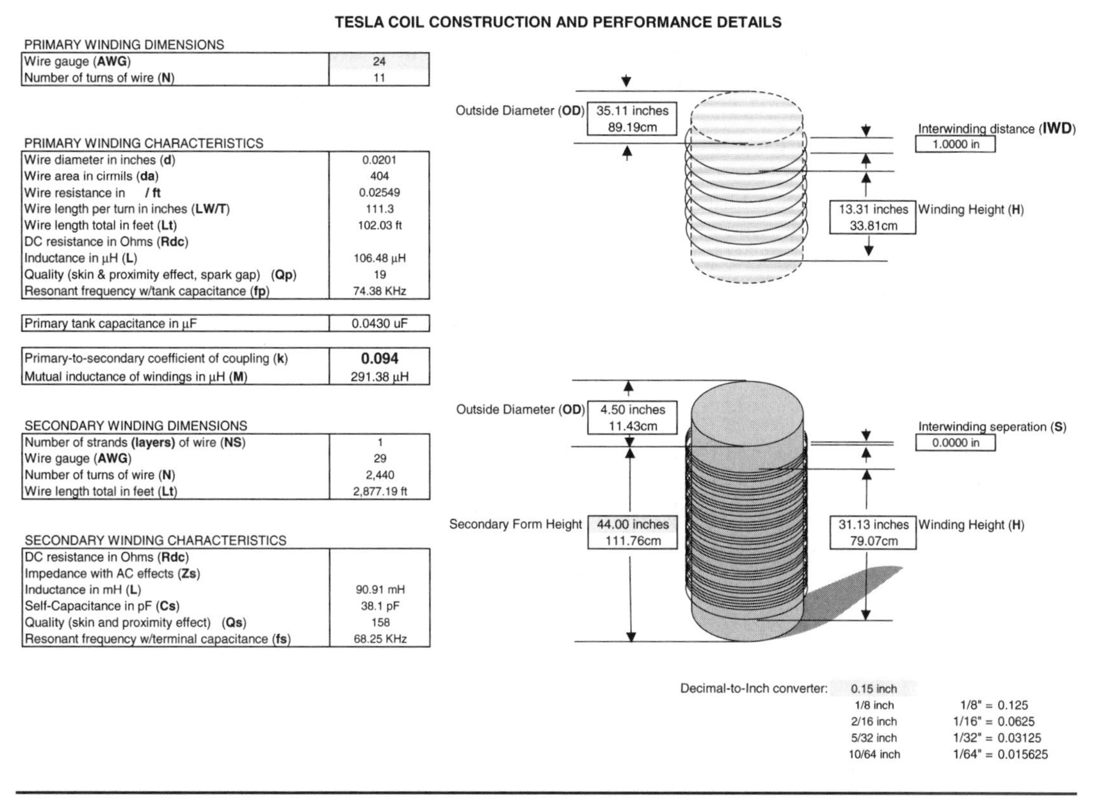

FIGURE 9-6 Summary of construction details for design calculations in Figure 2-7.

BPS and typically operates as a timing switch. It is more difficult to construct than the stationary gap but offers immediate rewards with increased coil performance and brighter spark output. Of course your controls will have to be capable of handling the increased power.

9.2.1 Determining the Operating Temperature of the Rotating and Stationary Electrodes in a Rotary Spark Gap

The rotating and stationary electrodes of the rotary spark gap must be able to handle the power levels in the primary circuit. As the power levels rise so does the power dissipated in the gap, with a corresponding rise in the operating temperature of the electrodes. The electrode operating temperature must be kept below the melting temperature of the material for obvious safety reasons. The electrode operating temperature should be determined to ensure it is well below the melting temperature of the material.

To calculate the electrode operating temperature, open the CH_6B.xls file, GAP TEMP worksheet (2). Whatever primary power is not transferred to the secondary and leakage inductance will be dissipated in the spark gap, assuming negligible losses in the tank capacitor and primary winding. This primary power was calculated in Chapters 2 or 6 and is entered in cell (B4) of the worksheet. The desired distance between the stationary and rotary electrodes is entered in cell (B3). Enter the required stationary electrode characteristics in cells (B8) through (B10) and the stationary electrode mounting characteristics in cells (B12) through (B16). If an optional heat sink is added to the stationary electrode mounting to further reduce the electrode operating temperature enter the required characteristics in cells (B18) through (B22). Enter the required rotary electrode characteristics in cells (B24) through (B27), and operating characteristics common to both stationary and rotary electrodes in cells (B29) through (B32).

The stationary and rotary electrodes are typically made from the same material. The most efficient and safest shape for a rotary spark gap electrode is the rod and will be assumed for all calculations. The formula for calculating the surface area of a cylinder is found in reference (6) and used to determine the surface area of a rod electrode:

$$SAe = \pi\, De\, Le + 2Aef \tag{9.1}$$

Where: SAe = Surface area of the electrode in inches = cell (F5), calculated value.
De = Diameter of the electrode in inches = cell (B8), enter value.
Le = Length of electrode in inches = cell (B9), enter value.
Aef = Surface area of electrode face in inches = calculated value from reference (5) for determining the surface area of a cylinder face:

$$Aef = \frac{\pi}{4} De^2$$

NOTE: *Surface area of the rotary electrodes (SAr) in cell (F12) and rotary electrode face (Arf) in cell (F11) are calculated using the same methodology.*

The mounting for the stationary electrode will follow either a cylindrical or rectangular shape. If the mounting is cylindrical enter a (1), or if rectangular enter a (2) in cell (B12). The surface area for a cylindrical stationary electrode mounting is:

$$SAm = \pi\, Dm Lm + 2\left(\frac{\pi}{4} Dm^2\right) \tag{9.2}$$

Where: SAm = Surface area of the stationary electrode mounting in inches = cell (F7), calculated value.
Dm = Diameter of the stationary electrode mounting in inches = cell (B13), enter value.
Lm = Length of stationary electrode mounting in inches = cell (B15), enter value.

If the stationary electrode mounting is rectangular the surface area is:

$$SAm = 4C\,Am Lm + 2C\,Am \tag{9.3}$$

Where: SAm = Surface area of the stationary electrode mounting in inches = cell (F7), calculated value.
CAm = Cross-sectional area of stationary electrode mounting in inches^2 = cell (B14), enter value.
Lm = Length of stationary electrode mounting in inches = cell (B15), enter value.

If a heat sink is mounted on the stationary electrode mounting the total surface area is increased and the mounting can dissipate more heat. The surface area for a finned heat sink is:

$$SAs = Ws\,Ls + 2(nf\,Ls\,Hf) \qquad (9.4)$$

Where: SAs = Surface area of the optional stationary electrode mounting heat sink in inches = cell (F9), calculated value.
Ws = Width of heat sink (perpendicular to fins) in inches = cell (B18), enter value.
Ls = Length of heat sink (parallel to fins) in inches = cell (B20), enter value.
nf = Number of fins used in heat sink = cell (B19), enter value.
Hf = Height of fins in inches = cell (B21), enter value.

The calculated surface area of the rotary electrodes in cell (F12) for the diameter entered in cell (B24) and length in cell (B25) uses equation (9.1), the same as the stationary electrode surface area.

The thermal conductivity of the rotating and stationary parts is determined for use in evaluating the conduction cooling mechanisms of the application. References (1) through (4) were used to construct the thermal properties of materials listed in Table 9-2. This table is used with a lookup function to derive the thermal conductivity when the desired material number in the left column of the table is entered into the following cells:

- The thermal conductivity of stationary electrode material (ke) in W/m-°C is derived in cell (F16) for the material number entered in cell (B10).
- The thermal conductivity of stationary electrode mounting material (kM) in W/m-°C is derived in cell (F18) for the material number entered in cell (B16).
- The thermal conductivity of stationary electrode material (kS) in W/m-°C is derived in cell (F20) for the material number entered in cell (B22).
- The thermal conductivity of rotary electrode material (kR) in W/m-°C is derived in cell (F23) for the material number entered in cell (B26).

CAUTION: *It should be noted that although the metals listed in the above table are found in publications for arc work and have been used as far back as early spark gap transmitters, they may present hazards to the user. Materials such as Beryllium Copper are particularly hazardous when grinding and the dust is inhaled. Even trace amounts on the hands can be ingested and present health hazards later. Consult the applicable Material Safety Data Sheet (MSDS) available from the supplier. The data sheets can also be found on dozens of free websites from an Internet search if not available from the supplier.*

The convection heat transfer coefficient (h) is also determined for the rotating and stationary parts for use in evaluating the convection cooling mechanisms of the application:

- The convection heat transfer coefficient of stationary parts (hS) in W/m^2-°C is derived in cell (F21) for the following gap characteristic entered in cell (B29):

  2.8 for vacuum or reduced pressure (70,000 ft altitude) gap environment is derived by entering (1) in cell (B29).

Material Number	Material Type	Melting point in °C	Melting point in °F	Thermal Conductivity (k) in W / m -°C
1	Aluminum	660°C	1,220.0°F	211
2	Brass (66% Cu, 34% Zn)	920°C	1,688.0°F	122
3	Copper	1,210°C	2,210.0°F	381
4	Beryllium Copper	1,800°C	3,272.0°F	105
5	Iron	1,535°C	2,795.0°F	67
6	Lead	327°C	620.6°F	34
7	Magnesium	651°C	1,203.8°F	158
8	Mercury	−39°C	−38.0°F	8.3
9	Nickel	1,455°C	2,651.0°F	90.6
10	Phosphor-bronze	1,050°C	1,922.0°F	69.2
11	Silver	961°C	1,760.9°F	415
12	Steel, SAE 1045 (0.5% C, balance Fe)	1,480°C	2,696.0°F	67
13	Steel, 18-8 stainless (0.1% C, 18% Cr, 8% Ni, balance Fe)	1,410°C	2,570.0°F	24
14	Titanium	1,800°C	3,272.0°F	15.7
15	Tungsten	3,370°C	6,098.0°F	197
16	Zinc	419°C	786.2°F	102.4
	Other Materials of Interest:			
	Air			0.0261
	Molybdenum	2,630°C	4,766.0°F	130
	Platinum	1,774°C	3,225.2°F	73.4
	Tin	232°C	449.4°F	66
	Fiberglass			0.048
	Polystyrene			0.15
	Epoxy			0.17–1.5
	Water			0.604
	Glass			1.7–3.4

TABLE 9-2 Thermal properties of materials suitable for use in rotary spark gaps.

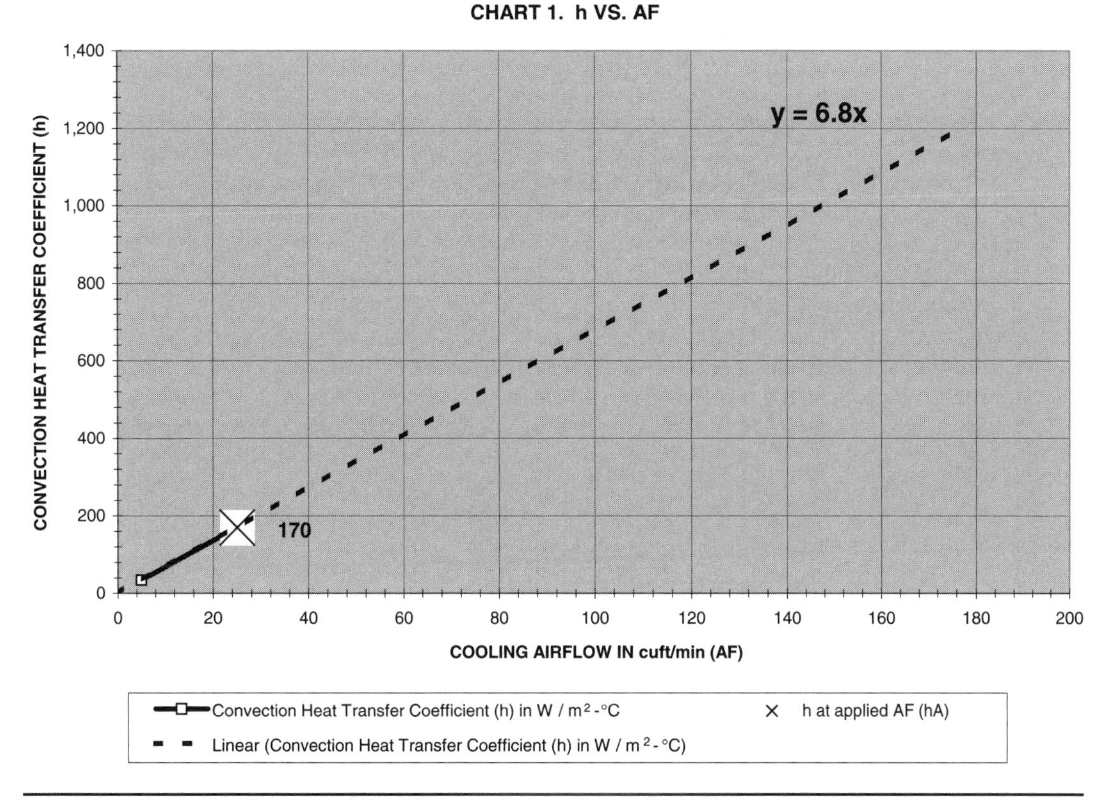

FIGURE 9-7 Convection heat transfer coefficient (*H*) vs cooling airflow (AF).

5.7 × 1.0 for standard pressure (sea level) and cooling airflow <1 cuft/min is derived by entering (2) in cell (B29) and <1.0 cuft/min in cell (B30).

6.8 × AFs for standard pressure and cooling airflow ≥1 cuft/min (AFs) is derived by entering (2) in cell (B29) and ≥1.0 cuft/min in cell (B30).

The convection heat transfer coefficients (h) are from reference (2).

NOTE: *Reference (2) provides a convection heat transfer coefficient (h) of 34 for cooling airflow (AF) of 5 cuft/min and an h of 170 for an AF of 25 cuft/min. These data were used to construct the graph shown in Figure 9-7. The linear curve fit formula shown in the figure is used to calculate h for the selected AF entered in cell (B30) by multiplying AF by 6.8 (6.8 × AFs or 6.8 × AFr).*

- The convection heat transfer coefficient of rotary electrodes (hR) in W/m^2-°C is derived in cell (F24) for the following gap characteristic entered in cell (B29):

  2.8 for vacuum or reduced pressure gap environment is derived by entering (1) in cell (B29).

6.8 × AFr for standard pressure and cooling airflow in cuft/min (AFr) is derived by entering (2) in cell (B29) and estimated airflow across rotary electrodes in cuft/min in cell (B27). Note the airflow is always >1.0 across the rotary electrodes.

There are several mechanisms in the rotary gap that help dissipate the heat generated during gap ionization. Cooling is provided when the heat is distributed through a larger surface area (conduction). Cooling is also provided through any interaction with airflow (convection). Some of this heat is conducted through the stationary electrode-to-the mounting where it is distributed throughout the larger surface area of the mounting providing cooling. If an optional heat sink is used there is also conduction through the mounting-to-heat sink, which provides additional cooling. The heat sink can provide further cooling when additional airflow is directed across it. Finally, the rotating electrodes are provided with cooling airflow as the rotor is spinning at a high rate of speed. All of these materials are involved in conduction and offer a thermal resistance to the transfer of heat. This thermal resistance is calculated using the previously determined thermal conductivity, the conduction path length and resulting surface area:

$$\theta = \frac{L}{k \bullet SA} \tag{9.5}$$

Where: θ = Thermal resistance to conduction in °C/W for rotary spark gap component listed below.
L = Length of conduction path in meters for rotary spark gap component listed below.
k = Thermal conductivity in W/m-°C for rotary spark gap component listed below.
SA = Surface area in meters for rotary spark gap component listed below.

Rotary spark gap component θ	L	k	SA
Air gap (θAd) in cell (F28)	Entered in cell (B3)	0.0261 from reference (2)	Calculated Aef value in cell (F4) from equation (9.1)
Stationary electrode (θsed) in cell (F32)	Entered in cell (B9)	ke derived in cell (F16)	Calculated Aef value in cell (F4) from equation (9.1)
Stationary mounting (θMd) in cell (F35)	Entered in cell (B15)	kM derived in cell (F18)	Calculated SAm value in cell (F7) from equation (9.2)
Optional heat sink (θSd) in cell (F38)	Entered in cell (B20)	kS derived in cell (F20)	Calculated SAs value in cell (F9) from equation (9.4)
Rotary electrode (θRed) in cell (F41)	Entered in cell (B25)	kR derived in cell (F23)	Calculated SAr value in cell (F12) from equation (9.1)

NOTE: *The thermal conductivity must be calculated in W/m-°C. All L dimensions are entered in inches and converted to meters in the corresponding cell in column (C) using a 0.0254 multiplier (inches × 2.54 = cm, 1 cm = 0.01 meters). All SA dimensions are calculated in inches and converted to meters in the corresponding cell in column (H) using a 0.0254 multiplier.*

Some of the heat produced in the spark gap is cooled through the surrounding air by convection. The faster the air flow the greater the cooling. Some cooling is provided to the rotary and stationary electrodes as the rotor spins. If enough electrical insulation and filtering can be provided to a cooling fan the additional airflow can be directed to the stationary electrodes, mounting, and optional heat sink. All of the mechanisms involved in convection offer a thermal resistance to the transfer of heat. This thermal resistance is calculated using the previously determined convection heat transfer coefficient and surface area:

$$\theta = \frac{1}{h \bullet SA} \tag{9.6}$$

Where: θ = Thermal resistance to convection in °C/W for rotary spark gap component listed below.
h = Convection heat transfer coefficient in W/m^2-°C for rotary spark gap component listed below.
SA = Surface area in meters for rotary spark gap component listed below.

Rotary spark gap component θ	h	SA
Air gap, stationary parts (θAv) in cell (F29)	hS derived in cell (F21)	Calculated Aef value in cell (F4) from equation (9.1)
Air gap, rotating parts (θRAv) in cell (F30)	hR derived in cell (F24)	Calculated Arf value in cell (F11) from equation (9.1)
Stationary electrode (θsev) in cell (F33)	hS derived in cell (F21)	Calculated Aef value in cell (F4) from equation (9.1)
Stationary mounting (θMv) in cell (F36)	hS derived in cell (F21)	Calculated SAm value in cell (F7) from equation (9.2)
Optional heat sink (θSv) in cell (F39)	hS derived in cell (F21)	Calculated SAs value in cell (F9) from equation (9.4)
Rotary electrode (θRev) in cell (F42)	hR derived in cell (F24)	Calculated SAr value in cell (F12) from equation (9.1)

NOTE: *The convection heat transfer coefficient must be calculated in W/m^2-°C. All SA dimensions are calculated in inches and converted to meters in the corresponding cell in column (H) using a 0.0254 multiplier.*

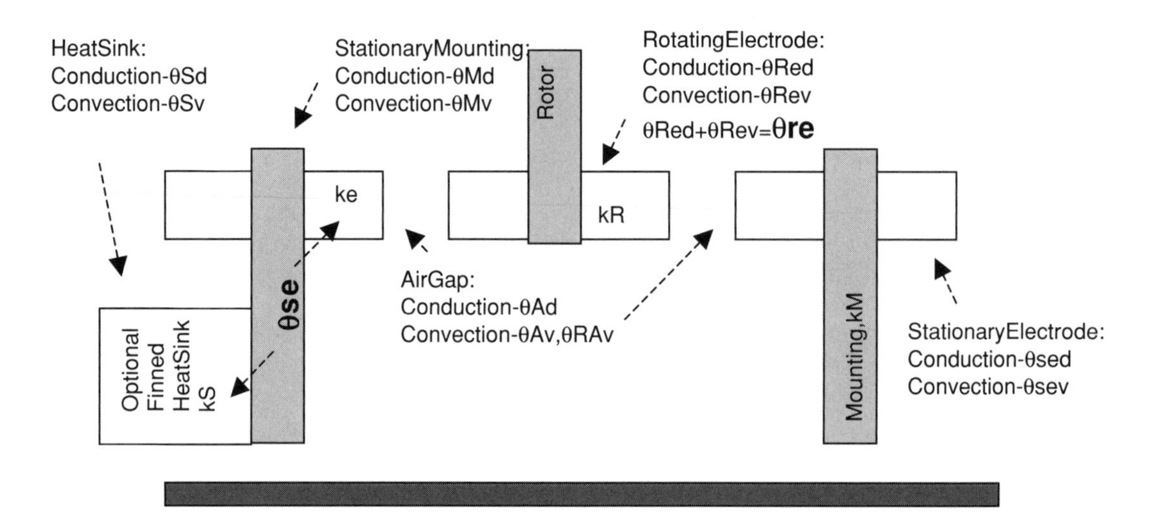

FIGURE 9-8 Conduction and convection mechanisms in a rotary gap.

The total thermal resistance is the sum of the conduction and convection thermal resistances. Refer to Figure 9-8 for an illustration of the conduction and convection mechanisms discussed. The thermal resistance for the stationary electrode conduction path-to-mounting-to-optional heat sink is θsed in cell (F32) + θMd in cell (F35) + θSd in cell (F38). The thermal resistance for the stationary electrode convection path is θMv in cell (F36) + θSv in cell (F39). The total thermal resistance of the stationary electrode conduction and convection mechanisms (θse) is shown in cell (F44).

The thermal resistance for the rotary electrode conduction path is θRed in cell (F41). The thermal resistance for the rotary electrode convection path is θRev in cell (F42). The total thermal resistance of the rotary electrode conduction and convection mechanisms (θre) is shown in cell (F45).

The power dissipated in the spark gap during ionization is calculated:

$$Pd = Pp(1 - \%P) \tag{9.7}$$

Where: Pd = Power dissipated in the spark gap during ionization in watts = cell (F47), calculated value.

Pp = Estimated load power of coil during operation in watts = cell (B4), enter value in kW. Converted to watts in cell (C4) using a 0.001 multiplier.

$\%P$ = Estimated primary power coupled to secondary (line power not dissipated in spark gap) in watts = cell (B32), enter estimated power in %.

The thermal rise above the mounting temperature for each of the two stationary electrodes is:

$$Trs = \theta se \bullet \frac{Pd}{2} \tag{9.8}$$

Where: Trs = Thermal rise above the mounting temperature for each of the two stationary electrodes in °C = cell (F48), calculated value.

θse = Total thermal resistance of the stationary electrode conduction and convection mechanisms in °C/W = cell (F44), summation of calculated values from equations (9.5) and (9.6).

Pd = Power dissipated in the spark gap during ionization in watts = cell (F47), calculated value from equation (9.7). This power is equally divided between the two stationary gaps.

The thermal rise above the mounting temperature for each of the rotary electrodes is:

$$Trr = \theta re \bullet \frac{Pd}{2} \tag{9.9}$$

Where: Trr = Thermal rise above the mounting temperature for each of the two stationary electrodes in °C = cell (F49), calculated value.

θre = Total thermal resistance of the rotary electrode conduction and convection mechanisms in °C/W = cell (F45), summation of calculated values from equations (9.5) and (9.6).

Pd = Power dissipated in the spark gap during ionization in watts = cell (F47), calculated value from equation (9.7). This power is equally divided between the two stationary gaps.

The resulting stationary electrode face temperature will be the hottest material temperature in the rotary spark gap application:

$$Ts = Tm + Trs \tag{9.10}$$

Where: Ts = Resulting stationary electrode face temperature in °C = cell (F50), calculated value.

Tm = Estimated ambient temperature of stationary electrode mounting or optional heat sink during operation in °C = cell (B31), enter value.

Trs = Thermal rise above the mounting temperature for each of the two stationary electrodes in °C = cell (F48), calculated value from equation (9.8).

The resulting electrode face temperature should be below the melting temperature of the material used in the stationary electrodes. The melting temperature for the beryllium copper electrodes is shown in cell (F51) and Table 9-2. The example shown does not have the desired conservancy therefore additional series stationary gaps were added to the application. The resulting operating temperature was decreased considerably from that shown in Figure 9-9 and resulted in temperatures well below the melting point of the material. To determine the effects of adding additional series gaps simply reduce the estimated primary power (cell B4) or increase the percentage of power coupled to the secondary (cell B32). If each series gap is adjusted to the same dimension as the rotary gaps the power dissipation is split equally among all of the gaps. Of course a test run will be required on the final application to ensure it still ionizes. If your electrodes are burning up or showing excessive early wear the electrode temperature is too high and the worksheet can be used to evaluate the effects of changing the design parameters before costly and time-consuming work is done.

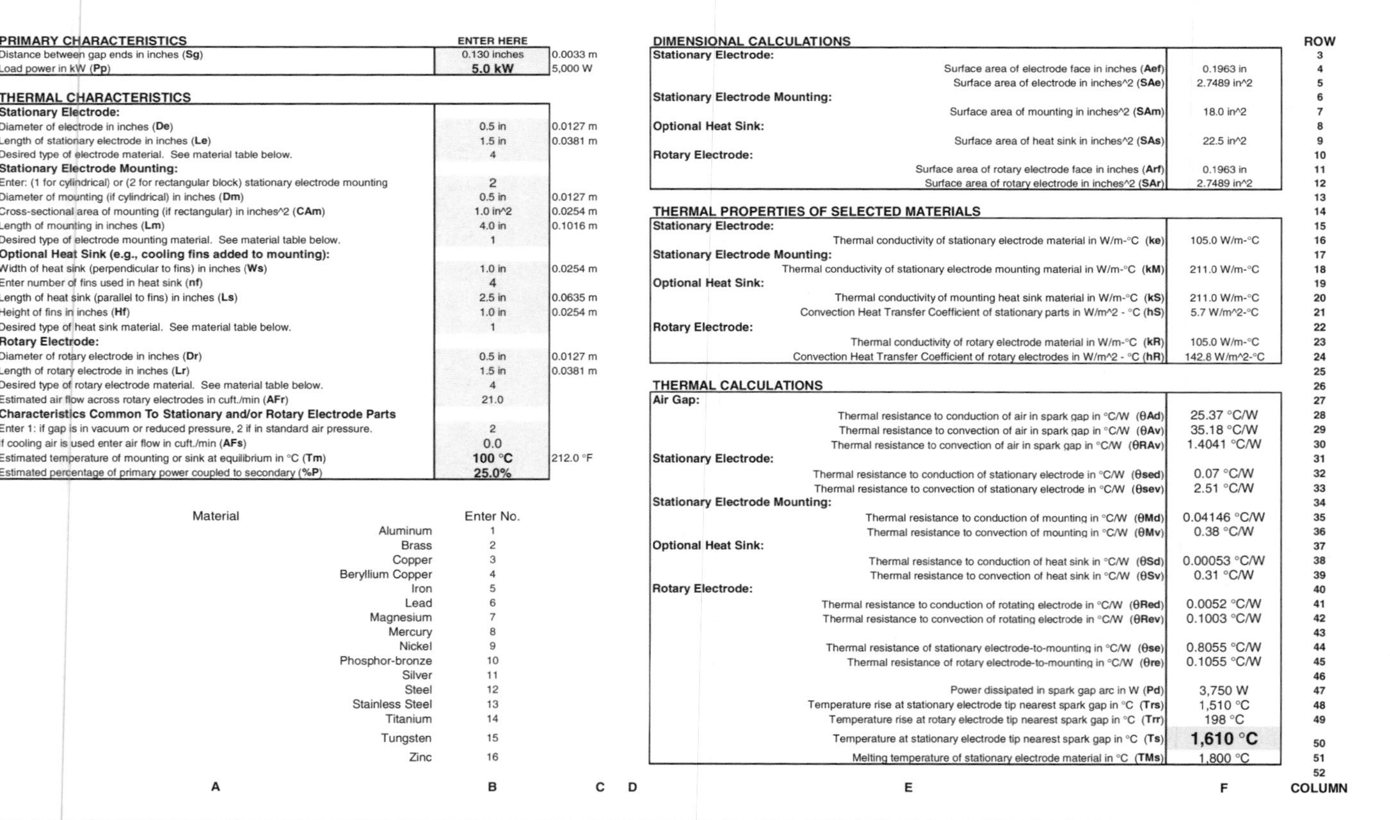

PRIMARY CHARACTERISTICS	ENTER HERE	
Distance between gap ends in inches (**Sg**)	0.130 inches	0.0033 m
Load power in kW (**Pp**)	**5.0 kW**	5,000 W

THERMAL CHARACTERISTICS		
Stationary Electrode:		
Diameter of electrode in inches (**De**)	0.5 in	0.0127 m
Length of stationary electrode in inches (**Le**)	1.5 in	0.0381 m
Desired type of electrode material. See material table below.	4	
Stationary Electrode Mounting:		
Enter: (1 for cylindrical) or (2 for rectangular block) stationary electrode mounting	2	
Diameter of mounting (if cylindrical) in inches (**Dm**)	0.5 in	0.0127 m
Cross-sectional area of mounting (if rectangular) in inches^2 (**CAm**)	1.0 in^2	0.0254 m
Length of mounting in inches (**Lm**)	4.0 in	0.1016 m
Desired type of electrode mounting material. See material table below.	1	
Optional Heat Sink (e.g., cooling fins added to mounting):		
Width of heat sink (perpendicular to fins) in inches (**Ws**)	1.0 in	0.0254 m
Enter number of fins used in heat sink (**nf**)	4	
Length of heat sink (parallel to fins) in inches (**Ls**)	2.5 in	0.0635 m
Height of fins in inches (**Hf**)	1.0 in	0.0254 m
Desired type of heat sink material. See material table below.	1	
Rotary Electrode:		
Diameter of rotary electrode in inches (**Dr**)	0.5 in	0.0127 m
Length of rotary electrode in inches (**Lr**)	1.5 in	0.0381 m
Desired type of rotary electrode material. See material table below.	4	
Estimated air flow across rotary electrodes in cuft./min (**AFr**)	21.0	
Characteristics Common To Stationary and/or Rotary Electrode Parts		
Enter 1: if gap is in vacuum or reduced pressure, 2 if in standard air pressure.	2	
If cooling air is used enter air flow in cuft./min (**AFs**)	0.0	
Estimated temperature of mounting or sink at equilibrium in °C (**Tm**)	**100 °C**	212.0 °F
Estimated percentage of primary power coupled to secondary (**%P**)	**25.0%**	

Material	Enter No.
Aluminum	1
Brass	2
Copper	3
Beryllium Copper	4
Iron	5
Lead	6
Magnesium	7
Mercury	8
Nickel	9
Phosphor-bronze	10
Silver	11
Steel	12
Stainless Steel	13
Titanium	14
Tungsten	15
Zinc	16

DIMENSIONAL CALCULATIONS		ROW
Stationary Electrode:		3
Surface area of electrode face in inches (**Aef**)	0.1963 in	4
Surface area of electrode in inches^2 (**SAe**)	2.7489 in^2	5
Stationary Electrode Mounting:		6
Surface area of mounting in inches^2 (**SAm**)	18.0 in^2	7
Optional Heat Sink:		8
Surface area of heat sink in inches^2 (**SAs**)	22.5 in^2	9
Rotary Electrode:		10
Surface area of rotary electrode face in inches (**Arf**)	0.1963 in	11
Surface area of rotary electrode in inches^2 (**SAr**)	2.7489 in^2	12
		13
THERMAL PROPERTIES OF SELECTED MATERIALS		14
Stationary Electrode:		15
Thermal conductivity of stationary electrode material in W/m-°C (**ke**)	105.0 W/m-°C	16
Stationary Electrode Mounting:		17
Thermal conductivity of stationary electrode mounting material in W/m-°C (**kM**)	211.0 W/m-°C	18
Optional Heat Sink:		19
Thermal conductivity of mounting heat sink material in W/m-°C (**kS**)	211.0 W/m-°C	20
Convection Heat Transfer Coefficient of stationary parts in W/m^2 - °C (**hS**)	5.7 W/m^2-°C	21
Rotary Electrode:		22
Thermal conductivity of rotary electrode material in W/m-°C (**kR**)	105.0 W/m-°C	23
Convection Heat Transfer Coefficient of rotary electrodes in W/m^2 - °C (**hR**)	142.8 W/m^2-°C	24
		25
THERMAL CALCULATIONS		26
Air Gap:		27
Thermal resistance to conduction of air in spark gap in °C/W (**θAd**)	25.37 °C/W	28
Thermal resistance to convection of air in spark gap in °C/W (**θAv**)	35.18 °C/W	29
Thermal resistance to convection of air in spark gap in °C/W (**θRAv**)	1.4041 °C/W	30
Stationary Electrode:		31
Thermal resistance to conduction of stationary electrode in °C/W (**θsed**)	0.07 °C/W	32
Thermal resistance to convection of stationary electrode in °C/W (**θsev**)	2.51 °C/W	33
Stationary Electrode Mounting:		34
Thermal resistance to conduction of mounting in °C/W (**θMd**)	0.04146 °C/W	35
Thermal resistance to convection of mounting in °C/W (**θMv**)	0.38 °C/W	36
Optional Heat Sink:		37
Thermal resistance to conduction of heat sink in °C/W (**θSd**)	0.00053 °C/W	38
Thermal resistance to convection of heat sink in °C/W (**θSv**)	0.31 °C/W	39
Rotary Electrode:		40
Thermal resistance to conduction of rotating electrode in °C/W (**θRed**)	0.0052 °C/W	41
Thermal resistance to convection of rotating electrode in °C/W (**θRev**)	0.1003 °C/W	42
		43
Thermal resistance of stationary electrode-to-mounting in °C/W (**θse**)	0.8055 °C/W	44
Thermal resistance of rotary electrode-to-mounting in °C/W (**θre**)	0.1055 °C/W	45
		46
Power dissipated in spark gap arc in W (**Pd**)	3,750 W	47
Temperature rise at stationary electrode tip nearest spark gap in °C (**Trs**)	1,510 °C	48
Temperature rise at rotary electrode tip nearest spark gap in °C (**Trr**)	198 °C	49
Temperature at stationary electrode tip nearest spark gap in °C (**Ts**)	**1,610 °C**	50
Melting temperature of stationary electrode material in °C (**TMs**)	1,800 °C	51
		52

A	B	C	D	E	F	COLUMN

FIGURE 9-9 Rotary gap electrode operating temperature for selected primary characteristics worksheet calculations.

The temperature of the arc at atmospheric pressure is quite high, typically reaching 3,777°C using copper electrodes and 5,877°C to 6,167°C for tungsten electrodes. However, the arc is not continuous. The arc is quenched in a short time period and is repeated at a rate equal to the BPS. The steady-state electrode temperature will typically be as characterized in the above calculations and much less than the arc temperature because the calculations use the average primary power.

9.2.2 Constructing the Rotary Spark Gap

The rotary gap design calculations are performed in CH_6A.xls file, BPS CALCULATOR worksheet (3) and detailed in Section 6.8. Once the desired rotary gap performance is calculated in the worksheet the rotational diameter of the electrodes, number and diameter of electrodes needed, and thickness of the rotor are defined. The remaining parameters are defined in the steps below and in Figure 9-10. I use the following techniques to make a rotary spark gap:

1. At your local plastic supplier or the next Hamfest, select a piece of plastic stock large enough to cut the desired diameter of the rotor. The thicker the piece the better; however, keep in mind the motor must have enough torque to turn the rotor. I recommend using at least 1/2 HP motors to ensure it starts-up quickly and does not load down with a 1″ thick × 6″ or larger diameter rotor.
2. Using the calculated rotor diameter (this is actually the diameter of the rotating electrodes) to achieve the desired BPS of the rotary gap, add enough extra material to the diameter to fasten the electrodes to the rotor. Step 7 details how the electrodes are secured to the rotor. If this is your first rotary gap, add 2″ to the calculated rotor diameter, which allows 1″ of additional material from the rotor's edge to the center of the electrodes. Cut the desired circle as carefully as possible with a router and circle jig. An alternate method is to draw a circle of the desired rotor diameter on the stock and carefully cut it out with a jig or sabre saw. The router makes the best circle in the stock. I use inexpensive 1/4″ fluted router bits and make 1/8″ deep cuts until the router bit penetrates the disc material. G-10 is time consuming material to work with and quickly takes the edge off your cutting tools. Do this in an area with ventilation and wear a filter mask as cutting produces a lot of fiber dust and fumes that should not be breathed.
3. If a router was used to cut the circle there is probably a hole in the center of the circle from using the circle jig. If a jig or saber type saw was used to cut the circle a hole needs to be placed in the center. Mark the circle with a center finder and use a drill press to make a small hole in the center. Enlarge the center hole to the diameter required to mount the hub being used. If the hub was made to slide back and forth along the motor shaft ensure the hole in the center of the rotor allows this movement after the hub is mounted to the rotor. This will make it easier to assemble and adjust the rotor after it is mounted to the shaft as it can be slid along the shaft.
4. Mount the hub to the rotor that will be used to secure the rotor to the motor's shaft. Ensure the hub is centered on the rotor stock and firmly secured with bolts. Remember, for most applications this assembly will be spinning at 1,800 to 3,600 RPM and *all*

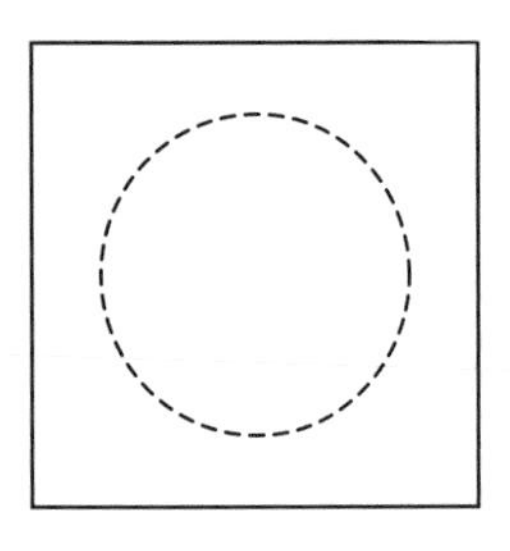

Step 1. Mark and cut circle from stock.

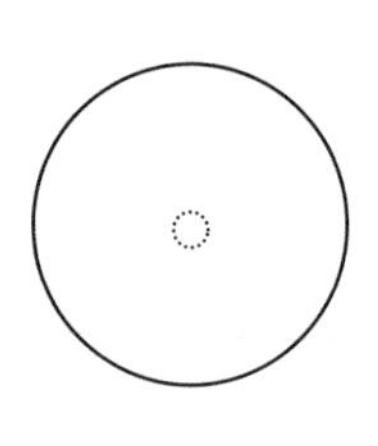

Step 2. Mark center of circle and drill.

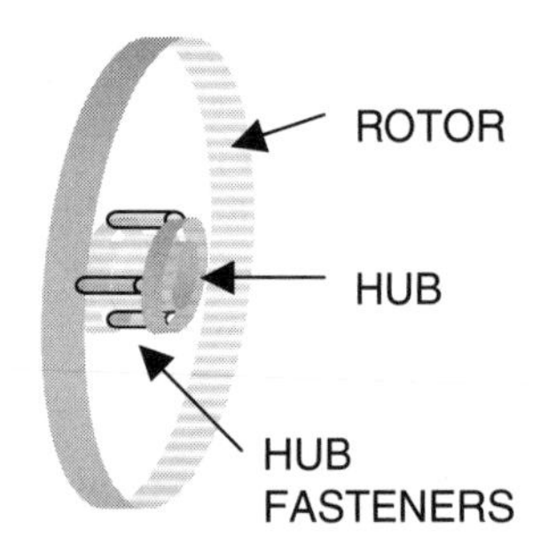

Step 3. Mount hub.

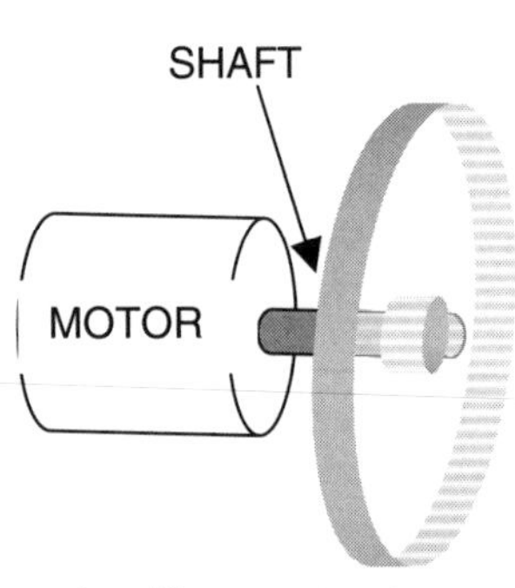

Step 4. Mount rotor to motor and secure.

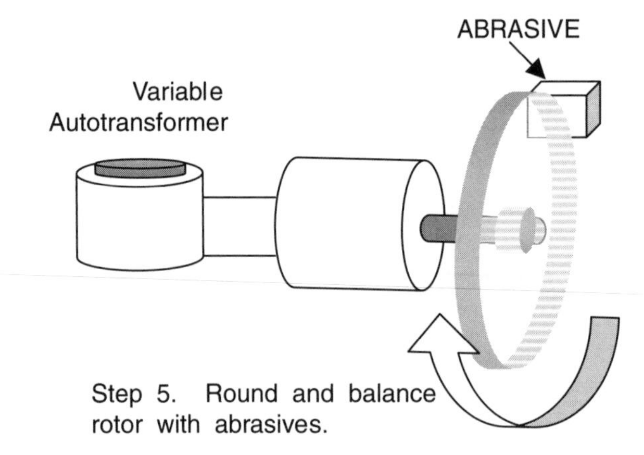

Step 5. Round and balance rotor with abrasives.

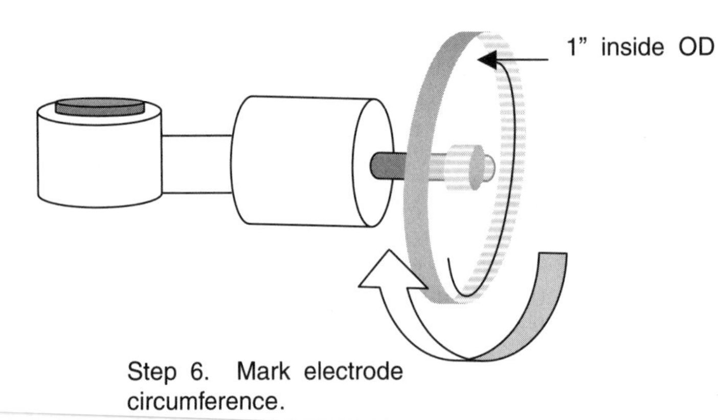

Step 6. Mark electrode circumference.

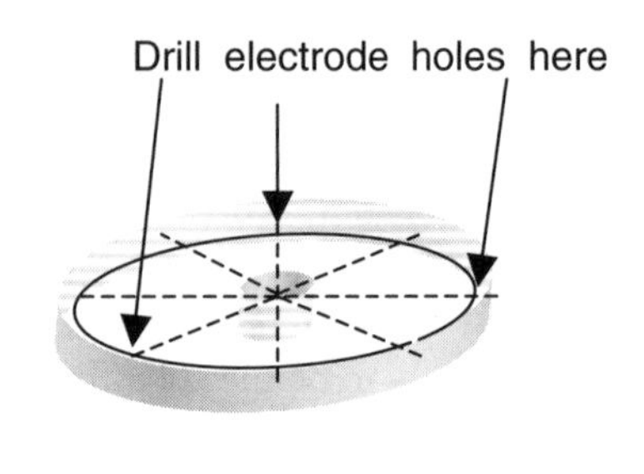

Step 7. Divide into equal sections.

Figure 9-10 Details of rotary spark gap construction.

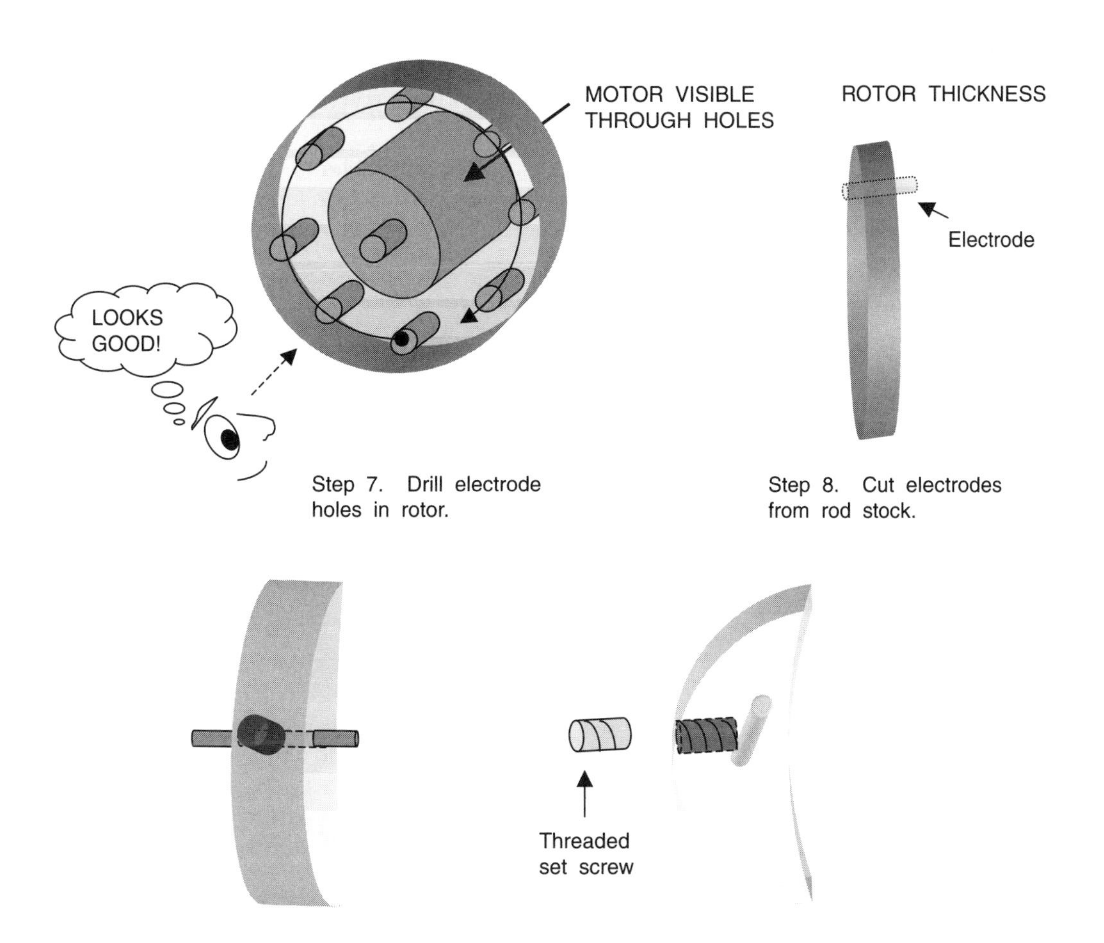

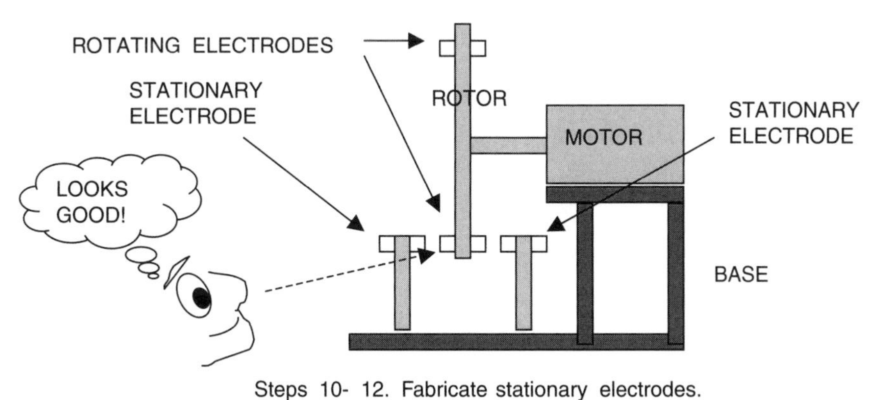

Steps 10- 12. Fabricate stationary electrodes.

FIGURE 9-10 Details of rotary spark gap construction. (*continued*)

parts need to stay on the rotor and the rotor on the motor! If you do not feel comfortable performing this step or you are unsure of what you are doing, seek assistance from a machinist or other experienced person.

5. Mount the rotor and hub onto the motor's shaft. Mount the motor to a solid surface with enough clearance to allow the rotor to turn. I prefer to construct the spark gap as a module and use plastic laminates for the base to mount the motor and stationary electrodes.
6. This next procedure may be met with some controversy, but it works well for me. Connect the motor wiring to a variable autotransformer and run it up *slowly* until the motor begins to turn. You probably won't get too much speed out of it at first as it may shake, rattle, and roll from being out of round (balance). Using a 40 or 60 grit abrasive cloth affixed to a sanding block, touch the rotating edge of the rotor. If you use a file or metal cutting edge for this step the laminate will quickly ruin the tool, abrasives are much cheaper. Increase the variable autotransformer input to increase the motor speed until the rotor turns at a moderate RPM when loaded with the sanding block. This will take just a few minutes of practice until it feels safe and a good cutting speed is reached. The abrasive will slowly cut down the rotor edge until it is balanced. Check the balance as you go by occasionally increasing the variable autotransformer input and RPM. Continue taking off material until the motor will spin at the maximum RPM (full line voltage applied from the variable autotransformer) with a minimum of vibration. Finish by smoothing the edge with 120 and 240 grit abrasive paper and polish with a paste abrasive. The rotor should now be smooth, rotate at the maximum motor speed with no perceptible vibration and remain securely fastened to the motor's shaft. As a bonus, you have just dynamically balanced the rotor to the motor that will be used to turn it. Remember to wear your mask and eye protection. This procedure is very messy so doing it outdoors results in the easiest cleanup.
7. Mark where to drill the holes for the electrodes. Run the motor one more time at a very slow RPM. Mark a circle on the face of the rotor (perpendicular to the shaft axis) by touching it with a marker or scribe that is securely fastened to the mounting plate (or carefully hold the marker still against the rotor face as it revolves). If this is your first rotary gap, place the marker 1″ inside of the outer circumference. This will leave 1″ of rotor material outside the electrode's center to fabricate the electrode fasteners. Remember the extra 2″ we added to the calculated rotor diameter in step 1? Glad we remembered that before we cut the circle!
8. Remove the rotor and hub from the motor shaft. Lay the rotor flat on a horizontal surface. Remove the hub from the rotor if necessary and reinstall to attach the rotor to the motor when required in the following steps. With a marker, divide the circle (marked in step 7) into as many equal parts as the number of electrodes selected in the BPS calculations. The rotor should now look like sliced pie. Where the circle from step 7 intersects with the pie slices is where the holes are drilled to insert the electrodes. In this manner the electrodes are equidistant on the rotor and cross the same point on the rotational axis. Select a drill bit of the same diameter as the electrode rod material. Drill the first electrode hole using a drill press. The holes must be exactly perpendicular to the rotor face so no hand-held wobbly drills. Remount the rotor on the motor shaft

and give it a spin by hand. Looking into the motor shaft a series of holes on the face of the rotor should provide a glimpse of the motor casing on the other side of the rotor, similar to a stroboscopic effect. If this does not occur, the holes are not drilled on the proper axis. See Figure 9-10 for orientation. If this is correct drill the remaining electrode holes.

9. Cut the electrode rod material to desired length. Repeat as many times as there are electrodes needed to produce the calculated BPS. Ensure that all electrodes are exactly the same length. The ends should be flat and polished. The electrodes will be centered on the rotor's width (thickness). If a desired length is unknown cut the rod length to the rotor thickness plus, say 0.5″ to 1.0″ extending on each side. For higher power levels the rod ends need to be further away from the rotor or the arc will carbonize the rotor material. Hard materials such as tungsten require a high-speed abrasive wheel (e.g., dremel tool) to cut. Ensure the electrodes will slide in their mounting holes in the rotor with little resistance. This may require enlarging the holes in the rotor slightly with an abrasive cloth to allow lateral movement of the electrodes for later adjustment.
10. Select Allen head (they have a recessed hexagonal slot for an Allen wrench) set screws to secure the electrodes in the rotor. The set screws should be of sufficient diameter to securely hold the electrode rod and course threaded for easier tapping of the laminate rotor material. Set screws of 1/4 to 3/8″ diameter are usually sufficient to hold an electrode that is less than 1″ in diameter. A fine thread will be difficult to cut in the laminate material and tends to disintegrate with use. Using a drill bit size specified for the tap being used, drill a straight hole for each set screw (use the drill press again) from the outside rotor edge (circumference) toward the rotor's center and perpendicular to the electrode axis. Drill far enough to reach the electrode hole and stop. Cut the threads using the tap. If the tap stops before the threads are fully cut you must drill out enough material on the other side of the electrode hole to finish cutting the threads. After the first hole is tapped you will have a pretty good idea what you are doing. Insert the electrodes into their holes and secure with the threaded set screws. Ensure all electrodes are centered in the disk.
11. Mount the rotor to the motor. Secure the motor to a permanent base or on your Tesla coil. Make sure there is clearance for the rotor to spin.
12. Now that the rotor is complete, the stationary electrodes (stators) are added. Select the same rod material for the stationary electrodes and fabricate metal mounting posts that will hold the stationary rod parallel to the rotating rods. Mount each post at least 2″ away from the rotor and motor case, the farther the better (too close and the high voltage will arc to the motor case or mounting structure). Use set screws to secure the electrodes to the posts. I prefer to mount the stationary electrodes at the outer periphery of the rotor circumference (as far away from the motor as you can get) to provide as much electrical insulation as possible.
13. Before energizing the motor, adjust the distance between the rotating electrodes and each of the stationary electrodes to 0.25″ to start. Spin the rotor by hand and ensure the electrodes do not touch each other. Each rotating electrode should pass the stationary electrodes with the same clearance. Be aware that any backlash (longitudinal play) in the motor shaft may require additional electrode clearance. *Ensure all mounting*

hardware is tightened. Slowly run the motor speed up to maximum with the variable autotransformer and ensure there is still adequate electrode clearance. If this goes well, adjust the stationary electrode gap to the desired clearance using the data in Figure 6-28 and the design calculations performed in Chapter 2 as a guide, and slowly run the motor up to speed with the variable autotransformer to ensure adequate clearance.

Check tightness of fasteners and security of moving parts before each coil run! For added safety a thread lock product can be used once all adjustments are finished. There is an optimum gap setting that can only be fine-tuned using the test run and adjust method.

9.3 Troubleshooting During the Test Run

Once the coil is built it must be tested. A spark will not always issue from the secondary when the power is applied due to some oversight or connection. Table 9-3 will assist in troubleshooting the coil when a spark is not evident on the secondary. The terms used in the table correspond to the drawing in Figure 2-1. A voltmeter is typically used to perform voltage checks to identify an open circuit. This can be safely done under typical conditions; however, exercise extreme caution when troubleshooting 240-V or higher line service. If you are not confident you can safely perform any of the steps seek qualified help.

9.4 Current-Limited Transformer Characteristics

Not all step-up transformers will drive a Tesla coil. When current-limited transformers are used, loading effects may inhibit spark gap ionization with no spark on the secondary. All neon sign transformers (NSTs) are current-limited transformers. References (7) through (9) were the sources used to characterize NST performance. NSTs are typically rated by the manufacturer for a maximum OPEN-CIRCUIT output voltage in volts rms and a maximum SHORT-CIRCUIT output current in mA rms. A 12 kV/30 mA rated NST will not supply 12 kV @ 30 mA. It can supply 12 kV to a very high impedance (nearly open-circuit), but as the load increases to draw 30 mA the output voltage decreases to a very low value. NSTs are designed to ionize the gas in an argon or neon filled tube. This requires a high voltage with little initial current. However, once ionized the gas tube requires very little voltage due to its low resistance (see Section 6.10), and if not limited would create a near short-circuit to the line. To prevent a short-circuit the NST includes a shunt in the core to increase the core's reluctance. This limits the short-circuit current to the rated value (e.g., 30 mA).

The output current vs. output voltage for a 12 kV/30 mA NST was measured from the output terminal-to-case (center tap) and is shown in Figure 9-11. The measured output voltage at the loaded output current is shown in the red trace and right hand *Y* axis while the output voltage in percent of the rated value is shown in the black trace and left hand *Y* axis. The measured voltage on the right hand *Y* axis is scaled from 0 to 6.5 kV to slightly offset it from the calculated left hand *Y* axis (0 to 100% corresponds to 0 to 6 kV) to display both the red

Trouble Symptom	Probable Cause	Test Procedure	Corrective Action
No meter indications on control panel.	No line power into control panel.	Ensure full line voltage is available at output of service panel, e.g., 120 V or 240 V.	If available continue to next step. If not, troubleshoot line service.
		Ensure full line voltage is available at input to control cabinet.	If available continue to next step. If not, repair open circuit.
	Open circuit in control panel.	Ensure full line voltage is available at output of control cabinet.	If available troubleshoot control cabinet metering. If not, trace voltage through relays, contactors, circuit protection, input filter, and variable autotransformer to locate and repair open circuit.
Good voltmeter but low ammeter indications on control panel.	Control circuits cannot support load.	Use a resistive type test load with an appropriate resistance value to ensure that the control cabinet can supply the required current at full output voltage.	If resistive load is supported with good meter indications resume test run and review calculations.If not, check all cabinet interconnections ensuring correctness and security. Pay close attention to variable autotransformer brushes and contactor/relay contacts.
Good voltmeter but high ammeter indications or low voltmeter and high ammeter indications on control panel.	Full or partial short circuit in control cabinet.	Disconnect load and run up line voltage using the variable autotransformer.	If high ammeter indications persist with no load there is a short circuit in cabinet. Troubleshoot and repair. If not, check indications using a resistive type test load with an appropriate resistance value.

Table 9-3 Trouble symptoms and corrective action for a coil with no output.

Trouble Symptom	Probable Cause	Test Procedure	Corrective Action
No spark issues from secondary. Visible rotary or stationary spark gap ionization (spark).	Primary oscillations are out of tune with resonant frequency of the secondary.	Connect primary tap one turn above or below the calculated number of turns required to tune the primary. Continue moving up or down one turn at a time if necessary.	DISCHARGE tank capacitor before handling (Section 5.12). Review calculations for accuracy. If correct, try a different primary turn and run test. If no spark issues after retuning continue to next step.
	Terminal capacitance too large for spark breakout.	Place a pointed director on terminal and rerun. If no spark issues move another *grounded* director within a few inches of the terminal. If no spark issues remove terminal and retune primary using different tap.	Tuning point is generally different with and without terminal capacitance. The grounded director checks for secondary output >30 kV (10 kV/in). If no spark issues continue to next step.
	Secondary winding not continuous or interwinding short.	Visually inspect winding for insulation anomalies or crossed windings. Perform continuity (resistance) check on secondary winding ensuring it is close to calculated DC resistance. Measure inductance ensuring it is close to calculated value.	If winding passes visual inspection, continuity and inductance tests continue to next step. If not, repair discontinuity, crossed winding or insulation, or rewind secondary (also use a new form).
	Primary-to-secondary coupling too tight.	Review calculations. Ensure accuracy and k value is below the critical coupling threshold for the number of turns used. If visible corona or secondary-to-primary breakdown is observed the coupling is too tight.	If calculations are correct try loosening the coupling by raising the base of the secondary a few inches at a time and run test.

TABLE 9-3 Trouble symptoms and corrective action for a coil with no output. (*continued*)

Trouble Symptom	Probable Cause	Test Procedure	Corrective Action
No spark issues from secondary. No visible rotary or stationary spark gap ionization (spark).	No variable output voltage to step-up transformer.	Ensure full line voltage is available to input of step-up transformer.	If available continue to next step. If not, repair open circuit.
	No step-up transformer output voltage or reduced output voltage.	With 1 V applied to the input of step-up transformer ensure the output voltage is increased according to the turns ratio. The full output voltage can be measured if the proper high-voltage metering is available, adequate insulation is used and caution is exercised.	DISCHARGE tank capacitor before handling (Section 5.12). If voltage is correct continue to next step. If not, replace step-up transformer. This occurs more often when using neon sign transformers (NSTs).
	Insulation breakdown between secondary windings of step-up transformer.	Use a resistive type test load with an appropriate resistance value to ensure the step-up transformer can supply the required current at full output voltage.	DISCHARGE tank capacitor before handling (Section 5.12). If resistive load is supported with good meter indications continue to next step. If not, replace step-up transformer. This occurs more often when using neon sign transformers (NSTs).
	Open or shorted tank capacitor or electrical overstress.	Ensure tank capacitor is functional.	DISCHARGE tank capacitor before handling (Section 5.12). Remove capacitor and test using appropriate test equipment. If good, reinstall and continue to next step. If not, replace.

TABLE 9-3 Trouble symptoms and corrective action for a coil with no output. *(continued)*

Trouble Symptom	Probable Cause	Test Procedure	Corrective Action
	Open tank circuit.	Ensure tank circuit connections are secure.	DISCHARGE capacitor before handling (Section 5.12). Inspect all interconnections in the tank circuit to ensure correctness and security. If good continue to next step. If not, repair.
	Excessive gap spacing.	Reduce gap spacing to achieve ionization.	DISCHARGE capacitor before handling (Section 5.12). Adjust the gap spacing to ionize at a lower voltage. Adjust in small steps until the gap fires. If ionization is not achieved perform above steps again until problem is found.

TABLE 9-3 Trouble symptoms and corrective action for a coil with no output. (*continued*)

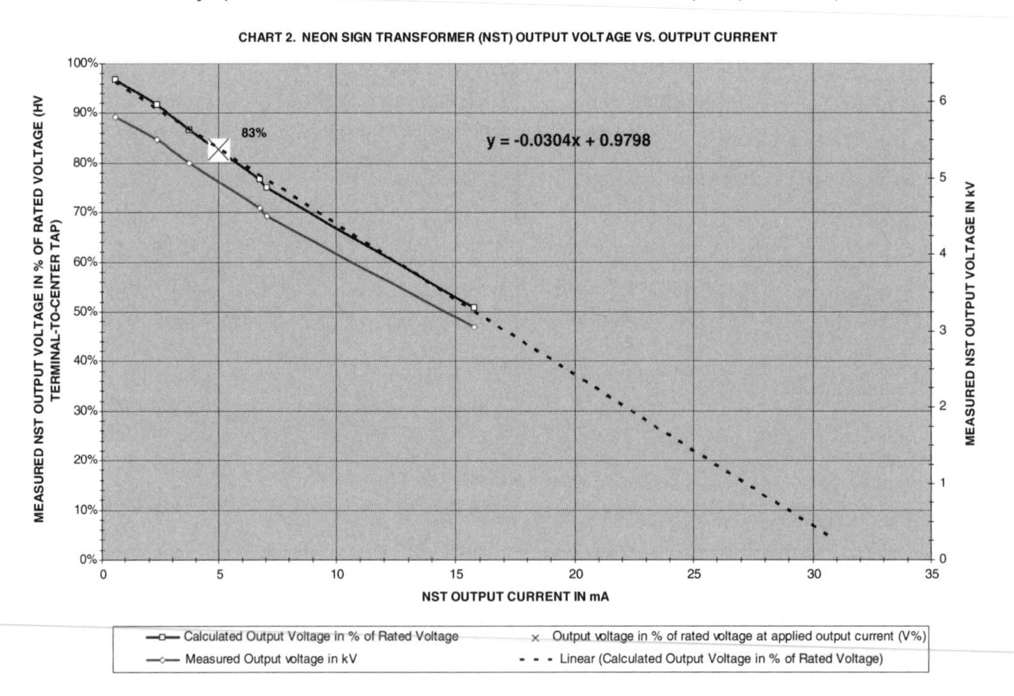

FIGURE 9-11 Measured 12 kV/30 mA NST output current vs. output voltage.

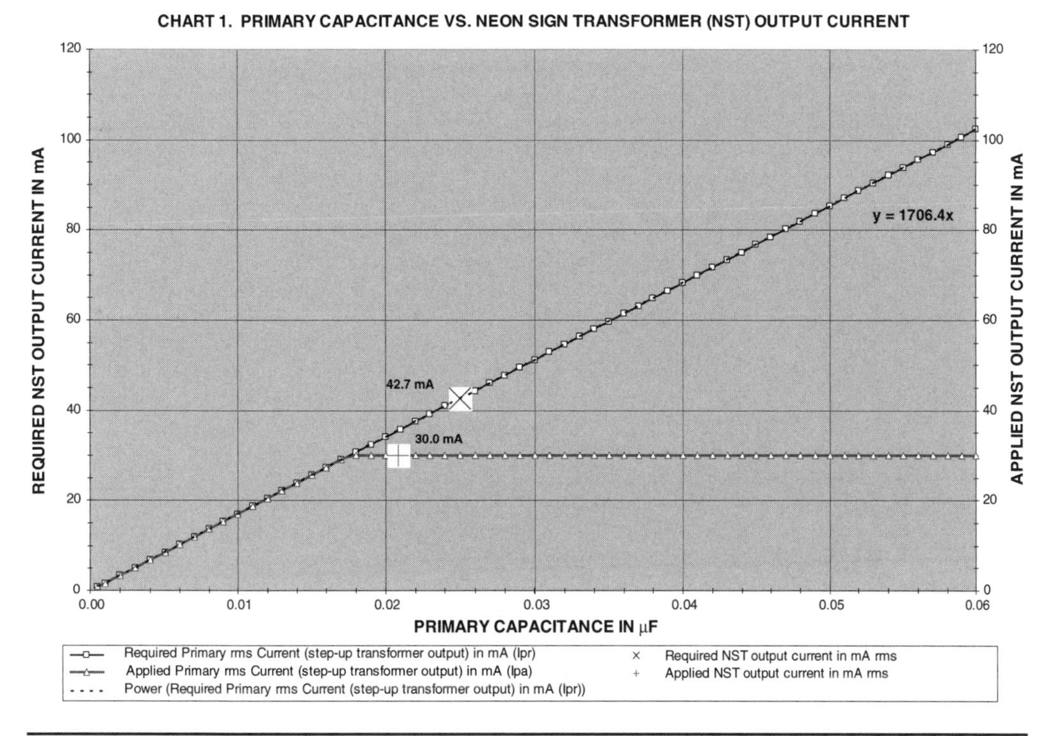

FIGURE 9-12 Single 12 kV/30 mA NST used to supply resonant tank circuit.

and black traces. With no load on the output of the NST (open-circuit) the output voltage is 6 kV from each output terminal to the center tap or 12 kV from output terminal-to-output terminal. This output voltage decreases linearly from the open-circuit voltage @ 0 mA to nearly 0 V at the rated output current. The output voltage in percent of the rated value is extended (extrapolated) from the measurements to the rated output current indicated by the dashed black trace and used to develop the linear curve fit formula shown in the figure. The linear curve fit illustrates the output voltage vs. output current relationship for any NST. If the NST output current is linearly scaled on the *X* axis (bottom) from 0 mA to rated value, and the output voltage linearly scaled on the *Y* axis from 0 V to rated value, a straight line can be drawn on the graph from rated voltage at 0 mA to 0 V at the rated current, which approximates the performance of any NST. For example, at an output current of 5 mA the rated 12 kV output is reduced to 83% or 12 kV × 0.83 = 10 kV and shown in the [X] in Figure 9-11. At approximately one half the rated output current (16 mA) the rated open-circuit output voltage is reduced 50% or 6 kV. As the output current approaches the rated short-circuit value the output voltage is very low (<10% of rated open-circuit output voltage). The NST must provide enough output current at an output voltage high enough to ionize the spark gap to be of any use in the design.

Veteran coiler Tom Vales provided a test circuit used in two performance runs as follows: the first run used a single 12 kV/30 mA NST and the second run used two parallel 9 kV/30 mA NSTs. The Tesla coil being driven used a tank capacitance of 0.025 μF with a 10-turn flat

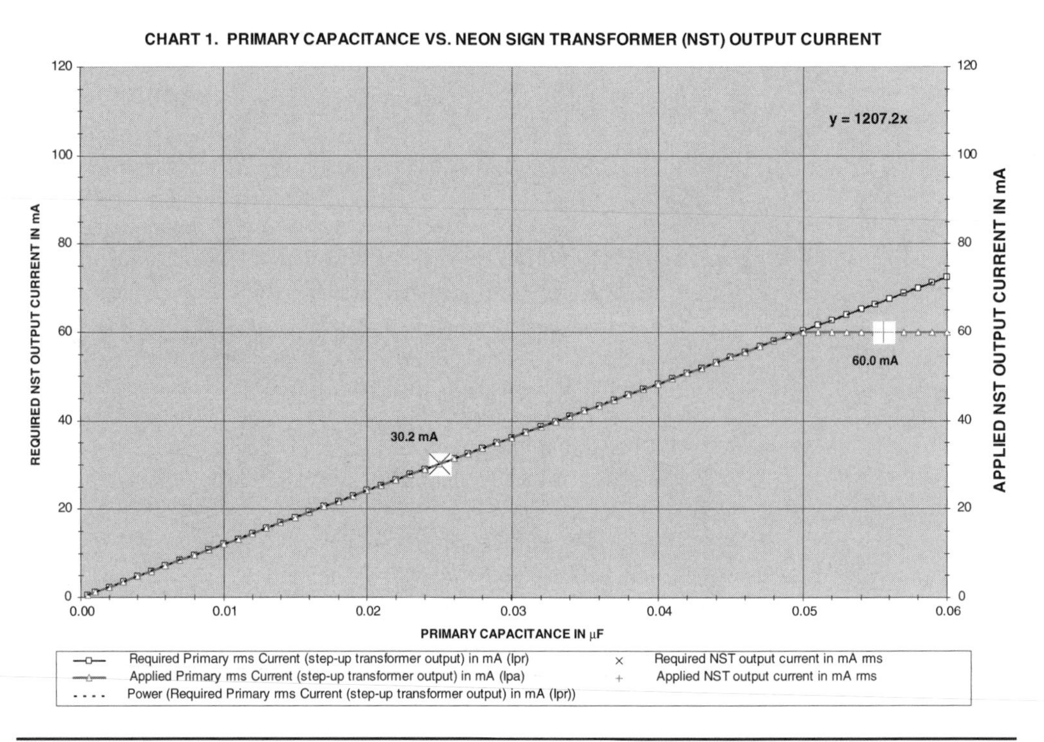

FIGURE 9-13 Two parallel 9 kV/30 mA NSTs used to supply resonant tank circuit.

Archemedes spiral and secondary winding resonance of approximately 185 kHz. The first run produced no output in the secondary as the spark gap never ionized. The second run produced gap ionization and a secondary output. A methodology was sought to determine what combination of NSTs are required to drive a resonant tank circuit.

The CH_6B.xls file, NST Current Output worksheet (4) was constructed to calculate and graph NST performance in a Tesla coil using primary capacitance values from 0.001 μF to 0.06 μF in the resonant circuit. The input parameters and calculations used in the worksheet and shown in Figure 9-14 are the same as described in Section 6.11 and Chapter 2 to determine the rms value of primary (step-up transformer output) current required for the applied primary characteristics. Figure 9-12 shows the performance when using a singe 12 kV/30 mA NST. The black trace plots the step-up transformer rms output current that is required by the resonant tank. The red trace plots the NST performance in the circuit. The [X] indicates the resonant tank requires 42.7 mA of rms current to supply a 0.025-μF tank capacitance with a 10-turn flat Archemedes spiral primary winding. The [+] indicates the NST can supply only 30 mA. Once the NST output current reaches its rated maximum of 30 mA rms (shown on the left or *Y* axis) the two traces divide. This point of division indicates a maximum primary capacitance of only 0.018 μF (shown on the bottom or *X* axis) can be used in this particular resonant tank circuit with a single 12 kV/30 mA NST. The 12 kV/30 mA NST would be overloaded using the 0.025-μF tank capacitance and no spark gap ionization would result. Once

PRIMARY CHARACTERISTICS

Row	A	B (ENTER HERE)	C
5	NST rated rms line voltage (**LV**)	120 V	100:1
6	NST Rated Output Current in mA rms (**irms**)	30.0 mA	0.030 A
7	NST Rated Output Voltage in kV rms (**vrms**)	12.0 kV	12000 V
8	Turns ratio of step-up transformer (**NT**)	**1:100**	NT = v / LV
9	Tank capacitance in μF (**Cp**)	**0.0250μF**	0.0000000250000 F
10	Primary winding DC resistance in Ω (**Rp**)	0.0100 Ω	
11	Calculated applied output voltage of step-up transformer in kVrms (**Vp**)	**12.00 KV**	16968 V

SPARK GAP CHARACTERISTICS

Row	A	B	C
14	Air Temperature in °C (**Ta**)	20.0 °C	68.0 °F
15	Air Pressure in torr (**AP**)	760	
16	Relative Humidity in Percent (**RH**)	50%	
17	Distance between gap ends in inches (**Sg**)	0.130 inches	0.33cm
18	Applied Overvoltage in percent (**Vo**)	0.0%	
19	Applied LINE frequency in Hz (**Lf**)	60 Hz	0.0167 sec
20	Enter Phase Shift of Synchronous Gap in Degrees (**PS**)	89.5	466.5311
21	Spark gap breaks per second (**BPS**)	120	0.008333 sec
22	Enter: (1 for linear) or (2 for exponential) gap material characteristics	2	
23	Minimum ionization current (**Imin**)	**1.0 A**	
24	Quench time in μs (**tQ**)	0.00 usec	0.000000 sec
25	Calculated breakdown voltage at applied positive alternation in kV (**BVp**)	**11.65 KV**	11647 V

Peak applied voltage (Vp) in cell (C9) must be greater than breakdown voltage (BVp) in cell (B22).

PRIMARY TUNING

Row	A	B	C
30	Enter: (1 for Archemedes Spiral) or (2 for Helical) Wound Primary	1	
31	Enter Inside Diameter of Archemedes Spiral (**ID**) or Outside Diameter of Helical Primary in inches (**OD**)	7.8 in	19.69cm
32	Interwinding Distance in inches (**IWD**)	0.750 in	1.91cm
33	Enter Total Number of Turns in Primary Winding (**Ttp**)	15	
34	Enter Angle of Inclination in° if using Archemedes Sprial (**θ**)	0.0 °	
35	Enter Desired Primary Turn Number Used To Tune (**Tp**)	**10**	
36	Enter Tuning Capacitance in μF (**Cpt**)	0.00000 μF	0.0000000000000 F
37	Enter Tuning Inductance in μH (**Lpt**)	0.0 uH	0.0000000000000 H
38	Calculated Primary Inductance in μH (**Lp**)	**37.89 uH**	0.00003789 H
39	Primary resonant frequency (**fP**)	163,518 Hz	fp=1/[2*π*sqrt([Lp+Lpt]*Cp)]

PRIMARY CALCULATIONS

Row	E	F
5	Find **ωp**	1027413
6	Find resonant oscillation time period in seconds (**tp**)	6.11554E-06
7	Find **tan ϕ** of primary oscillations	0.000000
8	Resonant primary impedance with S.G., w/o reflected sec (**Zpss**)	2.1795 Ω
9	Find primary Quality factor (**Qps**) with spark gap	17.78
10	Find decrement factor of primary (**δP**)	**0.17668**
11	Maximum primary winding voltage (**Vpp**)	2982 V
12	Peak primary tank current in Amps (**Ip**)	**69.4 A**
13	Find primary rms current (step-up transformer output) in Amps (**Ip**)	0.043 A
14	Lineal spark gap resistance in Ohms (**Rgl**)	2.1020 Ω
15	Exponential spark gap resistance in Ohms (**Rge**)	2.1695 Ω
17	Number of primary oscillations using solution in C74	**10**
18	Number of primary oscillations using solution in C74	**24**
19	Calculated primary oscillation time period (**tP**)	**146.7 μs**
21	Required ionization time of spark gap in air in μs (**ti**)	18.1 μs
22	Required primary oscillation time in μs (**tPR**)	**164.8 μs**
23	Minimum required operating time of spark gap in μs (**tO**)	0.000165 sec
25	Maximum useable tank capacitance in μF (**Cpm**)	**0.0208μF**
26	Peak primary tank current in Amps with maximum useable tank capacitance (**Ipm**)	**57.8 A**
27	Find available step-up transformer current output in Amps rms (**Ipa**)	**30.0 mA**
30	Find required step-up transformer current output in Amps rms (**Ipr**)	**42.7 mA**
31	Find Load power in watts (**Pp**)	**511.9W**
32	Find maximum rated Load power in watts (**Pr**)	**360.0W**
33	Find Line current in Amps rms (**IL**)	**4.3 A**
34	Find transformer load resistance in Ohms (**ZP**)	0.3 Ω

ROW 5–40; COLUMN A B C D E F

FIGURE 9-14 NST performance worksheet.

the NST is overloaded the output voltage drops rapidly to a very low value and the spark gap cannot be ionized with the low output voltage.

Figure 9-13 shows the performance results when the single 12 kV/30 mA NST was replaced with two parallel 9 kV/30 mA NSTs. Two significant changes result: the voltage applied to the tank capacitance is reduced, which lowers the peak primary current and the required rms primary current. Using the single 12 kV/30 mA NST required 42.7 mA rms of output current, which exceeds the available 30 mA NST rating. Changing to two parallel 9 kV/30 mA NSTs requires only 31.7 mA rms of output current (indicated by the [X], which is less than the available 60.0 mA NST rating shown by the [+] in Figure 9-13). A theoretical maximum tank capacitance of about 0.05 μF could be used with the two parallel 9 kV/30 mA NSTs. However, the output voltage of the NSTs would not be able to ionize the spark gap using this much capacitance and the actual maximum tank capacitance would be somewhere between 0.025 μF and 0.05 μF.

With each of the two parallel 9 kV/30 mA, NSTs supplying 15 mA, the output voltage of each will be at about 50% rated value (open-circuit output voltage). This may still be enough to ionize a small gap (<0.065″). This is why the second test run produced spark gap ionization and a secondary spark where the first run produced no visible ionization or spark.

When NSTs are selected for use as the step-up transformer in a Tesla coil design the resonant tank circuit should be evaluated to estimate the rms current required from the output of the step-up transformer. If an NST is used it must be able to supply enough voltage to ionize the spark gap at this rms output current required by the resonant primary circuit. The required current can be calculated and used to determine whether the NST can supply the requirements. If the coil is already built and does not perform as expected, this technique can be used to evaluate design changes.

References

1. *Reference Data for Radio Engineers*, 4th ed. ITT: 1956, pp. 45–46.
2. Abbott Electronics, Inc. website: www.abbottelectronics.com, Application Note: Thermal Management—Heat Sinking.
3. Hamilton Precision Metals website: http://www.hpmetals.com/thermCOND.asp, thermal conductivity tables.
4. Georgia State University's Department of Physics and Astronomy website: http://hyperphysics.phy-astr.gsu.edu/hbase/tables/thrcn.html, thermal conductivity tables.
5. *The VNR Concise Encyclopedia of Mathematics.* Van Nostrand Reinhold Company: 1977, p. 174.
6. Ibid, p. 191.
7. Allanson International, Inc. website: www.allanson.com. Lighting Electrics Group, Application Guidelines for Neon Transformers.
8. Universal Lighting Technologies, Inc. website: www.signasign.com. Neon Transformer Catalog and Application Data.
9. France, A Scott Fetzer Company website: www.franceformer.com. Frequently Asked Questions.

10

CHAPTER

Engineering Aids

The following sections were included to provide data and equations useful to the designer and will assist the coiler in understanding some of the key relationships and parameters associated with Tesla coils.

10.1 Frequency, Time, and Wavelength Relationships

To illustrate the relationship between frequency (f), time period (t) and wavelength (λ) in the repetitive waveforms found in AC and RF design, Figure 10-1 is presented. A log-log scaling is used to display a calculated wavelength (black trace) using equation (10.1) on the right-hand Y-axis and a calculated time period (red trace) using equation (10.2) on the left-hand Y-axis for the frequencies shown in the X-axis. Also read on the right-hand Y-axis is the one-half λ in the green trace and one-quarter λ in the blue trace. The scaling makes this relationship appear linear and more common sense. It will become intuitive with design practice.

Using the CH_10.xls file, f lambda t relationships worksheet (1) will calculate: the wavelength in feet and time period in μsec for any frequency in kHz entered in cell (B2), the frequency in kHz and wavelength in feet for any time period in μsec entered in cell (B6), and the frequency in kHz and time period in μsec for any wavelength in feet entered in cell (B10) (see App. B). Equations (10.1) and (10.2) show the relationship between these parameters. The graph shown in Figure 10-1 is displayed in Chart 1, rows 16 to 63 of the worksheet.

$$\lambda = \frac{c}{f} \tag{10.1}$$

Where: λ = Wavelength of frequency = Calculated wavelength in feet in column (G) and meters in column (I) for corresponding frequency in column (H).
c = Propagation speed of wavefront in free space (vacuum) = 9.84×10^8 feet/sec, or 2.998×10^8 meters/sec.
f = Frequency in Hz.

$$t = \frac{1}{f} \tag{10.2}$$

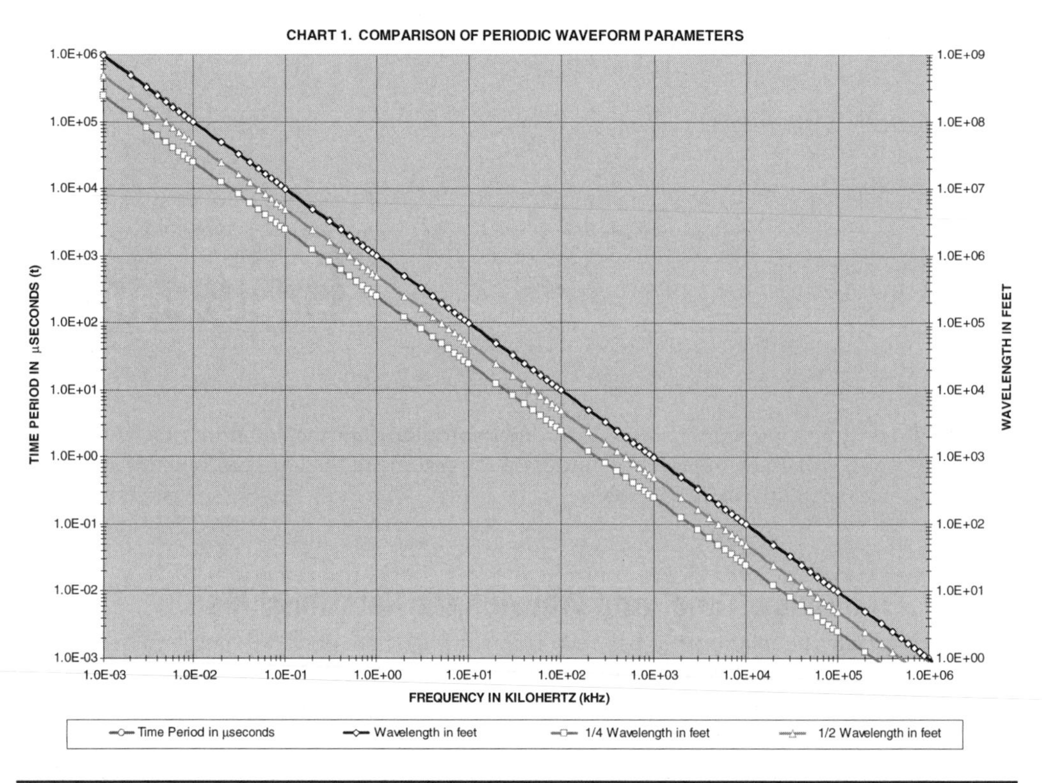

FIGURE 10-1 Comparison of periodic waveform parameters.

Where: t = Time period of resonant frequency in seconds = calculated value in column (F). Converted to μsec using a 1e6 multiplier.

f = Frequency in Hz.

Equation (10.3) calculates the energy level in a periodic waveform of specified frequency from reference (5). The energy level in electron volts for periodic waveforms is shown in Figure 10-2 at the corresponding frequency in the X-axis. The graph shown in the figure is displayed in Chart 2, rows 67 to 118 of the worksheet.

$$eV = \frac{12.4}{\lambda} \bullet 1{,}000 \tag{10.3}$$

Where: eV = Energy of periodic waveform in electron volts = calculated value in column (K). 1.0 keV = 12.4 / λ(in Å).

λ = Wavelength of frequency in angstroms (Å). Calculated wavelength in meters (m) in column (I) is converted to Å in column (J) using a 1e10 multiplier. 1.0 Å $= 10^{-8}$ cm. 1 m $= 10^{2}$ cm. 1.0 Å $= 10^{-10}$ m.

As the frequency increases in Figure 10-2 the higher the energy contained in the oscillations.

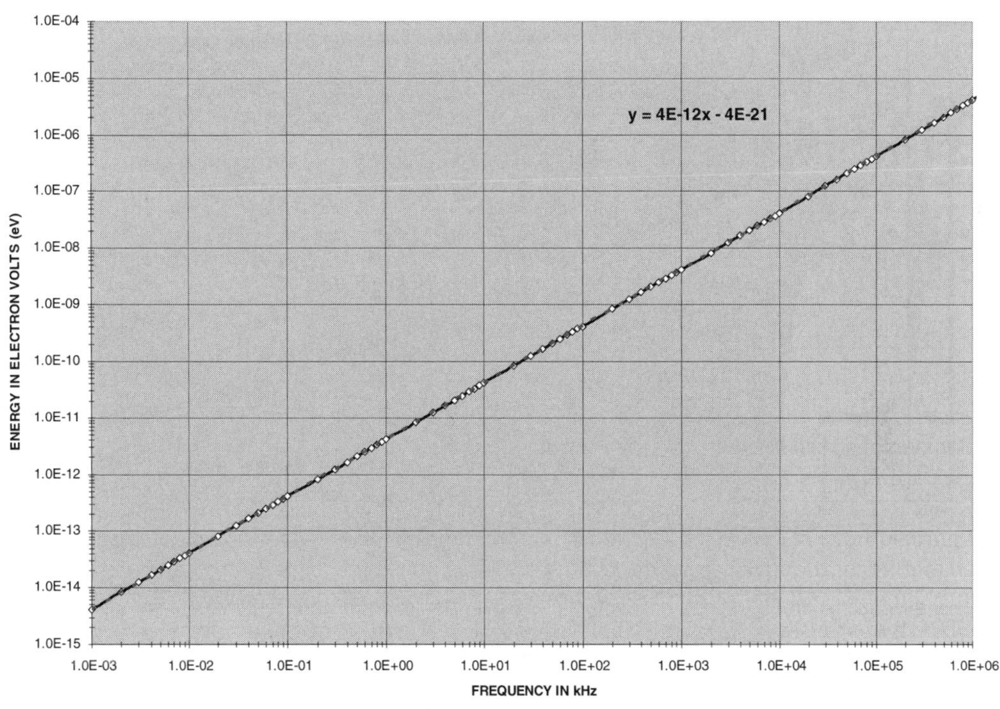

FIGURE 10-2 Energy levels in periodic waveforms.

10.2 Exponential Waveform Calculations

The exponential rise and decay of current in an inductor and voltage in a capacitor are illustrated when the instantaneous value of current or voltage is calculated for a span of time. From reference (1) the instantaneous value of current in an inductor and voltage in a capacitor during its rise (charge) for a given time period is calculated:

$$I, V = \text{Im}, Vm \times \left[1 - \in \left(\frac{-t}{tp}\right)\right] \tag{10.4}$$

Where: I, V = Instantaneous value of current in an inductor or voltage in a capacitor during its rise (charge) for the given time period (tp).

Im, Vm = Maximum value of current in an inductor or voltage in a capacitor at the end of charge time.

$\in$ = Exponential function (EXP).

t = Cumulative time from zero for instant of time (instantaneous) of interest.

tp = Total time period for rise of current in an inductor or voltage in a capacitor.

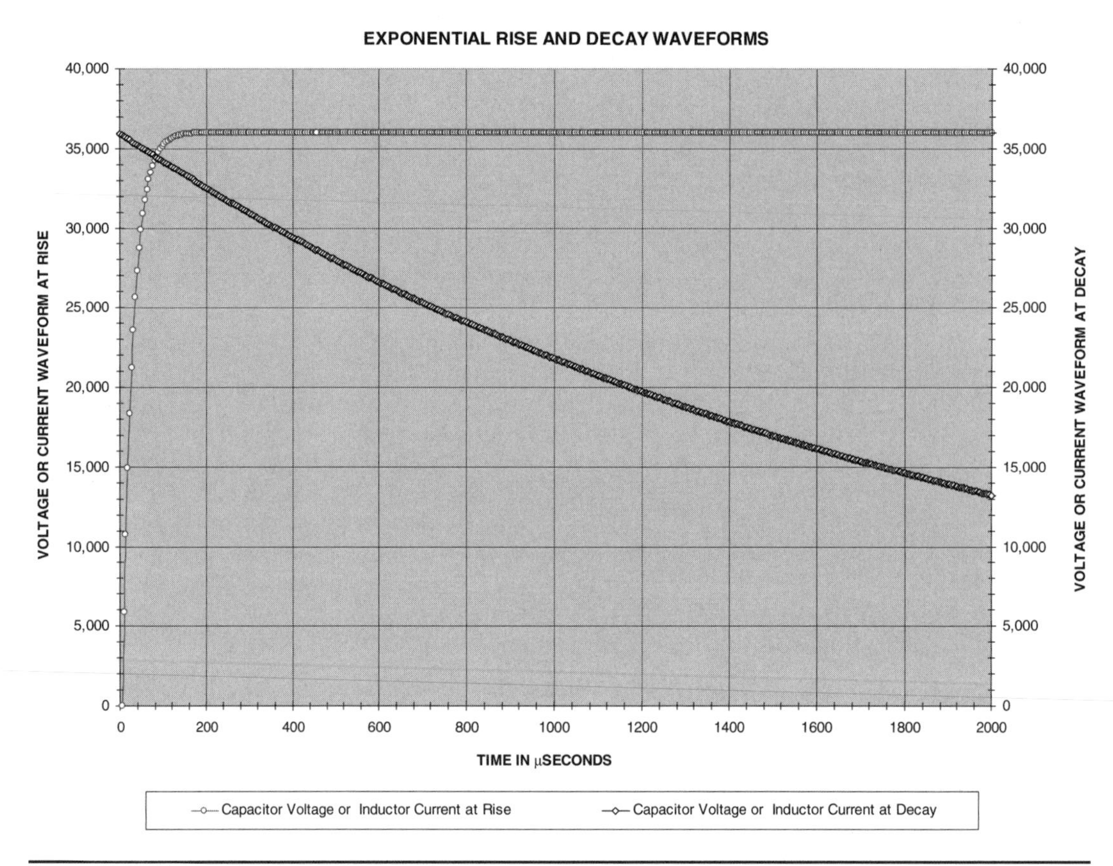

FIGURE 10-3 Exponential rise and decay of current or voltage in a reactive component.

The instantaneous value of current in an inductor and voltage in a capacitor during its decay (discharge) for a given time period is calculated:

$$I, V = Im, Vm \times \in \left(\frac{-t}{tp}\right) \tag{10.5}$$

Figure 10-3 displays the calculations for the charge and discharge of a reactive component using a charge time period of 25 μs and discharge time period of 2,000 μs. The discharge waveform in Figure 10-3 shows a similarity to the peak envelope of the secondary waveforms of Chapter 2. CH_10.xls file, Exponential Waveform worksheet (2) will produce an exponential waveform on the graph in rows 13 to 52 for any frequency, rise or fall time, and peak voltage entered in cells (B4) thru (B7).

Another Excel method to model an exponential voltage (V) rise in a circuit is to use the formula: $V \times$ EXP(δ). And to model a decay use the formula: $V \times$ EXP($-\delta$), where delta is the waveform decrement.

10.3 Calculating Output Voltage in an Iron Core Step-Up Transformer

To calculate the high-voltage output of your step-up transformer, open the CH_10.xls file, Line vs. HV Step-up worksheet (3). Enter the line voltage input in cell (B3) and the step-up ratio in cell (B4) or (B5). The step-up transformer HV output at the corresponding input voltage (adjustable with a variable autotransformer) is displayed on the graph in rows 7 to 58 and shown in Figure 10-4. The graph can be printed on magnetic sheet and placed on a metal

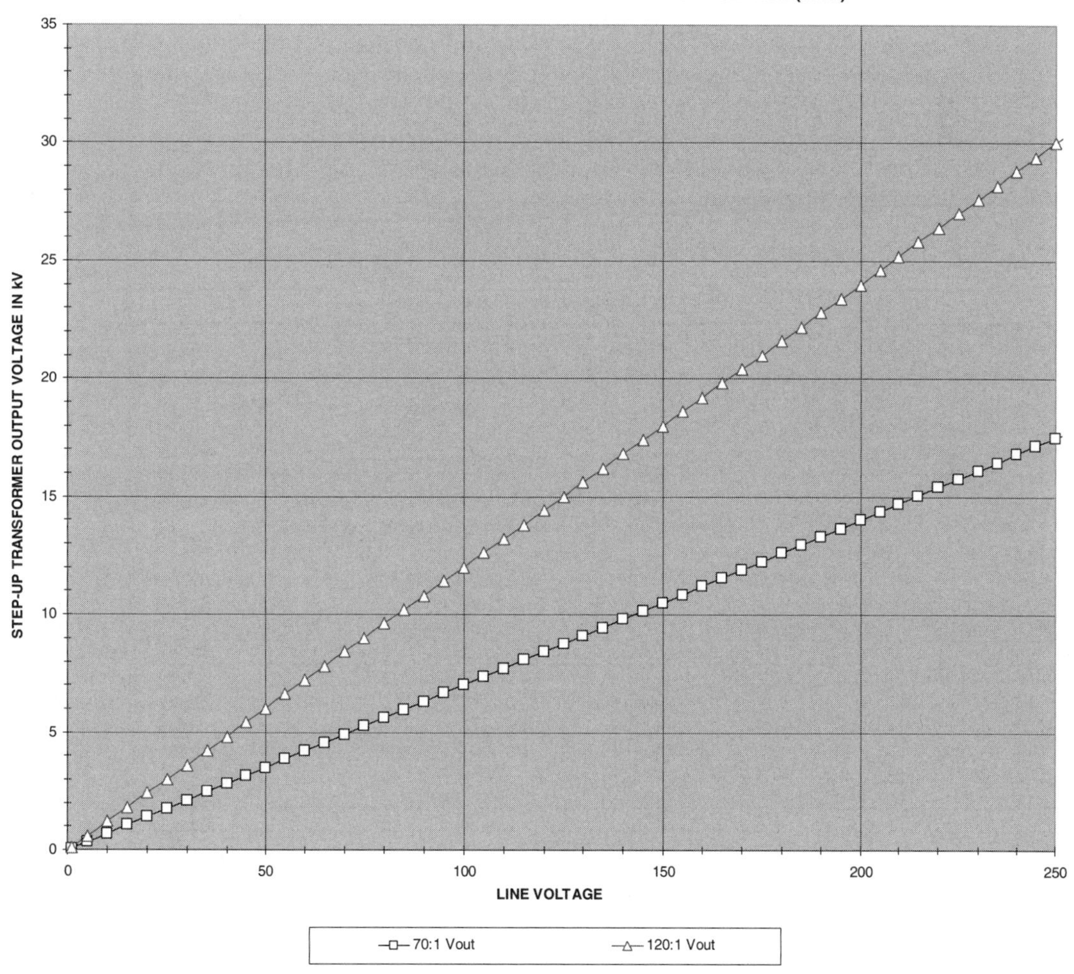

FIGURE 10-4 High-voltage output of 70:1 and 120:1 step-up transformers for variable autotransformer line voltages.

control cabinet for a professional-looking appearance. It provides quick reference for a selected line voltage for up to two step-up transformers. If only one transformer trace is desired enter a 0 step-up ratio in cell (B4) or (B5).

10.4 Decibel (dB) Conversions

When working with commercial EMI filters, Spice modeling, manufacturer's data, vacuum tubes, and other electronic parameters, it is often necessary to convert a voltage, current, or power to decibels (dB). It may also become necessary to convert dB to a voltage, current, or power. To assist with these calculations open the CH_10.xls file, dB conversion worksheet (4).

Figure 10-5 shows the worksheet construction using the conversion formulae shown in Table 10-1. Begin by entering the characteristic impedance of the system in cell (F2). Impedances were calculated for the spark gap coil in Chapter 2. For other applications the characteristic impedance may not be known. If so enter 50 ohms, which according to reference (2) is optimum for any RF system because it minimizes skin effect losses. To use these conversions the dB value from manufacturer data must be referenced to a dBm system where 0 dB = 1 mW. This is typically used throughout the manufacturing industry although some peculiar references such as dBW (0 db = 1 W) exist.

A	B	C	D	E	F	ROW
		Characteristic impedance of system in ohms:			50	2
Input:			Converts to:			3
						4
dBm	47.1		47.1	dBm		5
Volts			50.639	Volts rms	143.207Vp-p	6
Current(mA)			1012.8	mA		7
Power (mW)			51286.1384	mW		8
						9
Input:			Converts to:			10
						11
dBm			41.8	dBm		12
Volts			27.4	Volts rms	77.448Vp-p	13
Current(mA)			547.7	mA		14
Power (mW)	15000		15000.0	mW		15
						16
Input:			Converts to:			17
						18
dBm			47.0	dBm		19
Volts rms	50	141.400Vp-p	50.0	Volts rms	141.400Vp-p	20
Current(mA)			999.7	mA		21
Power (mW)			49984.9	mW		22
						23
Input:			Converts to:			24
						25
dBm			4.7	dBm		26
Volts			0.4	Volts rms	1.089Vp-p	27
Current(mA)	7.7		7.7	mA		28
Power (mW)			3.0	mW		29
A	B	C	D	E	F	COLUMN

FIGURE 10-5 DB Conversion worksheet.

Type Conversion	Conversion Units of Measurement	Conversion Formula
Power (*P*)	*P* in mW to dBm *P* in mW to *V* *P* in mW to *I* in mA	$10 \bullet \log P(\text{mW})$ $\sqrt{P \bullet R \bullet 1000}$ $P(\text{mW})/E$
dBm	dBm to *P* in mW dBm to *V* dBm to *I* in mA	$\log^{-1}\left(\frac{\text{dBm}}{10}\right)$ $\sqrt{R \bullet \log^{-1}\left(\frac{\text{dBm}}{10}\right)}$ $P(\text{mW})/E$
Voltage (*V*)	*V* to dBm *V* to *P* in mW *V* to *I* in mA	$10\log\left(\frac{E^2}{R} \bullet 1000\right)$ $\frac{E^2}{R} \bullet 1000$ $P(\text{mW})/E$
Current (*I*)	*I* in mA to *P* in mW *I* in mA to dBm *I* in mA to *V*	$\frac{I(\text{mA})^2 \bullet R}{1000}$ $10\log\left(\frac{I(\text{mA})}{1000}\right)^2 \bullet R$ $\frac{P(\text{mW})}{I(\text{mA})}$

Thanks are extended to Mike Ray for his assistance in developing the conversions shown in Table 10-1, which are compatible with Excel.

TABLE 10-1 Decibel (dB) conversion formulae.

Entering a dBm value in cell (B5) converts it to an rms voltage in cell (D6), peak-to-peak voltage in cell (F6), current in mA in cell (D7) and power in mW in cell (D8).

Entering a power in mW in cell (B15) converts it to an rms voltage in cell (D13), peak-to-peak voltage in cell (F13), current in mA in cell (D14), and dBm value in cell (D12).

Entering an rms voltage in cell (B20) converts it to an rms voltage in cell (C20), current in mA in cell (D21), power in mW in cell (D22), and dBm value in cell (D19).

Entering a current in mA in cell (B28) converts it to an rms voltage in cell (D27), peak-to-peak voltage in cell (F27), power in mW in cell (D29), and dBm value in cell (D26).

10.5 RMS and Average Equivalents of AC Waveforms

The formulae for calculating the root mean square (RMS) and average values of different AC waveforms are shown in Table 10-2 and found in references (3) and (4). The CH_10.xls file, RMS CONVERSIONS worksheet (7) can be used to calculate rms equivalent values for the waveform's peak voltage, "on" time, and repetition period. The worksheet is shown in Figure 10-6 with depictions of each type of waveform and associated parameters. Pulsed circuits such as Tesla coils produce waveform peaks much higher than values measured with meters; however,

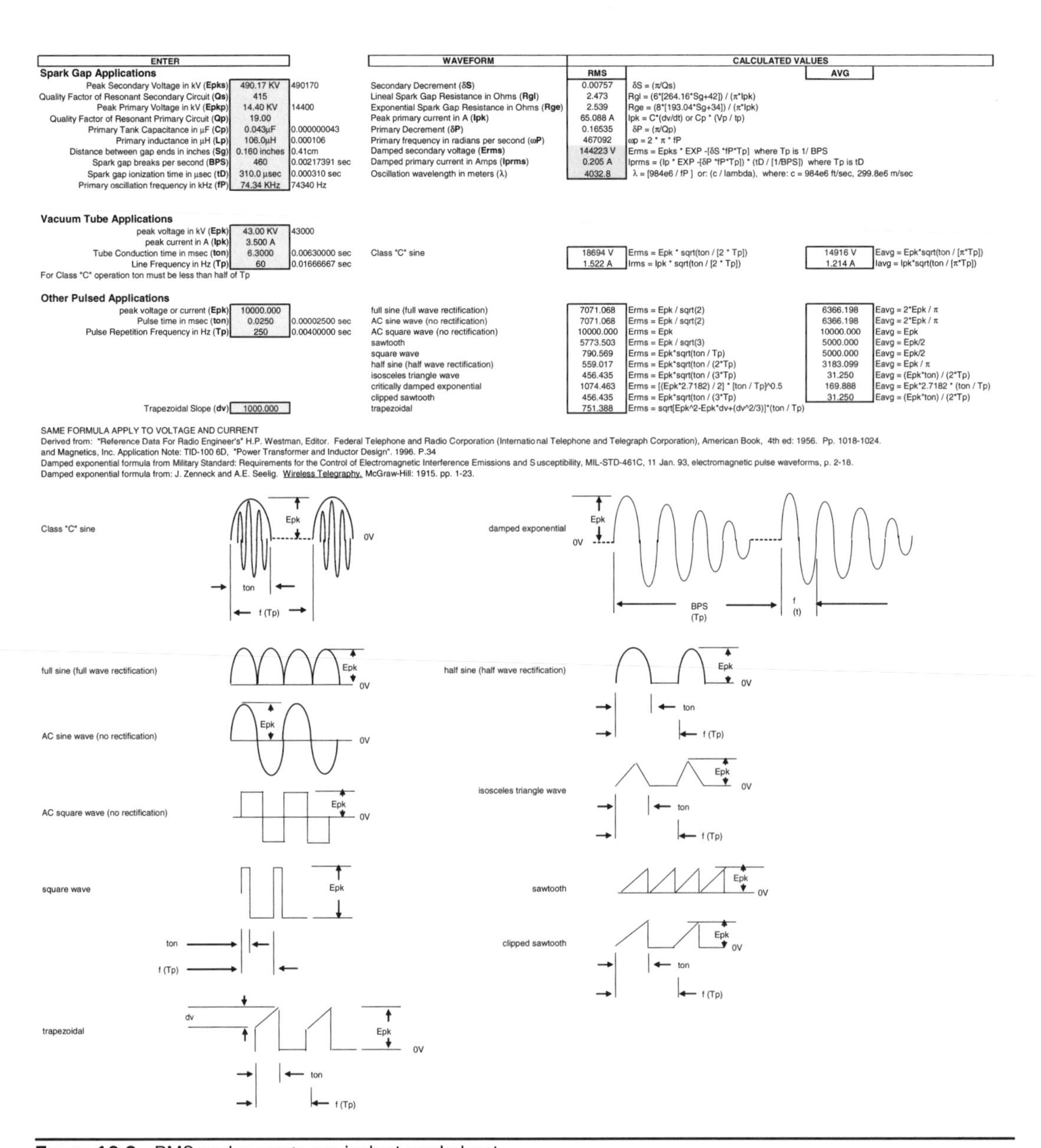

FIGURE 10-6 RMS and average equivalent worksheet.

the rms or average equivalent values can be calculated for comparison to measurements or for determining load and line characteristics. The same formulae are used for both voltage and current waveforms. The damped exponential waveform is typical of a spark gap Tesla coil and the class "C" sinewave is typical of a vacuum tube Tesla coil. Other waveforms encountered in pulsed circuits have been included.

AC Waveform	RMS Equivalent	Average Equivalent	
Critically damped exponential	$\frac{p2.7182}{2}\left(\frac{t}{Tp}\right)^{\frac{1}{2}}$	$\frac{p2.7182}{2}\left(\frac{t}{Tp}\right)$	*p* is peak *I* or *V* *t* is oscillation time period *Tp* is spark gap break rate
Class "C" sinewave	$p\sqrt{\frac{ton}{2Tp}}$	$p\sqrt{\frac{ton}{\pi Tp}}$	*p* is peak *I* or *V* ton is tube conduction time *Tp* is line time period (16.67ms for 60-Hz line)
Sinewave (sinusoidal) no rectification	$\frac{p}{\sqrt{2}}$ or $\frac{p}{1.414}$	$\frac{2p}{\pi}$	*p* is peak *I* or *V*
Sinewave (sinusoidal) full wave rectification	$\frac{p}{\sqrt{2}}$ or $\frac{p}{1.414}$	$\frac{2p}{\pi}$	*p* is peak *I* or *V*
Sinewave (sinusoidal) half wave rectification	$p\sqrt{\frac{ton}{2Tp}}$	$\frac{p}{\pi}$	*p* is peak *I* or *V* ton is 1/2 line time period (8.33ms for 60-Hz line) *Tp* is line time period (16.67ms for 60-Hz line)
Square wave no duty cycle or rectification	*p*	*p*	*p* is peak *I* or *V*
Square wave (pulse) duty cycled or rectified	$p\sqrt{\frac{ton}{Tp}}$	$\frac{p}{\pi}$	*p* is peak *I* or *V* ton is time period of pulse *Tp* is pulse repetition time period
Sawtooth	$\frac{p}{\sqrt{3}}$	$\frac{p}{2}$	*p* is peak *I* or *V*
Clipped sawtooth	$p\sqrt{\frac{ton}{3Tp}}$	$\frac{(p\bullet ton)}{(2Tp)}$	*p* is peak *I* or *V* ton is time period of ramp (sawtooth pulse) *Tp* is pulse repetition time period
Isosceles triangle	$p\sqrt{\frac{ton}{3Tp}}$	$\frac{(p\bullet ton)}{(2Tp)}$	*p* is peak *I* or *V* ton is time period of triangular pulse *Tp* is pulse repetition time period
trapezoidal	$\sqrt{\left(p^2 - p\bullet dv + \frac{dv^2}{3}\right)\bullet\frac{ton}{Tp}}$		*p* is peak *I* or *V* ton is time period of triangular pulse *Tp* is pulse repetition time period *dv* is current or voltage ramp value

Note: an exponent of (1/2) is the same as 0.5 or the $^2\sqrt{}$ function. $x^{(1/3)}$ is the same as $^3\sqrt{x}$ and $x^{(1/4)}$ is the same as $^4\sqrt{x}$. Exponents and roots are inversely proportional.

TABLE 10-2 RMS and average conversions for AC waveforms.

References

1. *Reference Data For Radio Engineers*. H.P. Westman, Editor. Federal Telephone and Radio Corporation (International Telephone and Telegraph Corporation), Fourth Ed, American Book: 1956, pp. 151–155.
2. Howard Johnson. *Why* 50 Ω. EDN magazine, Cahner's Business Information. Sept. 14, 2000, p. 30.
3. *Reference Data For Radio Engineers*. H.P. Westman, Editor. Federal Telephone and Radio Corporation (International Telephone and Telegraph Corporation), Fourth Ed, American Book: 1956, pp. 1018–1024.
4. Magnetics, Inc. Application Note: TID-100 6D. *Power Transformer and Inductor Design*. 1996, p. 34.
5. F.A. Mettler, M.D. and A.C. Upton, M.D. *Medical Effects of Ionizing Radiation*. Second Ed. W.B. Saunders Co. Div. of Harcourt Brace & Co.: 1995, pp. 1–4.

APPENDIX A

A Short Biography of Nikola Tesla

If you want to succeed, get some enemies.

Edison, as quoted in the *Ladies Home Journal*, April 1898

Much has been written on Tesla. Not all of it is accurate nor respectful of the man who almost single handedly invented the twentieth century. In 1895 any person who did not recognize the name Nikola Tesla would most likely have been dead. However, if a study was conducted today, most electrical engineers could not tell you one historic detail of Tesla's work, if they have even heard of him. In the old Serbo-Croatian language, Tesla translated to English is a tool used to square large logs into square timbers. In an attempt to appeal to your common sense, follow along with me as we highlight Tesla's life.

Born at midnight, July 10, 1856 in Smiljan (now Yugoslavia) he was an interesting and creative youth. He possessed a very divergent thinking style as well as a photographic memory. Very early in life he displayed cognitive abilities that most adults never achieve. He was a true child prodigy. His mind was able to visualize the mechanical details and electrical performance of an electrical circuit much like watching the Spice models in Chapter 8 run through a transient analysis on the computer. And that was just one of Tesla's amazing abilities.

As he attended the University in Prague the young Tesla was consumed by the thought of improving the dynamo. Working for the fledgling telephone company in Budapest, he patented several advancements in the art. It was during this time in 1882, still shy of his 26th birthday, that he first conceived the induction motor, AC generation and distribution, and the rotating magnetic field theory. In literally a microsecond burst of inventive genius, Mr. Tesla had solved what had eluded hundreds of inventors and engineers since Michael Faraday and Joseph Henry simultaneously discovered the coexisting electric and magnetic fields of current flow in 1830. His mind continued to solve problems around the clock, the solution often revealed with a blinding flash of light or intense imagery, possibly linked to his phenomenal photographic memory. In his mind he was able to design, construct, and run tests on any of his ideas. His concept for an AC power system was ridiculed in Europe. While later working for the Edison Company in Paris, Charles Batchelor (Edison's foremost manager) suggested to Mr. Tesla that he go to America and work for the Edison Company. There his ideas might be better received.

Mr. Tesla arrived in the United States in 1884 with only the change in his pocket and the clothes on his back. With a letter of introduction from Batchelor he secured a job working for Thomas Edison in New York. The letter stated "I know of two great men and you are one of them; the other is this young man." At this time in history there was only one method of power generation and distribution; direct current (DC). DC cannot be transformed to a higher voltage/lower current for distribution like alternating current (AC), because the current must constantly alternate in value to transfer power from the primary-to-secondary in a transformer. The Edison 110-VDC distribution system required much larger conductors to transmit the same power as an AC system and suffered from large line voltage drops with high power dissipation and heat. This understandably started many fires and was plagued with breakdowns. And the large diameter copper transmission lines were prohibitively expensive. The most pressing of problems with Edison's DC system was the inability of his engineers to connect more than one generator at a time to the distribution system. This limited his Pearl Street station to a service of only 400 electric lights. Edison and Tesla soon entered into an agreement which would award Tesla $50,000 for solving the single generator problem. Performing the work of several engineers and incorporating his own design improvements Mr. Tesla had the Edison plants running at top efficiency after a year of arduous and dedicated work. In typical fashion Edison did not honor the award and Tesla left his employ.

His ideas for improved lighting were soon formed into a small company, and patents issued. The investors quickly swindled the enterprise and Mr. Tesla was back in the street. Mr. Tesla survived a humiliating, impoverished existence digging ditches for laying Edison DC conduit and sewer lines for many months. He demonstrated his first "Egg of Columbus" to investors who quickly set up Tesla Company, Incorporated. In 1888 Tesla was invited to lecture to the American Institute of Electrical Engineers on the subject "A New System of Alternating Current Motors and Transformers." Now gaining momentum and attention with each passing day, George Westinghouse soon sought out Mr. Tesla. His ideas for an AC system were so promisingly received by Mr. Westinghouse that Tesla was immediately offered a contract for $1,000,000 and royalties of $1.00 per horsepower generated by the system, in exchange for the rights to his patents. Remember that Tesla had already advanced Edison's DC system to commercial success while in their employ. The war of the currents between DC and AC has been forgotten, but was one of the most exciting times in world history. At stake was the electrification of the world and the profits from doing so. Westinghouse engineers had by this time constructed a workable 133-Hz AC system in Great Barrington, Massachusetts. They did not however, have a working AC motor, which was the primary interest of George Westinghouse. The first successful AC power system using generators, distribution network, and motors was built near Telluride, Colorado in 1891. Telluride, high in the Rocky Mountains, needed power to operate its silver mines. It was becoming bankrupt trying to transport coal and timber by mule for the steam plants in use; and the AC system, though still unproven, looked promising to the developers. If you have ever tried to boil a pot of water at 10,000 feet of altitude you can appreciate the dilemma that Telluride faced with steam power. Tesla and Westinghouse were contracted to design a hydroelectric AC transmission system beginning at a river over two miles from the mines. It can still be seen in operation today. Other independent power projects were being developed in Europe, but the Colorado project preceded all of them. In the war of the currents, Tesla not only designed the winning AC system but also provided key improvements to the losing DC system. Now that is genius. It is a wonder his name is so obscured by history.

At the height of the current wars Westinghouse was awarded the contract to provide the AC power generation and lighting for the 1893 Columbian Exposition in Chicago, Illinois. The Edison Company would not sell Westinghouse their incandescent light nor grant a license to manufacture them. Westinghouse was forced to develop a different light that didn't infringe on the Edison patent. This was done in the nick of time and the successful lighting of the World's Fair of 1893 was instrumental in Westinghouse receiving the contract to build the AC hydroelectric generators at Niagara Falls, New York. The fair was the largest display of electrical lighting in the world at that time. As if that were not impressive enough, Tesla filled an entire hall with his electrical experiments and apparatus and demonstrated never-before-seen high-voltage effects, including drawing currents of hundreds of thousands of cycles per second at hundreds of thousands of volts through his body. He literally glowed from the violet streamers of corona. The public and press hailed him a "wizard."

After the World's Fair Tesla was hired as a consultant for Westinghouse's Niagara Falls contract. Unable to convince the Westinghouse engineers that 60-Hz generation was optimum for his polyphase system, Tesla left for Europe in 1893 to deliver a series of lectures before the Royal Society in London. The Westinghouse engineers didn't understand the new system, argued for 133 Hz, but finally adopted Tesla's 60 Hz. The 60 Hz system also provided an elegant time standard and was soon used to synchronize the hydroelectric generators to customer clocks with exceptional accuracy and reliability. The Tesla system operated most efficiently at 50 to 60 Hz and his induction motors did not like the 133 Hz. Mr. Tesla was well received by the Royal Society and his fame increased. His lectures demonstrated a new science. Resonant transformers of high frequency and high potential demonstrated an ability to transmit power and intelligence *without wires* (wireless). Although inspired by Hertz's earlier work, Tesla's improvements were original, possessing many key features of radio as shown in Figure A-1. Tesla not only lectured but displayed high-voltage radio-frequency effects in practical demonstrations with working equipment. Nothing else like this existed at the time. *Tesla possessed an exclusive knowledge of AC power distribution and radio.* After all, he was the inventor and discoverer. All others, wrongly claiming priority to Tesla's discoveries, were participants of his 1888 and 1893 lectures, or received the information second-hand from attendees. The high voltage "oscillation transformer" used in his lectures is now known as a "Tesla Coil" after its discoverer.

During the current wars both the Edison and Westinghouse companies were overextended trying to develop and promote their respective systems. The financier J.P. Morgan organized a takeover of both companies forming a power monopoly. Morgan did not own the companies outright but profited in arranging loans to overextended companies using investor capital. Morgan usually ended up with a good share of company stock from these transactions, therefore he would profit only as much as the company, and a monopoly can merely name its profits, especially one as important to twentieth century industry as power was. At the turn of the century business dealings such as these received no attention or were conducted privately, being hidden from the consumer. The Edison takeover was renamed General Electric (GE) but the Westinghouse takeover retained its name. For Westinghouse to continue the commercialization of the Tesla Polyphase System, it was necessary for Mr. Tesla to release Westinghouse from his promised arrangement for royalties of $1.00 per horsepower. This Tesla selflessly did with the comment "it will be my gift to the world." The new General Electric received the contract to build the distribution system for the Niagara project, but needed the Tesla patents to do so. This was not a problem as the Morgan monopoly now owned Westinghouse, which had exclusive rights to the Tesla patents. GE and Westinghouse had a patent exchange agreement where

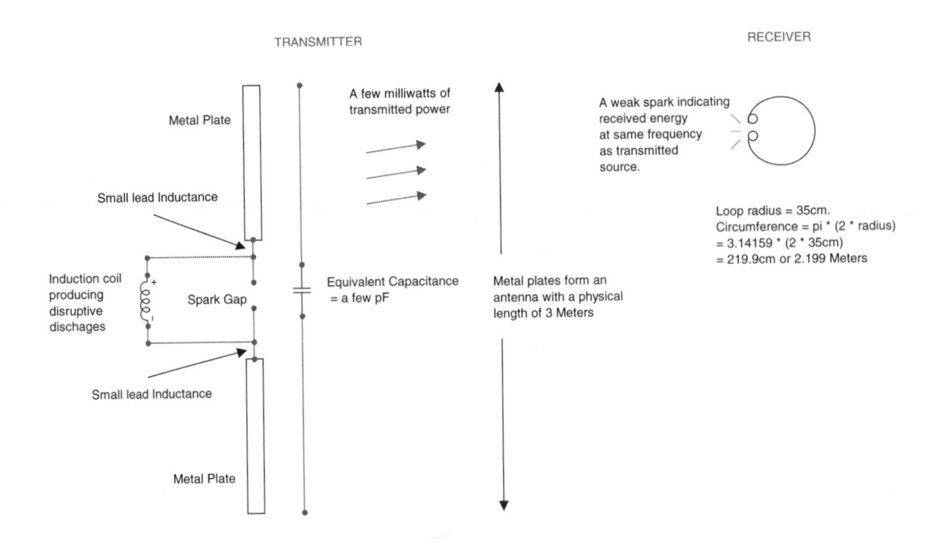

Simplified Hertz Circuit (1888)

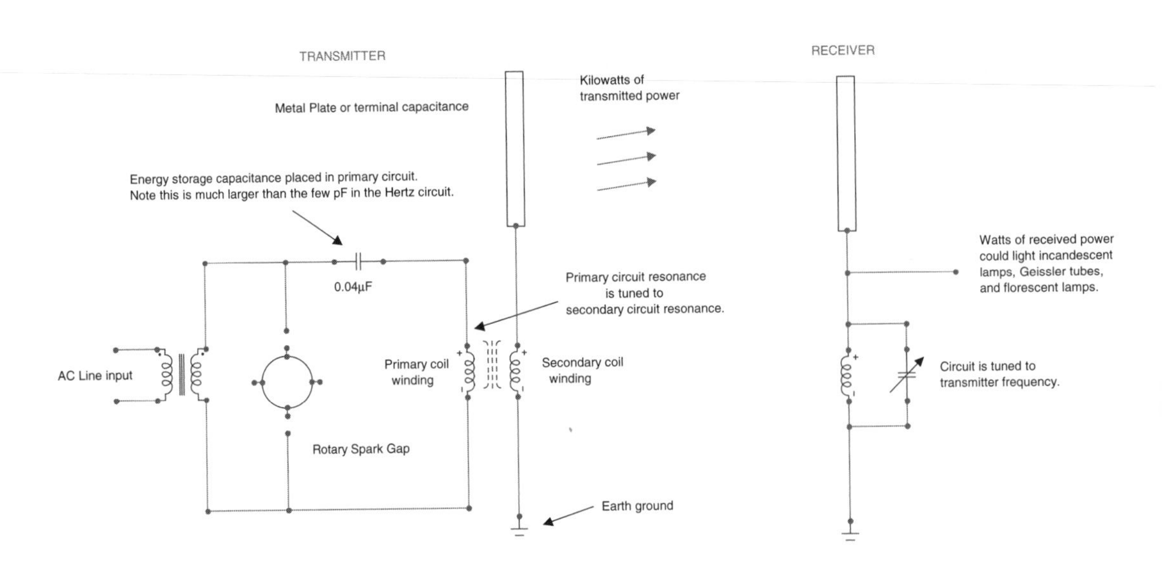

Simplified Tesla Circuit (1892 - 1893)

From: Corum, James F. and Kenneth L. Vacuum Tube Tesla Coils. Corum & Assoc: 1987. Pages X-2 thru X-6.
H. Hertz. Weidmann's Annalen, Vol. 31, 1887, Pg. 421 and Electric Waves, Page 40 is referenced.

FIGURE A-1 Tesla's early improvements to Hertz's radio work.

key patents were shared without risk of infringement. By 1895 the Niagara project's Adams Plant, containing the AC hydroelectric generators, was completed and the distribution system reached Buffalo by the end of 1896. On his return from the European lectures to Buffalo for the Niagara project dedication, Tesla had already conceived a *wireless* system of power generation and distribution that would make the newly completed Niagara project obsolete. Tesla's contribution to the first commercially successful AC system was already being overshadowed by the power industry. Thanks to commercial advertising and promotion the public associated AC power with Westinghouse and General Electric, not with Nikola Tesla. The Morgan companies found their own stars to steal the spotlight, one of the most successful was Charles Proteus Steinmetz at GE.

Tesla was now very famous and lived an extravagant lifestyle. Universities bestowed on him honorary doctoral degrees. He immersed himself in research at his New York City laboratory. His evenings were engaged socializing with New York's Fortune 500, partly to find investors for his new power scheme. He included as his friends the writer Mark Twain, the architect Stanford White, publisher of Century Magazine Robert Underwood Johnson, and the actress Sarah Bernhardt. John J. Astor invited him to occupy a penthouse in the Waldorf-Astoria, free of charge for the remainder of his life. His portraits of this period reveal a striking personage, so it is not hard to believe the immense success he enjoyed during this period. His research progressed until tragically, one evening, his laboratory burnt to the ground, the most probable cause being arson. This prompted him to move his research to Colorado Springs, Colorado.

The year 1899 was probably the most inventive period of his life but also the most mysterious. His published "Colorado Springs Notes" leave the reader with a limited view of Tesla's thoughts. He kept few notes during his lifetime. They were either an unnecessary expense of time or he had learned the danger of his competitors using his ideas without crediting the inventor or paying royalties on his patents. He may have intentionally left out important details and conclusions. Among his observations were earth resonance, generation of millions of volts with corresponding lightning discharge effects (lightning bolts of at least 50 feet evident in his published photographs and more than 100 feet by his own admission in reference 7), increased efficiency of his resonant transformer by adding an "extra coil," and perhaps strangest of all—reception of intelligent radio signals. Tesla attributed these to extra-terrestrial origin, not unlikely since his was the only work at the time whose scale could transmit a radio wave any considerable distance. On his return to New York in 1900, Tesla published a description of his work in Century Magazine with photographs of his famous Colorado Springs experiments. The world was overwhelmed. Not one other person had the slightest notion of what Tesla was doing. He was soon ridiculed for his extra-terrestrial observations. When Marconi announced a few decades later that he had received "space signals," he was hailed as a hero.

A very young Guglielmo Marconi had a wealthy father who was well connected in the Italian government. As he graduated from high school, Marconi pursued the new field of Radio. Beginning with Heinrich Hertz's revelation in 1888 the scientific world busily pursued what to do with the new science. Present at the European Tesla lectures and later visited by Tesla was Sir Oliver Lodge. Dr. Tesla gave to Lodge his entire lecture content along with working equipment of tuned circuits. Lodge later passed the equipment and knowledge to Marconi (reference 5). The young Marconi used the design without crediting its inventor to construct a radio transmitter. By December of 1901 Marconi, with almost unlimited funds from his father's Italian government connections, was credited with inventing "radio" when he succeeded in transmitting a wireless telegraphy signal across the Atlantic. I find it difficult to believe that such an inexperienced youth would be more instrumental in advancing a new science than a man practically twice his age with orders of magnitude more experience, who had already electrified the world with his polyphase AC system. And there are the lectures, which predate this by several years. I submit to your common sense who was the inventor of Radio.

While Marconi was working on crudely transmitting a carrier wave and trying to receive it at a distance using Dr. Tesla's ideas, Tesla worked on transmitting not just a carrier wave, but electric power. He visualized transmitting the Niagara Falls hydroelectric power (his invention) through the earth in a stationary wave to a distant point. This wave would transmit the full power of Niagara with negligible loss, compared to about 20 to 50% efficiency using the wire distribution system of today. The wave could also be modulated to transmit intelligence. He demonstrated these principles with his Madison Square Garden demonstration in 1896. Using remote radio control Dr. Tesla maneuvered a small boat, which he called an automaton, through a large pool. During the demonstration he would issue verbal commands and manipulating a control panel the boat would respond to his commands of "turn right" or "blink lights." Observers were so astonished they thought it magic, but Tesla clearly demonstrated to the public that he preceded all inventors in *wireless*. Dr. Tesla also demonstrated his automaton to the highest ranks of the U.S. Navy, but their shortsightedness could find no use for the invention. Perhaps the Navy was not shortsighted as they resurrected wireless remote control decades later at no cost and no credit to its discoverer. Better to wait fifty years and get it for free!

His ideas now took on substance. J. P. Morgan began to finance the ill-fated Wardenclyff Project in Shoreham, Long Island. Dr. Tesla's plans of *wireless* power distribution were eventually revealed to Morgan. Someone also pointed out to Morgan that the consumer could not be billed for the power received because there was no way to meter usage of *wireless* power. It should be obvious that these ideas were not well received by Morgan who was just beginning to squeeze out profits from his General Electric–Westinghouse power monopoly, with every major city in the world soliciting contracts for their own hydroelectric power plant. By 1908 the Wardenclyff Project was without funding and lost to creditors. It was too far ahead of its time.

Dr. Tesla retreated into a life of mysterious solitude in New York City. He continued to work out his power scheme and make new discoveries. He entered into many business ventures and worked as a consultant for industry, but he was never able to make his financial ends meet. While Morgan prospered from the Tesla Polyphase System monopoly, the inventor was evicted from his lodgings, unable to pay for them. Remember J. J. Astor's invitation for life in the Waldorf-Astoria penthouse? In keeping with Tesla's tragic life he lost these opulent lodgings as well as a dedicated investor when Astor went down with the Titanic. The industry quickly forgot its debt to the man who gave away $1.00 per horsepower (today the Canadian and American sides of Niagara alone produce 10,000,000 kilowatts of power which is more than 13,000,000 horsepower). Remember that hydroelectric power is virtually free once the generation plant, distribution system, and infrastructure are built and paid for. The early private power companies were small and could be paid off in just a few years, the remaining life of the system generated mostly profit. Also worth noting was the Morgan group's activities in the holding company scandal, which precipitated the crash of the stock market in October of 1929. At this time the power industry was a series of monopolies and the largest business in the nation. Morgan, Samuel Insull, and others created giant holding companies in the 1920s, which consolidated private and public power utilities. Stock and bond sales were inflated and capital pilfered to produce huge dividends, the overvalued stocks helping to bring down the stock market. As expected, those at the top lost little while the common stock holders would lose their savings and utility customers would pay increased rates while the bankrupt utilities were in receivership (through Morgan's banks). The National Electric Light Association (NELA) was the power industries' lobbying arm before the crash. To distract public attention to the ties between NELA and the power industry after the crash, they renamed the organization the Edison Electric Institute. Public furor lost momentum behind the Edison name.

Toward the end of his life the Yugoslavian government presented Dr. Tesla with an honorary income, which enabled the inventor to live those final years in relative comfort. He received nothing from his own adopted country, America. Dr. Tesla stated publicly his most valuable possession was his American citizenship papers. Industry and the press treated Dr. Tesla disrespectfully, which left an indelible mark on history. Today he is more often regarded as a crazy old man who played with pigeons than the inventor of twentieth century industry. Virtually every significant electrical invention of the twentieth century can be linked directly to Tesla's work in AC power and radio.

In 1943, the U.S. Supreme Court decided that Nikola Tesla had preceded all others in the invention of "Radio," and became the legal "Father of Radio." The outcome of the 40-year patent infringement was unknown by Dr. Tesla as he had died earlier in the year. His papers and personal effects were impounded by the U.S. government and not released until years after his death. There is no way to know what the government kept or what secrets died with Tesla, but they were his to keep. Tesla's critics contradict themselves when ignoring all the attention paid to him by the U.S. government during WWII. Too much attention if he were just an eccentric who played with pigeons.

References

1. Cheney, Margaret and Uth, Robert. *Tesla Master of Lightning*. Barnes & Noble Books, NY: 1999.
2. Seifer, Marc J. *Wizard: The Life and Times of Nikola Tesla. Biograph of a Genius*. Birch Lane Press: 1996.
3. Hunt, Inez and Draper, Wanetta. *Lightning in His Hand*. Omni Publication, Hawthorne, CA: 1964.
4. O'neill, John J. *Prodigal Genius: The Life Story of N. Tesla*. Ives Washburn, NY: 1944.
5. Lomas, Robert. *The Man Who Invented the Twentieth Century: Nikola Tesla, Forgotten Genius of Electricity*. Headline Book Publishing, London: 1999.
6. Richard Rudolph and Scott Ridley. *Power Struggle: The Hundred-Year War over Electricity*. Harper & Row, New York: 1986.
7. Nikola Tesla. *The Problem of Increasing Human Energy*. Century Magazine: 1900.

APPENDIX B

Index of Worksheets

Table B-1 is provided as an index to the worksheets used to perform the calculations in this design guide. You can find these worksheets at the companion Web site, which can be found at www.mhprofessional.com/tilbury/.

File	Worksheet Number	Worksheet Title	Worksheet Function
CH_2.xls	1	AWG vs VS (W_TERMINAL)	Calculate secondary voltage for selected AWG and operating characteristics of a spark gap coil.
	2	CALCULATION ERROR	Determine maximum calculation error.
	3	CONSTRUCTION SPECS	Observe Tesla coil construction and performance details.
CH_2A.xls	1	PRIMARY AND SECONDARY WAVEFORMS	Calculate and display line voltage and primary current and secondary voltage waveforms of a spark gap coil.
CH_3.xls	1	SPARK GAP COIL RESONANCE	Calculate resonant characteristics in a spark gap coil.
	2	RLC RESONANCE	Calculate resonant characteristics in a series RLC or parallel LC circuit.
	3	RECTANGULAR	Calculate and observe characteristics of periodic sinusoidal waveforms.
CH_4.xls	1	WIRE TABLE	Observe Magnet Wire Table.
	2	SKIN EFFECT	Calculate skin effect in coiled magnet wire.
	3	HELIX (magnet wire)	Calculate inductance and other parameters for a helical winding using magnet wire.

TABLE B-1 Worksheet index.

File	Worksheet Number	Worksheet Title	Worksheet Function
	4	HELIX (bare wire)	Calculate inductance and other parameters for a helical winding using bare or insulated wire.
	5	HELIX (Solenoid)	Calculate inductance and other parameters for a solenoid winding (or coil gun) using magnet wire.
	6	ARCHIMEDES (mag wire)	Calculate inductance and other parameters for an Archimedes Spiral winding using magnet wire.
	7	ARCHIMEDES (bare wire)	Calculate inductance and other parameters for an Archimedes Spiral winding using bare or insulated wire.
	8	Toroid (air core)	Calculate inductance and other parameters for a toroidal air core winding using magnet wire.
	9	Toroid (ferrite core)	Calculate inductance, and other parameters for a toroidal ferrite core winding using magnet wire.
	10	PRIMARY DIMENSIONS	Calculate primary form dimensions.
CH_4A.xls	1	MUTUAL INDUCTANCE	Calculate the mutual inductance of two coaxial windings that are concentric or non-concentric.
	2	k for $A = 0.1$	Calculate mutual inductance and coupling coefficient for coaxial coils wound on same form with primary-to-grid winding height ratio of 0.1.
	3	k for $A = 0.2$	Calculate mutual inductance and coupling coefficient for coaxial coils wound on same form with primary-to-grid winding height ratio of 0.2.

TABLE B-1 Worksheet index. (*continued*)

File	Worksheet Number	Worksheet Title	Worksheet Function
	4	k for $A = 0.3$	Calculate mutual inductance and coupling coefficient for coaxial coils wound on same form with primary-to-grid winding height ratio of 0.3.
	5	k for $A = 0.5$	Calculate mutual inductance and coupling coefficient for coaxial coils wound on same form with primary-to-grid winding height ratio of 0.5.
	6	k for $A = 0.7$	Calculate mutual inductance and coupling coefficient for coaxial coils wound on same form with primary-to-grid winding height ratio of 0.7.
	7	k for $A = 1.0$	Calculate mutual inductance and coupling coefficient for coaxial coils wound on same form with primary-to-grid winding height ratio of 1.0.
	8	H vs D	Calculate Q in a winding for selected height vs. diameter characteristics.
	9	B vs. H	Calculate hysteresis curve for selected spark gap coil characteristics.
	10	L per Turn	Calculate the inductance per turn of helically wound tuning coil.
	11	IRON CORE RELATIONSHIPS	Calculate relationships in an iron core transformer.
	12	AIR CORE RELATIONSHIPS	Calculate relationships in an air core transformer.
CH_5.xls	1	Dielectric Constants (k)	Observe Dielectric Constants Table.
	2	LEYDEN JAR	Calculate capacitance and other parameters for a Leyden jar type capacitor.
	3	PLATE	Calculate capacitance and other parameters for a Plate type capacitor.
	4	SPHERICAL	Calculate capacitance and other parameters for a Spherical type capacitor.

TABLE B-1 Worksheet index. (*continued*)

File	Worksheet Number	Worksheet Title	Worksheet Function
	5	TOROID	Calculate capacitance and other parameters for a Toroidal type capacitor.
	6	STRAY CAPACITANCE	Calculate terminal-to-ground plane capacitance and other stray capacitance values using the plate capacitor technique.
	7	CMAX vs STEP UP	Calculate maximum usable tank capacitor value for selected step-up transformer.
	8	FILM-FOIL OIL CAP	Calculate electrical stresses for film/paper-foil oil filled capacitors. Montena data.
	9	CERAMIC CAP	Calculate electrical stresses for ceramic capacitors. Sprague data.
	10	MICA CAP	Calculate electrical stresses for mica capacitors. Cornell-Dubilier data.
	11	PLASTIC PULSE CAP	Calculate electrical stresses for film (pulse) capacitors. Maxwell data.
	12	Ct vs fo	Calculate effects of terminal capacitance on secondary resonant frequency of a spark gap coil.
CH_6.xls	1	HVEF 1984	Observe breakdown voltage tables for spherical gap ends. Source: E. Kuffel and W.S. Zaengl. *High Voltage Engineering: Fundamentals*.
	2	SPHERICAL GAP ENDS	Observe breakdown voltage tables for spherical gap ends. Source: Naidu & Kamaraju, *High Voltage Engineering*.
	3	ROD GAP ENDS	Observe breakdown voltage tables for rod (flat) gap ends. Source: Naidu & Kamaraju, *High Voltage Engineering*.

TABLE B-1 Worksheet index. *(continued)*

File	Worksheet Number	Worksheet Title	Worksheet Function
	4	AIEE 1947	Observe breakdown voltage tables for spherical gap ends. Source: American Institute of Electric Engineers, 1947.
	5	WIRELESS TELEGRAPHY 1915	Observe breakdown voltage tables for spherical gap ends. Source: Zenneck, *Wireless Telegraphy*: 1915.
	6	PRESSURE CONVERSIONS	Convert air pressure units for different standards.
	7	ARC LENGTH	Calculate actual spark length resulting from an arc of travel between terminals.
	8	GAP RESISTANCE	Calculate ionized gap resistance for selected primary current, gap separation and electrode material. Source: Zenneck, *Wireless Telegraphy*: 1915.
	9	DECREMENT	Display damped oscillation for selected decrement value.
	10	IONIZATION CURRENT	Observe data for arc characteristics. Source: *Vacuum Metallurgy*. 1958.
	11	*Ea* vs. *I*	Calculate voltage drop vs. current in an ionized gap using Ayrton equation. Source: Hertha Ayrton. *The Electric Arc*: 1902.
CH_6A.xls	1	IONIZATION TIME	Calculate ionization characteristics of a spark gap coil.
	2	PRIMARY OSCILLATIONS	Calculate primary decrement characteristics of a spark gap coil.
CH_6B.xls	1	BPS CALCULATOR	Calculate breaks per second (BPS) and other rotary spark gap parameters.
	2	GAP TEMP	Calculate rotating and stationary electrode operating temperature for a rotary spark gap.

Table B-1 Worksheet index. *(continued)*

File	Worksheet Number	Worksheet Title	Worksheet Function
	3	VS vs SPARK LENGTH	Observe spark length data for high voltage terminal output of a Tesla coil.
	4	NST Current Output	Calculate output voltage vs. output current characteristics for a neon sign transformer.
CH_7.xls	1	RESISTIVE CURRENT LIMIT	Calculate limited output current vs. resistance setting for potentiometer used as series current limiter.
	2	INDUCTIVE CURRENT LIMIT	Calculate limited output current vs. amp setting for arc welders used as series current limiter.
	3	INTERCONNECT SKIN EFFECT	Calculate skin effect in solid, tubular, and rectangular conductors.
	4	INPUT FILTER	Calculate dB, voltage transients and filter attenuation in AC line.
	5	FUSING CURRENT OF WIRE	Calculate fusing current of wire (current required to melt wire).
	6	AUTOTRANSFORMERS	Calculate allowable overcurrent vs. time duration at selected ambient temperature in variable autotransformers.
	7	AMPACITY	Observe ampacity rating for selected AWG copper wire.
	8	CIRCUIT BREAKERS	Calculate derating factors in thermal circuit breakers.
	9	LOAD I vs HORSEPOWER	Observe load current vs. horsepower rating for AC and DC motors.
	10	RESISTOR DERATING	Calculate derated power dissipation for potentiometer used as series current limiter.
	11	GROUND RESISTANCE	Observe ground resistance calculations for selected schemes and number of electrodes
	12	FUSING CURRENT (Onderdonk)	Calculate fusing time and current of wire (current required to melt wire) using Onderdonk's equation.

TABLE B-1 Worksheet index. (*continued*)

File	Worksheet Number	Worksheet Title	Worksheet Function
CH_7A.xls	1	IRON CORE SATURATION	Calculate operating parameters and saturation threshold in an iron core used as a reactor.
	2	B vs H AIR CORE	Calculate magnetic flux density (*B*) vs. magnetizing force (*H*) in an air core resonant transformer.
CH_10.xls	1	F, lambda, T RELATIONSHIP	Calculate frequency, time, and wavelength conversions
	2	EXPONENTIAL WAVEFORM	Calculate parameters of an exponential waveform
	3	LINE VS. HV STEP UP	Calculate voltage output of step-up transformer for selected line voltages
	4	dB CONVERSIONS	Calculate dB, voltage, current, and power conversions
	5	Phase Angle	Calculate phase angle conversions
	6	POLAR-RECTANGULAR	Calculate polar-to-rectangular and rectangular-to-polar conversions
	7	RMS CONVERSIONS	Calculate rms and average equivalents for AC waveforms

Table B-1 Worksheet index. (*continued*)

C

APPENDIX

Metric Prefixes, Measurement Standards, and Symbols

Table C-1 lists the metric prefixes for the International System of Units (SI) approved by the National Institute of Standards and Technology (NIST). Those most common to electrical work

Decimal Number	Multiplier Factor	Name	Symbol
1,000,000,000,000,000,000,000,000.	10^{24}	yotta	Y
1,000,000,000,000,000,000,000.	10^{21}	zetta	Z
1,000,000,000,000,000,000.	10^{18}	exa	E
1,000,000,000,000,000.	10^{15}	peta	P
1,000,000,000,000.	10^{12}	tera	T
1,000,000,000.	10^{9}	giga	G
1,000,000.	10^{6}	mega	**M**
1,000.	10^{3}	kilo	**k**
100.	10^{2}	hecto	h
10.	10^{1}	deka	da
0.1	10^{-1}	deci	d
0.01	10^{-2}	centi	c
0.001	10^{-3}	milli	**m**
0.000001	10^{-6}	micro	μ
0.000000001	10^{-9}	nano	**n**
0.000000000001	10^{-12}	pico	**p**
0.000000000000001	10^{-15}	femto	f
0.000000000000000001	10^{-18}	atto	a
0.000000000000000000001	10^{-21}	zepto	z
0.000000000000000000000001	10^{-24}	yocto	y

TABLE C-1 Metric prefixes.

are emboldened. Expressing large values using a metric prefix is often easier than using the decimal number as illustrated in the table. Note the increasing or decreasing movement of three decimal places when moving up or down in the table. The multiplier factor corresponds to standard engineering notation.

Table C-2 lists the symbols used in this design guide and the electrical parameters they represent.

Symbol	Greek Alphabet Name	Electrical Parameter	See Section:
α	alpha	attenuation factor	2.3
β	beta	form factor	
δ	delta (lower case)	waveform decrement	2.2, 2.3
		skin depth	2.2, 4.7
ø	phi	angle	2.5
η	eta	efficiency	7.15
λ	lambda	wavelength	2.2
π	pi	pi = 3.14159...	
θ	theta	phase angle	2.8
ρ	rho	resistivity	7.11.1
σ	sigma	conductivity	7.11.1
ω	omega (lower case)	angular velocity (frequency in radians per second)	
Δ	delta (upper case)	change (maximum – minimum value)	
Ω	omega (upper case)	resistance in ohms	

TABLE C-2 Symbols.

Index

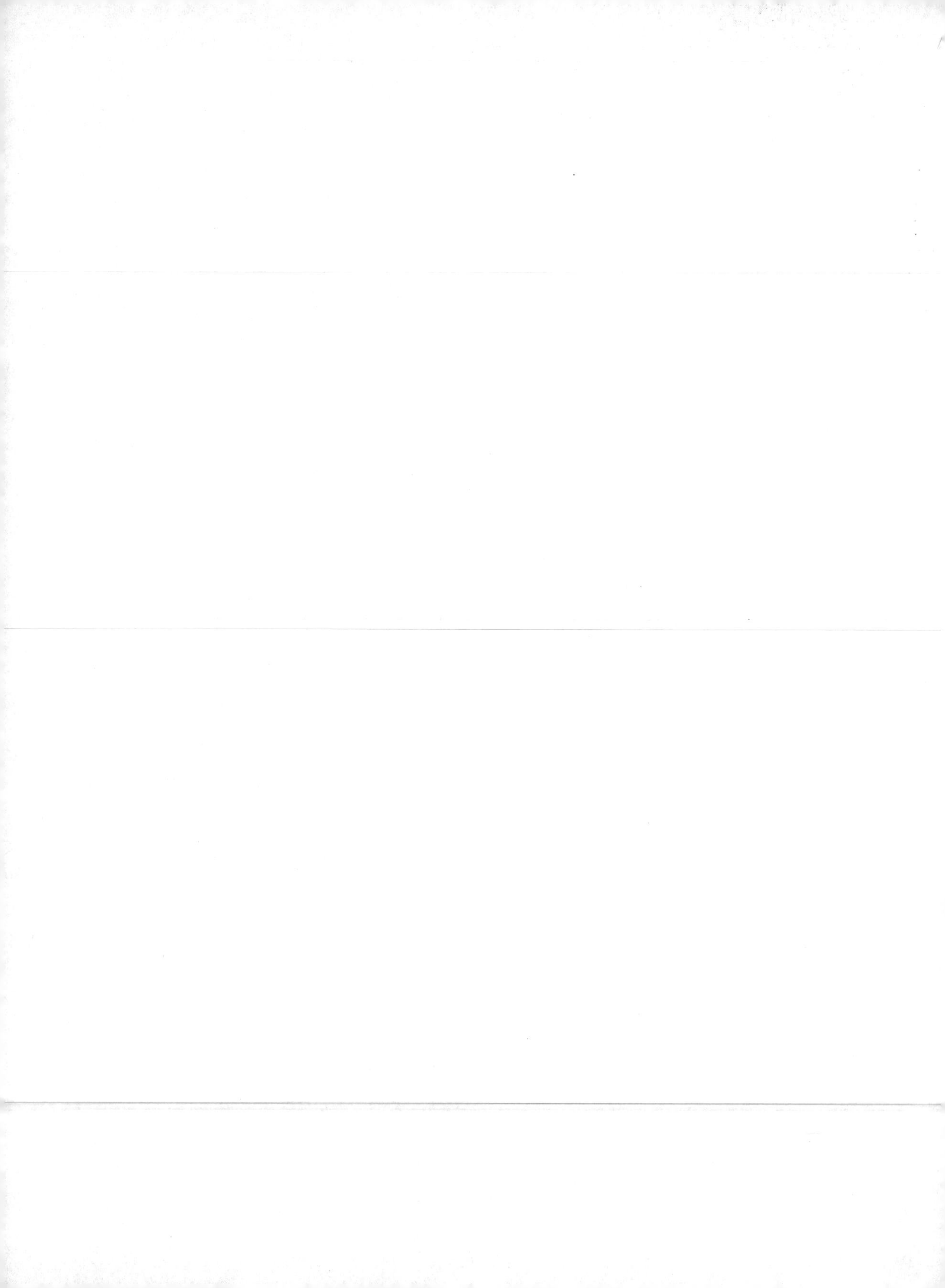